T0338299

**The Scaled Boundary
Finite Element Method**

The Scaled Boundary Finite Element Method

Introduction to Theory and Implementation

Chongmin Song
University of New South Wales
Sydney, Australia

Registered Office(s)
John Wiley & Sons, Inc., 111 River Street, Hoboken, NJ 07030, USA
John Wiley & Sons Ltd, The Atrium, Southern Gate, Chichester, West Sussex, PO19 8SQ, UK

Editorial Office
The Atrium, Southern Gate, Chichester, West Sussex, PO19 8SQ, UK

For details of our global editorial offices, customer services, and more information about Wiley products visit us at www.wiley.com.

Wiley also publishes its books in a variety of electronic formats and by print-on-demand. Some content that appears in standard print versions of this book may not be available in other formats.

Library of Congress Cataloging-in-Publication Data

Names: Song, Chongmin, author.
Title: The scaled boundary finite element method : introduction to theory and implementation / by Chongmin Song.
Description: Hoboken, New Jersey : John Wiley & Sons, 2018. | Includes bibliographical references and index. |
Identifiers: LCCN 2018003949 (print) | LCCN 2018010627 (ebook) | ISBN 9781119388463 (pdf) | ISBN 9781119388456 (epub) | ISBN 9781119388159 (cloth)
Subjects: LCSH: Finite element method. | Boundary element methods.
Classification: LCC QC20.7.F56 (ebook) | LCC QC20.7.F56 S66 2018 (print) | DDC 518/.25–dc23
LC record available at https://lccn.loc.gov/2018003949

Cover design: Wiley
Cover image: Courtesy of Chongmin Song

Printed in Singapore by C.O.S. Printers Pte Ltd
Set in 10/12pt Warnock by SPi Global, Pondicherry, India

10 9 8 7 6 5 4 3 2 1

To Feng, Helena and Sophie
for their support, patience
and many sacrifices

Contents

Preface

The finite element method is nowadays the most widely applied numerical method in computational stress analysis. Many finite element software packages are available commercially or in the public domain. This book presents the scaled boundary finite elements, a type of finite elements based on the scaled boundary technique. These elements are practically arbitrary polygons/polyhedra in shape, leading to a substantial reduction of the mesh generation burden. The semi-analytical representation of the displacement field in an element is advantageous when modelling singularity problems. Furthermore, the scaled boundary finite elements are seamlessly compatible with standard displacement-based finite elements. Incorporating these kinds of elements in a finite element software package has great potential in directly integrating geometric models and finite element analysis, in modelling singularity problems and crack propagation and in adaptive analysis.

A basic knowledge of the finite element method, linear algebra and ordinary differential equations at the level of typical undergraduate courses would be beneficial to the understanding of this book. The theoretical derivations are presented step-by-step in great detail and are thus self-contained.

This book should appeal to numerical analysts and software developers in various fields of engineering and science, such as civil engineering, mechanical engineering, computational mathematics and material science. This book can be used for an advanced course on finite elements and numerical methods.

Structure of the Book

This book consists of ten chapters covering the theory, computer implementation and application of the scaled boundary finite element method. This book commences with an overview of the scaled boundary finite element method in Chapter 1. The salient features of this method are illustrated with examples. The remaining chapters are organized into two Parts.

Part I, including Chapters 2–5, provides an introduction to the fundamental principles and the MATLAB programming of the scaled boundary finite element method. When selecting the contents, the primary consideration is simplicity, allowing the reader to follow and to learn. The theory is limited to the two-dimensional case. In developing the scaled boundary finite element formulations, only the simplest 2-node line element is addressed. All theoretical developments are explained step-by-step and with numerical examples. A computer program written in MATLAB is presented along with the theory and forms an integral part of the text. All the MATLAB functions and scripts necessary for the reader to reproduce all the numerical examples in this Part are provided.

- In Chapter 2, the key concepts of the scaled boundary finite element method are presented. The scaling requirement, scaled boundary coordinates and transformation are introduced. A detailed derivation of the scaled boundary finite element equation by applying the virtual work principle is provided.
- In Chapter 3, a solution procedure of the scaled boundary finite element equation is addressed. The quantities commonly employed in the finite element method, such as the stiffness matrix, the mass matrix and equivalent nodal force vector, are obtained. The shape functions of polygon elements are constructed, which allows the scaled boundary finite element method to be applied as a general finite element method. The static and dynamic analyses are demonstrated by examples.
- In Chapter 4, the automatic generation of polygon mesh is addressed. Examples are presented to demonstrate the salient features and to evaluate the accuracy and convergence of the scaled boundary finite element method.
- In Chapter 5, modelling considerations in applying scaled boundary finite element method are discussed. The guidelines that are not conventional or intuitive from the experience of the finite element analysis are discussed. Some of the applications for which the scaled boundary finite element method is advantageous are illustrated by examples. They provide a basis for establishing good practice in carrying out scaled boundary finite element analyses.

Part II, including Chapters 6–10, expands on the basics introduced in Part I to systematically develop the theory of the scaled boundary finite element method. The significant advantages that the scaled boundary finite element method offers in several types of challenging applications are demonstrated.

- In Chapter 6, the scaled boundary finite element equation in three dimensions is derived for elastodynamics by applying the Galerkin's weighted residual technique. The properties of the equation and the linear completeness of the solution are examined theoretically.
- In Chapter 7, a solution procedure of the scaled boundary finite element equation based on the matrix exponential function and block-diagonal Schur decomposition is presented. It provides a robust representation of singularities, to which the polynomial interpolations in standard finite elements are unsuitable, and contributes to the advantages of the scaled boundary finite element method in fracture analysis.
- In Chapter 8, the use of high-order elements is presented with the focus on the spectral elements. The scaled boundary finite element method is applied as high-order polygon/polyhedral elements of arbitrary edges, which eases the difficulties of spectral elements in modelling complex geometries. Numerical examples demonstrate that the spectral elements lead to exponential convergence.
- In Chapter 9, automatic mesh generation for use with the scaled boundary finite element method in two and three dimensions is addressed. Based on a quadtree/octree algorithm, mesh techniques are developed to handle CAD models and digital images. Fully automatic analyses of complex digital images of meso-structures of materials and STL models of statues are demonstrated with examples.

- In Chapter 10, the application of the scaled boundary finite element method to linear elastic fracture mechanics is covered. The stress singularities at crack tips and vertices of multi-material wedges are treated. Generalized stress intensity factors and the T-stress are evaluated. This approach does not require local mesh refinements, enrichments with analytical functions or special post-processing techniques. Crack propagation problems are conveniently handled owing to the simplicity in computing the stress intensity factors and the flexibility in meshing.

Two appendices are provided. Appendix A presents the governing equations of linear elasticity, and Appendix B introduces the matrix power function.

Accompanying Computer Program Platypus

A computer program Platypus implementing the scaled boundary finite element method for stress analysis in MATLAB is integrated into the text of the book. Platypus includes a mesh generator for use to discretize simple problem domains into meshes of polygon elements of arbitrary number of edges. The polygon elements are constructed by modelling the boundary of the polygon elements with 2-node line elements. After obtaining the element solutions using the scaled boundary finite element theory contained in this book, standard finite element procedures are followed to perform linear static and dynamic analyses. Basic post-processing functions are included. All the examples in Part I of this book are produced with Platypus.

The MATLAB functions are documented in detail. In particular, cross-references to the equations implemented in a function are provided within the function for the purpose of clear documentation. The MATLAB code will be helpful in understanding the theory presented in the text and provides an unambiguous interpretation of the mathematical theory and equations. The code can also be incorporated into a general finite element programme and forms the basis for developing more advanced versions.

The computer program Platypus is available at:

https://www.dropbox.com/sh/0nrc7tb9rjhfya3/AABHNwY5oo6SuLI2N4bnNh19a?dl=0

The above internet address can be acquired by scanning the QR-code below:

Electronic files on the installation and use of Platypus are found in the same place.

Acknowledgements

The scaled boundary finite element method originated in research led by my mentor Dr. John P. Wolf at the Swiss Federal Institute of Technology, in Lausanne. I gratefully acknowledge his significant scientific contributions. Without his leadership, scientific insight, strong support and trust, this research would not have been brought to fruition.

I would like to thank my collaborators and students at the University of New South Wales (UNSW Sydney) for their contributions to the work contained in this book. In particular, the efforts of Ean Tat Ooi, Albert Suputra, Carolin Birk, Hauke Gravenkamp, Yan Liu, Hossein Talebi, Sundararajan Natarajan, Hou Man, Mohammad Bazyar, Suriyon Prempramote, Irene Chiong, Morsaleen Chowdhury, Chao Li, Xiaojun Chen, Ke He, Tingsong Xiang, Junchao Wang, Lei Liu, Weiwei Xing, Junqi Zhang and Duc Tran are greatly appreciated.

I express my deepest gratitude to Professor Gao Lin at the Dalian University of Technology, China, Fellow of the Chinese Academy of Sciences, for his very significant contributions, encouragement and support throughout the development of the scaled boundary finite element method.

Some of the research leading to this book was financially supported by the Australia Research Council. This support is gratefully acknowledged.

Sydney, Australia   *Chongmin Song*
October, 2017

About the Companion Website

This book is accompanied by a companion website:

The URL is:
www.wiley.com/go/author/scaledboundaryfiniteelementmethod

The website includes Source codes

- Platypus

1

Introduction

1.1 Numerical Modelling

The advances in numerical modelling techniques and computer technology during the past few decades have transformed the analysis and design of many types of engineering structures. Today, numerical simulations as an integral part of Computer-Aided Engineering (CAE) are routinely performed in civil engineering, mechanical engineering, aerospace and other industries.

Many problems in engineering and science are formulated as field problems by mathematical models in terms of field variables, such as displacements, potentials, etc. The mathematical models consist of governing differential equations to describe physical laws (such as equilibrium and compatibility in a stress analysis), material constitutive models and boundary conditions enforced on the problem domains. Classical methods of engineering analysis are often based on analytical solutions to the mathematical model. The analytical solutions are expressed as mathematical functions and can be evaluated at any locations of interest. However, analytical solutions are available only for very simple problems. For many engineering problems, considerable simplifications have to be made in order to apply the classical analysis methods. This often leads to over-conservative designs.

Numerical modelling techniques have been developed to analyse complex engineering problems. In the numerical modelling of a field problem, the mathematical model is solved by a numerical method. A common feature of the numerical methods is the use of mesh discretization to divide a complex problem domain into a set of discrete pieces. Simple formulations are constructed on each piece and assembled together. The mathematical model of a field problem is approximated by a system of algebraic equations with unknowns at a finite number of discrete points. The solution generally converges to the exact solution with the increasing number of discrete points. Numerical methods rely on computers for number crunching.

The finite element method (Bathe, 1996; Cook et al., 2002; Reddy, 2005; Zienkiewicz et al., 2005, to name a few) is probably the most popular and powerful numerical method for the stress analysis of solids and structures. In the standard finite element method, a problem domain is discretized into a set of discrete subdomains called elements. The edges of the elements have to conform to the boundary of the problem domain. The shapes of the elements are limited to a few simple ones. They are triangles and quadrilaterals (i.e. polygons with 3 or 4 edges) in two dimensions, and tetrahedrons, pyramid,

The Scaled Boundary Finite Element Method: Introduction to Theory and Implementation,
First Edition. Chongmin Song.
© 2018 John Wiley & Sons Ltd. Published 2018 by John Wiley & Sons Ltd.
Companion website: www.wiley.com/go/author/scaledboundaryfiniteelementmethod

wedge and hexahedra (i.e. polyhedra with 4–6 faces) in three dimensions. These elementary shapes allow the development of simple element solution. Typically, the solution within an element is approximated locally by piecewise polynomials, which leads to extremely efficient procedures (e.g. for numerical integration) suited to number crunching by computers. Element formulations are constructed as algebraic equations by satisfying the governing differential equations in a weak form. Assembling the element equations according to the element connectivity leads to a global system of algebraic equations that can be solved by known solution techniques.

The finite element method is highly versatile and widely applicable in numerical modelling largely due to the simplicity in formulating element equations. Nowadays, many commercial software packages implementing the finite element method are available. Finite element analysis is widely performed in many disciplines of engineering and science. However, the conventional finite element method faces challenges when applied to certain classes of engineering problems. Examples pertinent to the contents of this book include the problems involving dynamic responses of unbounded domains, strain/stress singularities and moving boundaries. In addition, the simple shapes and formulations of finite elements place a heavy burden on mesh generation, which often requires frequent human interventions.

In the numerical model of a structure under an earthquake action, the supporting soil is often simplified as an unbounded domain extending to infinity in comparison with the dimension of the structure. A so-called radiation condition, which states that no energy be radiated from infinity towards the structure, has to be enforced. Direct application of the finite element method would be computationally expensive as the size of the finite elements is limited by the frequency/wavelength of interest. Various techniques such as the viscous boundary, viscous-spring boundary, etc. have been developed for use with finite elements to overcome this limitation. The boundary element method, which satisfies the radiation condition by a fundamental solution, is another attractive alternative in the modelling of unbounded domains.

In the application of linear elastic fracture mechanics to evaluate the propagation of a crack in a brittle material, the stress field around the crack tip is of primary interest. Based on the theory of elasticity, stress/strain singularities exist at the crack tip. The standard finite element analysis of cracks suffers from slow convergence as the polynomial basis functions of standard finite elements do not resemble the singular functions in the strain and stress solutions. The same happens when modelling interface cracks and multi-material corners of composite materials. To achieve accurate solutions, a large number of elements are required around a singularity point. This further increases the burden on mesh generation. Additional post-processing techniques, such as path-independent integrals, are often required to extract the stress intensity factors and other parameters for engineering design.

In engineering practice, structural designs are routinely performed on Computer-Aided Design (CAD) systems. The geometry of a design is commonly described by Non-Uniform Rational B-Splines (NURBS). To perform a finite element analysis, a mesh that conforms to the boundary of the geometric model has to be generated. For computational efficiency considerations, the element size gradation is indispensable. Automatic mesh generation from CAD models has been an active research subject and various meshing techniques have been developed (Watson, 1981; Yerry and Shephard, 1984; Löhner and Parikh, 1988; Blacker and Meyers, 1993, to name a few). Generally speaking, a triangular/tetrahedral mesh is easier to generate automatically than a quadrilateral/hexahedral mesh, while quadrilateral/hexahedral elements are more advantageous in a finite element analysis. Nowadays, an automatic mesh generator is an indispensable feature of any popular commercial finite element software package. However, the generation of a high-quality finite element mesh conforming to the boundary very often requires tedious human interventions in practical engineering computations. The process can be cumbersome and time-consuming. Furthermore, continuous remeshing is highly desirable in some types of analysis. Examples include the modelling of moving boundaries that result from crack propagation or phase transition in matter, adaptive analysis that aims to automatically achieve a specified accuracy in an optimal way, and topology optimization. Robust and efficient remeshing is a challenging task in a finite element analysis.

Various numerical methods have been proposed recently to overcome the above shortcomings of the finite element method in solving practical engineering problems. One major trend is to develop numerical methods that reduce the meshing burden, especially the human efforts and interventions required in the mesh generation. Many novel numerical methods have been proposed. Notable examples include the various meshfree methods (Lucy, 1977; Belytschko et al., 1994; Atluri and Zhu, 1998), extended/generalized finite element methods (Moës et al., 1999), the finite cell method (Parvizian et al., 2007), and isogeometric analysis (Hughes et al., 2005) to name a few. A comprehensive review of the recent rapid developments in this area is a daunting task beyond the scope of this book.

Advances in digital technology have led to the development and increasing popularity of other file formats than NURBS to describe the geometries of objects. Two notable examples are the STL (STereoLithography) format and the digital image.

1) The STL format is widely used in 3D printing and rapid prototyping (Bassoli et al., 2007; Rengier et al., 2010). It is now widely supported by CAD systems. The popularity of the STL format is largely attributed to its extreme simplicity: the boundary of a geometric model is represented by a list of unstructured triangular facets. However, this simple format allows the existence of ill-shaped, overlapped or self-intersecting facets and holes on the boundary. These defects, albeit small in size, are incompatible with the finite element method.
2) Digital images are increasingly being applied in engineering and science. Various digital imaging technologies, such as X-ray computed tomography (X-ray CT) scans, magnetic resonance imaging (MRI) and ultrasound have become available with a rapid increase in resolution. For example, the meso-structure of a material can be acquired non-destructively by X-ray CT. A digital image is composed of pixels in 2D and voxels in 3D. The geometry is represented by the colour intensity of the pixels/voxels. The algorithms developed for the automatic mesh generation of NURBS-based models are not directly applicable to STL models or digital images, which hinders the application of numerical methods.

The scaled boundary finite element method presented in this book aims to overcome some of the limitations and shortcomings of the standard finite element method mentioned above and to facilitate the expansion of computational mechanics into new areas. The scaled boundary finite element method was originally developed for the dynamic analysis of unbounded domains, where the radiation boundary condition is formulated at infinity (Wolf and Song, 1996; Bazyar and Song, 2008; Li et al., 2013b; Chen et al., 2015b). In this method, a problem domain is divided into subdomians satisfying the so-called scaling requirement (see Section 2.2.1 on page 31). Generally speaking, the shape of such a subdomain is more complex than that of a finite element. The scaled boundary finite element method adapts a semi-analytical approach in constructing an approximate solution of a subdomain. Only the boundary of the subdomain is discretized. The solution in the other direction is obtained analytically. This is different from the standard finite element technique that requires an *a priori* choice of the interpolation functions over the subdomain/element. This method was initially called the 'consistent infinitesimal finite-element method' as a cell of finite elements adjacent to the boundary is used in the original mechanically based derivation (Song and Wolf, 1995, 1996; Wolf and Song, 1995). The derivation of the scaled boundary finite element method based on the Galerkin weighted residual technique is reported in Song and Wolf (1997), where the name 'scaled boundary finite element method' was coined. A virtual work-based derivation is presented in Deeks and Wolf (2002d).

The major steps of the theoretical background of the scaled boundary finite element method for the modelling of unbounded domains are summarized as follows (Song and Wolf, 1997; Deeks, 2004). The boundary of an unbounded domain is discretized. In this aspect, this method is similar to the boundary element method (Brebbia et al., 1984; Dominguez, 1993) and leads to a reduction in spatial dimensions by one. By applying the finite element technique along the circumferential directions parallel to the boundary, the governing partial differential equations in the domain are transformed into ordinary differential equations with a radial coordinate extending to infinity as the independent variable. The ordinary differential equations can be solved analytically. This can be regarded as a semi-analytical (or semi-discrete) method as the solution is numerical in the circumferential directions and analytical in the radial direction. Using the semi-analytical solution, the radiation condition is enforced rigorously at infinity. Unlike the boundary element method, no fundamental solution is required. No singular integrals must be evaluated. General anisotropic materials can be analysed without increasing the computational effort. The results converge as the elements at the boundary are refined.

Further research has revealed that the scaled boundary finite element method is much more general than a niche technique to model unbounded domains. This method has since been developed into numerical techniques with attractive features for a broader range of applications, covering the evaluation of stress intensity factors and the T-stress at cracks and interface cracks (Song and Wolf, 2002; Song, 2005; Chidgzey and Deeks, 2005; Liu and Lin, 2007; Yang and Deeks, 2008; Bird et al., 2010; Zhong et al., 2014) and of edge singularities (Lindemann and Becker, 2002), elastodynamics (Song, 2009; Yang et al., 2011; Gravenkamp et al., 2012), adaptive analysis in statics (Deeks and Wolf, 2002b,c; Vu and Deeks, 2006, 2008) and dynamics (Yang et al., 2011), diffusion (Birk and Song, 2009), crack propagation (Yang, 2006; Yang and Deeks, 2007a; Zhu et al., 2013), electromagnetics (Liu et al., 2010), sloshing analysis of liquid tanks (Lin et al., 2015), fluid-structure interaction (Li, 2009; Wang et al., 2015), meshless method (Deeks and Augarde, 2005), 3D fracture problems (Mittelstedt and Becker, 2006; Goswami and Becker, 2012; Hell and Becker, 2015), isogeometric analysis (Klinkel et al., 2015; Chen et al., 2015a; Natarajan et al., 2015), plate bending (Man et al., 2012, 2013, 2014a), etc.

Recently, the scaled boundary finite element method has been applied to constructed polygon elements (Chiong et al., 2014). The shape functions of the polygon elements are determined from the scaled boundary finite element solution. Such a polygon element is interpreted as a type of finite element with an arbitrary number of edges. Furthermore, a polygon mesh is readily generated as a dual mesh of a triangular mesh. Thus, the scaled boundary finite element method can, in principle, be applied as widely as the finite element method. The polygon elements based on the scaled boundary finite element method are applied to the problems involving the fracture of functionally graded materials (Chiong et al., 2014), elasto-plasticity (Ooi et al., 2014), geometrically and physically nonlinearity (Behnke et al., 2014), crack propagation (Ooi et al., 2012b, 2013; Dai et al., 2015) and piezoelectric materials (Li et al., 2013a, 2014; Sladek et al., 2015, 2016). The procedure used to construct polygon elements in two dimensions is easily extended to construct polyhedral elements in three dimensions (Saputra et al., 2015). The polygon elements in two dimensions and polyhedral elements in three dimensions based on the scaled boundary finite element method are termed S-elements (see Section 1.2 on page 6).

The high degree of flexibility of the shape of the S-elements significantly simplifies the meshing process. Instead of using the dual mesh of a triangular mesh, a polygon mesh is generated by the simple and efficient quadtree algorithm (Man et al., 2014b; Ooi et al., 2015), which is one of the intermediate techniques for generating triangular meshes. The use of the S-elements with quadtree (in two dimensions) and octree (in three dimensions) is extended to stress analysis of geometric models provided as digital images (Saputra et al., 2015). The same procedure is further developed to analyse STL models (Chapter 9).

This book will present the scaled boundary finite element method in the context of a generally applicable numerical method with the emphasis on the following key features in applications:

1) Reducing the mesh generation burden. This is achieved by:
   a) A unified approach to the automatic mesh generation of geometric models in the formats of NURBS, STL and digital images.
   b) The direct connecting of non-matching meshes in domain decomposition and contact mechanics.
2) Semi-analytical solutions of singular stress fields, leading to accurate solutions at reduced mesh generation burden.
3) Compatibility with standard finite elements. The S-elements based on the scaled boundary finite element method can be coupled seamlessly with standard finite elements. The formulations can be simply implemented in an existing finite element computer program as another type of element (Zou et al., 2017).

The author strongly believes that the scaled boundary finite element method offers great potential in developing software for fully-automatic numerical modelling and simulation.

A brief summary of the concept of the scaled boundary finite element method is provided in Section 1.2. Some example applications are presented in Section 1.3 to illustrate the salient features of the scaled boundary finite element method.

1.2   Overview of the Scaled Boundary Finite Element Method

As mentioned previously (Section 1.1), a common feature of numerical methods is the use of mesh discretization to divide a complex problem domain into a mesh of discrete pieces. In the finite element method, the pieces are called finite elements or, simply, elements.

The modelling process of the scaled boundary finite element method is illustrated in Figure 1.1 using a two-dimensional problem. The problem domain in Figure 1.1a is divided into discrete pieces of simple geometry as shown in Figure 1.1b by introducing a net of non-overlapping internal lines. The pieces of simple geometry are termed S-domains. As will be explained in Section 2.2.1, the shape of an S-domain only needs to meet the scaling requirement, which can be briefly expressed as, that a scaling centre can be identified from which the whole boundary of the S-domain is directly visible. The scaling centres of the S-domains are indicated by the markers '⊕'. The boundary of the problems is broken into short segments by the internal lines. The lines on the boundary and the internal lines constitute the edges of the S-domains. The S-domains are seamlessly connected at their common edges, i.e. the internal lines. This process is similar to the construction of a geometrical model in a finite element pre-processor.

All the lines defining the S-domains are meshed into line elements in the next step of the modelling process. Figures 1.1c and 1.1d show two possible meshes. Since the lines are only connected at their ends, each line can be discretized independently of the others. In Figure 1.1c, the lines are divided into different numbers of 3-node line elements. The discretization of an S-domain is constructed from the discretization of the lines forming its edges (Section 2.2.2). The discretization in Figure 1.1d is obtained by local refinement of the three horizontal lines around S-domains 1 and 6 of the discretization in Figure 1.1c. Elements of different orders are used in the same mesh. The refinement

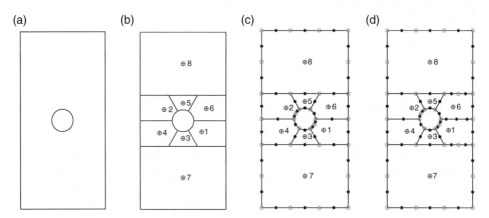

Figure 1.1 Discretization process of a rectangular plate with a circular hole in the scaled boundary finite element method. (a) Problem domain. (b) Discretization of the problem domain into S-domains. (c) Discretization of the boundaries of S-domains with 3-node line elements. The red nodes indicate the end nodes of line elements. (d) Modification of the boundary discretization of S-domains: Replacing the 3-node element at the common edge of S-domains 1 and 6 with a 5-node line element, and dividing the common edge between S-domains 1 and 7 and between S-domains 6 and 8 into two 3-node line elements.

(a)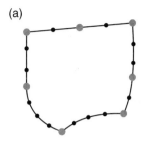

(b)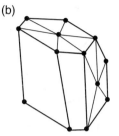

Figure 1.2 Examples of S-elements. (a) in two dimensions. (b) in three dimensions.

on one line affects only the two S-domains directly connected to it. As the S-domains are connected seamlessly by the internal lines, the displacement compatibility between two adjacent S-domains is enforced automatically.

A boundary-discretized S-domain is defined as an S-element.[1] From the point of view of domain discretization, the S-elements in a scaled boundary finite element analysis play the same role as finite elements in the finite element method. Generally speaking, an S-element can have any number of edges in two dimensions and any number of faces and edges in three dimensions. An edge or a face can also be divided into multiple elements. A two-dimensional S-element and a three-dimensional S-element are shown in Figure 1.2 as examples.

In a linear elastic analysis, the formulation of an S-element in the scaled boundary finite element method is fundamentally different from the finite element formulation in the following two aspects:

1) The discretization is only performed on the boundary of an S-element. In two dimensions, the boundary is divided into line elements as illustrated in Figure 1.2a. The line elements are modelled in a manner similar to one-dimensional finite elements (see Section 2.2.2). The compatibility between elements is enforced by simply joining their end nodes (shown by the larger dots in Figure 1.2a) together. Any type of displacement-based elements can be used and mixed together with other types. One edge of an S-element can also be divided into multiple line elements. In three dimensions (Figure 1.2b), the boundary of an S-element is divided into surface elements. The displacement compatibility is ensured by using the same discretization at the common edge of two connecting surface elements.

---

1 During the development of the scaled boundary finite element method, several other terms were used to refer to an S-domain or S-element. No distinction is made between an S-domain, a geometrical entity, and an S-element with the boundary discretization. Initially, the method was developed to model problems involving unbounded domains. In the substructure technique, the whole unbounded domain of a problem is considered as one S-domain and the term 'subdomain' was used. In Song (2004b), a dynamic fracture analysis is performed by dividing a problem domain into several pieces that are considerably more complex than finite elements. Each of the pieces is termed a 'super-element'. To further extend its application range, the scaled boundary finite method is applied to construct polygon elements, which permits automatic mesh generation. Elastoplastic analysis (Ooi et al., 2014), crack propagation (Ooi et al., 2013) and functionally-graded materials (Chiong et al., 2014) are addressed. In this context, the term 'scaled boundary polygon element' or simply 'scaled boundary polygon' is used. All these terms are appropriate for certain classes of applications of the scaled boundary finite element method, but not for others. Thus, the term 'S-element' is coined to describe an entity that satisfies the scaling requirement and is modelled by the scaled boundary finite element method with a boundary discretization.

 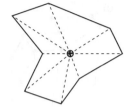 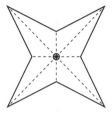

Figure 1.3 Examples of octagons that satisfy the scaling requirement.

2) The interior of the S-element is represented by scaling the boundary continuously towards a point called the scaling centre (see Section 2.2.3). No volume discretization is performed. The scaled boundary finite element method reduces the governing partial differential equations to a set of ordinary differential equations called the scaled boundary finite element equation (see Section 2.3). They represent a stronger equilibrium requirement than the finite element equations obtained from the volume discretization. Solving the ordinary differential equations analytically (see Section 3.3) leads to the stiffness matrix of the S-element and the displacement distribution inside the S-element. The solution has a richer basis than polynomials. For example, the stress singularity at a crack tip is represented analytically and accurately without any *a priori* knowledge or special treatments (Chapter 10).

A two-dimensional S-element can be regarded as a polygon element in the finite element sense. As long as the scaling requirement is satisfied, polygon elements of any number of sides (n−gons) can be constructed. The polygons can also be concave. Figure 1.3 shows three examples of octagons that meet the scaling requirement, and thus, can be used as S-elements in the scaled boundary finite element method. The generation of polygon meshes is closely related to the generation of triangular meshes. A simple function (see Section 4.3.3) to generate a polygon mesh from a triangular mesh is included as a part of the MATLAB code Platypus accompanying this book. The scaled boundary finite element method can be directly used with a polygon mesh generator to analyse problems of complex geometry.

The S-elements are much more flexible than finite elements to use in mesh generation owing to the following two features:

1) An S-element can assume more complex shapes than a finite element can. In Figure 1.1b, S-elements 1, 2, 4 and 6 have five edges. The shape of S-elements 7 and 8 is a rectangle, but the upper edge of S-element 7 and lower edge of S-element 8 are each composed of three independent line segments to be compatible with the adjacent S-elements. They can be regarded as polygons of six edges. The scaled boundary finite element method treats triangles and quadrilaterals (e.g. S-elements 3 and 5 in Figure 1.1b) as special cases. The boundary of an S-element can also be curved.

2) Only the boundary of an S-element is discretized. As depicted in Figure 1.1c, the edges of the S-elements are modelled with 3-node (quadratic) elements. No internal meshing of the S-element is required. Furthermore, since one edge can be shared by only two S-elements, the changes caused by any modifications or refinements of the mesh on one edge are limited to the two S-elements connected to the edge. Figure 1.1d illustrates the flexibility in meshing by modifying the mesh in Figure 1.1c.

The 3-node element at the common edge between S-elements 1 and 6 is replaced with a 5-node element. The elements on the common edges between S-elements 1 and 7 and between S-elements 6 and 8 are each divided into two 3-node elements. The modifications do not necessitate any other changes of the mesh. This feature is advantageous for mesh transition and local mesh refinement.

After the S-element formulations (including the stiffness matrix) are established, the rest of the solution procedure of a linear elastic problem follows that of the finite element method. The unknowns are defined at the nodes of the global mesh. The system of algebraic equations of the whole problem is obtained by assembling the stiffness matrices of the S-elements to satisfy the physical laws such as compatibility and equilibrium (see Section 3.5). The assemblage procedure is based on the connectivity of the S-elements to the global mesh and follows that of the standard finite element method by treating an S-element as a finite element. The finite element techniques for the enforcement of boundary conditions are directly applicable. The nodal displacements are obtained by solving the resulting system of algebraic equations.

The nodal solution of an S-element is extracted from the nodal solution of the global mesh based on the connectivity. The internal solution is obtained from the S-element formulation (Section 3.7). The stress field is expressed semi-analytically. The stresses at the boundary can be evaluated directly at the Gauss points of line elements. At a crack tip, the strain/stress singularity is represented by analytical functions, leading to a simple and accurate approach to determine the stress intensity factors (Chapter 10).

The polygon S-elements can be regarded as a type of generalized finite element of more complex geometries than standard elements. The internal displacement distribution over an S-element is obtained from the solution of its scaled boundary finite element equation. By expressing the internal displacements as functions of nodal displacements at the boundary, the shape functions of the S-element are obtained (Chiong et al., 2014). High-order elements are constructed using spectral elements on the boundary. A high-order mass matrix is developed in Song (2009) to model the high-frequency responses of a high-order polygon element without an internal mesh. An example of a shape function of a cracked S-element is shown in Figure 1.4. The availability of shapes functions permits the direct use of standard finite element procedures to broaden the applications of the scaled boundary finite element technique. Problems of functionally-graded materials are studied in Chiong et al. (2014). Elastoplastic analyses are performed in Ooi et al. (2014). A development to perform geometrically nonlinear analysis is reported in Behnke et al. (2014).

Indeed, the scaled boundary finite element method involves a few mathematical techniques (e.g. linear ordinary differential equations, the Schur decomposition and the matrix exponent) that are not commonly found in textbooks on the finite element method. The perceived complexity in mathematics could be considered a disadvantage of the scaled boundary finite element method in the context of the current state of computational solid mechanics, which is dominated by the finite element method. On the other hand, these mathematical techniques are well established. Their theory and computational implementation are covered by many mathematical textbooks. The solution of Euler-Cauchy ordinary differential equations is taught in engineering mathematics. The reader may refer to Kreyszig (2011) for details. The Schur decomposition and matrix exponent are explained in Golub and van Loan (1996). MATLAB functions are available for these operations.

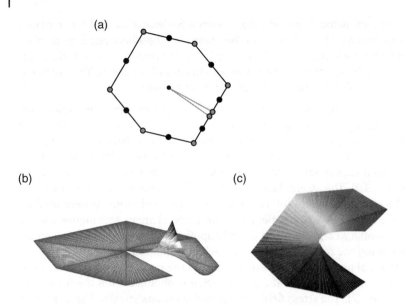

Figure 1.4 Example of shape functions of a cracked S-element. (a) Polygon S-element with edges discretized using quadratic elements. (b) Shape function for the crack mouth node. (c) Shape function near the crack tip, showing the $\sqrt{r}$ distribution. *Source*: Figure 5 in (Chiong et al., 2014).

1.3 Features and Example Applications of the Scaled Boundary Finite Element Method

In the scaled boundary finite element method, a problem domain is discretized into a mesh of S-elements. This is analogous to the discretization of the problem domain into finite elements in the finite element method. In comparison with standard finite elements, an S-element has the following two defining characteristics arising from its formulation:

1) A semi-analytical solution of field variables in an S-element. In a finite element, the solution of field variables is approximated by the polynomial interpolation of nodal values. In an S-element, the scaled boundary finite element solution is approximated by the polynomials in the circumferential directions parallel to the boundary. In the radial direction, the solution is obtained analytically and not limited to polynomials. Owing to the semi-analytical nature of the solution, the scaled boundary finite element method is advantageous in modelling problems involving infinity, such as unbound domains (where a dimension is infinite), and linear fracture problems (where the stresses tend to infinity at a crack tip).

2) A high degree of flexibility for mesh generation. The shape of an S-element in the scaled boundary finite element method needs to meet only the scaling requirement. It may assume more complex shapes than finite elements. While finite elements are limited to triangles and quadrilaterals in two dimensions, and tetrahedrons, pyramids, wedges and hexahedra in three dimensions, an S-element can be polygons and polyhedra of arbitrary number of edges and faces. In addition, only the boundary of the S-element is discretized. The domain of the S-element is obtained by scaling the

boundary. Owing to the flexibility in the shape of an S-element and the boundary discretization, the scaled boundary finite element method greatly reduces the mesh generation burden.

Several example applications for stress analysis are summarized in this section. The aim is to illustrate the salient features of the scaled boundary finite element method. No complete information is provided. Some of the examples will be detailed in later chapters, and some require further reading of the relevant references.

1.3.1 Linear Elastic Fracture Mechanics: Crack Terminating at Material Interface

A rectangular body with a crack terminating at a material interface is shown in Figure 1.5a with its dimensions. The inclination angle of the crack is α. The crack length is $a = b/\cos\alpha$. The two materials are orthotropic. The rectangular body is loaded by a uniform tension p at the lower and upper edges. Details of this example can be found in Example 10.5 on page 430.

This example illustrates the simplicity of the scaled boundary finite element method in modelling stress singularities in linear fracture mechanics. The whole problem is modelled by a single S-element. The scaling centre is selected at the crack tip. The mesh is shown in Figure 1.5b. The outer boundary of the rectangular body is discretized with 12 nine-node elements.

A unique feature of the scaled boundary finite element method is that the crack faces and material interface passing through the scaling centre are represented without discretization. At the crack mouth (Figure 1.5b), one node on each crack face is introduced and the two elements are not connected. The crack faces are generated by these two independent nodes at the crack mouth when scaling the boundary. At a material interface, the two elements (Figure 1.5b) are connected. Each of the elements is assigned to the corresponding material. The material interface is formed by scaling the common node.

Figure 1.5 Crack terminating at material interface. (a) Geometry. (b) Mesh.

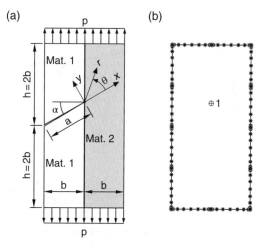

The scaled boundary finite element solution in the radial direction emanating from the crack tip, including the stress singularity, is expressed analytically as a power series. The generalized stress intensity factors and orders of singularity, which characterize the singular stress field, are determined directly from their definitions using standard stress recovery techniques.

A shape sensitivity analysis with respect to the crack inclination angle α is conveniently performed. It is sufficient to move the scaling centre with the crack tip on the material interface. The same boundary mesh can be used without modification. The generalized stress intensity factors and orders of singularity are plotted in Figure 1.6 for varying crack inclination angle α.

At the crack inclination angle $\alpha = 30°$, the variation of the generalized stress intensity factors with respect to the angular coordinate θ (see Figure 1.5) is shown in Figure 1.7. The result is useful in determining the direction of crack propagation.

The above scaled boundary finite element analysis is computationally very efficient. Using an in-house code written in MATLAB, the analysis takes less than 0.3s to complete on a Microsoft Surface Pro 3 laptop computer with an i7-4650U CPU.

(a)

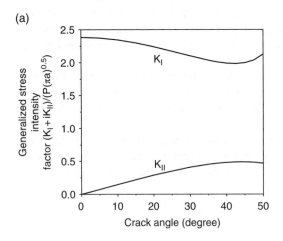

(b)

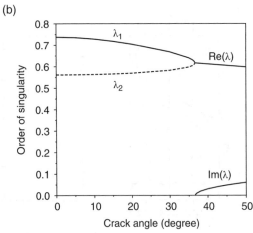

Figure 1.6 Crack terminating at material interface with varying crack angle α. (a) Generalized stress intensity factors. (b) Order of singularity.

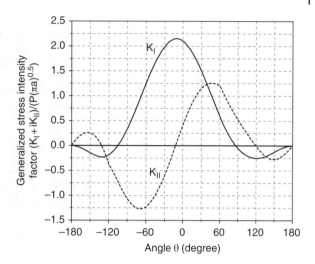

Figure 1.7 Crack terminating at material interface. Angular variation of generalized stress intensity factors at crack inclination angle $\alpha = 30°$.

The salient features of the scaled boundary finite element method in modelling singular stress field include:

1) Simplicity in mesh generation as described above. Furthermore, straight crack faces and material interfaces passing through the crack tip are not discretized.
2) The scaled boundary finite element solution is expressed analytically in the radial direction as a power series. The stress intensity factors, T-stresses and the coefficients of higher-order terms can be evaluated directly based on their definitions.
3) No fundamental solution or asymptotic expansion is required, which permits anisotropic materials and multi-material corners to be addressed without increasing the computational effort.
4) Not only real and complex power singularities but also power-logarithmic singularities can be modelled robustly.
5) The generalized stress intensity factors can be evaluated at any angular coordinate around the crack tip by using a stress recovery technique in the finite element method.

1.3.2 Automatic Mesh Generation Based on Quadtree/Octree

The scaled boundary finite element method is highly complementary with the quadtree/octree algorithm for mesh generation and remeshing. Their combination provides a promising technique for integrating geometric models and numerical analysis in a fully automatic manner (see Section 9.6 on page 378).

The quadtree/octree algorithm recursively subdivides square/cubic cells into smaller ones by halving the edge lengths. The algorithm is simple and highly efficient, but its direct application in the finite element analysis, where a cell is modelled by a square/cubic element, is limited mostly by two issues. One is the existence of hanging nodes between elements of different size. The other is the difficulty in representing a curved boundary by square/cubic elements.

These two issues are resolved straightforwardly in the scaled boundary finite element method. The analysis of a square body with multiple circular holes using a quadtree mesh

(a)   (b)

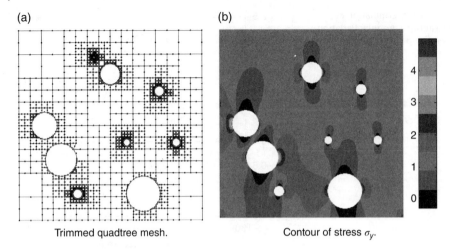

Trimmed quadtree mesh.   Contour of stress σ_y.

Figure 1.8 A square body with multiple holes under tension.

is shown in Figure 1.8 as an example. Each quadtree cell is modelled as an S-element. When the sizes of S-elements change, the common line between a larger S-element and the two adjacent smaller S-elements is divided into two line elements. The compatibility between the S-elements of different size is satisfied automatically and the hanging nodes are thus eliminated. At a curved boundary, a square element truncated by the boundary is directly modelled as a polygon S-element. Higher-order elements can be used to increase the accuracy in representing a curved boundary. Details of this example are described in Section 9.7.1 on page 384.

The analysis of a cube with one octant being removed is illustrated in Figure 1.9. The geometry of the cube and the boundary conditions (the default boundary condition is traction-free) are given in Figure 1.9a. An S-element mesh and the contour of the displacement in z-direction are depicted in Figure 1.9b. Smaller S-elements are employed around the edges of the cutout, where the stresses concentrate. The use of S-elements eliminates the hanging nodes between the octree cells of different size. Figure 1.9c presents the tetrahedral finite element mesh of similar element size. A convergence study is performed. The displacement error norms are shown in Figure 1.9d. The scaled boundary finite element method leads to more accurate results than the finite element method for both linear and quadratic elements.

1.3.3 Treatment of Non-matching Meshes

Non-matching meshes occur in domain decomposition and contact mechanics. Special techniques, often introducing additional unknown variables, are required in a finite element analysis to enforce the compatibility and equilibrium at the interfaces of non-matching meshes. Using the flexibility of S-elements in their shapes (i.e. an S-element may have any number of faces and edges, and only the boundary of an S-element is discretized), non-matching meshes can be converted to matching meshes (see Section 5.3).

A two-dimensional example is shown in Figure 1.10. Two rectangular domains are meshed independently by either standard finite elements or S-elements. To link the

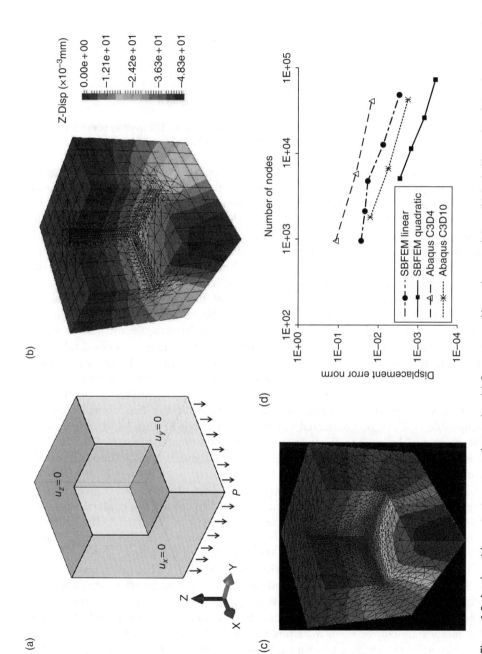

Figure 1.9 A cube with a cutout corner under tension. (a) Geometry and boundary conditions. (b) Scaled boundary finite element mesh and z-displacement contour. (c) Finite element mesh and z-displacement contour. (d) Convergence of displacements.

(a)   (b)

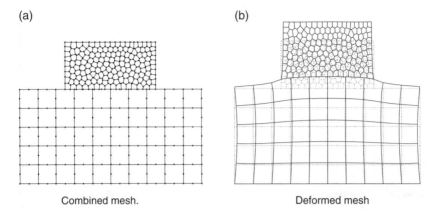

Combined mesh.   Deformed mesh

Figure 1.10 Modelling of two square domains meshed independently. The meshes of the two domains are linked at the interface by inserting the nodes of one mesh into the other.

non-matching meshes, the discretization at the interface is modified by including the nodes from both meshes. This operation breaks the discretization of both meshes at the interface into shorter line elements. The two-dimensional elements connected to the interface are replaced by S-elements with their edges formed by these shorter line elements. The two meshes have identical interface discretization and become compatible. Figure 1.10a depicts the combined mesh obtained by connecting the meshes of the two domains node-by-node. It is observed from the deformed mesh in Figure 1.10b that the displacements of the two domains are compatible.

The same approach is also applicable to treat three-dimensional non-matching meshes. An example is shown in Figure 1.11. The hexahedral meshes of the two domains in Figures 1.11a and 1.11c are generated independently and non-matching at the interface (shaded areas). The intersecting points (within a specified tolerance) of the interfaces meshes of the two domains are found (Figure 1.11b). On the interface, the quadrilateral face of a hexahedral element is divided into several surface elements. The hexahedral element is replaced by an S-element with the surface elements on the quadrilateral. The surface meshes on the interfaces of the two domains become compatible and can be connected simply node-to-node (Figure 1.11d). Figure 1.11e shows a part of the mesh to illustrate the interior of the combined mesh. A modal shape of free vibration is shown in Figure 1.11f. The displacements are continuous across the two domains.

Converting non-matching meshes to matching ones simplifies the contact analysis. The node-to-node scheme can always be employed to enforce the contact constrains. The analysis of the Hertz contact problem is depicted in Figure 1.12. The geometry, dimensions and boundary conditions are shown in Figure 1.12a. The Young's modulus of rectangular domain is chosen as 5000 times that of the semi-cylindrical domain to simulate a rigid supporting block. The Poisson's ratios of both domains are equal to $v = 0.3$. Coulomb friction with coefficient $\mu = 0$ and plain strain conditions are assumed. A uniform pressure $p = 1.3485\text{kN/m}$ is applied at the top surface of the semi-cylinder. The non-matching meshes of the two domains are shown in Figure 1.12b. The meshes are converted to matching meshes over the potential contact area. The distribution of the contact pressure is plotted in Figure 1.12c, where $p_{\max}$ and $b_{\max}$

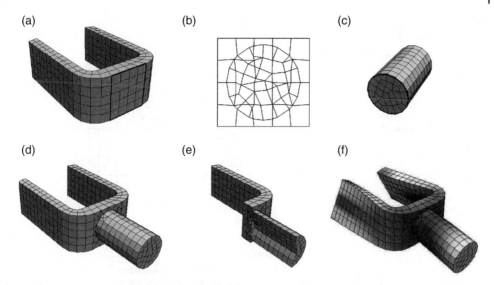

Figure 1.11 Connecting 3D non-matching mesh. (a) Mesh of domain 1. (b) Surface mesh (enlarged) of the interface, where the surface elements represent the boundary of the 3D S-elements connected to the interface. (c) Mesh of domain 2. (d) Combined mesh. (e) Interior view of the combined mesh. (f) A modal shape where the right end is fixed.

are maximum values of contact pressure and the half-width of the contact area. The results obtained from the scaled boundary finite element method converge to the exact solution as the mesh is refined. The contour plot of the vertical stress σ_{yy} is shown in Figure 1.12d (denoted as S_{22} in the legend).

1.3.4 Crack Propagation

Two commonly encountered obstacles in the simulation of a crack propagation problem by finite elements are the handling of the evolving crack path and the modelling of stress singularity at the crack tip. The scaled boundary finite element method provides an attractive alternative (Ooi et al., 2012b, 2013, 2015). Since an S-element may have any number of edges, the propagating crack path is represented by splitting each S-element on the path into two smaller ones. The singular stress field is modelled accurately by the S-element surrounding the crack tip (see Section 1.3.1 and Chapter 10). The fact that no local refinement of the S-element mesh is required around the crack tip contributes to reducing the meshing and re-meshing burden. The process requires only a minimal amount of remeshing and is easily automated (Section 10.7).

An example (Ooi et al., 2015) of the simulation of crack propagation is shown in Figure 1.13. The geometry, initial crack and dimensions (unit: mm) are shown in Figure 1.13a. The initial quadtree mesh is displayed in Figure 1.13b. The crack path is deflected towards the circular hole as shown in Figure 1.13c.

1.3.5 Adaptive Analysis

The semi-analytical solution of the scaled boundary finite element method also provides an *a posteriori* error indicator of the solution. In combination with the

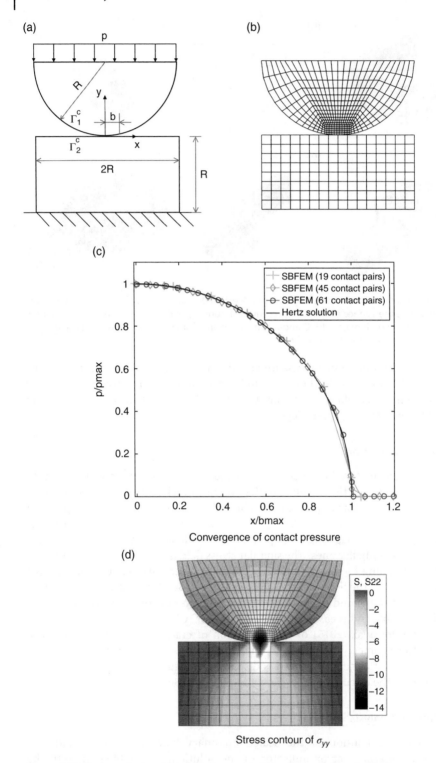

Figure 1.12 Hertz contact problem between a cuboid and a parallel semi-cylinder.

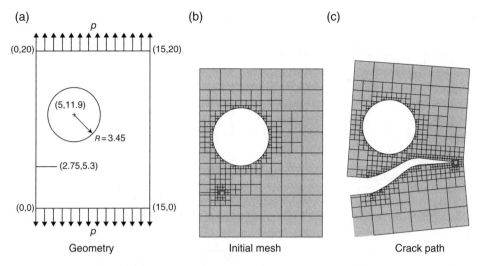

Figure 1.13 Crack path deflection by a circular hole.

quadtree mesh generation algorithm, a simple and convenient procedure for adaptive analysis is developed. The analysis of a mechanical part in Figure 1.14a is shown as an example. The initial mesh of S-elements is shown in Figure 1.14b. The boundaries of the S-elements are discretized by quadratic elements. The mesh is refined adaptively to reduce the error in strain energy. The meshes of the first four refinements are shown in Figures 1.14c to 1.14f. Note that the mesh of the last (4th) refinement in Figure 1.14f is finer around the stress concentrators (small holes, reentrant corners and corners at the fixed boundary) than elsewhere. The quadtree mesh results in rapid local refinement.

1.3.6 Transient Wave Scattering in an Alluvial Basin

The scaled boundary finite element method is developed for the direct time-domain analysis of wave scattering problems in Bazyar and Song (2017). The analysis of wave scattering in a five-layer alluvial basin is shown in Figure 1.15. The site conditions are illustrated in Figure 1.15a. A near field enclosing the alluvial basin is selected and discretized by a quadtree mesh of S-elements (Figure 1.15b). The remaining far field is modelled as an unbounded S-domain (see point 3b on page 36). A spatially and temporally local approximation is used. The radiation condition is satisfied rigorously. The seismic inputs are considered incident fields of obliquely plane waves. The equivalent boundary tractions are applied at the interface between the near field and the far field. The transient responses at points B and D indicated in Figure 1.15 are plotted in Figures 1.15c and d, respectively.

1.3.7 Automatic Image-based Analysis

Quadtree/octree algorithm is commonly used in image processing. Quadtree/octree cells are generated by recursively subdividing one cell into four/eight smaller ones of equal size until only one material exists in one cell. This algorithm provides a

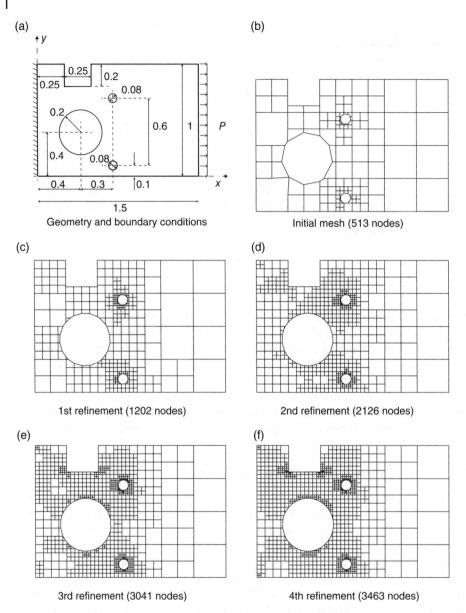

Figure 1.14 Adaptive analysis of a mechanical part.

fully-automatic and highly-efficient approach for the generation of S-element mesh from digital images (see Section 9.3).

1.3.7.1 Two-dimensional Elastoplastic Analysis of Cast Iron

The digital image of the microstructure of a nodular cast iron in Collini (2005) is shown in Figure 1.16a. The size of the image is 454×343 pixels. Three constituent materials are present in the image. The black represents the graphite, and the dark grey represents the

(a)

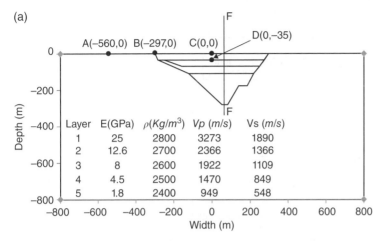

Geometry and material property

(b)

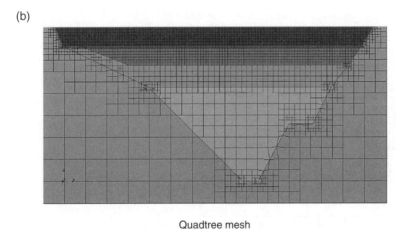

Quadtree mesh

Figure 1.15 Transient response of an alluvial basin subject to a plane *SV*−wave propagating with incident angle of 30°.

pearlite, and light grey represents the ferrite. The graphite is usually considered as a void, because it has a much lower Young's modulus than the ferrous matrix, and the interface bond between the nodule and its encapsulating cavity is relatively weak.

A quadtree mesh is shown in Figure 1.16b. The materials are assigned to the quadtree cells based on their grey levels. The size of the smallest quadtree cells is chosen as 1 pixel and, thus, the full resolution of the image is preserved. The size of the largest cells is 16 pixels. The ratio of the number of quadtree cells to the total number of pixels of the image is approximately 13%.

The quadtree cells are directly modelled as S-elements of 4–8 edges. An elastoplastic analysis of the specimen under uniaxial tension in the horizontal direction is performed. The von Mises stress and equivalent plastic strain are depicted in Figures 1.16c and d, respectively. As the graphite behaves like voids in the continuum material matrix, stress concentrations originated at the boundaries of the voids perpendicular to the

(c)

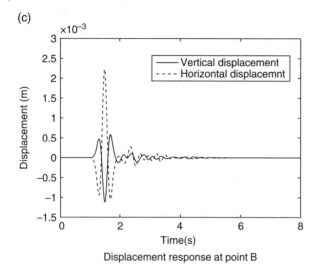

Displacement response at point B

(d)

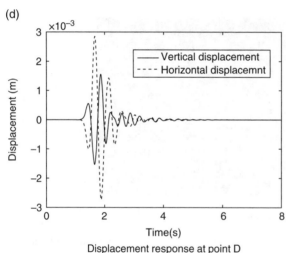

Displacement response at point D

Figure 1.15 (*Continued*)

loading direction. The plastic strain is localized in the ferrite phase because of the lower yield stress. The stress and strain concentration bands appear at roughly 45 degrees with respect to the loading axis. The slip bands of stress and strain concentrations are connecting the voids and indicate possible fracture initiation locations (Collini, 2005).

1.3.7.2 Three-dimensional Concrete Specimen

A three-dimensional image-based analysis of a concrete specimen is illustrated in Figure 1.17. The digit image in Figure 1.17a consists of $256 \times 256 \times 256$ (i.e. 16,777,216) voxels. The black and grey phases represent the aggregates and mortar, respective. The octree mesh of S-elements shown in Figure 1.17b is generated fully automatically. The resolution of the mesh is 1 voxel, leading to 3,002,329 S-elements.

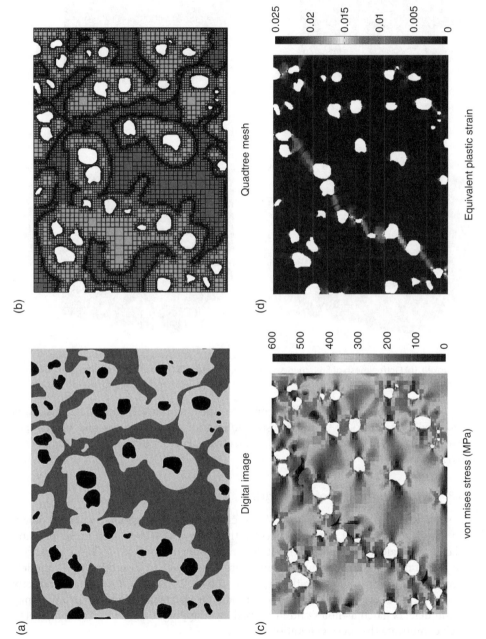

(a)

Digital image

(b)

Quadtree mesh

(c)

von mises stress (MPa)

600
500
400
300
200
100
0

(d)

Equivalent plastic strain

0.025
0.02
0.015
0.01
0.005
0

Figure 1.16 Two-dimensional image-based elastoplastic analysis of cast iron.

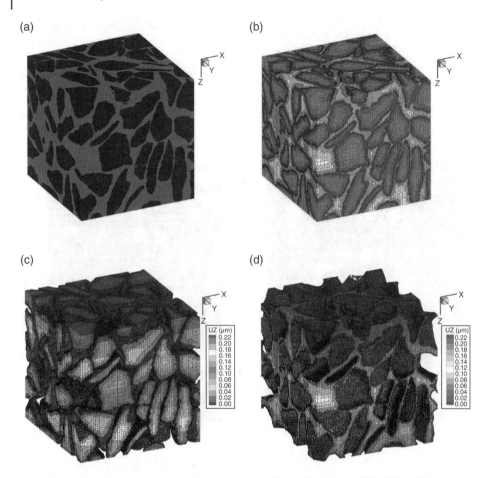

Figure 1.17 Three-dimensional image-based analysis of concrete specimen. (a) Digital image (256^3 voxel). (b) Octree mesh of S-elements. (c) Contour of z-displacement of aggregates. (d) Contour of z-displacement of mortar.

A linear elastic analysis of the concrete specimen under uniaxial tension is performed. The contours of the z-displacements of the aggregates and mortar are shown in Figure 1.17c and Figure 1.17d, respectively. Further details of this example are described in Section 9.5.2.

1.3.8 Automatic Analysis of STL Models

The standard tessellation language (STL) describes the surface geometry of a 3D object by a list of unstructured triangular facets. This format is widely used in 3D printing and computer-aided design (CAD) systems and supported by many software packages. Common finite element mesh generators are developed to discretize CAD models represented as Non-uniform rational Basis splines (NURBS). They do not directly handle STL models. Furthermore, an STL model for visualization and 3D printing may contain defects (such as holes and ill-shaped, overlapping, or self-intersecting facets), which complicate the mesh generation process.

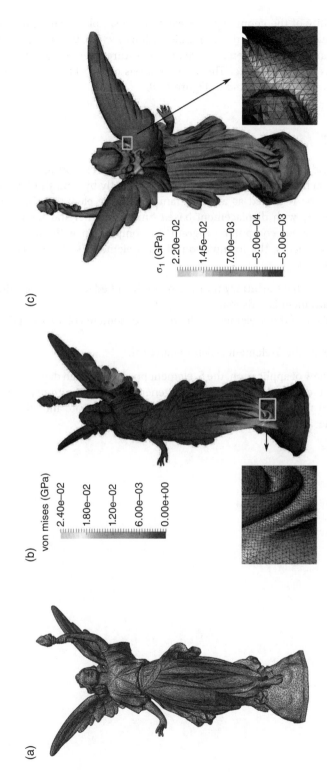

Figure 1.18 Stress analysis of STL model of Lucy. (a) STL model. (b) von Mises stress. (c) First principal stress.

Using the flexibility of the geometry of S-elements, a simple meshing procedure of STL models is developed based on the octree algorithm (Section 9.6 on page 378). After the mesh resolution, maximum element size and curvature tolerance are specified, the meshing process is fully automatic. The stress analysis of an STL model of the Lucy statue under self-weight is illustrated in Figure 1.18.

1.4 Summary

In the scaled boundary finite element method, a problem can be decomposed into several domains. Each domain is discretized independently by a mesh of S-elements. The S-element can be regarded as another family of finite elements. In fact, it can be coupled seamlessly with displacement-based finite elements and incorporated in an existing finite element computer program. In comparison with the formulation of standard finite elements, the formulation of an S-element is characterized by the following three features

1) The S-element may have arbitrary number of faces and edges on the condition that the scaling requirement is satisfied.
2) Only the boundary of the S-element is divided into elements of a spatial dimension lower by one.
3) The solution inside the S-element is semi-analytical.

From the point of view of application, the S-element possesses a high degree of flexibility in its shape, which alleviates the burden of mesh generation of the finite element method. The semi-analytical solution of the S-element provides an accurate and efficient representation of singularities as occurring at crack tips.

Part I

Basic Concepts and MATLAB Implementation of the Scaled Boundary Finite Element Method in Two Dimensions

Everything should be made as simple as possible, but not simpler.

—Albert Einstein

The above famous quote by Albert Einstein provided valuable philosophical guidance on our research on the development of the scaled boundary finite element method. Although what is 'as simple as possible' is problem-dependent and subjective, we strive to develop and apply simple approaches (albeit which may be unfamiliar at the first sight) to solve challenging problems in numerical modelling and simulation.

The same philosophy is followed in writing this part of the present book. For the sake of simplicity, the contents are mostly limited to what are essential to understand the scaled boundary finite element method. This part is focused on two-dimensional stress analysis. It is built upon the 2-node isoparametric line element shown in Figure I.1 which is arguably the simplest element in the finite element method.

To model two-dimensional problems, several 2-node line elements are connected sequentially to form the edges of a polygon as shown in Figure I.2a. The only requirement on the shape of the polygon is that a so-called scaling centre (i.e. a point from where all the elements are directly visible) can be identified. This type of polygons is known in mathematics as star-shaped polygons (Stewart and Tall, 1983). The scaling centre is indicated by the symbol $\oplus$ in Figure I.2. The polygon (the grey area in Figure I.2) is obtained by scaling the line elements with respect to the scaling centre. There is no limitation on the number of line elements to use to define the polygon edges. One edge of a polygon can also be formed by more than one element as depicted in Figure I.2b. Additionally, it is possible for the line elements to form an open loop as illustrated in Figure I.2c. Scaling the elements results in a polygon with two extra straight edges connecting the scaling centre to the nodes at the two ends of the loop.

A polygon obtained by scaling the line elements is referred to as an S-element in this book. The element formulations of a polygonal S-element will be established in this Part. The role of the S-elements in the scaled boundary finite element method is the same as that of the finite elements in the finite element method.

The S-elements provide a high degree of flexibility in mesh generation. Three examples are illustrated in Figure I.3. The first example (Figure I.3a) is a mesh generated by Poly-Mesher (Talischi et al., 2012). The second example (Figure I.3b) is a mesh converted

The Scaled Boundary Finite Element Method: Introduction to Theory and Implementation,
First Edition. Chongmin Song.
© 2018 John Wiley & Sons Ltd. Published 2018 by John Wiley & Sons Ltd.
Companion website: www.wiley.com/go/author/scaledboundaryfiniteelementmethod

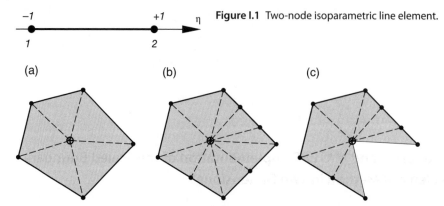

Figure I.1 Two-node isoparametric line element.

Figure I.2 Construction of polygons using 2-node line elements. (a) A pentagon enclosed by a closed loop with one element per edge. (b) A pentagon enclosed by a closed loop with varying number of elements per edge. (c) A polygon formed by an open loop of elements.

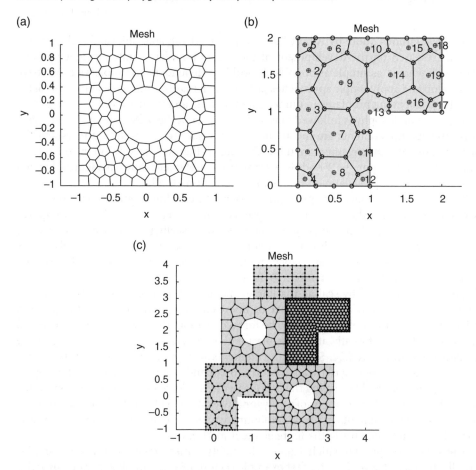

Figure I.3 Examples of meshes of S-elements. (a) A square with a circular hole. (b) An L-shaped panel. S-element 13 surrounding the re-entrant corner is formed by scaling an open loop of line elements (see Figure I.2c). (c) A system composed of multi-domains. The meshes of the individual domains are generated independently and connected at their interfaces by subdividing the edges of the S-elements connected to the interfaces.

from a triangular mesh. In this mesh, S-element 13 is obtained by scaling an open loop of line elements and each edge is modelled by two line elements. This unique feature of the S-element provides an efficient and accurate technique to model the stress singularity at the re-entrant corner (see Section 4.4.3.1 on page 203). The last example (Figure I.3c) shows a mesh of a mosaic of five domains (see Example 5.2 on page 231). The meshes of all the S-elements are generated independently and are connected together at their inter-faces. The interface edges of the polygon mesh of a domain are subdivided to include the interface nodes of the mesh of the adjacent domain. The displacement compatibility is ensured without additional treatments.

In the application of the scaled boundary finite element method, the process is inverted. A problem domain is decomposed into several smaller domains. Each of the smaller domains is discretized into a mesh of S-elements. Two adjacent S-elements are connected by their common edge. The edges of the S-elements are modelled by one or more line elements. The global system of algebraic equations is obtained by assembling the property matrices of the S-elements.

A computer program named Platypus[1] is provided with this book. Platypus per-forms two-dimensional stress analysis. It also includes a mesh generator of polygon S-elements. The boundaries of an S-element are divided into the 2-node line elements. The program is written in MATLAB and detailed documentation is included in the code. All the MATLAB functions and scripts that are used to produce the numerical examples in this part of the book are also provided. They aim to help the readers become familiar with the scaled boundary finite element method by following through the execution step-by-step. The example files can also serve as templates to assist the users in modelling problems of their own choice.

Part I of the book is organized as follows. In Chapter 2, the formulations of S-elements are derived. This corresponds to the construction of element formulations in the finite element method. In Chapter 3, the solution procedures of the scaled boundary finite element equations are developed. The stiffness matrix, mass matrix and displacement field of an S-element are obtained. In Chapter 4, techniques for the automatic genera-tion of polygon S-element meshes and numerical examples of the scaled boundary finite element analysis are presented. In Chapter 5, some modelling issues and salient features of the scaled boundary finite element method are discussed.

---

1 The platypus is an Australian native animal. It has many unlikely features and is considered to be one of nature's most unusual animals. When the first specimen of a platypus was sent to Great Britain for examination, scientists thought that it was an elaborate hoax. The platypus is described on pages 227–232 in Volume 10 of *Naturalist's Miscellany* by George Shaw (the document is available at http://www .biodiversitylibrary.org/item/124636#page/236/mode/2up). Many interesting and amazing facts on platypus are available on the internet (http://www.nationalgeographic.com/animals/mammals/p/platypus/, https://en .wikipedia.org/wiki/Platypus, http://mentalfloss.com/article/63062/10-curious-and-quirky-platypus-facts, to name a few sites). The platypus is featured on the reverse of the Australian 20c coin. It is the animal emblem of the state of New South Wales. 'Syd' the platypus was one of the three mascots of the Sydney 2000 Olympics.

2

Basic Formulations of the Scaled Boundary Finite Element Method

2.1 Introduction

This chapter presents the basic concepts of the scaled boundary finite element method for two-dimensional stress analysis. The theoretical formulation and implementation are mostly limited to those of simplistic polygonal S-elements (see, for example, Figure 1.3). The most basic 2-node line element (see Figure I.1) is used for the discretization of the edges of an S-element. The scaled boundary finite element formulation for an S-element is developed. From the point of view of development process, this is equivalent to the development of element formulation in the finite element method.

2.2 Modelling of Geometry in Scaled Boundary Coordinates

A key concept of the scaled boundary finite element method is to devise the scaled boundary coordinates to describe the geometry of an S-domain. The role of the scaled boundary coordinates is similar to that of the reference coordinates in the isoparametric finite elements. This coordinate system involves the discretization of the boundary of the S-domain, leading to an S-element. This allows a semi-analytical solution to be obtained.

Same as in the finite element method, the directions of displacements and forces remain in the Cartesian coordinates so that the assembly of S-elements can be conveniently formulated to satisfy the compatibility and the equilibrium.

2.2.1 S-domains: Scaling Requirement on Geometry, Scaling Centre and Scaling of Boundary

In this section, only the geometry of an S-domain is addressed. To guarantee that the so-called scaled boundary transformation introduced in Section 2.2.3 below is unique, the geometry of the S-domain has to meet a so-called *scaling requirement: there exists a region from which every point on the boundary is directly visible, or in other words, there exists a region that has direct line of sight of the whole boundary.* The scaling requirement can also be stated inversely as that there exists a region that is directly visible from any point on the boundary. A point called the *scaling centre* is selected in this region.

The Scaled Boundary Finite Element Method: Introduction to Theory and Implementation,
First Edition. Chongmin Song.
© 2018 John Wiley & Sons Ltd. Published 2018 by John Wiley & Sons Ltd.
Companion website: www.wiley.com/go/author/scaledboundaryfiniteelementmethod

The scaling requirement on the geometry of a domain is illustrated in Figure 2.1. The domain V shown in Figure 2.1a satisfies the scaling requirement as a point from where every point on the boundary S (the bold solid line) is directly visible can be identified. Such a point is selected as the scaling centre O. The direct lines of sight from the scaling centre to the corners of the domain are drawn as dashed lines. Note that the boundary does not have to be convex and the number of edges is not limited. Figure 2.1b shows an example of a domain that does not satisfy the scaling requirement. No point can be found that has direct visibility of the three points A, B and C on the boundary S at the same time. To apply the scaled boundary finite element method, such a domain has to be subdivided into several smaller ones that satisfy the scaling requirement.

In the scaled boundary finite element method, an S-domain is described by continuously scaling its boundary with respect to a scaling centre. Figure 2.2 depicts the operation to generate the S-domain V by scaling its boundary S. When the boundary is scaled by a factor less than 1, an internal curve similar to the boundary is obtained. Two such internal curves corresponding to scaling factors 0.8 and 0.6, respectively, are shown in Figure 2.2 as thin solid lines. When the scaling is performed continuously from the boundary (with a scaling factor 1) to the scaling centre (with a scaling factor 0), the whole domain is covered. This process is analogous to covering a surface by continuously moving a line. When the scaling requirement is satisfied, a point in the domain is covered once and once only in the scaling process.

The scaling operation to describe an S-domain is addressed in mathematical terms. As shown in Figure 2.3, a system of Cartesian coordinates x, y is defined. Its origin is selected at the scaling centre O. This choice simplifies the derivation of equations, and does not lead to loss of generality. The position vector of a point (x, y) in the S-domain

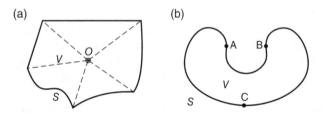

(a)   (b)

Figure 2.1 Illustration of the scaling requirement. (a) A bounded domain V that satisfies the scaling requirement. The scaling centre is indicated by the marker $\oplus$ from where every point on the boundary S (bold solid line) is directly visible. The dashed lines show examples of direct lines of sight. Such a domain will be referred to as an S-domain. (b) A bounded domain V that does not satisfy the scaling requirement.

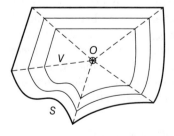

Figure 2.2 Representation of an S-domain V by scaling its boundary S with respect to the scaling centre O selected inside the domain. The dash lines indicate the lines of sight from the scaling centre. The thin lines indicate two typical internal curves resulting from scaling the boundary.

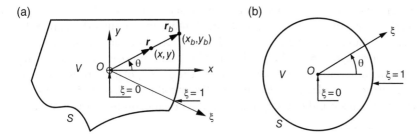

Figure 2.3 (a) Scaling of the boundary of an S-domain using the radial coordinate ξ ($\xi = 1$ at the boundary and $\xi = 0$ at the scaling centre) as the scaling factor. (b) An S-domain is transformed to a unit circular domain in the system of radial coordinate ξ and angular coordinate θ.

(Figure 2.3a) is expressed as

$$\mathbf{r} = x\mathbf{i} + y\mathbf{j} \tag{2.1}$$

where $\mathbf{i}$ and $\mathbf{j}$ are unit vectors along the x and y directions, respectively. The angular coordinate (i.e. the angle between the x–axis and the vector $\mathbf{r}$) is denoted as θ.

Along the radial line connecting the scaling centre and the point (x, y), the point (x_b, y_b) on the boundary is addressed. The point (x, y) and the point (x_b, y_b) have the same angular coordinate θ. The position vector $\mathbf{r}_b$ is expressed as

$$\mathbf{r}_b = x_b\mathbf{i} + y_b\mathbf{j} \tag{2.2}$$

When the scaling requirement is satisfied, no two points on the boundary have the same value of angular coordinate θ. The boundary can be defined by a single-valued function $r_b(\theta)$. It is often more flexible to describe a boundary of complex shape by two single-valued parametric functions $x_b(\theta)$ and $y_b(\theta)$

$$x_b = x_b(\theta) \tag{2.3a}$$
$$y_b = y_b(\theta) \tag{2.3b}$$

where the angular coordinate θ is used as the parametric variable. This parametric description will be used in the scaled boundary coordinates (Section 2.2.3).

A radial coordinate ξ emanating from the scaling centre O is introduced as the scaling factor (Figure 2.3a). The radial coordinate ξ is selected as $\xi = 0$ at the scaling centre and $\xi = 1$ on the boundary. The scaling requirement ensures that the radial coordinate ξ and the angular coordinate θ are not parallel.

As shown in Figure 2.3a, scaling the point (x_b, y_b) to the scaling centre O generates a radial line connecting the two points and passing through the point (x, y). The point (x, y) can thus by obtained by scaling the point (x_b, y_b) on the boundary with the radial coordinate ξ as

$$\mathbf{r} = \xi\mathbf{r}_b \tag{2.4}$$

The scaling operation is expressed in the Cartesian coordinates as

$$x = \xi x_b \tag{2.5a}$$
$$y = \xi y_b \tag{2.5b}$$

Equation (2.5) can be regarded as a coordinate transformation between (x, y) and (ξ, θ). As shown in Figure 2.3b, an S-domain is transformed into a unit circular domain in the system of radial coordinate ξ and angular coordinate θ. The boundary is specified by a constant radial coordinate $\xi = 1$. This choice simplifies the enforcement of boundary conditions in solving differential equations, much like the use of polar coordinates in the solution of problems on circular domains. In fact, the coordinate system $(\xi,\ \theta)$ resembles the polar coordinates r and θ. Using Eq. (2.5), the radial coordinate r is written as

$$r = \xi r_b = \xi \sqrt{x_b^2 + y_b^2} \tag{2.6a}$$

where

$$r_b = \sqrt{x_b^2 + y_b^2} \tag{2.6b}$$

is the radial coordinate on boundary. The angular coordinate θ is expressed as

$$\theta = \arctan \frac{y_b}{x_b} \tag{2.6c}$$

with the principal value $-\pi < \theta \le \pi$. The scaling requirement guarantees that the angle θ is a unique valued function along the boundary.

It is worthwhile noting the following remarks on the scaling requirement and the S-domain.

1) Star-shaped domains in mathematics (Stewart and Tall, 1983) are S-domains. A domain is star-shaped if there exists a point such that a straight line segment connecting this point with any other point in the domain is contained in the domain. In two and three dimensions, this criterion is equivalent to having direct lines of sight of the whole boundary from a point. Figure 2.4a depicts an example of star-shaped polygonal domains in computational geometry (Preparata and Shamos, 1985). The lines of sight are shown by dashed lines. The set of points that have direct line of sight of the whole boundary is called the kernel of the star-shaped polygon domain as shown by the shaded area in Figure 2.4b. Therefore, a star-shaped domain meets the scaling requirement and the scaling centre is selected within the kernel. A star-shaped domain is not necessarily convex. As shown in Figure 2.4, some parts of the boundary can be concave. All convex polygons are star-shaped. A convex

(a)   (b)

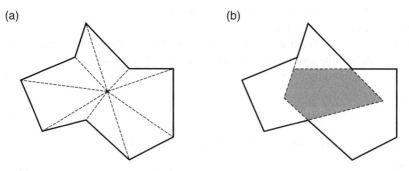

Figure 2.4 Star-shaped polygonal domain. (a) The dashed lines are the straight line segments that connect a point with the vortices of the polygonal domain. (b) The shaded region shows the kernel of the star-shaped polygonal domain.

Figure 2.5 An unbounded S-domain. The unbounded domain *V* is covered by scaling its boundary *S* with respect to the scaling centre *O* selected outside of the domain. The thin lines indicate internal curves resulting from scaling the boundary with a factor larger than 1. The dashed lines indicate the lines of sight from the scaling centre.

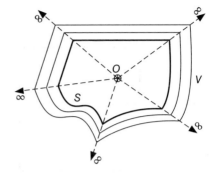

polygon coincides with its own kernel, i.e. the whole boundary of a convex polygon is directly visible from any point in the polygon.

On the other hand, an S-domain may not be a star-shaped domain. Some of these cases are given below.

2) The scaled boundary technique is also applicable to the modelling of unbounded (or exterior) domains. An unbounded domain is depicted in Figure 2.5. The scaling centre *O* is selected outside of the unbounded domain. Since the whole boundary is directly visible from the scaling centre, the unbounded domain is an S-domain, but not a star-shaped domain. The unbounded S-domain *V* is represented by scaling its boundary *S* with a scaling factor varying continuously from 1 at the boundary to infinity. The two internal curves obtained by scaling the boundary with the scaling factors 1.1 and 1.2, respectively, are shown in Figure 2.5 as thin solid lines.

3) Two useful extensions to the above base cases of forming S-domains by scaling their boundaries are described briefly in the following and will be addressed in detail in the subsequent chapters.

   a) The first is the modelling of problems with a straight edge crack and material interfaces as illustrated in Figure 2.6. Figure 2.6a is an edge-cracked square. The scaling centre is selected at the crack tip. The two straight crack faces passing through the scaling centre are not included in the scaling process and are denoted as 'side-face'. The rest of the boundary is referred to as the 'defining curve', which is not closed. The S-domain of the cracked square is obtained by continuously scaling the defining curve with respect to the scaling centre. The side-faces (crack faces) are formed in the scaling process by the two ends of the defining curve.

   A corner formed by three materials occupying the domains V_1, V_2 and V_3 is shown in Figure 2.6b. The scaling centre is selected at the vertex. The defining curves for the three domains are plotted by bold solid lines. They are jointed at their common ends denoted by circles. The three domains are obtained by continuously scaling the defining curves and are modelled as one S-domain. The side-faces and material interfaces are formed in the scaling process by the ends of the defining curves.

   As will be demonstrated in Example 3.9 and in Chapter 10, the scaled boundary finite element method leads to a semi-analytical solution of the strain/stress singularity at the crack tip and multi-material corners. Accurate results can be obtained without a local mesh refinement or special element, which represents

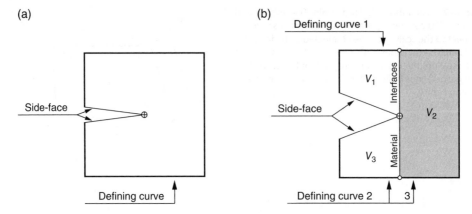

Figure 2.6 (a) Modelling of an edge-cracked square by scaling the defining curve. The two side-faces (crack faces) are formed by scaling the two ends of the defining curve. (b) Modelling of a three-material corner by scaling the defining curves. The two side-faces and the material interfaces are formed by scaling the ends of the defining curves.

one of the advantages of the scaled boundary finite element method in computational fracture mechanics.

b) The second extension is the modelling of a half-plane with an excavation as shown in Figure 2.7. The problem domain is modelled as one S-domain. The scaling centre is selected outside of the domain and on the extension of the free surface of the half-plane. The free surface is not included in the scaling process and denoted as 'side-face'. The excavation line is referred to as the defining curve and scaled with respect to the scaling centre. The S-domain with the two side-faces representing the free surface are formed by scaling the defining curve. A semi-analytical solution satisfying the boundary conditions at infinity can be obtained without discretizing the half-plane.

4) The visibility of the boundary from the scaling centre is determined by the angle between the boundary and the line of sight from the scaling centre. When the angle is a right angle, the visibility is the highest. When the angle is very small (or very close to 180°) like those ones shaded in Figure 2.8a, the low visibility of the boundary from the scaling centre will affect the numerical accuracy of the scaled boundary transformation and the scaled boundary finite element analysis. This is analogous to the effect of elements with small internal angles on the accuracy of a finite element analysis. To improve the visibility of the boundary, i.e. to avoid very small angles between

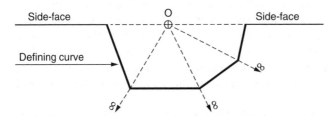

Figure 2.7 Modelling of an excavation in a half-plane as an S-domain by scaling the defining curve. The two side-faces (free surface) are formed by scaling the two ends of the defining curve.

(a)   (b)

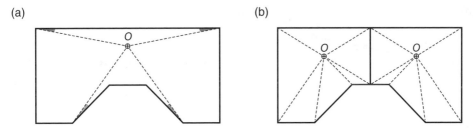

Figure 2.8 (a) An S-domain with small angles between the boundary and the lines of sight from the scaling centre, as indicated by the shaded areas, leading to low visibility of the boundary. (b) Subdivision of the S-domain to increase angles between the boundary and the lines of sight from the scaling centre and, thus, the visibility of boundary.

the lines of sight and the boundary, an S-domain can be subdivided as shown in Figure 2.8b.

2.2.2   S-elements: Boundary Discretization of S-domains

In practical applications of the scaled boundary finite element method, the boundary of an S-domain can be any general shape satisfying the scaling requirement. Generally speaking, the solution of a problem is only feasible in semi-analytical form with a piece-wise discretized description of the boundary. As shown in Figure 2.9, the boundary S of the S-domain V is divided into line elements. The large dots indicate the end nodes of elements. The reference coordinate of the line elements follows the counter-clockwise direction around the scaling centre (i.e. the right-hand rule). The S-domain and its boundary discretization define an S-element. The solution inside the S-element will be obtained analytically, leading to a semi-analytical procedure.

Standard isoparametric formulations of the finite element method apply to describe the geometry of the elements on the boundary. The Cartesian coordinates of the nodes of an element on boundary are arranged in $\{x\}$, $\{y\}$. The geometry and displacements of the isoparametric line element are interpolated using the same shape functions from their nodal values.

A 2-node element is shown in Figure 2.10a. It is the simplest element and will be used in all the examples in this chapter. The nodal coordinate vectors of the element

Figure 2.9 An S-element obtained by discretizing the boundary S of S-domain V. Displacement-based elements of different orders are used. The direction of the elements has to follow the counter-clockwise direction around the scaling centre.

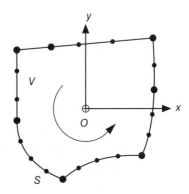

(a)

(b)

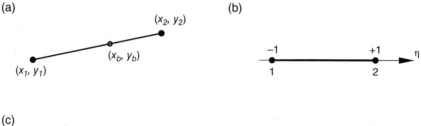

(c)

Figure 2.10 Two-node line element on boundary. The element direction must follow the counter-clockwise direction around the scaling centre. (a) Physical element. A point (x_b, y_b) on the element is obtained by interpolating nodal coordinates, see Eq. (2.17). (b) Parent element in natural coordinate η. (c) Shape functions in natural coordinate η.

are equal to

$$\{x\} = [\, x_1 \ x_2 \,]^T \tag{2.7a}$$

$$\{y\} = [\, y_1 \ y_2 \,]^T \tag{2.7b}$$

The coordinates of an arbitrary point on the element (on the boundary) are denoted as x_b, y_b.

The parent element in the natural coordinate η (ξ has been used for the radial coordinate) is illustrated in Figure 2.10b. It has a length of 2 units. The natural coordinates of nodes 1 and 2 are equal to -1 and $+1$, respectively. The linear interpolation of, for example, the x–coordinate on the 2-node element is expressed as

$$x_b(\eta) = a_0 + a_1\eta \tag{2.8}$$

The interpolation constants a_0 and a_1 are determined by formulating Eq. (2.8) at the two nodes, leading to

$$x_1 = x_b(-1) = a_0 + a_1 \times (-1) \tag{2.9a}$$

$$x_2 = x_b(+1) = a_0 + a_1 \times (+1) \tag{2.9b}$$

The solution of Eq. (2.9) is expressed as

$$a_0 = \frac{1}{2}(x_1 + x_2) \tag{2.10a}$$

$$a_1 = \frac{1}{2}(x_2 - x_1) \tag{2.10b}$$

Using Eq. (2.10), Eq. (2.8) is written as

$$x_b(\eta) = \frac{1}{2}(x_1 + x_2) + \frac{1}{2}(x_2 - x_1)\eta = \frac{1}{2}(1 - \eta)x_1 + \frac{1}{2}(1 + \eta)x_2 \tag{2.11}$$

The interpolation in Eq. 2.11 is expressed in terms of the nodal coordinates as

$$x_b(\eta) = N_1(\eta)x_1 + N_2(\eta)x_2 \tag{2.12}$$

where the shape functions are equal to

$$N_1(\eta) = \frac{1}{2}(1 - \eta) \tag{2.13a}$$

$$N_2(\eta) = \frac{1}{2}(1 + \eta) \tag{2.13b}$$

The two shape functions $N_1(\eta)$ and $N_2(\eta)$ are plotted in Figure 2.10c. It is easy to verify that the shape functions possess the following properties:

- Kronecker delta functions

$$N_i(\eta_j) = \delta_{ij} = \begin{cases} 1 & \text{when} \quad i = j \\ 0 & \text{when} \quad i \neq j \end{cases} \tag{2.14}$$

- Partition of unity

$$\sum_i N_i(\eta) = 1 \tag{2.15}$$

The shape functions are expressed in matrix form as

$$[N(\eta)] = \left[\frac{1}{2}(1 - \eta) \quad \frac{1}{2}(1 + \eta) \right] \tag{2.16}$$

The interpolation of coordinate x_b in Eq. (2.12) and, similarly, the interpolation of coordinate y_b are written as

$$x_b = [N]\{x\} \tag{2.17a}$$
$$y_b = [N]\{y\} \tag{2.17b}$$

where, for conciseness, the arguments η have been omitted from the coordinates $x_b = x_b(\eta)$, $y_b = x_b(\eta)$ and the shape functions $[N] = [N(\eta)]$.

A 3-node line element is shown Figure 2.11a. The nodal coordinate vectors are expressed as

$$\{x\} = [\, x_1 \quad x_2 \quad x_3 \,]^T \tag{2.18a}$$
$$\{y\} = [\, y_1 \quad y_2 \quad y_3 \,]^T \tag{2.18b}$$

The parent element in the natural coordinate η is shown in Figure 2.11b. The shape functions are equal to

$$[N(\eta)] = \left[\frac{1}{-2}\eta(1 - \eta) \quad 1 - \eta^2 \quad \frac{1}{2}\eta(1 + \eta) \right] \tag{2.19}$$

and plotted in Figure 2.10c. They have the properties of the Kronecker delta and satisfy the partition of unity.

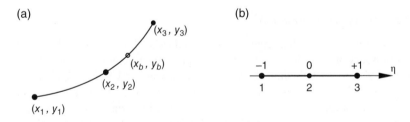

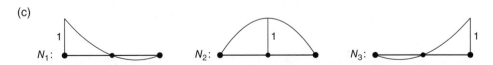

Figure 2.11 Three-node line element on boundary. The element direction must follow the counter-clockwise direction around the scaling centre. (a) Physical element. A point (x_b, y_b) on the element is obtained by interpolating nodal coordinates, see Eq. (2.17). (b) Parent element in natural coordinate η. (c) Shape functions in natural coordinate η.

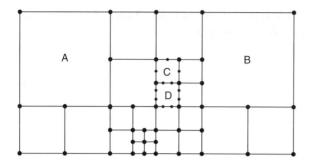

Figure 2.12 A quadtree mesh illustrating the simplicity of mesh generation and remeshing using S-elements. This mesh of S-elements satisfies displacement compatibility requirement.

The only requirement for the boundary discretization as shown in Figure 2.9 is that the displacement is continuous at the end nodes where the elements are connected. As long as this requirement is satisfied, the elements can be of any type and order. They can also be mixed together in one S-element. In addition, one edge of an S-element can have more than one element As one edge is shared by two S-elements only, compatibility is automatically satisfied. Practically, any displacement-based elements can be used.

The fact that an S-element allows a flexible boundary discretization reduces significantly the burden on conventional finite element mesh generation and remeshing. This is demonstrated by the simple quadtree mesh shown in Figure 2.12. A quadtree mesh is highly efficient and suitable for adaptive analysis, but its application in finite element analysis is greatly hindered by the presence of hanging nodes causing displacement incompatibility (Zienkiewicz et al., 2005). When a quadtree cell is modelled as one S-element in the scaled boundary finite element method, this difficulty is avoided by subdividing an edge at a hanging node into two or more line elements. For example, the lower and right edges of S-element A in Figure 2.12 are divided into two 2-node line elements so that the compatibility with the adjacent S-elements of a smaller size is satisfied. Similarly, the left edge of S-element B is divided into three 2-node line elements by the two hanging nodes. S-elements C and D depict the mesh refinement by increasing the element order (*p*-refinement). The compatibility with adjacent S-elements is satisfied automatically on the common edges. The refinement of one S-element will affect at most the four adjacent S-elements.

2.2.3   Scaled Boundary Transformation

As explained in Section 2.2.1, the S-domain is described by scaling the boundary. For the discretized boundary of an S-element, the scaling operation applies to individual elements. As shown in Figure 2.13, the line element S^e covers a sector V^e of the S-element when it is scaled towards the scaling centre. The S-element is the assembly of all the sectors resulting from scaling the individual elements on the boundary. Therefore, one sector covered by the scaling of a single line element is addressed in the derivation of the scaled boundary finite element equation. The equations for the S-element are obtained by enforcing the compatibility and equilibrium between the sectors.

2.2.3.1   Scaled Boundary Coordinates

The boundary scaling discussed in Section 2.2.1 is applied to a single line element on the boundary. The process is depicted in Figure 2.14 using a 3-node element. Scaling the point (x_b, y_b) on the line element leads to a radial line. The coordinates of a point (x, y)

Figure 2.13 Representation of an S-element by scaling the line elements on boundary.

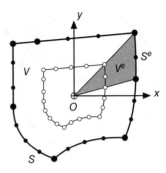

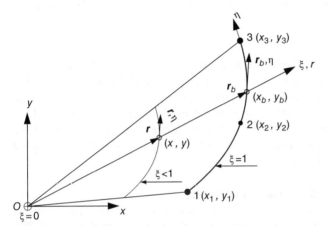

Figure 2.14 Scaled boundary coordinates defined by scaling a line element at the boundary.

along the radial line and inside the domain are obtained by substituting Eq. (2.17) into Eq. (2.5)

$$x = \xi [N(\eta)]\{x\} \tag{2.20a}$$
$$y = \xi [N(\eta)]\{y\} \tag{2.20b}$$

ξ, η are called *the scaled boundary coordinates* in two dimensions. ξ is a (dimensionless) radial coordinate. η is the circumferential coordinate. They form a right-hand coordinate system. Equation (2.20) defines the *scaled boundary transformation*, which transforms the coordinates between x, y and ξ, η. The scaling requirement ensures the uniqueness of the transformation.

The scaled boundary transformation can be regarded as a semi-analytical approach to transform an S-element into a circular domain. For example, the boundary S of the S-element V in Figure 2.13 is transformed into a circle described by a constant radial coordinate $\xi = 1$ as illustrated in Figure 2.15 (see also Figure 2.3). The S-element is specified by $0 \leq \xi \leq 1$.

The circumferential direction, parallel to the boundary, around the scaling centre O is described by the element number of the scaled element and the local coordinate η as in one-dimensional finite elements. A straight line passing through the scaling centre O in x, y coordinates remains as a straight line and is described by a constant η. The angular

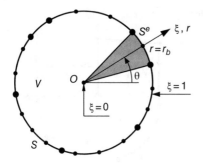

Figure 2.15 Representation of the polygonal domain shown in Figure 2.13 in the scaled boundary coordinates.

coordinate θ in Eq. (2.6c) is expressed as a function of the circumferential coordinate η

$$\theta(\eta) = \arctan \frac{y_b(\eta)}{x_b(\eta)} \tag{2.21}$$

which is independent of the radial coordinate ξ. Since every point on the boundary of an S-domain is directly visible, the function $\theta(\eta)$ is single-valued. The element number and the circumferential coordinate η of the element can be regarded as a discrete representation of the angular coordinate θ. An element on the boundary in Figure 2.13 is transformed into an arc defined by $\xi = 1$ and $-1 \le \eta \le 1$ in Figure 2.15. The part of the domain covered by scaling an element on the boundary in Figure 2.13 is a sector defined by $0 \le \xi \le 1$ and $-1 \le \eta \le 1$ in Figure 2.15.

2.2.3.2 Coordinate Transformation of Partial Derivatives

The scaled boundary transformation in Eq. (2.20) is similar to the coordinate transformation in isoparametric finite elements (Cook et al., 2002; Zienkiewicz et al., 2005). The function $\xi[N(\eta)]$ is equivalent to the shape functions of an isoparametric finite element from the viewpoint of coordinate transformation. The procedure for coordinate transformation in the formulation of isoparametric finite elements is followed to perform the scaled boundary transformation.

The transformation of partial derivatives of the spatial dimensions is addressed. In the governing differential equations (Section A.2 on page 454), the displacement field is expressed in the Cartesian coordinates x and y. The partial derivatives with respect to x and y are required. In the scaled boundary finite element method (as in the isoparametric finite element formulation), the displacement field is written in the scaled boundary coordinates ξ and η (see Eq. (2.78) in Section 2.4). The partial derivatives with respect to the Cartesian coordinates x and y are not available explicitly and are obtained from those with respect to the scaled boundary coordinates ξ and η.

Applying the chain rule, the partial differential operators with respect to the scaled boundary coordinates ξ and η are expressed as[1]

$$\frac{\partial}{\partial \xi} = \frac{\partial}{\partial x}\frac{\partial x}{\partial \xi} + \frac{\partial}{\partial y}\frac{\partial y}{\partial \xi} \tag{2.22a}$$

$$\frac{\partial}{\partial \eta} = \frac{\partial}{\partial x}\frac{\partial x}{\partial \eta} + \frac{\partial}{\partial y}\frac{\partial y}{\partial \eta} \tag{2.22b}$$

---

1 More details of the derivation of the transformation of spatial derivative are presented in Section 6.4 for the three-dimensional case.

It is expressed in the matrix form as

$$
\left\{ \begin{array}{c} \dfrac{\partial}{\partial \xi} \\[2mm] \dfrac{\partial}{\partial \eta} \end{array} \right\} = [J] \left\{ \begin{array}{c} \dfrac{\partial}{\partial x} \\[2mm] \dfrac{\partial}{\partial y} \end{array} \right\}
\tag{2.23}
$$

with the Jacobian matrix defined as

$$
[J] = \begin{bmatrix} x_{,\xi} & y_{,\xi} \\ x_{,\eta} & y_{,\eta} \end{bmatrix}
\tag{2.24}
$$

A comma followed by a subscript is used to denote partial differentiation with respect to the variable in the subscript. The derivatives with respects to x, y are transformed into those with respect to ξ, η by inverting Eq. (2.23), resulting in

$$
\left\{ \begin{array}{c} \dfrac{\partial}{\partial x} \\[2mm] \dfrac{\partial}{\partial y} \end{array} \right\} = [J]^{-1} \left\{ \begin{array}{c} \dfrac{\partial}{\partial \xi} \\[2mm] \dfrac{\partial}{\partial \eta} \end{array} \right\}
\tag{2.25}
$$

The partial derivatives in the Jacobian matrix (Eq. (2.24)) are obtained from Eq. (2.20) (or equivalently Eqs. (2.5) and (2.17)) as

$$
x_{,\xi} = x_b = [N]\{x\}
\tag{2.26a}
$$
$$
x_{,\eta} = \xi x_{b,\eta} = \xi [N]_{,\eta}\{x\}
\tag{2.26b}
$$
$$
y_{,\xi} = y_b = [N]\{y\}
\tag{2.26c}
$$
$$
y_{,\eta} = \xi y_{b,\eta} = \xi [N]_{,\eta}\{y\}
\tag{2.26d}
$$

Substituting Eq. (2.26) into Eq. (2.24) and separating the coordinates ξ from η yields

$$
[J] = \mathrm{diag}(1,\ \xi)[J_b]
\tag{2.27}
$$

where $[J_b]$ is the Jacobian matrix at the boundary ($\xi = 1$)

$$
[J_b] = \begin{bmatrix} x_b & y_b \\ x_{b,\eta} & y_{b,\eta} \end{bmatrix}
\tag{2.28}
$$

It is a function of η and depends on the geometry of the element only. Substituting Eq. (2.27) into Eq. (2.25) results in

$$
\left\{ \begin{array}{c} \dfrac{\partial}{\partial x} \\[2mm] \dfrac{\partial}{\partial y} \end{array} \right\} = [J_b]^{-1} \left\{ \begin{array}{c} \dfrac{\partial}{\partial \xi} \\[2mm] \dfrac{1}{\xi}\dfrac{\partial}{\partial \eta} \end{array} \right\}
\tag{2.29}
$$

The inverse of the Jacobian at the boundary $[J_b]$ is written as

$$
[J_b]^{-1} = \frac{1}{|J_b|} \begin{bmatrix} y_{b,\eta} & -y_b \\ -x_{b,\eta} & x_b \end{bmatrix}
\tag{2.30}
$$

The determinant of $[J_b]$ is expressed as

$$
|J_b| = x_b y_{b,\eta} - y_b x_{b,\eta}
\tag{2.31}
$$

Substituting Eq. (2.30) into Eq. (2.29) results in the transformation of the differential operators

$$
\begin{Bmatrix} \dfrac{\partial}{\partial x} \\[2mm] \dfrac{\partial}{\partial y} \end{Bmatrix} = \frac{1}{|J_b|} \begin{Bmatrix} y_{b,\eta} \\ -x_{b,\eta} \end{Bmatrix} \frac{\partial}{\partial \xi} + \frac{1}{|J_b|}\frac{1}{\xi} \begin{Bmatrix} -y_b \\ x_b \end{Bmatrix} \frac{\partial}{\partial \eta}
\tag{2.32}
$$

The partial derivatives with respect to ξ and η are separated to different terms.

2.2.3.3 Geometrical Properties in Scaled Boundary Coordinates

The geometrical properties of an S-element, which are required for the derivation of the scaled boundary finite element equation in the subsequent sections, are formulated in the scaled boundary coordinates.

The boundary of the S-element is considered. The part of the boundary represented by a line element is depicted in Figure 2.14. The position vector of a point (x_b, y_b) is given in Eq. (2.2). Its tangential vector is expressed as

$$
\mathbf{r}_{b,\eta} = x_{b,\eta}\mathbf{i} + y_{b,\eta}\mathbf{j}
\tag{2.33}
$$

It is shown that the determinant of the Jacobian on boundary, $|J_b|$ in Eq. (2.31), is equal to the area of the parallelogram formed by vectors $\mathbf{r}_b$ and $\mathbf{r}_{b,\eta}$

$$
\begin{aligned}
|\mathbf{r} \times \mathbf{r}_{,\eta}| &= \begin{vmatrix} 1 & 1 & \mathbf{k} \\ x_b & y_b & 0 \\ x_{b,\eta} & y_{b,\eta} & 0 \end{vmatrix} \\
&= x_b y_{b,\eta} - y_b x_{b,\eta} \\
&= |J_b|
\end{aligned}
\tag{2.34}
$$

where $\mathbf{k}$ is the unit vector along z (perpendicular to $x - y$ plane) direction. The vector $\mathbf{r}_b$ represents the line of sight from the scaling centre to the point (x_b, y_b) and the vector $\mathbf{r}_{b,\eta}$ the tangential direction of the boundary. When the scaling requirement is met, the area of the parallelogram will not be equal to 0. Since the scaled boundary coordinate system follows the right-hand rule, $|J_b|$ is always positive and the Jacobian matrix $[J_b]$ is invertible. The scaled boundary transformation is thus well defined. In a scaled boundary finite element analysis, the angle between the vector $\mathbf{r}_b$ and vector $\mathbf{r}_{b,\eta}$ should not be too small (Figure 2.8) so that the determinant $|J_b|$ is not close to 0 and the Jacobian matrix $[J_b]$ is well-conditioned.

A sector covered by scaling an element at the boundary as depicted in Figure 2.16 is considered. Several coordinate curves with constant values of scaled boundary coordinates ξ and η are plotted in Figure 2.16. Those with constant values of the radial coordinate ξ, denoted as S_r, are shown as solid lines, and these with constant values of the circumferential coordinate η, denoted as S_c, are shown as dashed lines.

A coordinate curve S_c with a constant ξ is considered. Its tangential vector $\mathbf{r}_{,\eta}$, shown in Figure 2.16, is equal to the derivative of the position vector (Eq. (2.1)) with respect to η. Using Eq. (2.4), it is written as

$$
\mathbf{r}_{,\eta} = \xi \mathbf{r}_{b,\eta}
\tag{2.35}
$$

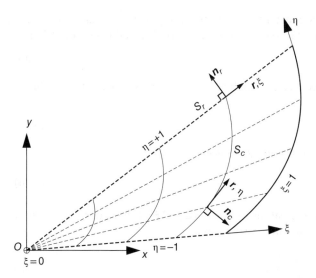

Figure 2.16 A sector of an S-element covered by scaling an element at the boundary.

where $\mathbf{r}_{b,\eta}$ (Eq. (2.33)) is the tangential vector at the boundary ($\xi = 1$). The magnitude of the tangential vector $\mathbf{r}_{,\eta}$ (Eq. (2.35)) is expressed as

$$|\mathbf{r}_{,\eta}| = \xi |\mathbf{r}_{b,\eta}| \tag{2.36}$$

where the magnitude of the tangential vector at the boundary $|\mathbf{r}_{b,\eta}|$ (Eq. (2.33)) is equal to

$$|\mathbf{r}_{b,\eta}| = \sqrt{\left(x_{b,\eta}\right)^2 + \left(y_{b,\eta}\right)^2} \tag{2.37}$$

An infinitesimal length on S_c is equal to (Eq. (2.36))

$$dS_c = |\mathbf{r}_{,\eta}| d\eta = \xi |\mathbf{r}_{b,\eta}| d\eta \tag{2.38}$$

Substituting Eq. (2.33) into Eq. (2.35) results in

$$\mathbf{r}_{,\eta} = \xi x_{b,\eta}\mathbf{i} + \xi y_{b,\eta}\mathbf{j} \tag{2.39}$$

which can also be obtained directly by differentiating Eq. (2.1) with respect to η and using Eq. (2.5). The unit outward normal vector $\mathbf{n}_c$, shown in Figure 2.16, and the tangential vector of a coordinate curve S_c form a right-hand coordinate system. The unit outward normal vector is obtained using Eq. (2.39) as

$$\mathbf{n}_c = \frac{y_{b,\eta}}{|\mathbf{r}_{b,\eta}|}\mathbf{i} - \frac{x_{b,\eta}}{|\mathbf{r}_{b,\eta}|}\mathbf{j} \tag{2.40}$$

It is independent of ξ, i.e., the unit outward normal vector at a given η is the same for any coordinate curve S_c. For later use, the unit outward normal vector $\mathbf{n}_c$ is expressed in matrix form as

$$\{n_c\} = \left\{ \begin{array}{c} n_{cx} \\ n_{cy} \end{array} \right\} = \frac{1}{|\mathbf{r}_{b,\eta}|} \left\{ \begin{array}{c} y_{b,\eta} \\ -x_{b,\eta} \end{array} \right\} \tag{2.41}$$

A coordinate curve S_r with a constant η is a radial line passing through the scaling centre (Figure 2.16). Its tangential (direction) vector, $\mathbf{r}_{,\xi}$, is obtained from Eq. (2.4) as

$$\mathbf{r}_{,\xi} = \mathbf{r}_b \qquad (2.42)$$

with the position vector $\mathbf{r}_b$ given in Eq. (2.2). It is expressed in the Cartesian coordinate system as

$$\mathbf{r}_{,\xi} = x_b \mathbf{i} + y_b \mathbf{j} \qquad (2.43)$$

Its unit outward normal vector $\mathbf{n}_r$ (Figure 2.16) is expressed as

$$\mathbf{n}_r = -\frac{y_b}{r_b}\mathbf{i} + \frac{x_b}{r_b}\mathbf{j} \qquad (2.44)$$

where r_b is the magnitude of the position vector $\mathbf{r}_b$ (Eq. (2.6b)). For later use, the unit outward normal vector $\mathbf{n}_r$ is written in matrix form as

$$\{n_r\} = \begin{Bmatrix} n_{rx} \\ n_{ry} \end{Bmatrix} = \frac{1}{r_b} \begin{Bmatrix} -y_b \\ x_b \end{Bmatrix} \qquad (2.45)$$

For infinitesimal changes $d\xi$, $d\eta$ of the scaled boundary coordinates, the signed infinitesimal volume dV is equal to the magnitude of the cross product of the infinitesimal increments of the tangential vectors (Eqs. (2.39) and (2.43))

$$dV = |(\mathbf{r}_{,\xi} \, d\xi) \times (\mathbf{r}_{,\eta} \, d\eta)|$$

$$= \begin{vmatrix} 1 & 1 & \mathbf{k} \\ x_b & y_b & 0 \\ \xi x_{b,\eta} & \xi y_{b,\eta} & 0 \end{vmatrix} d\xi d\eta$$

$$= \xi(x_b y_{b,\eta} - y_b x_{b,\eta}) d\xi d\eta \qquad (2.46)$$

Using Eq. (2.31), Eq. (2.46) is expressed as

$$dV = \xi |J_b| d\xi d\eta \qquad (2.47)$$

Note that $|J_b|$ is positive.

■ Example 2.1   Two-node Element: Scaled Boundary Transformation

A 2-node line element with the nodal coordinates (x_1, y_1) and (x_2, y_2) is shown in Figure 2.17. Perform the scaled boundary transformation of sector covered by scaling this element.

1) Boundary discretization

Note that the nodes are numbered in such a manner that the natural coordinate η of the element (see Figure 2.17) follows the counter-clockwise direction around the scaling centre O. The nodal coordinates are arranged as

$$\{x\} = \begin{Bmatrix} x_1 \\ x_2 \end{Bmatrix} \qquad (2.48a)$$

$$\{y\} = \begin{Bmatrix} y_1 \\ y_2 \end{Bmatrix} \qquad (2.48b)$$

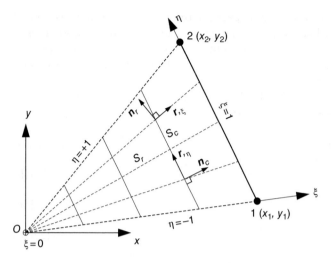

Figure 2.17 Scaled boundary coordinate transformation of a 2-node line element.

Substituting the nodal coordinate vectors $\{x\}$, $\{y\}$ and the shape functions in Eq. (2.16) into Eq. (2.17), the geometry of the line element is given in Cartesian coordinates by

$$x_b = [N]\{x\} = \frac{1}{2}(1-\eta)x_1 + \frac{1}{2}(1+\eta)x_2 = \bar{x} + \frac{1}{2}\Delta_x\eta \qquad (2.49a)$$

$$y_b = [N]\{y\} = \frac{1}{2}(1-\eta)y_1 + \frac{1}{2}(1+\eta)y_2 = \bar{y} + \frac{1}{2}\Delta_y\eta \qquad (2.49b)$$

with the abbreviations

$$\Delta_x = x_2 - x_1 \qquad (2.50a)$$

$$\Delta_y = y_2 - y_1 \qquad (2.50b)$$

and

$$\bar{x} = \frac{1}{2}\left(x_1 + x_2\right) \qquad (2.51a)$$

$$\bar{y} = \frac{1}{2}\left(y_1 + y_2\right) \qquad (2.51b)$$

For later use, differentiating Eq. (2.49) leads to

$$x_{b,\eta} = \frac{1}{2}\Delta_x \qquad (2.52a)$$

$$y_{b,\eta} = \frac{1}{2}\Delta_y \qquad (2.52b)$$

2) Scaled boundary coordinates
   Substituting Eq. (2.49) into Eq. (2.5) results in

$$x = \xi x_b = \xi\left(\bar{x} + \frac{1}{2}\Delta_x\eta\right) \qquad (2.53a)$$

$$y = \xi y_b = \xi\left(\bar{y} + \frac{1}{2}\Delta_y\eta\right) \qquad (2.53b)$$

This equation defines the transformation between the Cartesian coordinates x, y and the scaled boundary coordinates ξ, η. As shown in Figure 2.17, coordinate curves S_c with constant ξ are parallel to the line element. Coordinate curves S_r with constant η are radial lines connecting the scaling centre O and the points on the line element.

3) Coordinate transformation

To obtain the Jacobian matrix defined in Eq. (2.24), the partial derivatives of x, y with respect to ξ, η are evaluated from Eq. (2.53)

$$x_{,\xi} = x_b = \bar{x} + \frac{1}{2}\Delta_x \eta \tag{2.54a}$$

$$y_{,\xi} = y_b = \bar{y} + \frac{1}{2}\Delta_y \eta \tag{2.54b}$$

$$x_{,\eta} = \xi x_{b,\eta} = \frac{1}{2}\Delta_x \xi \tag{2.54c}$$

$$y_{,\eta} = \xi y_{b,\eta} = \frac{1}{2}\Delta_y \xi \tag{2.54d}$$

leading to

$$[J] = \mathrm{diag}(1,\ \xi)[J_b] \tag{2.55}$$

where the Jacobian matrix at the boundary is expressed as

$$[J_b] = \begin{bmatrix} \bar{x} + \frac{1}{2}\Delta_x\eta & \bar{y} + \frac{1}{2}\Delta_y\eta \\ \frac{1}{2}\Delta_x & \frac{1}{2}\Delta_y \end{bmatrix} \tag{2.56}$$

The same result can also be obtained by directly substituting Eq. (2.49) and Eq. (2.52) into Eq. (2.28). The determinant of the Jacobian matrix on the boundary (Eq. (2.31)) is expressed, after subtracting the second row multiplied with η from the first row, as

$$|J_b| = \frac{1}{2}\begin{vmatrix} \bar{x} & \bar{y} \\ \Delta_x & \Delta_y \end{vmatrix} = \frac{1}{2}\left(\bar{x}\Delta_y - \bar{y}\Delta_x\right) \tag{2.57}$$

Using Eqs. (2.50) and (2.51), Eq. (2.57) is simplified as

$$|J_b| = \frac{1}{2}\begin{vmatrix} x_1 & y_1 \\ x_2 & y_2 \end{vmatrix} = \frac{1}{2}(x_1 y_2 - x_2 y_1) \tag{2.58}$$

$|J_b|$ is equal to the area of the triangle covered by scaling the 2-node element (Figure 2.17). To ensure that the Jacobian matrix is well conditioned, small angles between the boundary and the line of sight from the scaling centre are to be avoided (Figure 2.8).

The inverse of the Jacobian $[J_b]$ (Eq. (2.30)) is expressed as

$$[J_b]^{-1} = \frac{1}{|J_b|}\begin{bmatrix} \frac{1}{2}\Delta_y & -\left(\bar{y} + \frac{1}{2}\Delta_y\eta\right) \\ \frac{1}{2}\Delta_x & \bar{x} + \frac{1}{2}\Delta_x\eta \end{bmatrix} \tag{2.59}$$

It provides the transformation of the partial derivatives in the Cartesian coordinates to those in the scaled boundary coordinates (Eq. (2.32))

$$
\left\{ \begin{array}{c} \dfrac{\partial}{\partial x} \\[2mm] \dfrac{\partial}{\partial y} \end{array} \right\} = \dfrac{1}{|J_b|} \left\{ \begin{array}{c} \dfrac{1}{2}\Delta_y \\[2mm] \dfrac{1}{2}\Delta_x \end{array} \right\} \dfrac{\partial}{\partial \xi} + \dfrac{1}{|J_b|}\dfrac{1}{\xi} \left\{ \begin{array}{c} -\left(\bar{y} + \dfrac{1}{2}\Delta_y\eta\right) \\[2mm] \bar{x} + \dfrac{1}{2}\Delta_x\eta \end{array} \right\} \dfrac{\partial}{\partial \eta}
\tag{2.60}
$$

4) Geometry properties of coordinate transformation

The tangential vector of a coordinate curve S_c with a constant ξ (Figure 2.17) is expressed as (Eq. (2.35) with Eqs. (2.33) and (2.52))

$$
\mathbf{r}_{,\eta} = \frac{1}{2}\xi(\Delta_x \mathbf{i} + \Delta_y \mathbf{j})
\tag{2.61}
$$

On the boundary ($\xi = 1$), it is equal to

$$
\mathbf{r}_{b,\eta} = \frac{1}{2}(\Delta_x \mathbf{i} + \Delta_y \mathbf{j})
\tag{2.62}
$$

The infinitesimal length of S_c equals (Eqs. (2.38) and (2.62))

$$
dS_c = \frac{1}{2}\xi\sqrt{\Delta_x^2 + \Delta_y^2}\,d\eta
\tag{2.63}
$$

It is easily verified that integrating over $-1 \leq \eta \leq +1$ leads to the length of a coordinate curve $\xi\sqrt{\Delta_x^2 + \Delta_y^2}$.

Substituting Eq. (2.62) into Eq. (2.40), the unit outward normal vector $\mathbf{n}_c$ of the coordinate curve S_c is expressed as

$$
\mathbf{n}_c = \frac{\Delta_y}{\sqrt{\Delta_x^2 + \Delta_y^2}}\mathbf{i} - \frac{\Delta_x}{\sqrt{\Delta_x^2 + \Delta_y^2}}\mathbf{j}
\tag{2.64}
$$

The unit outward normal vector $\mathbf{n}_r$ of a coordinate curve S_r is

$$
\mathbf{n}_r = -\frac{y_b}{r_b}\mathbf{i} + \frac{x_b}{r_b}\mathbf{j}
$$

with x_b and y_b given in Eq. (2.49) and r_b in Eq. (2.6b). The equations of the unit outward normal vectors $\mathbf{n}_c$ and $\mathbf{n}_r$ can be verified by inspecting Figure 2.17.

The infinitesimal area dV is obtained by substituting Eq. (2.58) into Eq. (2.47)

$$
dV = \frac{1}{2}(x_1 y_2 - x_2 y_1)\xi\, d\xi\, d\eta
\tag{2.65}
$$

Integrating Eq. (2.65) over the domain defined by $0 \leq \xi \leq 1$ and $-1 \leq \eta \leq +1$ results in $(x_1 y_2 - x_2 y_1)/2$, which is the area of the triangular sector covered by scaling the 2-node element.

◼

2.3 Governing Equations of Linear Elasticity in Scaled Boundary Coordinates

The spatial coordinates of governing equations for two-dimensional (2D) linear elasticity found in Section A.2 are transformed into the scaled boundary coordinates. The directions of displacement components $\{u\}$, and strain components $\{\varepsilon\}$ and stress components $\{\sigma\}$ are retained in the original Cartesian coordinate directions. This is analogous to the procedure for developing isoparametric elements in the standard finite element method (Cook et al., 2002; Zienkiewicz et al., 2005).

The spatial derivatives are transformed from the Cartesian coordinates x, y into the scaled boundary coordinates ξ, η in Eq. (2.23). Substituting this equation into the differential operator for 2D elasticity (Eq. (A.31)) and grouping the terms according to the partial derivatives results in

$$[L] = \begin{bmatrix} \dfrac{\partial}{\partial x} & 0 \\[2mm] 0 & \dfrac{\partial}{\partial y} \\[2mm] \dfrac{\partial}{\partial y} & \dfrac{\partial}{\partial x} \end{bmatrix} = [b_1]\frac{\partial}{\partial \xi} + \frac{1}{\xi}[b_2]\frac{\partial}{\partial \eta} \tag{2.66}$$

where the matrices

$$[b_1] = \frac{1}{|J_b|}\begin{bmatrix} y_{b,\eta} & 0 \\ 0 & -x_{b,\eta} \\ -x_{b,\eta} & y_{b,\eta} \end{bmatrix} \tag{2.67a}$$

$$[b_2] = \frac{1}{|J_b|}\begin{bmatrix} -y_b & 0 \\ 0 & x_b \\ x_b & -y_b \end{bmatrix} \tag{2.67b}$$

are introduced. Note that $[b_1]$ and $[b_2]$ depend only on the geometry of the element at the boundary and are independent of ξ. It is easy to verify from Eq. (2.67) that the following identity between $[b_1]$ and $[b_2]$ exists

$$(|J_b|[b_2])_{,\eta} = -|J_b|[b_1] \tag{2.68}$$

This will be used later in the derivation of the scaled boundary finite element equation.

Substituting Eq. (2.66) into Eq. (A.5), the strains are expressed in the scaled boundary coordinates as

$$\{\varepsilon\} = [b_1]\frac{\partial \{u\}}{\partial \xi} + \frac{1}{\xi}[b_2]\frac{\partial \{u\}}{\partial \eta} \tag{2.69}$$

The directions of the strain and stress components are in the Cartesian coordinates. The stress-strain relationship (Eq. (A.11)) and the elasticity matrix $[D]$ are not affected by the transformation of the spatial coordinates.

The surface tractions on the coordinate curves S_c and S_r are determined using Eq. (A.36) and the unit outward normal vectors in the scaled boundary coordinates. On a coordinate curve S_c (parallel to the boundary) with a constant radial coordinate ξ, the unit outward normal is given in Eq. (2.41). Replacing n_x and n_y in Eq. (A.36) with n_{cx} and n_{cy}, respectively, the surface traction $\{t_c\} = \{t_c(\eta)\}$ is expressed as

$$\{t_c\} = \frac{1}{|\mathbf{r}_{b,\eta}|} \begin{bmatrix} y_{b,\eta} & 0 & -x_{b,\eta} \\ 0 & -x_{b,\eta} & y_{b,\eta} \end{bmatrix} \{\sigma\} \tag{2.70}$$

Comparing Eq. (2.70) to Eq. (2.67a), it is identified that

$$\{t_c\} = \frac{|J_b|}{|\mathbf{r}_{b,\eta}|} [b_1]^T \{\sigma\} \tag{2.71}$$

applies.

Similarly, the surface tractions $\{t_r\} = \{t_r(\xi)\}$ on a coordinate curve S_r (a radial line passing through the scaling centre) are determined by substituting the components of the unit outward normal vector given in Eq. (2.45) into Eq. (A.36) and leads to

$$\{t_r\} = \frac{1}{r_b} \begin{bmatrix} -y_b & 0 & x_b \\ 0 & x_b & -y_b \end{bmatrix} \{\sigma\} \tag{2.72}$$

Replacing the matrix on the right-hand side of Eq. (2.72) using Eq. (2.67b) results in

$$\{t_r\} = \frac{|J_b|}{r_b} [b_2]^T \{\sigma\} \tag{2.73}$$

2.4 Semi-analytical Representation of Displacement and Strain Fields

The equations of equilibrium are expressed by substituting Eq. (2.66) into Eq. (A.10) as

$$[b_1]^T \frac{\partial \{\sigma\}}{\partial \xi} + \frac{1}{\xi} [b_2]^T \frac{\partial \{\sigma\}}{\partial \eta} = \{0\} \tag{2.74}$$

The displacement field in an S-element is represented semi-analytically in the scaled boundary finite element method. In the circumferential direction, the solution is given at discrete nodal points as in 1D finite elements. In the radial direction, the solution is obtained analytically.

The radial lines passing through the scaling centre O and a node at the boundary as shown in Figure 2.18a are addressed. The lines are numbered sequentially corresponding to the numbering of nodes at the boundary. Along a radial line i, the circumferential coordinate η is constant and nodal displacement functions $u_{ix}(\xi)$ and $u_{iy}(\xi)$ are introduced. They analytically describe the variation of the displacement components. Their directions are along the Cartesian coordinates x and y. All the nodal displacement functions of an S-element (Figure 2.18a) are assembled to construct a vector of nodal displacement functions

$$\{u(\xi)\} = [u_{1x}(\xi) \quad u_{1y}(\xi) \quad u_{2x}(\xi) \quad u_{2y}(\xi) \quad \dots \quad u_{ix}(\xi) \quad u_{iy}(\xi) \quad \dots]^T \tag{2.75}$$

On the boundary ($\xi = 1$), the nodal displacement functions are equal to the nodal displacements denoted as $\{d\}$

$$\{d\} = \{u(\xi = 1)\} \tag{2.76}$$

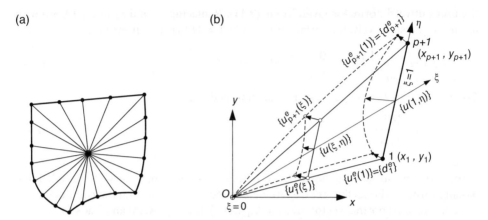

Figure 2.18 Semi-analytical representation of displacement field in an S-element by interpolating nodal displacement functions element-by-element independently. (a) Displacement functions are introduced along radial lines connecting the scaling centre and nodes at the boundary. (b) Displacement field in a sector covered by scaling one line element on boundary. The dashed lines show the deformed shapes. Displacements along the circumferential direction η are obtained by interpolating the nodal displacement functions.

The displacement field is obtained by interpolating the displacement functions in the circumferential direction η.

Similar to the formulation of standard displacement-based finite elements, the interpolation and other operations are performed element-by-element and independently of each other. A sector covered by the scaling of one isoparametric displacement-based p-th order element at the boundary is illustrated in Figure 2.18b, where only the two end nodes and one middle node are shown. The nodes are numbered sequentially from 1 to $p+1$ following the counter-clockwise direction. The element and a coordinate curve with constant radial coordinates ξ are plotted. The nodal displacement functions related to all the nodes of this element are assembled in a vector $\{u^e(\xi)\} = [u^e_{1x}(\xi) \quad u^e_{1y}(\xi) \quad \ldots \quad u^e_{(p+1)x}(\xi) \quad u^e_{(p+1)y}(\xi)]^T$, where the superscript e denotes functions of one element. The displacement functions $\{u^e_1(\xi)\} = [u^e_{1x}(\xi) \quad u^e_{1y}(\xi)]^T$ and $\{u^e_{p+1}(\xi)\} = [u^e_{(p+1)x}(\xi) \quad u^e_{(p+1)y}(\xi)]^T$ at the two end nodes of the element are depicted by dashed lines in Figure 2.18. The displacement functions, $\{u^e(\xi)\}$, of a line element are related to the displacement functions, $\{u(\xi)\}$, of the S-element via the element connectivity. The assembling of displacement functions of the line elements to form the displacement functions of the S-element is expressed as

$$\{u(\xi)\} = \sum_e \{u^e(\xi)\} \tag{2.77}$$

where the symbol $\sum_e$ indicates standard finite element assembly of all the line elements at the boundary.

The displacements $\{u\} = \{u(\xi,\eta)\} = [u_x(\xi,\eta) \quad u_y(\xi,\eta)]^T$ at a point (ξ,η) inside a sector are obtained by interpolating the displacement functions $\{u^e(\xi)\}$. At the specified value of the radial coordinate ξ, the displacement functions $u^e_{ix}(\xi)$ and $u^e_{iy}(\xi)$, where

$i = 1, 2, \ldots, p + 1$, are evaluated. The displacements in the x- and y-directions at the specified circumferential coordinate η are interpolated independently

$$u_x = u_x(\xi, \eta) = \sum_{i=1}^{p+1} N_i u_{ix}^e(\xi) \tag{2.78a}$$

$$u_y = u_y(\xi, \eta) = \sum_{i=1}^{p+1} N_i u_{iy}^e(\xi) \tag{2.78b}$$

where $N_i = N_i(\eta)$ are the shape functions of the p-th order element. The deformed shapes of the element and coordinate curve are shown by dashed lines in Figure 2.18b.

Equation (2.78) is expressed in matrix form as

$$\{u\} = \{u(\xi, \eta)\} = [N_u]\{u^e(\xi)\} \tag{2.79}$$

with

$$[N_u] = \begin{bmatrix} N_1 & 0 & N_2 & 0 & \cdots & N_{p+1} & 0 \\ 0 & N_1 & 0 & N_2 & \cdots & 0 & N_{p+1} \end{bmatrix} \tag{2.80}$$

Substituting the displacement field in Eq. (2.79) into Eq. (2.69), the strain field $\{\varepsilon\} = \{\varepsilon(\xi, \eta)\}$ is expressed in the scaled boundary coordinates as

$$\{\varepsilon\} = [b_1][N_u]\{u^e(\xi)\},_\xi + \frac{1}{\xi}[b_2][N_u],_\eta \{u^e(\xi)\} \tag{2.81}$$

The strain-displacement matrices $[B_1] = [B_1(\eta)]$ and $[B_2] = [B_2(\eta)]$ are introduced

$$[B_1] = [b_1][N_u] \tag{2.82a}$$

$$[B_2] = [b_2][N_u],_\eta \tag{2.82b}$$

They depend on only the geometry of the line element. Using Eq. (2.82), the strain field in Eq. (2.81) is rewritten as

$$\{\varepsilon\} = [B_1]\{u^e(\xi)\},_\xi + \frac{1}{\xi}[B_2]\{u^e(\xi)\} \tag{2.83}$$

It is expressed semi-analytically using the displacement functions.

The stress field $\{\sigma\} = \{\sigma(\xi, \eta)\}$ follows from Eqs. (A.11) and (2.83) as

$$\{\sigma\} = [D]\left([B_1]\{u^e(\xi)\},_\xi + \frac{1}{\xi}[B_2]\{u^e(\xi)\}\right) \tag{2.84}$$

The surface tractions on a coordinate curve S_c with a constant radial coordinate ξ and a coordinate curve S_r with a constant radial coordinate η are given in Eqs. (2.71) and (2.73), respectively. Using Eq. (2.84), they can also be expressed in terms of the displacement functions.

2.5 Derivation of the Scaled Boundary Finite Element Equation by the Virtual Work Principle

The scaled boundary finite element equation is derived by converting the governing partial differential equations to a weak form in the circumferential direction only. This leads to a system of ordinary differential equations with the radial coordinate as the independent variable. For a linear elastostatic problem without the presence of any body force, the scaled boundary finite element equation is a system of homogeneous Euler-Cauchy ordinary differential equations and can be solved analytically by following standard procedures (Kreyszig, 2011). The scaled boundary finite element equation can be derived by applying either the Galerkin weighted residual method (Song and Wolf, 1997) or the virtual work principle (Deeks and Wolf, 2002d). Other techniques, such as the principle of stationary potential energy, should also be possible. In this section, the derivation based on the virtual work principle is presented.

2.5.1   Virtual Displacement and Strain Fields in Scaled Boundary Coordinates

In the principle of virtual work, also known as the principle of virtual displacements, a virtual displacement field $\{\delta u\} = \{\delta u(\xi, \eta)\} = [\delta u_x(\xi, \eta), \; \delta u_y(\xi, \eta)]^T$ is postulated. The visual displacement field satisfies the displacement boundary conditions and does not alter the stress field and external loads. It is represented analogously to the real displacement field in Section 2.4.

Virtual displacement functions $\{\delta u(\xi)\}$ are introduced on the radial lines connecting the scaling centre and the nodes at the boundary of the S-element. The value of the virtual displacement functions at the boundary ($\xi = 1$) is denoted as

$$\{\delta d\} = \{\delta u(\xi = 1)\} \tag{2.85}$$

The virtual displacement field $\{\delta u\}$ is obtained by interpolating the virtual displacement functions element-by-element as in Eq. (2.79) using the shape functions $[N_u]$ of a line element.

$$\{\delta u\} = [N_u]\{\delta u^e(\xi)\} \tag{2.86}$$

The virtual strain field $\{\delta \varepsilon\} = \{\delta \varepsilon(\xi, \eta)\} = [\delta \varepsilon_x(\xi, \eta), \; \delta \varepsilon_y(\xi, \eta), \; \delta \gamma_{xy}(\xi, \eta)]^T$ produced by the virtual displacement field is determined in the same way as the real strain field. For a sector covered by scaling a line element at the boundary (Figure 2.18). Using Eqs. (2.86) and (2.69), the virtual strain field is expressed as

$$\{\delta \varepsilon\} = [B_1]\{\delta u^e(\xi)\}_{,\xi} + \frac{1}{\xi}[B_2]\{\delta u^e(\xi)\} \tag{2.87}$$

where $[B_1]$ and $[B_2]$ are given in Eq. (2.82).

2.5.2   Nodal Force Functions

Corresponding to the nodal displacement functions, the nodal force functions are introduced on the radial lines connecting the scaling centre and the nodes on the boundary to represent the stress field.

A coordinate curve S_c defined by scaling a line element at the boundary with a constant ξ is shown in Figure 2.16. The normal vector $\mathbf{n}_c$ (Eq. (2.40)) is the outward normal vector of the sector covered by scaling the curve S_c to the scaling centre O. The surface tractions $\{t_c\}$ on curve S_c resulting from the stress field $\{\sigma\}$ are given in Eq. (2.71). For the sector covered by scaling the line element at the boundary, the nodal force functions $\{q^e(\xi)\}$ are defined as being equivalent in the sense of virtual work to the surface tractions $\{t_c\}$

$$\{\delta u^e(\xi)\}^T \{q^e(\xi)\} = \int_{S_c} \{\delta u\}^T \{t_c\} dS_c \tag{2.88}$$

Substituting the virtual displacement field $\{\delta u\}$ (Eq. (2.86)) into Eq. (2.88) and considering the arbitrariness of the virtual displacement functions $\{\delta^e u(\xi)\}$, the nodal force functions are expressed as

$$\{q^e(\xi)\} = \int_{S_c} [N_u]^T \{t_c\} dS_c \tag{2.89}$$

Substituting the surface traction (Eq. (2.71)) and the infinitesimal surface dS_c (Eq. (2.38)) into Eq. (2.89) yields

$$\{q^e(\xi)\} = \int_{-1}^{+1} [N_u]^T \frac{|J_b|}{|\mathbf{r}_{b,\eta}|} [b_1]^T \{\sigma\} \xi |\mathbf{r}_{b,\eta}| d\eta$$

$$= \int_{-1}^{+1} [N_u]^T [b_1]^T \{\sigma\} \xi |J_b| d\eta \tag{2.90}$$

Using Eq. (2.82a), Eq. (2.90) is rewritten as

$$\{q^e(\xi)\} = \xi \int_{-1}^{+1} [B_1]^T \{\sigma\} |J_b| d\eta \tag{2.91}$$

The nodal force vectors of individual line elements are assembled to form the nodal force vector of the S-element

$$\{q(\xi)\} = \sum_e \{q^e(\xi)\} \tag{2.92}$$

The nodal force functions $\{q(\xi)\}$ of the S-element at a specified radial coordinate ξ are similar to the nodal forces of an element in the standard finite element method. In particular, the nodal forces of the S-element (Figure 2.9) are given by $\{q(\xi = 1)\}$, i.e. the nodal force functions $\{q(\xi)\}$ at the boundary $\xi = 1$.

2.5.3 The Scaled Boundary Finite Element Equation

The virtual work principle is applied to a single S-element. A pentagon S-element is depicted in Figure 2.19 as an example. The boundary of the S-element is discretized with line elements. The nodal displacements and nodal forces are shown in Figure 2.19a and Figure 2.19b, respectively. In each sector defined by scaling a line element towards the scaling centre, the (virtual) displacement and strain fields are described in Section 2.4. For conciseness in the derivation of the scaled boundary finite element equation, only concentrated nodal forces are considered in this section (The surface tractions and body loads are addressed in Sections 3.4.2 and 3.9, respectively.)

The principle of virtual work is expressed for the S-element as

$$\underbrace{\int_V \{\delta\varepsilon\}^T \{\sigma\} dV}_{U_\varepsilon} = \{\delta d\}^T \{F\} \tag{2.93}$$

(a)   (b)

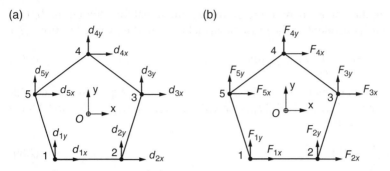

Figure 2.19 Nodal displacement and nodal forces of a pentagon S-element. (a) Nodal displacements. (b) Nodal forces.

where $\{\sigma\}$ is the stress field, and $\{F\}$ consists of the nodal forces (Figure 2.19b). The left-hand side is the virtual increment of internal strain energy, and the right-hand side the virtual increment of the work done by the external forces.

The virtual increment of the strain energy of the S-element at the left-hand side of Eq. (2.93) is evaluated sector-by-sector corresponding to the line elements at the boundary. Substituting the virtual strain field in Eq. (2.87) for a sector V^e into U_ε in Eq. (2.93) results in

$$U_\varepsilon = \underbrace{\sum \int_{V^e} \{\delta u^e(\xi)\}_{,\xi}^T [B_1]^T \{\sigma\} \mathrm{d}V}_{U_{\varepsilon I}} + \underbrace{\sum \int_{V^e} \{\delta u^e(\xi)\}^T [B_2]^T \{\sigma\} \frac{1}{\xi} \mathrm{d}V}_{U_{\varepsilon II}} \tag{2.94}$$

The two terms on the right-hand side of this equation are considered separately in the following.

The first term on the right-hand side of Eq. (2.94) is expressed, using Eq. (2.47) for $\mathrm{d}V$ in the scaled boundary coordinates, as

$$U_{\varepsilon I} = \sum \int_0^1 \{\delta u^e(\xi)\}_{,\xi}^T \int_{-1}^{+1} [B_1]^T \xi \{\sigma\} |J_b| \mathrm{d}\eta \mathrm{d}\xi \tag{2.95}$$

Replacing the summation with the finite element assembly, it is rewritten as

$$U_{\varepsilon I} = \int_0^1 \{\delta u(\xi)\}_{,\xi}^T \sum_e \xi \int_{-1}^{+1} [B_1]^T \{\sigma\} |J_b| \mathrm{d}\eta \mathrm{d}\xi \tag{2.96}$$

where the assembly of the virtual displacement vector $\{\delta u(\xi)\}$ of the S-element follows from Eq. (2.77). Considering Eq. (2.91), Eq. (2.96) is expressed as

$$U_{\varepsilon I} = \int_0^1 \{\delta u(\xi)\}_{,\xi}^T \sum_e \{q^e(\xi)\} \mathrm{d}\xi \tag{2.97}$$

Introducing the assembly of the nodal force functions in Eq. (2.92) leads to

$$U_{\varepsilon I} = \int_0^1 \{\delta u(\xi)\}_{,\xi}^T \{q(\xi)\} \mathrm{d}\xi \tag{2.98}$$

Applying integration by parts with respect to ξ to Eq. (2.98) results in

$$U_{\varepsilon I} = \left(\{\delta u(\xi)\}^T \{q(\xi)\} \right)_0^1 - \int_0^1 \{\delta u(\xi)\}^T \{q(\xi)\}_{,\xi} \, d\xi \tag{2.99}$$

At the lower limit $\xi = 0$, the first term on the left-hand side vanishes. Considering Eq. (2.85) at the upper limit of the first term, Eq. (2.99) is rewritten as

$$U_{\varepsilon I} = \{\delta d\}^T \{q(\xi = 1)\} - \int_0^1 \{\delta u(\xi)\}^T \{q(\xi)\}_{,\xi} \, d\xi \tag{2.100}$$

The second term on the right-hand side of Eq. (2.94) is also evaluated sector-by-sector

$$U_{\varepsilon II} = \sum \int_0^1 \{\delta u^e(\xi)\}^T \int_{-1}^{+1} [B_2]^T \{\sigma\} |J_b| d\eta d\xi \tag{2.101}$$

Replacing the summation with finite element assembly, Eq. (2.101) is expressed as

$$U_{\varepsilon II} = \int_0^1 \{\delta u(\xi)\}^T \sum_e \int_{-1}^{+1} [B_2]^T \{\sigma\} |J_b| d\eta d\xi \tag{2.102}$$

The virtual increment of the strain energy in Eq. (2.94) is obtained by substituting $U_{\varepsilon I}$ in Eq. (2.100) and $U_{\varepsilon II}$ in Eq. (2.102) as

$$U_{\varepsilon} = \{\delta d\}^T q(\xi = 1) - \int_0^1 \{\delta u(\xi)\}^T \left(\{q(\xi)\}_{,\xi} - \sum_e \int_{-1}^{+1} [B_2]^T \{\sigma\} |J_b| d\eta \right) d\xi \tag{2.103}$$

Substituting Eq. (2.103) back into Eq. (2.93), the complete virtual work equation of the S-element is obtained

$$\{\delta d\}^T q(\xi = 1) - \int_0^1 \{\delta u(\xi)\}^T \left(\{q(\xi)\}_{,\xi} - \sum_e \int_{-1}^{+1} [B_2]^T \{\sigma\} |J_b| d\eta \right) d\xi$$
$$= \{\delta d\}^T \{F\} \tag{2.104}$$

This equation is to be satisfied for all admissible $\{\delta u(\xi)\}$ (on the boundary $\{\delta d\} = \{\delta u(\xi = 1)\}$ in Eq. (2.85) applies) in strong form along the radial direction ξ. This yields two conditions. At the boundary (boundary conditions),

$$\{F\} = \{q(\xi = 1)\} \tag{2.105}$$

has to hold. In the domain ($0 \leq \xi < 1$), the integrand is set to zero leading to

$$\{q(\xi)\}_{,\xi} = \sum_e \int_{-1}^{+1} [B_2]^T \{\sigma\} |J_b| d\eta \tag{2.106}$$

where the nodal force functions are related to stresses $\{\sigma\}$ (Eq. (2.91)). This equation can be regarded as the equations of equilibrium expressed in a weak form in the circumferential direction η. When Eq. (2.106) is satisfied, the virtual increment of the strain energy, Eq. (2.103) is simply equal to (Eq. 2.105)

$$U_{\varepsilon} = \{\delta d\}^T \{q(\xi = 1)\} \tag{2.107}$$

i.e. the work done by the nodal forces at the boundary.

After replacing the stresses $\{\sigma\}$ using Eq. (2.84), Eqs. (2.91) and (2.106) are expanded and rearranged as

$$\{q(\xi)\} = \sum_e \xi \int_{-1}^{+1} [B_1]^T [D] \left([B_1]\{u^e(\xi)\}_{,\xi} + \frac{1}{\xi}[B_2]\{u^e(\xi)\} \right) |J_b| d\eta \qquad (2.108)$$

and

$$\{q(\xi)\}_{,\xi} = \sum_e \int_{-1}^{+1} [B_2]^T [D] \left([B_1]\{u^e(\xi)\}_{,\xi} + \frac{1}{\xi}[B_2]\{u^e(\xi)\} \right) |J_b| d\eta \qquad (2.109)$$

respectively. The integrations with respect to η are separated from the nodal displacement functions, which depend on ξ only. The following coefficient matrices are introduced

$$[E_0] = \sum_e [E_0^e]; \qquad \text{with} \quad [E_0^e] = \int_{-1}^{+1} [B_1]^T [D][B_1]|J_b| d\eta \qquad (2.110a)$$

$$[E_1] = \sum_e [E_1^e]; \qquad \text{with} \quad [E_1^e] = \int_{-1}^{+1} [B_2]^T [D][B_1]|J_b| d\eta \qquad (2.110b)$$

$$[E_2] = \sum_e [E_2^e]; \qquad \text{with} \quad [E_2^e] = \int_{-1}^{+1} [B_2]^T [D][B_2]|J_b| d\eta \qquad (2.110c)$$

It is worthwhile noting that:

1) The coefficient matrices $[E_0]$, $[E_1]$ and $[E_2]$ of the S-element are obtained by assembling the element coefficient matrices $[E_0^e]$, $[E_1^e]$ and $[E_2^e]$ element-by-element according to the element connectivity.
2) The element coefficient matrices are evaluated on the individual line element at the boundary. They depend on only the geometry of the boundary, as represented by the line element.
3) The integration can be performed numerically in the same way as the computation of the stiffness matrix of one-dimensional finite elements. The rules for the choice of the order of integration quadrature in the conventional finite element method are equally applicable.
4) The relationship of the element coefficient matrices to the stiffness matrix of a two-dimensional element is shown in Section 5.2.3, Wolf and Song (1996).
5) Both $[E_0]$ and $[E_2]$ are symmetric. It is also shown later in Section 6.10.1 that $[E_0]$ is positive definite.

Equations (2.108) and (2.109) are simplified, by using the coefficient matrices in Eq. (2.110), as

$$\{q(\xi)\} = [E_0]\xi\{u(\xi)\}_{,\xi} + [E_1]^T\{u(\xi)\} \qquad (2.111a)$$

$$\xi\{q(\xi)\}_{,\xi} = [E_1]\xi\{u(\xi)\}_{,\xi} + [E_2]\{u(\xi)\} \qquad (2.111b)$$

The nodal force functions $\{q(\xi)\}$ can be eliminated by substituting Eq. (2.111a) into Eq. (2.111b) leading to the scaled boundary finite element equation in displacement

$$[E_0]\xi^2\{u(\xi)\}_{,\xi\xi} + ([E_0] + [E_1]^T - [E_1])\xi\{u(\xi)\}_{,\xi} - [E_2]\{u(\xi)\} = 0 \qquad (2.112)$$

It is a system of second-order ordinary differential equations with the dimensionless radial coordinate ξ as the independent variable. In the derivation, the governing partial differential equations of linear elasticity are weakened in the circumferential direction in the manner of the finite element method, while the strong form remains in the radial direction.

◼ Example 2.2 Two-node Element: Element Coefficient Matrices

Evaluate the element coefficient matrices and the equivalent nodal forces of the 2-node element shown in Figure 2.17 in Example 2.1, where the scaled boundary transformation has been performed.

1) Substituting Eqs. (2.49) and (2.52) into Eq. (2.67) leads to

$$[b_1] = \frac{1}{|J_b|}[C_1] \tag{2.113a}$$

$$[b_2] = \frac{1}{|J_b|}[C_2] - \frac{\eta}{|J_b|}[C_1] \tag{2.113b}$$

with the abbreviations

$$[C_1] = \frac{1}{2}\begin{bmatrix} \Delta_y & 0 \\ 0 & -\Delta_x \\ -\Delta_x & \Delta_y \end{bmatrix} \tag{2.114a}$$

$$[C_2] = \begin{bmatrix} -\bar{y} & 0 \\ 0 & \bar{x} \\ \bar{x} & -\bar{y} \end{bmatrix} \tag{2.114b}$$

The matrices $[b_1]$ and $[b_2]$ can also be obtained by using their definitions in Eq. (2.66) and the transformation of derivatives from Cartesian coordinates to the scaled boundary coordinates given in Eq. (2.60) in Example 2.1.

2) Using Eq. (2.16), the shape function matrix (Eq. (2.80)) of the 2-node element is expressed as

$$[N_u] = \frac{1}{2}\begin{bmatrix} 1-\eta & 0 & 1+\eta & 0 \\ 0 & 1-\eta & 0 & 1+\eta \end{bmatrix}$$

$$= \frac{1}{2}\begin{bmatrix} 1 & 0 & 1 & 0 \\ 0 & 1 & 0 & 1 \end{bmatrix} + \frac{\eta}{2}\begin{bmatrix} -1 & 0 & 1 & 0 \\ 0 & -1 & 0 & 1 \end{bmatrix} \tag{2.115}$$

The derivative of the shape function matrix with respect to the natural coordinate η of the parent element is obtained as

$$[N_u]_{,\eta} = \frac{1}{2}\begin{bmatrix} -1 & 0 & 1 & 0 \\ 0 & -1 & 0 & 1 \end{bmatrix} \tag{2.116}$$

3) Substituting Eqs. (2.113), (2.115) and (2.116) into Eq. (2.82), the strain-displacement matrices are express as

$$[B_1] = [b_1][N_u]$$

$$= \frac{1}{2|J_b|}\big[[C_1]\,[C_1]\big] + \frac{\eta}{2|J_b|}\big[-[C_1]\,[C_1]\big] \tag{2.117a}$$

$$[B_2] = [b_2][N_u]_{,\eta}$$

$$= \frac{1}{2|J_b|}\big[-[C_2]\,[C_2]\big] - \frac{\eta}{2|J_b|}\big[-[C_1]\,[C_1]\big] \tag{2.117b}$$

4) The following abbreviations will be introduced below to simplify nomenclature

$$[Q_0] = \frac{1}{2|J_b|}[C_1]^T[D][C_1] \tag{2.118a}$$

$$[Q_1] = \frac{1}{2|J_b|}[C_2]^T[D][C_1] \tag{2.118b}$$

$$[Q_2] = \frac{1}{2|J_b|}[C_2]^T[D][C_2] \tag{2.118c}$$

Substituting Eq. (2.117a) into Eq. (2.110a) and using $\int_{-1}^{+1} 1 \times d\eta = 0$, the element coefficient matrix $[E_0^e]$ is obtained

$$[E_0^e] = \int_{-1}^{+1} [B_1]^T[D][B_1]|J_b|d\eta$$

$$= \int_{-1}^{+1} \frac{1}{2|J_b|}\big[\,[C_1]\ [C_1]\,\big]^T [D]\frac{1}{2|J_b|}\big[\,[C_1]\ [C_1]\,\big]\,|J_b|d\eta$$

$$+ \int_{-1}^{+1} \frac{\eta}{2|J_b|}\big[-[C_1]\ [C_1]\big]^T [D]\frac{\eta}{2|J_b|}\big[-[C_1]\ [C_1]\big]\,|J_b|d\eta$$

$$= \frac{1}{2}\begin{bmatrix} [Q_0] & [Q_0] \\ [Q_0] & [Q_0] \end{bmatrix}\times 2 + \frac{1}{2}\begin{bmatrix} [Q_0] & -[Q_0] \\ -[Q_0] & [Q_0] \end{bmatrix}\times\frac{2}{3}$$

$$= \frac{2}{3}\begin{bmatrix} 2[Q_0] & [Q_0] \\ [Q_0] & 2[Q_0] \end{bmatrix} \tag{2.119a}$$

Similarly, the coefficient matrix $[E_1^e]$ is obtained by substituting Eq. (2.117) into Eq. (2.110b)

$$[E_1^e] = \int_{-1}^{+1} [B_2]^T[D][B_1]|J_b|d\eta$$

$$= \int_{-1}^{+1} \frac{1}{2|J_b|}\big[-[C_2]\ [C_2]\big]^T [D]\frac{1}{2|J_b|}\big[\,[C_1]\ [C_1]\,\big]\,|J_b|d\eta$$

$$- \int \frac{\eta}{2|J_b|}\big[-[C_1]\ [C_1]\big]^T [D]\frac{\eta}{2|J_b|}\big[-[C_1]\ [C_1]\big]\,|J_b|d\eta$$

$$= \frac{1}{2}\begin{bmatrix} -[Q_1] & -[Q_1] \\ [Q_1] & [Q_1] \end{bmatrix}\times 2 - \frac{1}{2}\begin{bmatrix} [Q_0] & -[Q_0] \\ -[Q_0] & [Q_0] \end{bmatrix}\times\frac{2}{3}$$

$$= \begin{bmatrix} -[Q_1] & -[Q_1] \\ [Q_1] & [Q_1] \end{bmatrix} - \frac{1}{3}\begin{bmatrix} [Q_0] & -[Q_0] \\ -[Q_0] & [Q_0] \end{bmatrix} \tag{2.119b}$$

and the coefficient matrix $[E_2^e]$ by substituting Eq. (2.117b) into Eq. (2.110c)

$$[E_2^e] = \int_{-1}^{+1} [B_2]^T[D][B_2]|J_b|d\eta$$

$$= \int_{-1}^{+1} \frac{1}{2|J_b|}\big[-[C_2]\ [C_2]\big]^T [D]\frac{1}{2|J_b|}\big[-[C_2]\ [C_2]\big]\,|J_b|d\eta$$

$$+ \int_{-1}^{+1} \frac{\eta}{2|J_b|}\big[-[C_1]\ [C_1]\big]^T [D]\frac{\eta}{2|J_b|}\big[-[C_1]\ [C_1]\big]\,|J_b|d\eta$$

$$= \frac{1}{2} \begin{bmatrix} [Q_2] & -[Q_2] \\ -[Q_2] & [Q_2] \end{bmatrix} \times 2 + \frac{1}{2} \begin{bmatrix} [Q_0] & -[Q_0] \\ -[Q_0] & [Q_0] \end{bmatrix} \times \frac{2}{3}$$

$$= \begin{bmatrix} [Q_2] & -[Q_2] \\ -[Q_2] & [Q_2] \end{bmatrix} + \frac{1}{3} \begin{bmatrix} [Q_0] & -[Q_0] \\ -[Q_0] & [Q_0] \end{bmatrix} \tag{2.119c}$$

The MATLAB code, as a part of the accompanying computer program Platypus, to compute the coefficient matrix of a 2-node line element is presented in Section 2.6. It is used in the subsequent sections for the static analysis of 2D problems.

Example 2.3   A square S-element is shown in Figure 2.20. Each edge is modelled by 1 line element. The dimensions, nodal numbers and the line element numbers (in circle) are shown in Figure 2.20. The elasticity constants are: Young's modulus E and Poisson's ratio $v = 0$. Considering the plane stress states, determine the element coefficient matrices $[E_0^e]$, $[E_1^e]$ and $[E_2^e]$ for Element 2 using the equations derived for 2-node line element in Example 2.2.

1) The scaling centre is selected at the centre of the square S-element. The origin of the Cartesian coordinates is placed at the scaling centre. The nodal coordinates and connectivity of the line elements at the boundary of the S-element are listed in the tables below. Note that the line elements follow the counter-clockwise direction around the scaling centre.

Nodal coordinates		
Node	x(m)	y(m)
1	−1	−1
2	1	−1
3	1	1
4	−1	1

Element connectivity		
Element	Node 1	Node 2
1	1	2
2	2	3
3	3	4
4	4	1

2) For Element 2, the coordinates of the two nodes are equal to

$$x_1 = 1; \quad y_1 = -1;$$
$$x_2 = 1; \quad y_2 = 1$$

Figure 2.20 A square S-element (Unit: meter).

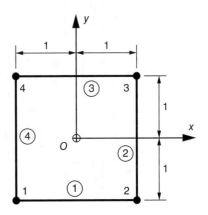

Eqs. (2.50) and (2.51) lead to

$$\Delta_x = 0; \quad \Delta_y = 2;$$
$$\bar{x} = 1; \quad \bar{y} = 0$$

The determinant of the Jacobian matrix at the boundary (Eq. (2.58)) is equal to

$$|J_b| = \frac{1}{2}(1 \times 1 - 1 \times (-1)) = 1$$

The following two matrices are obtained from Eq. (2.114)

$$[C_1] = \begin{bmatrix} 1 & 0 \\ 0 & 0 \\ 0 & 1 \end{bmatrix}; \quad [C_2] = \begin{bmatrix} 0 & 0 \\ 0 & 1 \\ 1 & 0 \end{bmatrix}$$

3) The elasticity matrix in plane stress state is expressed as (Eq. (A.42))

$$[D] = \frac{E}{2} \begin{bmatrix} 2 & 0 & 0 \\ 0 & 2 & 0 \\ 0 & 0 & 1 \end{bmatrix}$$

4) Using $[C_1]$, $[C_2]$ and $[D]$ above, Equation (2.118) leads to

$$[Q_0] = \frac{1}{2 \times 1} \begin{bmatrix} 1 & 0 & 0 \\ 0 & 0 & 1 \end{bmatrix} \frac{E}{2} \begin{bmatrix} 2 & 0 & 0 \\ 0 & 2 & 0 \\ 0 & 0 & 1 \end{bmatrix} \begin{bmatrix} 1 & 0 \\ 0 & 0 \\ 0 & 1 \end{bmatrix}$$

$$= \frac{E}{4} \begin{bmatrix} 2 & 0 & 0 \\ 0 & 0 & 1 \end{bmatrix} \begin{bmatrix} 1 & 0 \\ 0 & 0 \\ 0 & 1 \end{bmatrix}$$

$$= \frac{E}{4} \begin{bmatrix} 2 & 0 \\ 0 & 1 \end{bmatrix}$$

$$[Q_1] = \frac{1}{2 \times 1} \begin{bmatrix} 0 & 0 & 1 \\ 0 & 1 & 0 \end{bmatrix} \frac{E}{2} \begin{bmatrix} 2 & 0 & 0 \\ 0 & 2 & 0 \\ 0 & 0 & 1 \end{bmatrix} \begin{bmatrix} 1 & 0 \\ 0 & 0 \\ 0 & 1 \end{bmatrix}$$

$$= \frac{E}{4} \begin{bmatrix} 0 & 0 & 1 \\ 0 & 2 & 0 \end{bmatrix} \begin{bmatrix} 1 & 0 \\ 0 & 0 \\ 0 & 1 \end{bmatrix}$$

$$= \frac{E}{4} \begin{bmatrix} 0 & 1 \\ 0 & 0 \end{bmatrix}$$

$$[Q_2] = \frac{1}{2 \times 1} \begin{bmatrix} 0 & 0 & 1 \\ 0 & 1 & 0 \end{bmatrix} \frac{E}{2} \begin{bmatrix} 2 & 0 & 0 \\ 0 & 2 & 0 \\ 0 & 0 & 1 \end{bmatrix} \begin{bmatrix} 0 & 0 \\ 0 & 1 \\ 1 & 0 \end{bmatrix}$$

$$= \frac{E}{4} \begin{bmatrix} 0 & 0 & 1 \\ 0 & 2 & 0 \end{bmatrix} \begin{bmatrix} 0 & 0 \\ 0 & 1 \\ 1 & 0 \end{bmatrix}$$

$$= \frac{E}{4} \begin{bmatrix} 1 & 0 \\ 0 & 2 \end{bmatrix}$$

The coefficient matrices of Element 2 are obtained using Eq. (2.119) as

$$[E_0^e] = \frac{E}{6}\begin{bmatrix} 4 & 0 & 2 & 0 \\ 0 & 2 & 0 & 1 \\ 2 & 0 & 4 & 0 \\ 0 & 1 & 0 & 2 \end{bmatrix}$$

$$[E_1^e] = \frac{E}{4}\begin{bmatrix} 0 & -1 & 0 & -1 \\ 0 & 0 & 0 & 0 \\ 0 & 1 & 0 & 1 \\ 0 & 0 & 0 & 0 \end{bmatrix} - \frac{1}{3} \times \frac{E}{4}\begin{bmatrix} 2 & 0 & -2 & 0 \\ 0 & 1 & 0 & -1 \\ -2 & 0 & 2 & 0 \\ 0 & -1 & 0 & 1 \end{bmatrix}$$

$$= \frac{E}{12}\begin{bmatrix} -2 & -3 & 2 & -3 \\ 0 & -1 & 0 & 1 \\ 2 & 3 & -2 & 3 \\ 0 & 1 & 0 & -1 \end{bmatrix}$$

$$[E_2^e] = \frac{E}{4}\begin{bmatrix} 1 & 0 & -1 & 0 \\ 0 & 2 & 0 & -2 \\ -1 & 0 & 1 & 0 \\ 0 & -2 & 0 & 2 \end{bmatrix} + \frac{1}{3} \times \frac{E}{4}\begin{bmatrix} 2 & 0 & -2 & 0 \\ 0 & 1 & 0 & -1 \\ -2 & 0 & 2 & 0 \\ 0 & -1 & 0 & 1 \end{bmatrix}$$

$$= \frac{E}{12}\begin{bmatrix} 5 & 0 & -5 & 0 \\ 0 & 7 & 0 & -7 \\ -5 & 0 & 5 & 0 \\ 0 & -7 & 0 & 7 \end{bmatrix}$$

2.6 Computer Program Platypus: Coefficient Matrices of an S-element

MATLAB functions for computing the coefficient matrices of an S-element in the computer program Platypus accompanying this book are listed in this section. The use of the functions is demonstrated by examples. The boundary of the S-element is divided into 2-node line elements. The equations for computing the element coefficient matrices are obtained by performing the integrations analytically in Example 2.1. The origin of the coordinates has to be placed at the scaling centre in these functions. This requirement can be easily met by a translation of the coordinate system as shown later in Code List 3.2.

2.6.1 Element Coefficient Matrices of a 2-node Line Element

The function `EleCoeff2NodeEle.m` computes the element coefficient matrices $[E_0^e]$, $[E_1^e]$ and $[E_2^e]$ of a 2-node line element. It is listed below in Code List 2.1. The function is documented with comments in the list. The equation numbers in the comments refer to the equations in the text of this book. Additional explanations and an example to use this function are provided after the code list.

Code List 2.1: Element coefficient matrices

```
1   function [ e0, e1, e2, m0 ] = EleCoeff2NodeEle( xy, mat)
2   %Coefficient matrices of a 2-node line element
3   %
4   %  Inputs:
5   %    xy(i,:)      - coordinates of node i (orgin at scaling centre).
6   %                   The nodes are numbered locally within
7   %                   each line element
8   %    mat          - material constants
9   %       mat.D     - elasticity matrix
10  %       mat.den   - mass density
11  %
12  %  Outputs:
13  %     e0, e1, e2, m0  - element coefficient matrices
14
15  dxy = xy(2,:)-xy(1,:);  % ......................... [Δx, Δy], Eq. (2.50)
16  mxy = sum(xy)/2;  % .................................. [x̄, ȳ], Eq. (2.51)
17  a = xy(1,1)*xy(2,2)-xy(2,1)*xy(1,2);  % ............. a = 2|Jb|, Eq. (2.58)
18  if a < 1.d-10
19      disp('negative area (EleCoeff2NodeEle)');
20      pause
21  end
22  C1 = 0.5*[ dxy(2) 0; 0 -dxy(1); -dxy(1) dxy(2)];  % ....... Eq. (2.114a)
23  C2 = [-mxy(2) 0; 0 mxy(1); mxy(1) -mxy(2)];  % ........... Eq. (2.114b)
24  Q0 = 1/a*(C1'*mat.D*C1);  % ............................... Eq. (2.118a)
25  Q1 = 1/a*(C2'*mat.D*C1);  % ............................... Eq. (2.118b)
26  Q2 = 1/a*(C2'*mat.D*C2);  % ............................... Eq. (2.118c)
27  e0 =  2/3*[2*Q0 Q0; Q0 2*Q0];  % .......................... Eq. (2.119a)
28  e1 = -1/3*[ Q0 -Q0; -Q0  Q0] + [-Q1 -Q1;  Q1 Q1];  % ...... Eq. (2.119b)
29  e2 =  1/3*[ Q0 -Q0; -Q0  Q0] + [ Q2 -Q2; -Q2 Q2];  % ...... Eq. (2.119c)
30
31  % ...   mass coefficent matrix, Eq. (3.112)
32  m0 = a*mat.den/6*[ 2 0  1 0; 0 2  0 1; 1 0  2 0; 0 1 0 2 ];
33
34  end
```

The inputs are the nodal coordinates (argument xy) of the two nodes of the line element and the material properties (argument mat) given by the elasticity matrix (field D) and mass density (field den). The two nodes are numbered locally within the element (Figure 2.10 on page 38) and the direction from node 1 to node 2 has to follow the counter-clockwise direction around the scaling centre (Figure 2.9). The x and y coordinates of a node are stored as one row, i.e. the first row of xy is $[x_1, y_1]$ and the second one $[x_2, y_2]$.

The coefficient $[M_0^e]$, which is related to the mass matrix, is also calculated (see the code starting from Line 31). It is given in Eq. (3.98) for the dynamic analysis in Section 3.10.

■ **Example 2.4**   Use the function `EleCoeff2NodeEle` in Code List 2.1 on page 64 to compute the element coefficient matrices $[E_0]$, $[E_1]$ and $[E_2]$ of the four 2-node line elements in the square S-element in Example 2.3 on page 61. Assume the Young's modulus is equal to $E = 10\,\mathrm{GPa}$.

The following MATLAB script calls the function `EleCoeff2NodeEle` to calculate the coefficient matrices of Element 1

```
%% Compute element coefficient matrices
% nodal coordinates of element given as [x1 y1; x2 y2]
xy = [-1 -1; 1 -1] % Element 1
% xy = [ 1 -1; 1 1] % Element 2
% xy = [ 1 1; -1 1] % Element 3
% xy = [-1 1; -1 -1] % Element 4

% elascity matrix (plane stress).
% E: Young's modulus; p: Poisson's ratio
ElasMtrx = @(E, p) E/(1-p^2)*[1 p 0; p 1 0; 0 0 (1-p)/2];
mat.D = ElasMtrx(10, 0); % E in GPa
mat.den = 2; % mass density in Mg per cubic meter

[ e0, e1, e2 ] = EleCoeff2NodeEle( xy, mat)
```

The above MATLAB script can be obtained by scanning the QR-code to the right.

The coordinates of the 4 nodes and the connectivity of the 4 line elements of the square S-element are given in Example 2.3. Nodes 1 and 2 of line element 1 are Nodes 1 (-1, -1) and 2 (1, -1) of the square S-element, respectively.

The elasticity matrix of the plane stress conditions is calculated by a so-called anonymous function `ElasMtrx`. It takes the Young's modulus E and Poisson's ratio p as inputs. The mass density `den` is set to zero as the mass matrix is not used in this example.

The MATLAB output is listed below

```
xy =
    -1      -1
     1      -1
```

```
e0 =
    3.3333         0    1.6667         0
         0    6.6667         0    3.3333
    1.6667         0    3.3333         0
         0    3.3333         0    6.6667

e1 =
   -0.8333         0    0.8333         0
    2.5000   -1.6667    2.5000    1.6667
    0.8333         0   -0.8333         0
   -2.5000    1.6667   -2.5000   -1.6667

e2 =
    5.8333         0   -5.8333         0
         0    4.1667         0   -4.1667
   -5.8333         0    5.8333         0
         0   -4.1667         0    4.1667
```

Element 2 is addressed. Nodes 1 and 2 of Element 2 are Nodes 2 $(1, -1)$ and 3 $(1, 1)$ of the square S-element, respectively. To obtain the coefficient matrices of Element 2, it is sufficient to replace the nodal coordinates of Element 1 in the previous MATLAB script with those of Element 2:

```
xy = [ 1 -1; 1 1] % Element 2
```

The MATLAB outputs are as follows:

```
xy =
     1    -1
     1     1

e0 =
    6.6667         0    3.3333         0
         0    3.3333         0    1.6667
    3.3333         0    6.6667         0
         0    1.6667         0    3.3333

e1 =
   -1.6667   -2.5000    1.6667   -2.5000
         0   -0.8333         0    0.8333
    1.6667    2.5000   -1.6667    2.5000
         0    0.8333         0   -0.8333

e2 =
    4.1667         0   -4.1667         0
         0    5.8333         0   -5.8333
   -4.1667         0    4.1667         0
         0   -5.8333         0    5.8333
```

Note that the same results of the coefficient matrices in Example 2.3 are obtained. Similarly, the coefficient matrices of Elements 3 and 4 are obtained by inputting their nodal coordinates

```
xy = [ 1 1; -1 1] % Element 3
```

for Element 3, and

```
xy = [-1 1; -1 -1] % Element 4
```

for Element 4. For the sake of brevity, only the result of the matrix $[E_0^e]$, which is used in the next section to illustrate the assembling process, is given below. The MATLAB output for Element 3 is

```
e0 =
      3.3333            0      1.6667            0
           0      6.6667            0      3.3333
      1.6667            0      3.3333            0
           0      3.3333            0      6.6667
```

and for Element 4 is
```
e0 =
      6.6667            0      3.3333            0
           0      3.3333            0      1.6667
      3.3333            0      6.6667            0
           0      1.6667            0      3.3333
```

2.6.2 Assembly of Coefficient Matrices of an S-element

The coefficient matrices of an S-element are obtained by assembling those of the line elements. The process is identical to the assembly of the element stiffness matrices of one-dimensional elements in the standard finite element method. The MATLAB function for assembly is listed below in Code List 2.2, followed by further explanations and an example.

Code List 2.2: Assembly of Coefficient Matrices of an S-element

```
1   function [ E0, E1, E2, M0 ] = SElementCoeffMtx(xy, conn, mat)
2   %Coefficient matrix of an S-element
3   %
4   % Inputs:
5   %    xy(i,:)      - coordinates of node i (orgin at scaling centre)
6   %                   The nodes are numbered locally within
7   %                   an S-element starting from 1
8   %    conn(ie,:) - local connectivity matrix of line element ie
9   %                   in the local nodal numbers of an S-element
10  %    mat          - material constants
11  %       mat.D      - elasticity matrix
12  %       mat.den    - mass density
13  %
14  % Outputs:
15  %    E0, E1, E2   - coefficient matrices of S-element
16
17  nd = 2*size(xy,1); % number of DOFs at boundary (2 DOFs per node)
18  % ...  initializing variables
19  E0 = zeros(nd, nd);
20  E1 = zeros(nd, nd);
21  E2 = zeros(nd, nd);
22  M0 = zeros(nd, nd);
23  for ie = 1:size(conn,1)  % ............. loop over elements at boundary
24      xyEle = xy(conn(ie,:),:);  % ...... nodal coordinates of an element
25      % ...  get element coefficient matrices of an element
26      [ ee0, ee1, ee2, em0 ] = EleCoeff2NodeEle(xyEle, mat);
27      % ...  local DOFs (in S-element) of an element
28      d = reshape([2*conn(ie,:)-1; 2*conn(ie,:)], 1, []);
29      % ...  assemble coefficient matrices of S-element
30      E0(d,d) = E0(d,d) + ee0;
```

```
31        E1(d,d)  = E1(d,d) + ee1;
32        E2(d,d)  = E2(d,d) + ee2;
33        M0(d,d)  = M0(d,d) + em0;
34   end
35
36   end
```

The inputs of this function include the nodal coordinates (argument xy), the connectivity table (argument conn) of the line elements and the same material constants as in Code List 2.1 on page 64. The coordinates (x_i, y_i) of node i are stored as the i-th row of the argument xy. The two nodes of a line element are stored as one row of the argument conn. This function loops over all the line elements to compute element coefficient matrices (by calling the function in Code List 2.1 on page 64) and assembles them to form the coefficient matrices of the S-element. At each node, two degrees of freedom (DOF) are considered. The vector d contains the connectivity between the DOFs of a 2-node line element and the DOFs of the S-element. It is constructed by the code on Line 28, which is explained in the following example. The assembling is performed according to the connectivity of DOFs.

Example 2.5 Use the function in Code List 2.2 to compute the coefficient matrices $[E_0]$, $[E_1]$ and $[E_2]$ of the square S-element in Example 2.4 on page 65. Assume the Young's modulus is equal to $E = 10\,\text{GPa}$ and Poisson's ratio $v = 0$.

The following MATLAB script calls the function SElementCoeffMtx to calculate the coefficient matrices of the S-element.

```
%% Compute coefficient matrices of square S-element
xy = [-1 -1; 1 -1; 1 1; -1 1]
conn = [1:4; 2:4 1]'
% elascity matrix (plane stress).
% E: Young's modulus; p: Poisson's ratio
ElasMtrx = @(E, p) E/(1-p^2)*[1 p 0; p 1 0; 0 0 (1-p)/2];
mat.D = ElasMtrx(10, 0); % E in GPa
mat.den = 2; % mass density in Mg per cubic meter
[ E0, E1, E2 ] = SElementCoeffMtx(xy, conn, mat)
```

The above MATLAB script can be obtained by scanning the QR-code to the right.

The nodal coordinates are displayed as

```
xy =
    -1      -1
     1      -1
     1       1
    -1       1
```

and the connectivity table as

```
conn =
     1      2
     2      3
     3      4
     4      1
```

The assembling process is detailed for Element 4, which corresponds to ie=4 in the for-loop. The two nodes of Element 4 is given by

```
conn(ie,:)
ans =
     4      1
```

The nodal coordinates of an element are extracted from the nodal coordinates of the S-element (Line 24) as

```
xyEle =
    -1       1
    -1      -1
```

and used to compute the element coefficient matrices (see Example 2.4).

Within an S-element, the DOFs are numbered locally starting from the first node of the S-element. The two DOFs of the i-th node of the S-element are thus $(2i - 1)$ for u_x and $2i$ for u_y. The connectivity of the DOFs of a line element is constructed in Line 28. The DOFs for u_x and u_y of the nodes of the element are formed, respectively, as the first and second rows of a matrix

```
[2*conn(ie,:)-1; 2*conn(ie,:)]
ans =
     7      1
     8      2
```

It is reshaped column-wise into one row as

```
d =
     7      8      1      2
```

Line 30 assembles the entries of an element coefficient matrix $[E_0^e]$ to the entries of the same DOFs in the coefficient matrix $[E_0]$ of the S-element according to the connectivity of the DOFs. For example, E0(d,d) is expanded as:

$$
\text{Element 4:} \quad \text{E0(d,d)} \Rightarrow
\begin{bmatrix}
E0(7,7) & E0(7,8) & E0(7,1) & E0(7,2) \\
E0(8,7) & E0(8,8) & E0(8,1) & E0(8,2) \\
E0(1,7) & E0(1,8) & E0(1,1) & E0(1,2) \\
E0(2,7) & E0(2,8) & E0(2,1) & E0(2,2)
\end{bmatrix}
$$

to which the coefficient matrix $[E_0^e]$ of Element 4 is added to.

Following the same steps, the connectivity of the DOFs of the other three elements are obtained as:

Element 1:   d = [1   2   3   4]

Element 2:   d = [3   4   5   6]

Element 3:   d = [5   6   7   8]

Their element coefficient matrices $[E_0^e]$ are assembled, respectively, to the following entries of the coefficient matrix of the S-element:

Element 1:
$$\begin{bmatrix} E0(1,1) & E0(1,2) & E0(1,3) & E0(1,4) \\ E0(2,1) & E0(2,2) & E0(2,3) & E0(2,4) \\ E0(3,1) & E0(3,2) & E0(3,3) & E0(3,4) \\ E0(4,1) & E0(4,2) & E0(4,3) & E0(4,4) \end{bmatrix}$$

Element 2:
$$\begin{bmatrix} E0(3,3) & E0(3,4) & E0(3,5) & E0(3,6) \\ E0(4,3) & E0(4,4) & E0(4,5) & E0(4,6) \\ E0(5,3) & E0(5,4) & E0(5,5) & E0(5,6) \\ E0(6,3) & E0(6,4) & E0(6,5) & E0(6,6) \end{bmatrix}$$

Element 3:
$$\begin{bmatrix} E0(5,5) & E0(5,6) & E0(5,7) & E0(5,8) \\ E0(6,5) & E0(6,6) & E0(6,7) & E0(6,8) \\ E0(7,5) & E0(7,6) & E0(7,7) & E0(7,8) \\ E0(8,5) & E0(8,6) & E0(8,7) & E0(8,8) \end{bmatrix}$$

Using the element coefficient matrices $[E_0^e]$ obtained in Example 2.4 on page 65 for the four elements, the coefficient matrix of the S-element is obtained as

$$[E_0] = \begin{bmatrix} 10.00 & 0.00 & 1.67 & 0.00 & 0.00 & 0.00 & 3.33 & 0.00 \\ 0.00 & 10.00 & 0.00 & 3.33 & 0.00 & 0.00 & 0.00 & 1.67 \\ 1.67 & 0.00 & 10.00 & 0.00 & 3.33 & 0.00 & 0.00 & 0.00 \\ 0.00 & 3.33 & 0.00 & 10.00 & 0.00 & 1.67 & 0.00 & 0.00 \\ 0.00 & 0.00 & 3.33 & 0.00 & 10.00 & 0.00 & 1.67 & 0.00 \\ 0.00 & 0.00 & 0.00 & 1.67 & 0.00 & 10.00 & 0.00 & 3.33 \\ 3.33 & 0.00 & 0.00 & 0.00 & 1.67 & 0.00 & 10.00 & 0.00 \\ 0.00 & 1.67 & 0.00 & 0.00 & 0.00 & 3.33 & 0.00 & 10.00 \end{bmatrix}$$

In the same way, the coefficient matrices $[E_1]$ and $[E_2]$ of the S-element can be determined by performing the assembly. For the sake of brevity, the intermediate results are omitted. Only the final results are provided below:

$$[E_1] = \begin{bmatrix} -2.50 & 2.50 & 0.83 & 0.00 & 0.00 & 0.00 & 1.67 & 2.50 \\ 2.50 & -2.50 & 2.50 & 1.67 & 0.00 & 0.00 & 0.00 & 0.83 \\ 0.83 & 0.00 & -2.50 & -2.50 & 1.67 & -2.50 & 0.00 & 0.00 \\ -2.50 & 1.67 & -2.50 & -2.50 & 0.00 & 0.83 & 0.00 & 0.00 \\ 0.00 & 0.00 & 1.67 & 2.50 & -2.50 & 2.50 & 0.83 & 0.00 \\ 0.00 & 0.00 & 0.00 & 0.83 & 2.50 & -2.50 & 2.50 & 1.67 \\ 1.67 & -2.50 & 0.00 & 0.00 & 0.83 & 0.00 & -2.50 & -2.50 \\ 0.00 & 0.83 & 0.00 & 0.00 & -2.50 & 1.67 & -2.50 & -2.50 \end{bmatrix}$$

$$[E_2] = \begin{bmatrix} 10.00 & 0.00 & -5.83 & 0.00 & 0.00 & 0.00 & -4.17 & 0.00 \\ 0.00 & 10.00 & 0.00 & -4.17 & 0.00 & 0.00 & 0.00 & -5.83 \\ -5.83 & 0.00 & 10.00 & 0.00 & -4.17 & 0.00 & 0.00 & 0.00 \\ 0.00 & -4.17 & 0.00 & 10.00 & 0.00 & -5.83 & 0.00 & 0.00 \\ 0.00 & 0.00 & -4.17 & 0.00 & 10.00 & 0.00 & -5.83 & 0.00 \\ 0.00 & 0.00 & 0.00 & -5.83 & 0.00 & 10.00 & 0.00 & -4.17 \\ -4.17 & 0.00 & 0.00 & 0.00 & -5.83 & 0.00 & 10.00 & 0.00 \\ 0.00 & -5.83 & 0.00 & 0.00 & 0.00 & -4.17 & 0.00 & 10.00 \end{bmatrix}$$

3

Solution of the Scaled Boundary Finite Element Equation by Eigenvalue Decomposition

The scaled boundary finite element equation (Eq. (2.112)) is a system of Euler-Cauchy equations (Kreyszig, 2011). It is also closely related to the algebraic Riccati equation (Laub, 1979; Golub and van Loan, 1996) and the solution of quadratic eigenvalue problems (Tisseur and Meerbergen, 2001). A large volume of literature, ranging from standard textbooks to recent journal articles, exists on the solution procedure for systems of Euler-Cauchy equations or the closely-related algebraic Riccati equations. In this chapter, the simplest eigenvalue method is presented and some important conclusions necessary for the numerical implementation are simply stated without proofs. A numerically robust solution procedure based on the block-diagonal Schur decomposition and the theoretical background related to the solution procedures are presented in Chapter 7.

3.1   Solution Procedure for the Scaled Boundary Finite Element Equations in Displacement

A solution procedure suitable for numerical computation is to convert the second-order ordinary differential equations in Eq. (2.112) to first-order ones. The number of degrees of freedom at the boundary of the S-element is equal to the number of equations in Eq. (2.112) and denoted as n. The reduction of the order of the differential equations from two to one doubles the number of equations to $2n$.

The system of $2n$ first-order ordinary differential equations can be formulated directly from Eqs. (2.111a) and (2.111b). The derivative of displacement functions with respect to the radial coordinate ξ is obtained from Eq. (2.111a) as

$$\xi\{u(\xi)\}_{,\xi} = [E_0]^{-1}\{q(\xi)\} - [E_0]^{-1}[E_1]^T\{u(\xi)\} \tag{3.1}$$

Eliminating $\xi\{u(\xi)\}_{,\xi}$ from Eq. (2.111b) results in

$$\xi\{q(\xi)\}_{,\xi} -[E_1][E_0]^{-1}\{q(\xi)\} + [E_1][E_0]^{-1}[E_1]^T\{u(\xi)\} + [E_2]\{u(\xi)\} = 0 \tag{3.2}$$

Introducing the variable $\{X(\xi)\}$ that consists of the n nodal displacement functions and the n force functions

$$\{X(\xi)\} = \left\{ \begin{array}{c} \{u(\xi)\} \\ \{q(\xi)\} \end{array} \right\} \tag{3.3}$$

The Scaled Boundary Finite Element Method: Introduction to Theory and Implementation, First Edition. Chongmin Song.
© 2018 John Wiley & Sons Ltd. Published 2018 by John Wiley & Sons Ltd.
Companion website: www.wiley.com/go/author/scaledboundaryfiniteelementmethod

Equations. (3.1) and (3.2) are reformulated as a system of $2n$ equations

$$\xi\{X(\xi)\}_{,\xi} = [Z_p]\{X(\xi)\} \tag{3.4}$$

with the coefficient matrix[1] $[Z_p]$ specified in

$$[Z_p] = \begin{bmatrix} -[E_0]^{-1}[E_1]^T & [E_0]^{-1} \\ [E_2] - [E_1][E_0]^{-1}[E_1]^T & [E_1][E_0]^{-1} \end{bmatrix} \tag{3.5}$$

As $[E_0]$ and $[E_2]$ are symmetric, $[Z_p]$ is a Hamiltonian matrix satisfying

$$([J_{2n}][Z_p])^T = [J_{2n}][Z_p] \tag{3.6}$$

where the matrix $[J_{2n}]$ of order $2n$ is defined as ($[I]$ is a $n \times n$ identity matrix)

$$[J_{2n}] = \begin{bmatrix} 0 & [I] \\ -[I] & 0 \end{bmatrix} \tag{3.7}$$

A well-known and simple procedure to solve Eq. (3.4) is to postulate that its solution is of the form of a power function

$$\{X(\xi)\} = \xi^\lambda\{\phi\} \tag{3.8}$$

Substituting the trial solution (Eq. (3.8)) into Eq. (3.4) results in

$$\lambda\xi^\lambda\{\phi\} = [Z_p]\xi^\lambda\{\phi\} \tag{3.9}$$

For this equation to hold at any ξ, the following condition has to be satisfied

$$[Z_p]\{\phi\} = \lambda\{\phi\} \tag{3.10}$$

which is simply the eigenvalue problem of $[Z_p]$ with the eigenvalue λ and the corresponding eigenvector $\{\phi\}$. Alternative, Eq. (3.10) can be written as

$$\left([Z_p] - \lambda[I]_{2n}\right)\{\phi\} = 0 \tag{3.11}$$

where $[I]_{2n}$ is an identity matrix of order $2n$.

The eigenvalue decomposition in Eq. (3.10) is expressed in matrix form as

$$[Z_p][\Phi] = [\Phi]\text{diag}(\lambda) \tag{3.12}$$

where the eigenvector matrix $[\Phi]$ consists of the $2n$ eigenvectors

$$[\Phi] = [\{\phi_1\} \quad \{\phi_2\} \quad \cdots \quad \{\phi_i\} \quad \cdots \quad \{\phi_{2n}\}] \tag{3.13}$$

The diagonal matrix of the eigenvalues is written as

$$\text{diag}(\lambda) = \text{diag}(\lambda_1 \quad \lambda_2 \quad \cdots \quad \lambda_i \quad \cdots \quad \lambda_{2n}) \tag{3.14}$$

Many mathematical libraries to perform the eigenvalue decomposition exist, for example, the function eig in MATLAB, the subroutine DGEEV in LAPACK (Anderson, 2005), to name a few.

---

1 The scaled boundary finite element method was originally developed for the modelling of unbounded (exterior) domains. For convenience in notations, $[Z] = -[Z_p]$ was used so that the positive eigenvalues correspond to the solution for an unbounded domain and the negative eigenvalues to the solution for a bounded (interior) domain. This book mostly is concerned with the modelling of bounded domains. It is more convenient to use $[Z_p]$ instead of $[Z]$. The solution for a bounded domain is obtained using the positive eigenvalues of $[Z_p]$.

The general solution of Eq. (3.4) is then expressed as a linear combination of the trial solutions corresponding to the $2n$ eigenvalues and eigenvectors of $[Z_p]$

$$\{X(\xi)\} = \sum_{i=1}^{2n} c_i \xi^{\lambda_i} \{\phi_i\}$$

$$= c_1 \xi^{\lambda_1} \{\phi_1\} + c_2 \xi^{\lambda_2} \{\phi_2\} + \dots + c_i \xi^{\lambda_i} \{\phi_i\} + \dots + c_{2n} \xi^{\lambda_{2n}} \{\phi_{2n}\} \qquad (3.15)$$

where c_i are the integration constants and dependent on the boundary conditions. Using Eqs. (3.3) and (3.15), and partitioning an eigenvector $\{\phi_i\}$ into two sub-vectors $\{\phi_i^{(u)}\}$ and $\{\phi_i^{(q)}\}$ of the same size n

$$\{\phi_i\} = \left\{ \begin{array}{c} \{\phi_i^{(u)}\} \\ \{\phi_i^{(q)}\} \end{array} \right\} \qquad (3.16)$$

the solutions for nodal displacement and force functions are expressed as

$$\{u(\xi)\} = \sum_{i=1}^{2n} c_i \xi^{\lambda_i} \{\phi_i^{(u)}\} \qquad (3.17a)$$

$$\{q(\xi)\} = \sum_{i=1}^{2n} c_i \xi^{\lambda_i} \{\phi_i^{(q)}\} \qquad (3.17b)$$

$\{\phi_i^{(u)}\}$ can be regarded as the displacement modes and c_i the modal participation factors. The eigenvalues and eigenvectors of a real matrix can occur as complex conjugate pairs. The solution of the ordinary differential equation is always real.

As shown later in Section 6.10.3, one property of a Hamiltonian matrix is that if λ (the index i is omitted for simplicity) is an eigenvalue, $-\lambda$ is also an eigenvalue. In other words, the eigenvalues of a Hamiltonian matrix occur in pairs of $(\lambda, -\lambda)$. As the matrix $[Z_p]$ is real, complex eigenvalues occur as conjugate pairs. Therefore, if a complex eigenvalue $\lambda = \lambda_R + i\lambda_I$ (where i is the imaginary unit, $\lambda_R = \mathrm{Re}(\lambda)$ and $\lambda_I = \mathrm{Im}(\lambda)$ are the real and imaginary parts, respectively) exists, $\lambda_R - i\lambda_I$, $-\lambda_R + i\lambda_I$ and $-\lambda_R - i\lambda_I$ are also eigenvalues.

For convenience, the $2n$ eigenvalues are sorted into the descending order of their real parts, i.e. $\mathrm{Re}(\lambda_1) \geq \mathrm{Re}(\lambda_2) \geq \dots \geq \mathrm{Re}(\lambda_{2n})$. The eigenvalues are separated into three groups according to the signs of their real part. The behaviour of the terms in the general solution (Eq. (3.15)) corresponding to each of the group is examined at the scaling centre ($\xi = 0$):

1) Eigenvalues with positive real parts, i.e. $\mathrm{Re}(\lambda_i) > 0$: $(n-2)$ eigenvalues (λ_i, $i = 1, 2, \dots, n-2$) belong to this group. The power function $\xi^{\lambda_i} = 0$ applies at the scaling centre $\xi = 0$.

2) Eigenvalues of zeros. Four eigenvalues (λ_i, $i = n-1, n, n+1, n+2$) are equal to 0. The power function $\xi^{\lambda_i} = \xi^0 = 1$ applies. Two of them, λ_{n-1} and λ_n, correspond to the two terms describing the translational rigid motions of the S-element. The entries of $\{\phi_{n-1}^{(u)}\}$ and $\{\phi_n^{(u)}\}$ can be set according to $u_x = 1$, $u_y = 0$ and $u_x = 0$, $u_y = 1$, respectively. As rigid body motions do not produce forces, $\{\phi_{n-1}^{(q)}\} = \{\phi_n^{(q)}\} = 0$ applies.[2]

_____

2 In early works on the scaled boundary finite element method, the zero eigenvalues are avoided by adding a small positive value to the diagonal entries of the coefficient matrix $[E_2]$. The two eigenvalues corresponding

3) Eigenvalues with negative real parts, i.e. $\text{Re}(\lambda_i) < 0$: $(n-2)$ eigenvalues $(\lambda_i, i = n + 3, n+4, \ldots, 2n)$ belong to this group. The power function $\xi^{\lambda_i} \to \infty$ applies at the scaling centre $\xi = 0$. As the displacements have to remain finite inside the S-element, the integration constants have to be equal to zero, i.e. $c_i = 0$ applies for $i = n + 3, n+4, \ldots, 2n$.

The solution for a bounded S-element is then written as

$$\{u(\xi)\} = \sum_{i=1}^{n} c_i \xi^{\lambda_i} \{\phi_i^{(u)}\}$$

$$= c_1 \xi^{\lambda_1} \{\phi_1^{(u)}\} + c_2 \xi^{\lambda_2} \{\phi_2^{(u)}\} + \ldots + c_i \xi^{\lambda_i} \{\phi_i^{(u)}\} + \ldots + c_n \xi^{\lambda_n} \{\phi_n^{(u)}\} \quad (3.18a)$$

$$\{q(\xi)\} = \sum_{i=1}^{n} c_i \xi^{\lambda_i} \{\phi_i^{(q)}\}$$

$$= c_1 \xi^{\lambda_1} \{\phi_1^{(q)}\} + c_2 \xi^{\lambda_2} \{\phi_2^{(q)}\} + \ldots + c_i \xi^{\lambda_i} \{\phi_i^{(q)}\} + \ldots + c_n \xi^{\lambda_n} \{\phi_n^{(q)}\} \quad (3.18b)$$

The remaining integration constants c_i, $(i = 1, 2, \ldots, n)$ are to be determined from the boundary condition at $\xi = 1$.

For convenience, the first n eigenvalues and eigenvectors are assembled into matrices. The eigenvalues form a diagonal matrix

$$\langle \lambda_b \rangle = \text{diag}(\lambda_1, \lambda_2, \ldots, \lambda_i, \ldots, \lambda_n) \quad (3.19)$$

The eigenvector matrix is constructed by combining the corresponding eigenvectors as columns (in other words, the i-th eigenvector is the i-th column)

$$\begin{bmatrix} [\Phi_b^{(u)}] \\ [\Phi_b^{(q)}] \end{bmatrix} = \begin{bmatrix} \{\phi_1^{(u)}\} & \{\phi_2^{(u)}\} & \cdots & \{\phi_i^{(u)}\} & \cdots & \{\phi_n^{(u)}\} \\ \{\phi_1^{(q)}\} & \{\phi_2^{(q)}\} & \cdots & \{\phi_i^{(q)}\} & \cdots & \{\phi_n^{(q)}\} \end{bmatrix} \quad (3.20)$$

It is partitioned into two blocks corresponding to the nodal displacement and force modes. Using Eqs. (3.19) and (3.20), Eq. (3.18) is written in matrix form as

$$\{u(\xi)\} = [\Phi_b^{(u)}] \xi^{\langle \lambda_b \rangle} \{c\} \quad (3.21a)$$

$$\{q(\xi)\} = [\Phi_b^{(q)}] \xi^{\langle \lambda_b \rangle} \{c\} \quad (3.21b)$$

where the vector $\{c\}$ assembles the integration constants

$$\{c\} = [c_1, c_2, \ldots, c_n]^T \quad (3.22)$$

Eliminating the integration constants from Eq. (3.21),

$$\{q(\xi)\} = [\Phi_b^{(q)}][\Phi_b^{(u)}]^{-1} \{u(\xi)\} \quad (3.23)$$

is obtained.

At the boundary ($\xi = 1$) of the S-element, as shown in Figure 2.19, the nodal forces $\{F\} = \{q(\xi = 1)\}$ (Eq. (2.105)) and the nodal displacements $\{d\} = \{u(\xi = 1)\}$

---

to the translational rigid motions become small positive numbers and retained in the solution for a bounded domain, while the other two eigenvalues of zero turn negative.

(Eq. (2.76)) apply. Formulating Eq. (3.23) at $\xi = 1$, the nodal force-displacement relationship of the S-element is expressed as

$$\{F\} = [K]\{d\} \tag{3.24}$$

where the stiffness matrix of the S-element is equal to

$$[K] = [\Phi_b^{(q)}][\Phi_b^{(u)}]^{-1} \tag{3.25}$$

It is shown in Section 6.12 that the stiffness matrix is symmetric.

For later use, the virtual increment of the strain energy is expressed by using Eqs. (2.107) and (3.23) at $\xi = 1$ with Eqs. (3.25) and (2.76) as

$$U_\varepsilon = \{\delta d\}^T [K]\{d\} \tag{3.26}$$

3.2 Pre-conditioning of Eigenvalue Problems

The coefficient matrices of a line element at the boundary, and thus the coefficient matrices of an S-element, depend on the elasticity matrix and are proportional to the Young's modulus (see Eqs. (2.110) and (A.42)) of the material. The four submatrices of matrix $[Z_p]$ in Eq (3.5) do not have the same dimension. For instance, $[E_0]^{-1}$ is inversely proportional to the Young's modulus, while $[E_2]$ is proportional to the Young's modulus. Depending on the unit chosen for the Young's modulus in an analysis, $[E_0]^{-1}$ and $[E_2]$ may differ by several orders of magnitude, leading to poor numerical accuracy of eigenvalues and eigenvectors. This numerical issue is overcome by preconditioning the coefficient matrices with a symmetric matrix $[P]$ ($[P] = [P]^T$)

$$[\hat{E}_0] = [P][E_0][P] \tag{3.27a}$$

$$[\hat{E}_1] = [P][E_1][P] \tag{3.27b}$$

$$[\hat{E}_2] = [P][E_2][P] \tag{3.27c}$$

A simple choice of $[P]$ is to use the diagonal entries $E_{0(i,i)}$ ($i = 1, 2, \ldots, n$) of the coefficient matrix $[E_0]$

$$[P] = \mathrm{diag}\left(\frac{1}{\sqrt{E_{0(i,i)}}}\right) \qquad (i = 1, 2, \ldots, n) \tag{3.28}$$

so that the diagonal entries of the transformed matrix $[\hat{E}_0]$ (Eq. (3.27a)) are equal to 1. Similar to the eigenvalue problem in Eq. (3.8) with the matrix $[Z_p]$ in Eq (3.5), the following eigenvalue problem is formulated

$$[\hat{Z}_p]\{\hat{\phi}\} = \hat{\lambda}\{\hat{\phi}\} \tag{3.29}$$

with the pre-conditioned coefficient matrix

$$[\hat{Z}_p] = \begin{bmatrix} -[\hat{E}_0]^{-1}[\hat{E}_1]^T & [\hat{E}_0]^{-1} \\ [\hat{E}_2] - [\hat{E}_1][\hat{E}_0]^{-1}[\hat{E}_1]^T & [\hat{E}_1][\hat{E}_0]^{-1} \end{bmatrix} \tag{3.30}$$

In the computer implementation, this pre-conditioned eigenvalue problem is solved for the eigenvalues $\hat{\lambda}$ and eigenvectors $\{\hat{\phi}\}$.

Substituting Eq. (3.27) into Eq. (3.30) and considering

$$[\hat{E}_0]^{-1} = [P]^{-1}[E_0]^{-1}[P]^{-1} \tag{3.31}$$

results in

$$
\begin{aligned}
[\hat{Z}_p] &= \begin{bmatrix} -[P]^{-1}[E_0]^{-1}[E_1]^T[P] & [P]^{-1}[E_0]^{-1}[P]^{-1} \\ [P]([E_2] - [E_1][E_0]^{-1}[E_1]^T)[P] & [P][E_1][E_0]^{-1}[P]^{-1} \end{bmatrix} \\
&= \begin{bmatrix} [P]^{-1} & 0 \\ 0 & [P] \end{bmatrix} [Z_p] \begin{bmatrix} [P] & 0 \\ 0 & [P]^{-1} \end{bmatrix}
\end{aligned}
\tag{3.32}
$$

Using Eq. (3.32), Eq. (3.29) is rewritten as

$$[Z_p] \begin{bmatrix} [P] & 0 \\ 0 & [P]^{-1} \end{bmatrix} \{\hat{\phi}\} = \hat{\lambda} \begin{bmatrix} [P] & 0 \\ 0 & [P]^{-1} \end{bmatrix} \{\hat{\phi}\} \tag{3.33}$$

Comparing Eq. (3.33) to Eq. (3.10), the eigenvalues and eigenvectors of the original eigenvalue problem in Eq. (3.10) are obtained from those of Eq. (3.33) as

$$\lambda = \hat{\lambda} \tag{3.34}$$

$$\{\phi\} = \begin{bmatrix} [P] & 0 \\ 0 & [P]^{-1} \end{bmatrix} \{\hat{\phi}\} \tag{3.35}$$

An eigenvector is normalized based on the upper half of the eigenvector, i.e. the displacement mode $\{\phi^{(u)}\}$. The largest entry of $\{\phi^{(u)}\}$ in absolute value, i.e. the infinity norm, is expressed as

$$\| \phi^{(u)} \|_\infty = \max(|\phi_1^{(u)}|, \ |\phi_2^{(u)}|, \ \dots, |\phi_n^{(u)}|) \tag{3.36}$$

The eigenvector is normalized as

$$\{\hat{\phi}^{(u)}\} = \frac{1}{\| \phi^{(u)} \|_\infty} \{\phi^{(u)}\} \tag{3.37}$$

After the normalization, $\| \hat{\phi}^{(u)} \|_\infty = 1$ applies.

3.3 Computer Program Platypus: Solution of the Scaled Boundary Finite Element Equation of a Bounded S-element by the Eigenvalue Method

In the accompanying computer program Platypus, the solution of the scaled boundary finite element equation of a bounded S-element ($0 \le \xi \le 1$) is performed by the function SElementSlnEigenMethod.m in Code List 3.1 on page 79. The inputs to this function are the coefficient matrices $[E_0]$, $[E_1]$ and $[E_2]$ of the S-element computed with the function SElementCoeffMtx.m in Code List 2.2 on page 67. The static stiffness matrix (argument K) of the bounded S-element is obtained. The eigenvalues (argument d) and the upper part of the eigenvector matrix (argument v) are also output for later use in computing the displacement field and post-processing. The function is documented in the code listing with reference to equations in the previous sections.

Code List 3.1: S-element solution by eigenvalue method

```
1   function [K, d, v11, M] = SElementSlnEigenMethod(E0, E1, E2, M0)
2   % Stiffness matrix of an S-Element
3   %
4   % Inputs:
5   %      E0, E1, E2, M0  - coefficient matrices of S-Element
6   %
7   % Outputs:
8   %      K       - stiffness matrix of S-Element
9   %      d       - eigenvalues
10  %      v11     - upper half of eigenvectors (displacement modes)
11  %      M       - mass matrix of S-Element
12
13  nd = size(E0,1); % number of DOFs of boundary nodes
14  id = 1:nd;       % index of selected eigenvalues (working variable)
15
16  % ...  Preconditioning
17  Pf = 1./sqrt(abs(diag(E0))); P  = diag(Pf);  % .............. Eq. (3.28)
18  E0 = P*E0*P; E1 = P*E1*P; E2 = P*E2*P;  % .................... Eq. (3.27)
19  % ...  Construct [Ẑ_p], see Eq. (3.30)
20  m = E0\[-E1' eye(nd)];
21  Zp = [m; E2+E1*m(:,id) -m(:,id)'];
22  % ...  eigenvalues and eigenvectors of [Ẑ_p], see Eq. (3.29)
23  [v, d] = eig(Zp);  % ............................. v̂ : eigenvector matrix
24  d = diag(d);  % ..................... eigenvalues stored as a column vector
25  % ...  index for sorting eignvalues in descending order of real part.
26  [~, idx] = sort(real(d),'descend');
27  % ...  select eigenvalues and eigenvectors for solution of bounded domain
28  d = d(idx(id));  % ............... select the first half of sorted eigenvalues
29  v = v(:, idx(id));  % ............. select the corresponding eigenvectors
30  v = diag([Pf;1./Pf])*v;  % .................................... Eq. (3.35)
31  % ...  modes of translational rigid body motion, see Item  2 on page 75
32  d(end-1:end) = 0;  % ...................... set last two eigenvalues to zero
33  v(:,end-1:end) = 0;  % ................... set last two eigenvectors to zero
34  v(1:2:nd-1,end-1) = 1;  % .................... set $u_x = 1$ in $\{\phi_{n-1}^{(u)}\}$
35  v(2:2:nd,end) = 1;  % .......................... set $u_y = 1$ in $\{\phi_n^{(u)}\}$
36  % ...  normalization of eigenvectors, see Eq. (3.37)
37  v = bsxfun( @rdivide, v, max(abs(v(1:nd-2,:))) );
38  v11 = v(id, :); v11inv = inv(v11);
39  K  = real(v(nd+id, :)*v11inv);  % ......... stiffness matrix, see Eq. (3.25)
40
41  % ...  mass matrix;
42  if nargout > 3
43      M0 = v11'*M0*v11;  % .................................... Eq. (3.103)
```

am is a square matrix with all columns being the vector of eigenvalues. The entry (i, j) equals λ_i. The entry (i, j) of am' equals λ_j.

```
44      am = d(:,ones(1,nd));
45      M0 = M0./(2+am+am');  % .... the entry $(i, j)$ of am+am' equals to $\lambda_i + \lambda_j$
```

```
46      M = real(v11inv'*M0*v11inv);  % ...................... Eq. (3.105)
47  end
48  end
```

■ **Example 3.1** Compute the solution of the scaled boundary finite element equation for the square S-element in Example 2.4 on page 65. Assume the Young's modulus is equal to $E = 10\,\text{GPa}$ and Poisson's ratio $v = 0$.

The coefficient matrices $[E_0]$, $[E_1]$ and $[E_2]$ of this problem have been determined in Example 2.5 on page 68. The MATLAB script is directly used by appending a line calling the function in MATLAB List 3.1.

```
%% Solution of a square S-element
xy = [-1 -1; 1 -1; 1 1; -1 1]
conn = [1:4 ; 2:4 1]'
% elascity matrix (plane stress).
% E: Young's modulus; p: Poisson's ratio
ElasMtrx = @(E, p) E/(1-p^2)*[1 p 0; p 1 0; 0 0 (1-p)/2];
mat.D = ElasMtrx(10, 0); % E in GPa
mat.den = 2; % mass density in Mg per cubic meter
[ E0, E1, E2 ] = SElementCoeffMtx(xy, conn, mat)
[ K, d, v ] = SElementSlnEigenMethod(E0, E1, E2)
```

The above MATLAB script can be obtained by scanning the QR-code to the right.

The S-element has eight DOFs on the boundary. An eigenvalue problem of a 16×16 matrix is solved. The 16 eigenvalues are equal to 1.915, 1.915, 1.000, 1.000, 1.000, 1.000, 0.000, 0.000, 0.000, 0.000, −1.000, −1.000, −1.000, −1.000, −1.915 and −1.915. The first eight eigenvalues are selected for the solution of the bounded S-element (Eq. (3.19))

$$\langle \lambda_b \rangle = \text{diag}(\ 1.915 \quad 1.915 \quad 1.000 \quad 1.000 \quad 1.000 \quad 1.000 \quad 0.000 \quad 0.000\ )$$

As shown in Eq. (3.18a), each pair of eigenvalue and eigenvector forms a term $\xi^{\lambda_i}\{\phi_i^{(\text{u})}\}$ of the solution of the displacement field of the S-element. The six eigenvalues with a positive real part lead to finite displacements in the S-element. There are four eigenvalues that are equal to 1. The four corresponding eigenvectors describe linear variations of displacement. They are the linear combinations of the rigid body rotation and the three constant strain deformations (see Section 6.11). The two eigenvalues of zero correspond to the translational rigid body motions, since $\xi^0 = 1$ applies.

The size of the eigenvector matrix in the solution of the square S-element is 16×8. The upper half of the eigenvector matrix, representing the displacement modes, is obtained as

$$
[\Phi_b^{(u)}] = \begin{bmatrix}
-1.000 & -0.087 & -1.000 & 0.187 & 0.018 & -0.342 & 1.000 & 0.000 \\
0.174 & 0.971 & 0.236 & -0.796 & 0.671 & -0.658 & 0.000 & 1.000 \\
1.000 & -0.256 & -0.101 & 1.000 & -0.326 & -1.000 & 1.000 & 0.000 \\
0.174 & -1.000 & -0.236 & -0.770 & -1.000 & 0.050 & 0.000 & 1.000 \\
-1.000 & -0.087 & 1.000 & -0.187 & -0.018 & 0.342 & 1.000 & 0.000 \\
0.174 & 0.971 & -0.236 & 0.796 & -0.671 & 0.658 & 0.000 & 1.000 \\
1.000 & -0.256 & 0.101 & -1.000 & 0.326 & 1.000 & 1.000 & 0.000 \\
0.174 & -1.000 & 0.236 & 0.770 & 1.000 & -0.050 & 0.000 & 1.000
\end{bmatrix}
$$

and the lower part of the eigenvector matrix, i.e. the nodal force modes, is equal to

$$
[\Phi_b^{(q)}] = \begin{bmatrix}
-4.574 & 0.386 & -6.067 & -1.164 & 5.129 & -1.838 & 0.000 & 0.000 \\
-0.000 & 4.507 & -1.573 & -4.928 & 1.759 & -8.168 & 0.000 & 0.000 \\
4.574 & -0.386 & 2.921 & 6.968 & 1.684 & -8.416 & 0.000 & 0.000 \\
0.000 & -4.507 & 1.573 & -10.732 & -5.054 & 2.086 & 0.000 & 0.000 \\
-4.574 & 0.386 & 6.067 & 1.164 & -5.129 & 1.838 & 0.000 & 0.000 \\
-0.000 & 4.507 & 1.573 & 4.928 & -1.759 & 8.168 & 0.000 & 0.000 \\
4.574 & -0.386 & -2.921 & -6.968 & -1.684 & 8.416 & 0.000 & 0.000 \\
0.000 & -4.507 & -1.573 & 10.732 & 5.054 & -2.086 & 0.000 & 0.000
\end{bmatrix}
$$

The last two columns of $[\Phi_b^{(q)}]$ corresponding to the two translational rigid body motions are equal to zero. A eigenvalue decomposition of $[\Phi_b^{(q)}]$ reveals that three zero-eigenvalues exist. The third one corresponds to the rigid body rotation (see Section 6.11).

Note that the eigenvectors are not uniquely defined. A linear combination of the eigenvectors having the same eigenvalue is also an eigenvector. Depending on the hardware and the version of MATLAB, the above eigenvector matrix may vary in value.

The stiffness matrix of the square S-element is equal to

$$
[K] = \begin{bmatrix}
4.894 & 1.250 & -2.394 & -1.250 & -2.606 & -1.250 & 0.106 & 1.250 \\
1.250 & 4.894 & 1.250 & 0.106 & -1.250 & -2.606 & -1.250 & -2.394 \\
-2.394 & 1.250 & 4.894 & -1.250 & 0.106 & -1.250 & -2.606 & 1.250 \\
-1.250 & 0.106 & -1.250 & 4.894 & 1.250 & -2.394 & 1.250 & -2.606 \\
-2.606 & -1.250 & 0.106 & 1.250 & 4.894 & 1.250 & -2.394 & -1.250 \\
-1.250 & -2.606 & -1.250 & -2.394 & 1.250 & 4.894 & 1.250 & 0.106 \\
0.106 & -1.250 & -2.606 & 1.250 & -2.394 & 1.250 & 4.894 & -1.250 \\
1.250 & -2.394 & 1.250 & -2.606 & -1.250 & 0.106 & -1.250 & 4.894
\end{bmatrix}
$$

Note that the stiffness matrix is symmetric.

🖥 **Example 3.2**   A regular pentagon S-element is shown in Figure 3.1. The vertices are on a circle with a radius of 1 m. Each edge is modelled with a 2-node line element. The nodal numbers and the line element numbers (in circle) are shown in Figure 3.1. The elasticity constants are: Young's modulus $E = 10\,\text{GPa}$ and Poison's ratio $v = 0$. Consider plane stress states. Use the eigenvalue method to determine the solution of the scaled boundary finite element equation of the S-element.

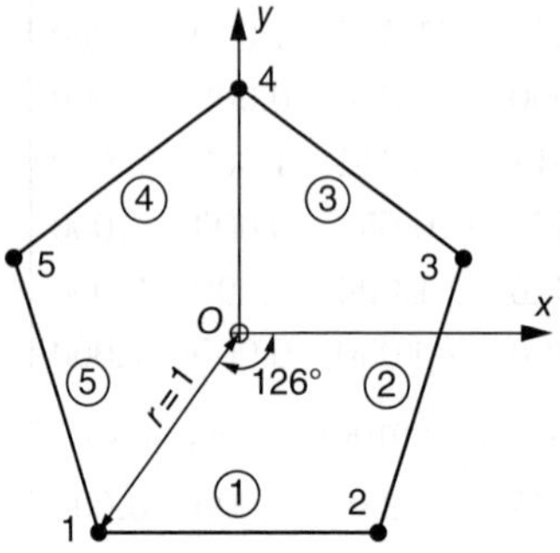

Figure 3.1 A regular pentagon S-element (Unit: meter).

The difference in the angular coordinate between two adjacent nodes is equal to $360°/5 = 72°$. The first node is at an angle of $-126°$. The nodal coordinates of i-th node are equal to

$$x = \cos(-126° + (i-1) \times 72°)$$
$$y = \sin(-126° + (i-1) \times 72°)$$

The nodal coordinates and the connectivity of the line elements are given in the tables below:

	Nodal coordinates			Element connectivity		
Node	x(m)	y(m)		Element	Node 1	Node 2
1	−0.5878	−0.8090		1	1	2
2	0.5878	−0.8090		2	2	3
3	0.9511	0.3090		3	3	4
4	0	1		4	4	5
5	−0.9511	0.3090		5	5	1

The following MATLAB script inputs the nodal coordinates, connectivity table and material constants and calls the MATLAB functions to compute the coefficient matrices and solution of the S-element. The script can easily be modified to compute regular polygon S-elements of other number of sides.

```
%% Solution of pentagon S-element
xy = [cosd(-126:72:180); sind(-126:72:180)]';
conn = [1:5; 2:5 1]';
% elasticity matrix (plane stress).
% E: Young's modulus; p: Poisson's ratio
```

```
ElasMtrx = @(E, p) E/(1-p^2)*[1 p 0; p 1 0; 0 0 (1-p)/2];
mat.D = ElasMtrx(10, 0); % E in GPa
mat.den = 2; % mass density in Mg per cubic meter
[ E0, E1, E2 ] = SElementCoeffMtx(xy, conn, mat);
[ K, d, v ] = SElementSlnEigenMethod(E0, E1, E2)
```

The above MATLAB script can be obtained by scanning the QR-code to the right.

There are 10 DOFs on the boundary of the pentagon. The 10 eigenvalues for the solution of the bounded domain are obtained as

$$\langle \lambda_b \rangle = \text{diag}(2.19 \quad 2.19 \quad 2.17 \quad 2.17 \quad 1.00 \quad 1.00 \quad 1.00 \quad 1.00 \quad 0.00 \quad 0.00)$$

Again, there are four eigenvalues equal to 1 and two eigenvalues equal to 0.

The eigenvector matrix is separated into the upper part representing the nodal displacement modes

$$[\Phi_1^{(u)}] = \begin{bmatrix}
0.10 & -1.00 & 0.88 & -0.38 & 0.86 & 0.87 & -0.79 & -0.27 & 1.00 & 0.00 \\
-0.71 & 0.32 & -0.23 & 0.73 & -0.09 & 0.08 & -0.30 & 0.98 & 0.00 & 1.00 \\
-0.25 & 0.97 & -0.58 & -0.56 & 0.76 & -0.21 & 0.34 & 0.07 & 1.00 & 0.00 \\
0.54 & 0.61 & 0.24 & -0.73 & 0.03 & 0.58 & -0.11 & 0.64 & 0.00 & 1.00 \\
0.63 & -0.16 & 0.32 & 0.44 & -0.39 & -1.00 & 1.00 & 0.32 & 1.00 & 0.00 \\
-0.52 & -0.95 & -1.00 & 0.34 & 0.11 & 0.28 & 0.23 & -0.59 & 0.00 & 1.00 \\
-1.00 & -0.24 & 0.32 & -1.00 & -1.00 & -0.40 & 0.28 & 0.12 & 1.00 & 0.00 \\
-0.12 & 0.65 & 0.53 & 0.07 & 0.04 & -0.41 & 0.25 & -1.00 & 0.00 & 1.00 \\
0.52 & 0.43 & -0.58 & 0.33 & -0.23 & 0.75 & -0.83 & -0.24 & 1.00 & 0.00 \\
0.81 & -0.63 & -0.71 & -0.56 & -0.09 & -0.53 & -0.07 & -0.03 & 0.00 & 1.00
\end{bmatrix}$$

and the lower part representing the nodal force modes

$$[\Phi_1^{(q)}] = \begin{bmatrix}
1.15 & -7.92 & 3.15 & -0.57 & 3.92 & 5.06 & -7.06 & -0.98 & 0.00 & 0.00 \\
-5.35 & 2.84 & -0.00 & 2.97 & 2.21 & 3.08 & -3.16 & 8.17 & 0.00 & 0.00 \\
-1.70 & 7.79 & -2.55 & -1.28 & 2.98 & -5.22 & 3.69 & 2.28 & 0.00 & 0.00 \\
3.92 & 5.04 & 1.85 & -2.73 & -2.80 & 3.19 & -0.71 & 7.22 & 0.00 & 0.00 \\
4.67 & -0.88 & 0.97 & 2.64 & -2.08 & -8.29 & 9.34 & 2.39 & 0.00 & 0.00 \\
-4.36 & -7.55 & -3.00 & 1.46 & -3.95 & -1.10 & 2.72 & -3.70 & 0.00 & 0.00 \\
-8.11 & -1.92 & 0.97 & -3.00 & -4.27 & 0.10 & 2.08 & -0.80 & 0.00 & 0.00 \\
-0.84 & 4.62 & 3.00 & 0.38 & 0.36 & -3.87 & 2.39 & -9.51 & 0.00 & 0.00 \\
3.98 & 2.92 & -2.55 & 2.20 & -0.56 & 8.35 & -8.05 & -2.89 & 0.00 & 0.00 \\
6.63 & -4.95 & -1.85 & -2.07 & 4.17 & -1.29 & -1.24 & -2.17 & 0.00 & 0.00
\end{bmatrix}$$

The eigenvector matrix is not unique. A different eigenvector matrix may be obtained depending on the hardware and software.

The stiffness matrix of the pentagon S-element is obtained as

$$
[K] =
\begin{bmatrix}
4.89 & 0.90 & -1.95 & -1.36 & -1.88 & -0.73 & -0.82 & -0.73 & -0.23 & 1.91 \\
0.90 & 5.48 & 1.36 & -0.05 & -1.08 & -1.29 & -0.39 & -2.36 & -0.80 & -1.77 \\
-1.95 & 1.36 & 4.89 & -0.90 & -0.23 & -1.91 & -0.82 & 0.73 & -1.88 & 0.73 \\
-1.36 & -0.05 & -0.90 & 5.48 & 0.80 & -1.77 & 0.39 & -2.36 & 1.08 & -1.29 \\
-1.88 & -1.08 & -0.23 & 0.80 & 5.95 & 0.56 & -1.30 & -0.45 & -2.54 & 0.17 \\
-0.73 & -1.29 & -1.91 & -1.77 & 0.56 & 4.41 & 2.26 & -0.71 & -0.17 & -0.64 \\
-0.82 & -0.39 & -0.82 & 0.39 & -1.30 & 2.26 & 4.23 & 0.00 & -1.30 & -2.26 \\
-0.73 & -2.36 & 0.73 & -2.36 & -0.45 & -0.71 & 0.00 & 6.13 & 0.45 & -0.71 \\
-0.23 & -0.80 & -1.88 & 1.08 & -2.54 & -0.17 & -1.30 & 0.45 & 5.95 & -0.56 \\
1.91 & -1.77 & 0.73 & -1.29 & 0.17 & -0.64 & -2.26 & -0.71 & -0.56 & 4.41
\end{bmatrix}
$$

◧

3.4 Assembly of S-elements and Solution of Global System of Equations

The equation of an S-element (Eq. (3.24)) is in the same form as the equation of a displacement-based finite element. Standard finite element procedures are followed to assemble the equations of all S-elements in a mesh and to perform the global analysis.

3.4.1 Assembly of S-elements

The stiffness matrices of the S-elements are assembled based on the connectivity of the S-elements to form the global stiffness matrices (subscript 'G' for global)

$$
[K_G] = \sum_{Se} [K]
$$

where $\sum_{Se}$ denotes the finite element assembly of the S-elements. The nodal displacements of the global system are assembled as

$$
\{d_G\} = \sum_{Se} \{d\}
$$

The assembly of the nodal forces are in equilibrium with the external loads

$$
\sum_{Se} \{F\} = \{F_G\}
$$

where $\{F_G\}$ stands for the load vector. It includes the contribution from all external loads. Concentrated nodal forces are directly assembled to the load vector. Surface tractions are considered below. Body loads are addressed in Section 3.9.

After the assembling process, the global system of equations is expressed as

$$[K_G]\{d_G\} = \{F_G\} \tag{3.38}$$

For simplicity in notation, the subscript G will be omitted from the variables of the global system wherever it would not cause confusion with the corresponding variable of the S-element equations.

3.4.2 Surface Tractions

The surface tractions are applied to the line elements on the edges of S-elements. One line element subject to a surface traction is illustrated in Figure 3.2. The nodal forces $\{F_T^e\}$ are equivalent to the surface tractions in the sense of virtual work

$$\{\delta d^e\}^T\{F_T^e\} = \int_{S^e} \{\delta u\}^T\{\bar{t}(\eta)\}dS \tag{3.39}$$

where $\{\delta u\}$ is the virtual displacement field (Eq. 2.86) and $\{\delta d^e\}$ the virtual nodal displacements at the nodes of the line element.

The virtual displacement field is obtained by interpolating the nodal values (Eq. (2.86)). Considering Eq. (2.85) at the boundary, the virtual displacements are expressed as

$$\{\delta u\} = [N_u]\{\delta d^e\} \tag{3.40}$$

Substituting Eq. (3.40) into Eq. (3.39) and considering the arbitrariness of the virtual nodal displacement $\{\delta d^e\}$, the equivalent nodal forces are equal to

$$\{F_T^e\} = \int_{S^e} [N_u]^T\{\bar{t}(\eta)\}dS \tag{3.41}$$

Substituting Eq. (2.38) at $\xi = 1$ and Eq. (2.37) into $\{F_T^e\}$ in Eq. (3.41) lead to

$$\{F_T^e\} = \int_{-1}^{+1} [N_u]^T\{\bar{t}(\eta)\} \sqrt{\left(x_{b,\eta}\right)^2 + \left(y_{b,\eta}\right)^2}\,d\eta \tag{3.42}$$

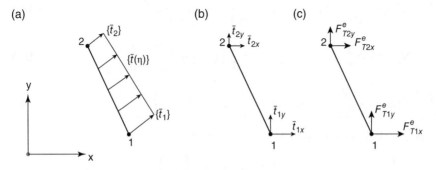

Figure 3.2 (a) Surface traction applied to a 2-node line element. (b) Components of nodal values of surface traction. (c) Equivalent nodal forces.

This is the same equation used in standard finite elements to determine the nodal forces equivalent to the surface traction. The integration over a line element can be evaluated by Gauss quadratures.

The contribution of the surface tractions to the load vector ($\{F_G\}$ in Eq. (3.38)) is obtained by assembling the equivalent nodal forces $\{F_T^e\}$ of the individual line elements.

For the 2-node line element shown in Figure 3.2, the integration in Eq. (3.42) can be performed analytically. The nodal vector of the surface traction is expressed as (Figure 3.2b)

$$\{\bar{t}\} = [\bar{t}_{1x}, \bar{t}_{1y}, \bar{t}_{2x}, \bar{t}_{2y}]^T \tag{3.43}$$

The surface traction is interpolated using the shape functions (Eq. (2.16))

$$\{\bar{t}(\eta)\} = [N_u]\{\bar{t}\} \tag{3.44}$$

Using Eq. (3.44) and

$$\sqrt{\left(x_{b,\eta}\right)^2 + \left(y_{b,\eta}\right)^2} = \frac{1}{2}\sqrt{\Delta_x^2 + \Delta_y^2} \tag{3.45}$$

obtained from Eq. (2.52), Eq. (3.42) are written as

$$\{F_T^e\} = \frac{1}{2}\sqrt{\Delta_x^2 + \Delta_y^2} \int_{-1}^{+1} [N_u]^T [N_u] d\eta \{\bar{t}\} \tag{3.46}$$

Using Eq. (2.80) for a 2-node element

$$[N_u]^T[N_u] = \begin{bmatrix} N_1^2 & 0 & N_1N_2 & 0 \\ 0 & N_1^2 & 0 & N_1N_2 \\ N_1N_2 & 0 & N_2^2 & 0 \\ 0 & N_1N_2 & 0 & N_2^2 \end{bmatrix} \tag{3.47}$$

is obtained with the shape functions given in Eq. (2.16). Considering the following integrations

$$\int_{-1}^{+1} N_1^2 d\eta = \frac{1}{4}\int_{-1}^{+1} (1-\eta)^2 \, d\eta = -\frac{1}{4}\frac{1}{3}(1-\eta)^3\Big|_{-1}^{+1} = \frac{2}{3} \tag{3.48a}$$

$$\int_{-1}^{+1} N_1 N_2 d\eta = \frac{1}{4}\int_{-1}^{+1} (1-\eta^2) \, d\eta = \frac{1}{4}\left(\eta - \frac{1}{3}\eta^3\right)\Big|_{-1}^{+1} = \frac{1}{3} \tag{3.48b}$$

$$\int_{-1}^{+1} N_2^2 d\eta = \frac{1}{4}\int_{-1}^{+1} (1+\eta)^2 \, d\eta = \frac{1}{4}\frac{1}{3}(1+\eta)^3\Big|_{-1}^{+1} = \frac{2}{3} \tag{3.48c}$$

the equivalent nodal forces (Figure 3.2c) are obtained as

$$\{F_T^e\} = \begin{Bmatrix} F_{1x}^e \\ F_{1y}^e \\ F_{2x}^e \\ F_{2y}^e \end{Bmatrix} = \frac{1}{6}\sqrt{\Delta_x^2 + \Delta_y^2} \begin{bmatrix} 2 & 0 & 1 & 0 \\ 0 & 2 & 0 & 1 \\ 1 & 0 & 2 & 0 \\ 0 & 1 & 0 & 2 \end{bmatrix} \begin{Bmatrix} \bar{t}_{1x} \\ \bar{t}_{1y} \\ \bar{t}_{2x} \\ \bar{t}_{2y} \end{Bmatrix} \tag{3.49}$$

When uniform surface tractions $\bar{t}_{1x} = \bar{t}_{2x} = \bar{t}_x$ and $\bar{t}_{1y} = \bar{t}_{2y} = \bar{t}_y$ are prescribed, the equivalent nodal forces in Eq. (3.49) are equal to

$$\{F_T^e\} = \begin{Bmatrix} F_{1x}^e \\ F_{1y}^e \\ F_{2x}^e \\ F_{2y}^e \end{Bmatrix} = \frac{1}{2}\sqrt{\Delta_x^2 + \Delta_y^2} \begin{Bmatrix} \bar{t}_x \\ \bar{t}_y \\ \bar{t}_x \\ \bar{t}_y \end{Bmatrix} \tag{3.50}$$

3.4.3 Enforcing Displacement Boundary Conditions

The prescribed nodal displacements are enforced by the following conceptual procedure. The nodal displacement vector $\{d\}$ is partitioned into the unknown displacements $\{d_1\}$ and the prescribed displacements $\{d_2\}$. The nodal load vector $\{F\}$ and stiffness matrix $[K]$ are partitioned conformably. Equation (3.38) is expressed as (omitting the subscript G)

$$\begin{bmatrix} [K_{11}] & [K_{12}] \\ [K_{21}] & [K_{22}] \end{bmatrix} \begin{Bmatrix} \{d_1\} \\ \{d_2\} \end{Bmatrix} = \begin{Bmatrix} \{F_1\} \\ \{F_2\} \end{Bmatrix} \tag{3.51}$$

For a well-posed problems, the external forces in $\{F_1\}$ have to be known, and the forces in $\{F_2\}$ are unknown support reactions. Moving the term associated with the prescribed displacements $\{d_2\}$ to the right-hand side of Eq. (3.51) leads to

$$\begin{bmatrix} [K_{11}] \\ [K_{21}] \end{bmatrix} \{d_1\} = \begin{Bmatrix} \{F_1\} \\ \{F_2\} \end{Bmatrix} - \begin{bmatrix} [K_{12}] \\ [K_{22}] \end{bmatrix} \{d_2\} \tag{3.52}$$

The unknown displacements $\{d_1\}$ are obtained by solving the first row block of Eq. (3.52)

$$[K_{11}]\{d_1\} = \{F_1\} - [K_{12}]\{d_2\} \tag{3.53}$$

The support reactions are determined from the second row block of Eq. (3.51) as

$$\{F_2\} = \begin{bmatrix} [K_{21}] & [K_{22}] \end{bmatrix} \begin{Bmatrix} \{d_1\} \\ \{d_2\} \end{Bmatrix} \tag{3.54}$$

The solution of the global system of equations leads to the nodal displacements at all the nodes. The nodal displacements of the S-elements can be extracted based on the connectivity of the S-elements.

3.5 Computer Program Platypus: Assembly and Solution

3.5.1 Assembly of Global Stiffness Matrix

The function SBFEMAssembly.m for the assembly of global stiffness and mass matrices of S-elements is listed in Code List 3.2. The structure of this function is introduced briefly below. The use of it is demonstrated in Example 3.3 on page 90.

The input consists of the nodal coordinates (argument: coord), the global connectivity of the line elements forming the S-elements (argument: sdConn), the coordinates of the scaling centre (argument: sdSC) of the S-elements and the material constants (argument: mat). The format of the input is shown in Example 3.3. The output includes the solution of the S-elements (argument: `sdSln`) stored for post-processing purpose and the stiffness and mass matrices of the global system (argument: `K` and `M`). This function has two parts. The first part computes the solution of all the S-elements. The second part assembles the global stiffness and mass matrices.

This function can handle S-elements defined by a polygon of arbitrary number of edges (see, e.g. Figure 2.2) and S-elements with cracks and notches such as the one in Figure 2.6. The line elements form a closed loop in the first case, but an open loop in the second. To accommodate both types of S-elements, an S-element is defined by a global connectivity table of the line elements (variable `sdConn`). When analysing an S-element, the nodes of the S-element are numbered locally. Compared with coding of standard finite elements, an extra operation to build a local connectivity table of the line elements is needed to acquire the ability to model cracks. This segment of code starts from Line 22 and is explained in Example 3.3 on page 90.

Before an S-element is solved, its coordinates are translated to have the origin of the local coordinate system at the scaling centre (see Line 25).

These results of the S-element solution that will be used for post-processing are stored in the variable `sdSln` (see Line 34).

The part of code on assembling the global stiffness and mass matrices (starting from Line 40) follows the standard finite element procedure. Sparse storage of MATLAB is adopted.

Code List 3.2: Global stiffness matrix

```
1   function [sdSln, K, M] = SBFEMAssembly(coord, sdConn, sdSC, mat)
2   %Assembly of global stiffness and mass matrices
3   %
4   %Inputs:
5   %   coord(i,:)    - coordinates of node i
6   %   sdConn{isd,:}(ie,:)  - S-element conncetivity. The nodes of
7   %                          line element ie in S-element isd.
8   %   sdSC(isd,:)   - coordinates of scaling centre of S-element isd
9   %   mat           - material constants
10  %       mat.D     - elasticity matrix
11  %       mat.den   - mass density
12  %
13  %Outputs:
14  %   sdSln         - solutions for S-element
15  %   K             - global stiffness matrix
16  %   M             - global mass matrix
17
18  % ...  Solution of subdomains
19  Nsd = length(sdConn);  % ........................ number of S-elements
20  sdSln = cell(Nsd,1);   % ................... store solutions for S-elements
21  for isd = 1:Nsd  % .............................. loop over S-elements
```

sdNode contains global nodal numbers of the nodes in an S-element. Vector `ic` maps the global connectivity to the local connectivity of the S-element

```
22      [sdNode, ~, ic] = unique(sdConn{isd}(:));  % . remove duplicats
23      sdNode = reshape(sdNode,1,[]);  % ........ nodes stored in one row
```

```
24      xy = coord(sdNode,:); % nodal coordinates
25      % ... transform coordinate origin to scaling centre
26      xy = bsxfun(@minus, xy, sdSC(isd,:));
27      % ... line element connectivity in local nodal numbers of an S-element
28      LConn = reshape(ic, size(sdConn{isd}) );
29      % ... compute S-element coefficient matrices
30      [ E0, E1, E2, M0 ] = SElementCoeffMtx(xy, LConn, mat);
31      % ... compute solution for S-element
32      [ K, d, v, M] = SElementSlnEigenMethod(E0, E1, E2, M0);
33
34      % ... store S-element data and solution
35      sdSln{isd}= struct('xy',xy, 'sc',sdSC(isd,:),  ...
36              'conn',LConn, 'node',sdNode, ...
37              'K',K, 'M',M, 'd',d, 'v', v);
38  end
39
40  % ... Assembly
41  % ... sum of entries of stiffness matrices of all S-elements
42  ncoe = sum(cellfun(@(x) numel(x.K), sdSln));
43  % ... initializing non-zero entries in global stiffness and mass matrix
44  K = zeros(ncoe,1); M = zeros(ncoe,1);
45  % ... rows and columns of non-zero entries in global stiffness matrix
46  Ki = K; Kj = K;
47  StartInd = 1; % ......... starting position of an S-element stiffness matrix
48  for isd = 1:length(sdSln); % .................... loop over S-Elements
49      % ... global DOFs of nodes in an S-element
50      dof = reshape([2*sdSln{isd}.node-1;2*sdSln{isd}.node], [], 1);
51      Ndof = length(dof); % ............ number of DOFs of an S-element
52      % ... row and column numbers of stiffness coefficients of an S-element
53      sdI = repmat(dof,1, Ndof); sdJ = sdI';
54
55      % ... store stiffness, row and column indices
56      EndInd = StartInd + Ndof^2 - 1; % .............. ending position
57      K (StartInd:EndInd) = sdSln{isd}.K(:);
58      M (StartInd:EndInd) = sdSln{isd}.M(:);
59      Ki(StartInd:EndInd) = sdI(:);
60      Kj(StartInd:EndInd) = sdJ(:);
61
62      StartInd = EndInd + 1; % ........... increment the starting position
63  end
64  % ... form global stiffness matrix
65  K = sparse(Ki,Kj,K); % ..... form global stiffness matrix in sparse storage
66  K = (K+K')/2; % ................................... ensure symmetry
67  % ... form global mass matrix
68  M = sparse(Ki,Kj,M); % ........ form global mass matrix in sparse storage
69  M = (M+M')/2; % ................................... ensure symmetry
70
71  end
```

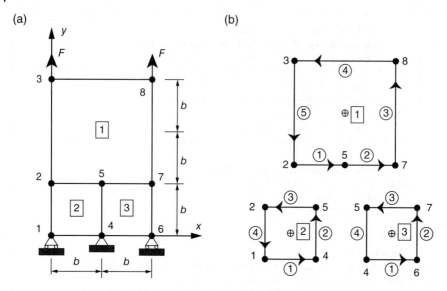

Figure 3.3 Modelling of a rectangular body by 3 S-elements. (a) Mesh. (b) Isolated S-elements with line elements defining the boundary.

▇ Example 3.3 Rectangular Body Under Uniaxial Tension: Assembly of Global Equations and Solution

A rectangular body is modelled by 3 S-elements as shown in Figure 3.3a. The dimensions (Unit: m) are indicated in Figure 3.3 with $b = 1$. The material constant are Young's modulus $E = 10$ GPa and Poisson's ratio $v = 0.25$. The input of the S-element data and the assembly of the global stiffness matrix are illustrated.

This problem is defined in the MATLAB script `Exmpl3SElements.m` listed below (The entry of boundary conditions starting from Line 17 is not used in this example and will be addressed later in Example 3.4 on page 102). The units are m for length and kN for force.

MATLAB script: A rectangular body modelled with 3 S-elements: uniaxial tension

```
1
2   % ... Mesh
3   % ... nodal coordinates. One node per row [x y]
4   coord = [0 0; 0  1; 0 3; 1 0; 1 1; 2 0; 2 1; 2 3];
```

Input S-element connectivity as a cell array (One S-element per cell). In a cell, the connectivity of line elements is given by one element per row [Node-1 Node-2].

```
5   sdConn = { [2 5; 5 7; 7 8; 8 3; 3 2];  % .................. S-element 1
6              [1 4; 4 5; 5 2; 2 1];  % ...................... S-element 2
7              [4 6; 6 7; 7 5; 5 4] };  % ...................... S-element 3
8   % ...  coordinates of scaling centres of S-elements.
9   sdSC = [ 1 2; 0.5 0.5; 1.5 0.5];  % .............. one S-element per row
```

```
10
11  % ...  Materials
12  % ...  E: Young's modulus; p: Poisson's ratio
13  ElasMtrx = @(E, p) E/(1-p^2)*[1 p 0;p 1 0;0 0 (1-p)/2];
14  mat.D = ElasMtrx(10E6, 0.25); % ........................... E in KPa
15  mat.den = 2; % ................................. mass density in Mg/m³
16
17  % ...  Boundary conditions
18  % ...  nodal forces. One force component per row: [Node Dir F]
19  BC_Frc = [3  2  1E3; 8  2  1E3]; % ....................... forces in KN
20  % ...  assemblage external forces
21  ndn = 2; % ........................................... 2 DOFs per node
22  NDof = ndn*size(coord,1); % ......................... number of DOFs
23  F = zeros(NDof,1); % .. initializing right-hand side of equation [K]{u} = {F}
24  F = AddNodalForces(BC_Frc, F); % ............ add prescribed nodal forces
25  % ...  displacement constraints. One constraint per row: [Node Dir Disp]
26  BC_Disp = [1 2 0; 4 1 0; 4 2 0; 6 2 0];
```

The node numbers are shown in Figure 3.3a. The nodal coordinates are stored in the variable `coord` sequentially starting from the first node. The coordinates x, y of one node occupy one row.

The S-element numbers are shown in the rectangular boxes. To illustrate the input data of the S-elements, the 3 S-elements are isolated in Figure 3.3b. The boundary of an S-element is described by line elements and the element numbers are shown within the circles. The line elements are numbered locally within each S-element. The sequence of local element numbering is of no significance. The connectivity of an S-element is defined by the connectivities of the line elements. It is stored in the cell array `sdConn` as a cell of a matrix (see Line 5 in the above MATLAB script). The cell variable of MATLAB can accommodate S-elements with varying number of line elements. When defining the line elements, the direction of an element from its first node to its second node (see Figure 2.10) has to follow the counter-clockwise direction with respect to the scaling centre as indicated in Figure 3.3b. The nodal numbers of one line element occupy one row.

The compatibility between the S-elements are satisfied automatically. For example, Line 25 in Figure 3.3a is shared by S-elements 1 and 2. Locally, this is modelled by Element 1 in S-element 1 and Element 3 in S-element 2. These two line elements have the same nodes and shape functions and the displacements are comparable along this line. Note that Element 1 in S-element 1 is defined by [2 5] while Element 3 in S-element 2 by [5 2]. The inverse of the sequence is necessary so that the line elements follow the counter-clockwise direction with respect to the scaling centres of the S-elements that they belong to.

The x, y coordinates of the scaling centres of the S-elements are stored in the matrix `sdSC` row-by-row. Typically, the geometrical centre of an S-element is chosen as the scaling centre. There are situations where it is advantageous to place the scaling centre at other locations (Figures 2.6 and 2.7).

The analysis is performed using the following MATLAB script to call the MATLAB function `SBFEMAssembly.m` in Code List 3.2 on page 88.

```
% input problem definition
Exmpl3SElements

% solution of S-elements and assemblage of global stiffness
% and mass matrices
[sdSln, K, M] = SBFEMAssembly(coord, sdConn, sdSC, mat);
```

The above MATLAB script can be obtained by scanning the QR-code to the right.

When analysing an S-element, the nodes connected to the S-element are numbered locally. The global connectivity table of the line elements for the 3 S-elements is

sdConn{1} =		sdConn{2} =		sdConn{3} =	
2	5	1	4	4	6
5	7	4	5	6	7
7	8	5	2	7	5
8	3	2	1	5	4
3	2				

The MATLAB function unique (Line 22) finds the global nodal number of all the nodes connected to this S-element but without the duplicated entries:

```
sdNode =    2    3    5    7    8    %S-element 1
sdNode =    1    2    4    5         %S-element 2
sdNode =    4    5    6    7         %S-element 3
```

The default behaviour of the unique function is to sort the nodal number in ascending order. As the sequence of local node numbering is of no significance, this default sequence is adopted. The connectivity of the local nodes to the global nodes is shown Table 3.1, which is similar to the element connectivity table of standard finite elements.

Table 3.1 Example 3.3 Nodal connectivity of S-elements.

S-element	Node 1	Node 2	Node 3	Node 4	Node 5
1	2	3	5	7	8
2	1	2	4	5	
3	4	5	6	7	

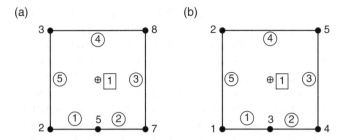

Figure 3.4 Example 3.3 Connectivity of line element in S-element 1. (a) Global nodal numbers. (b) Local nodal numbers.

Within an S-element, the connectivity of the line elements in global nodal numbers is mapped to the connectivity in the local nodal numbers. The connectivity of S-element 1 is depicted in Figure 3.4, where Figure 3.4a shows the global nodal numbers and Figure 3.4b the local nodal numbers. The `unique` function also returns an index vector `ic`, which maps the global nodal numbers to the local nodal numbers of the S-element. The local connectivity table of the line elements is built by reshaping the index vector `ic`:

```
%S-element 1        %S-element 2        %S-element 3
LConn =             LConn =             LConn =
    1    3              1    3              1    3
    3    4              3    4              3    4
    4    5              4    2              4    2
    5    2              2    1              2    1
    2    1
```

The result can easily be verified for S-element 1 by inspecting Figure 3.4b.

In the derivation of the scaled boundary finite element equation, it is assumed that the origin of the coordinate system is at the scaling centre (Section 2.2.2 and Figure 2.9). To satisfy this requirement, the local nodal coordinates of an S-element are obtained by subtracting the coordinates of the scaling centre from the global nodal coordinates of the S-element (Line 25) and storing in the variable `xy`. The local coordinates of the 3 S-elements are:

```
%S-element 1        %S-element 2        %S-element 3
xy =                xy =                xy =
    -1    -1            -0.5    -0.5          -0.5    -0.5
    -1     1            -0.5     0.5          -0.5     0.5
     0    -1             0.5    -0.5           0.5    -0.5
     1    -1             0.5     0.5           0.5     0.5
     1     1
```

After gathering the local nodal coordinates and connectivity, the S-element solutions can be obtained by calling the function `SBFEMAssembly.m`, as in Examples 3.1 and 3.2. The results are stored in the variable `sdSln`. The stiffness matrices (Unit: KPa.) of the S-elements are in the field `K` of the variable `sdSln`. The stiffness matrix of S-element 1 is equal to

$$
10^6
\begin{bmatrix}
5.54 & 0.98 & 1.14 & 0.01 & -2.69 & 0.02 & -1.85 & 0.33 & -2.14 & -1.34 \\
0.98 & 3.88 & -0.99 & -1.35 & 1.35 & -1.06 & -0.33 & -0.16 & -1.01 & -1.31 \\
1.14 & -0.99 & 4.69 & -1.67 & -1.00 & 1.31 & -2.14 & 1.01 & -2.69 & 0.33 \\
0.01 & -1.35 & -1.67 & 4.69 & 0.65 & -2.67 & 1.34 & -1.31 & -0.33 & 0.65 \\
-2.69 & 1.35 & -1.00 & 0.65 & 7.38 & -0.00 & -2.69 & -1.35 & -1.00 & -0.65 \\
0.02 & -1.06 & 1.31 & -2.67 & 0.00 & 7.45 & -0.02 & -1.06 & -1.31 & -2.67 \\
-1.85 & -0.33 & -2.14 & 1.34 & -2.69 & -0.02 & 5.54 & -0.98 & 1.14 & -0.01 \\
0.33 & -0.16 & 1.01 & -1.31 & -1.35 & -1.06 & -0.98 & 3.88 & 0.99 & -1.35 \\
-2.14 & -1.01 & -2.69 & -0.33 & -1.00 & -1.31 & 1.14 & 0.99 & 4.69 & 1.67 \\
-1.34 & -1.31 & 0.33 & 0.65 & -0.65 & -2.67 & -0.01 & -1.35 & 1.67 & 4.69
\end{bmatrix}
$$

S-elements 2 and 3 have the same stiffness matrix

$$
10^6
\begin{bmatrix}
4.69 & 1.67 & 0.64 & 0.33 & -2.69 & -0.33 & -2.64 & -1.67 \\
1.67 & 4.69 & -0.33 & -2.69 & 0.33 & 0.64 & -1.67 & -2.64 \\
0.64 & -0.33 & 4.69 & -1.67 & -2.64 & 1.67 & -2.69 & 0.33 \\
0.33 & -2.69 & -1.67 & 4.69 & 1.67 & -2.64 & -0.33 & 0.64 \\
-2.69 & 0.33 & -2.64 & 1.67 & 4.69 & -1.67 & 0.64 & -0.33 \\
-0.33 & 0.64 & 1.67 & -2.64 & -1.67 & 4.69 & 0.33 & -2.69 \\
-2.64 & -1.67 & -2.69 & -0.33 & 0.64 & 0.33 & 4.69 & 1.67 \\
-1.67 & -2.64 & 0.33 & 0.64 & -0.33 & -2.69 & 1.67 & 4.69
\end{bmatrix}
$$

The assembly of the global stiffness matrix is performed by the second part of the function SBFEMAssembly.m (starting from Line 40 in Code List 3.2) in the same manner as in the standard finite element method. The connectivity of the local DOFs of an S-element to the global DOFs are obtained from the nodal connectivity table (Table 3.1) as

	Connectivity of DOFs									
Local DOF	1	2	3	4	5	6	7	8	9	10
S-element 1	3	4	5	6	9	10	13	14	15	16
S-element 2	1	2	3	4	7	8	9	10		
S-element 3	7	8	9	10	11	12	13	14		

Only the square block of the global stiffness matrix operating on Nodes 1-5 is shown below

$$
10^6
\begin{bmatrix}
4.69 & 1.67 & 0.64 & 0.33 & 0.00 & 0.00 & -2.69 & -0.33 & -2.64 & -1.67 \\
1.67 & 4.69 & -0.33 & -2.69 & 0.00 & 0.00 & 0.33 & 0.64 & -1.67 & -2.64 \\
0.64 & -0.33 & 10.23 & -0.69 & 1.14 & 0.01 & -2.64 & 1.67 & -5.38 & 0.36 \\
0.33 & -2.69 & -0.69 & 8.57 & -0.99 & -1.35 & 1.67 & -2.64 & 1.02 & -0.42 \\
0.00 & 0.00 & 1.14 & -0.99 & 4.69 & -1.67 & 0.00 & 0.00 & -1.00 & 1.31 \\
0.00 & 0.00 & 0.01 & -1.35 & -1.67 & 4.69 & 0.00 & 0.00 & 0.65 & -2.67 \\
-2.69 & 0.33 & -2.64 & 1.67 & 0.00 & 0.00 & 9.38 & 0.00 & 1.28 & -0.00 \\
-0.33 & 0.64 & 1.67 & -2.64 & 0.00 & 0.00 & 0.00 & 9.38 & 0.00 & -5.38 \\
-2.64 & -1.67 & -5.38 & 1.02 & -1.00 & 0.65 & 1.28 & 0.00 & 16.76 & -0.00 \\
-1.67 & -2.64 & 0.36 & -0.42 & 1.31 & -2.67 & -0.00 & -5.38 & -0.00 & 16.83
\end{bmatrix}
$$

3.5.2 Assembly of Load Vector

Various external loadings can be included in the external load vector $\{F\}$ at the right-hand side of Eq. (3.38), although only concentrated nodal forces and surface tractions are considered in this book.

The function `AddNodalForces.m` in the Code List 3.3 assembles concentrated nodal forces into the load vector stored in the array variable F. The concentrated forces are added to the corresponding entries of F on input. The result is stored in F for output. To facilitate data input for meshes with a large number of nodes, the prescribed nodal forces are entered into the matrix variable BC_Frc. The forces are also resolved to the x- and y-components. One force component is specified as one row in BC_Frc. The 3 entries in a row are the nodal number, direction of the force (1 for x-direction and 2 for y-direction) and the value of the force. The prescribed nodal forces are taken as zeros if they are not included in the input.

Code List 3.3: Assembly of load vector from concentrated nodal forces

```
1   function [ F ] = AddNodalForces(BC_Frc, F)
2   %Assembly of prescribed nodal forces to load vector
3
4   %Inputs:
5   %   BC_Frc(i,:)    - one force component per row [Node Dir F]
6   %   F              - nodal force vector
7   %
8   %Output:
9   %   F              - nodal force vector
10
11  ndn = 2;  % ......................................... 2 DOFs per node
12  if ~isempty(BC_Frc)
13      fdof = (BC_Frc(:,1)-1)*ndn + BC_Frc(:,2);  % ............. DOFs
14      F(fdof) = F(fdof) + BC_Frc(:,3);  % ........... accumulate forces
15  end
16
17  end
```

The function `addSurfTraction.m` in Code List 3.4 assembles the equivalent nodal forces of the prescribed surface tractions. The surface tractions are applied to the edges of the S-elements. The edges subject to tractions are defined by the two input arguments, coord and edge. The variable coord stores the global nodal coordinates. One row of the variable edge stores the two global nodal numbers of one edge. The four surface traction components at the two nodes of the edge (see Figure 3.2) are stored in the corresponding column of the input argument trac. If trac contains one column only, all the edges have the same surface tractions given by single column of trac. The equivalent nodal forces are accumulated in the input/output argument F.

Code List 3.4: Assembly of equivalent nodal forces of surface traction

```
1   function [ F ] = addSurfTraction(coord, edge, trac, F)
2   %Assembly of surface tractions as equivalent nodal forces
```

```
3    % to load vector
4    %
5    %Inputs:
6    %   coord(i,:)    - coordinates of node i
7    %   edge(i,1:2)   - the 2 nodes of edge i
8    %   trac          - surface traction
9    %       when trac has only one column
10   %           trac(1:4) - surface traction at the 2 nodes of all edges
11   %       when trac has more than one column
12   %           trac(1:4,i) - surface traction at the 2 nodes of edge i
13   %
14   %Outputs:
15   %   F             - global load vector
16
17   if size(trac,2) == 1  % ...... expand uniform surface traction to all edges
18       trac = trac(:, ones(1,length(edge)));
19   end
20
21   % ...  equivalent nodal forces
22   fmtx = [ 2 0 1 0; 0 2 0 1; 1 0 2 0; 0 1 0 2];  % ...... see Eq. (3.49)
23   edgeLen = sqrt(sum( ( coord(edge(:,2),:) - ...
24                         coord(edge(:,1),:) ).^2, 2 ));  % edge length
25   nodalF = 1/6*fmtx*trac.*edgeLen(:,ones(1,4))';  % ........ Eq. (3.49)
26
27   % ...  assembly of nodal forces
28   for ii = 1:size(edge,1)
29       dofs = [2*edge(ii,1)-1 2*edge(ii,1) ...
30               2*edge(ii,2)-1 2*edge(ii,2)];
31       F(dofs) = F(dofs) + nodalF(:,ii);
32   end
33
34   end
```

3.5.3 Solution of Global System of Equations

A simple function SolverStatics.m is written to perform 2D static analysis by solving the global system of equations (Eq. (3.38)). It is provided in Code List 3.5. The input arguments are the global stiffness matrix K, the displacement boundary conditions BC_Disp and the nodal load vector F.

The prescribed nodal displacements are resolved to the DOFs in the x- and y-directions. The prescribed displacement of one DOF is specified in the variable BC_Disp by one row consisting of the nodal number, the direction of the DOF (1 for x-displacement and 2 for y-displacement) and the amount of displacement.

The output consists of the nodal displacements (argument: U) and nodal forces (argument: F). The nodal forces $\{F\}$ at all the nodes, instead of the support reactions $\{F_2\}$ only, are determined from the displacements using Eq. (3.51) in the following function SolverStatics.m for the purpose of verification of results.

Code List 3.5: SBFE analysis of 2D Statics

```
1   function [d, F] = SolverStatics(K, BC_Disp, F)
2   %2D Static Analysis by the Scaled Boundary Finite Element Method
3   %
4   %Inputs:
5   %  K          - static stiffness matrix
6   %  BC_Disp    - prescribed displacements in rows of
7   %                 [Node, Direction (=1 for x; =2 for y), Displacement]
8   %  F          - external load vector
9   %
10  %Outputs:
11  %  d          - nodal displacements
12  %  F          - external nodal forces including support reactions
13
14  ndn = 2;  % ........................................ 2 DOFs per node
15  NDof = size(K,1);
16  d = zeros(NDof,1);  % .............. Initialization of nodal displacements
17
18  %  ...  enforcing displacement boundary condition
19  %  ...  initialization of unconstrained (free) DOFs with unknown displace-
    ments
20  FDofs  = 1:NDof;
21  if ~isempty(BC_Disp)
22      %  ...  constrained DOFs with prescribed displacements
23      CDofs = (BC_Disp(:,1)-1)*ndn + BC_Disp(:,2);
24      FDofs(CDofs) = [];  % .................... remove constrained DOFs
25      F = F - K(:,CDofs)*BC_Disp(:,3);  % .................... Eq. (3.52)
26      d(CDofs) = BC_Disp(:,3);  % ........ store prescribed displacements
27  end
28
29  %  ...  displacement of free DOFs, see Eq. (3.53)
30  d(FDofs) = K(FDofs,FDofs)\F(FDofs);
31  %  ...  external forces, see Eq. (3.51)
32  F = K*d;
33
34  end
```

3.5.4 Utility Functions

A function PlotSBFEMesh.m is provided in Code List 3.6 to plot a simple mesh of polygon S-elements to aid the input of mesh data. The inputs of this function are the nodal coordinates (argument: coord) and the element connectivity of the S-elements (argument: sdConn). Labelling of nodes and additional controls and features are specified as fields of the structure variable opt. Detailed explanations on the options are given in the list of codes.

The S-elements can be filled with a specified colour (option opt.fill). The filling is performed by breaking an S-element into triangles formed by the scaling centre and 2-node line elements at the boundary of the S-element. The MATLAB patch function is called to plot the triangles filled with the colour.

The mesh is plotted as a net of lines defining the edges of the polygon S-elements. The lines are formed by the line elements, but duplicate entries are deleted. The `plot` function draws all the lines with the specified options.

Code List 3.6: Plot mesh of polygon S-elements

```
1  function PlotSBFEMesh(coord, sdConn, opt)
2  %Plot polygon S-element mesh
3  %
4  %Inputs:
5  %   coord(i,:)            - coordinates of node i
6  %   sdConn{isd}(ie,:)     - S-element connectivity. The nodes of
7  %                           line element ie in S-element isd.
8  %   plot options:
9  %     opt=struct('LineSpec','-b', 'sdSC', sdSC, 'LabelSC', 14, ...
10 %             'fill', [0.9 0.9 0.9], 'PlotNode', 1, 'LabelNode', 14, ...
11 %             'MarkerSize', 6, 'BC_Disp',BC_Disp, 'BC_Frc',BC_Frc);
12 %   where the options are:
13 %       opt.sdSC     : scaling centres of S-elements.
14 %       opt.LabelSC  : If specified, plot a marker at
15 %                          the scaling centre
16 %                          = 0, do not label S-element
17 %                          > 0, show S-element number
18 %                          If > 2, it also specifies the font size.
19 %       opt.fill     : =[r g b]. Fill an S-element with colour.
20 %                          opt.sdSC has also to be given.
21 %       opt.LineSpec : LineSpec of 'plot' function in Matlab
22 %       opt.LineWidth: LineWidth of 'plot' function in Matlab
23 %       opt.PlotNode : = 0, do not plot node symbol; otherwise, plot
24 %       opt.LabelNode: = 0, do not label nodes;
25 %                          > 0, show nodal number. If opt.LabelNode > 5,
26 %                              it specifies the font size.
27 %                          < 0, draw a marker only
28 %       opt.MarkerSize: marker size of nodes and scaling centres
29 %       opt.BC_Disp   : if specified, plot a marker at a node
30 %                          with prescribed displacement(s)
31 %       opt.BC_Frc    : if specified, plot a marker at a node
32 %                          with applied force(s)
33
34 LineWidth = 1;
35 LineSPec = '-k';
36 % use specified LineSpec if present
37 if nargin > 2
38     if isfield(opt, 'LineSpec') && ~isempty(opt.LineSpec)
39         LineSPec = opt.LineSpec;
40     end
41     if isfield(opt, 'LineWidth') && ~isempty(opt.LineWidth)
42         LineWidth = opt.LineWidth;
43     end
44 end
45
46 nsd = length(sdConn); % ........................ number of S-elements
```

```
47
48   % ...  fill S-elements by treating scaling centre and an edge as a triangle
49   meshEdge = vertcat(sdConn{:});
50   if nargin > 2
51       if isfield(opt, 'sdSC') && ~isempty(opt.sdSC)
52           if isfield(opt, 'fill') && ~isempty(opt.fill)
53               p = [coord; opt.sdSC]; % ........................ points
54               nNode = length(coord); % .............. number of nodes
55               % ...  appending scaling centres
56               nEdge = cellfun(@length, sdConn);
57               % ...  initilization of array of scaling centre
58               cnt = zeros(sum(nEdge),1);
59               ib = 1; % ................................. starting index
60               for ii = 1:nsd;
61                   ie = ib-1 + nEdge(ii); % .............. ending index
62                   cnt(ib:ie) = nNode + ii; % .......... scaling centre
63                   ib = ie + 1;
64               end
65               t = [meshEdge cnt]; % ........................ triangles
66               patch('Faces',t,'Vertices',p, ...
67                   'FaceColor',opt.fill,'LineStyle','none');
68           end
69       end
70   end
71
72   % ...  plot mesh
73   meshEdge = vertcat(sdConn{:});
74   meshEdge = unique(sort(meshEdge,2),'rows');
75   hold on
76   X = [coord(meshEdge(:,1),1)'; coord(meshEdge(:,2),1)'];
77   Y = [coord(meshEdge(:,1),2)'; coord(meshEdge(:,2),2)'];
78   plot(X,Y,LineSPec,'LineWidth',LineWidth);
79   xlabel('x'); ylabel('y'); % .............................. label axes
80
81   % ...  apply plot options
82   if nargin > 2
83       if isfield(opt, 'MarkerSize')
84           markersize = opt.MarkerSize;
85       else
86           markersize = 5;
87       end
88       if isfield(opt, 'sdSC') && ~isempty(opt.sdSC)
89           if isfield(opt, 'LabelSC') && ~isempty(opt.LabelSC)
90               % ...  plot scaling centre
91               plot(opt.sdSC(:,1), opt.sdSC(:,2), 'r+', ...
92                   'MarkerSize', markersize, 'LineWidth', LineWidth);
93               plot(opt.sdSC(:,1), opt.sdSC(:,2), 'ro', ...
94                   'MarkerSize', markersize, 'LineWidth', LineWidth);
95               if opt.LabelSC > 1
96                   if opt.LabelSC > 5
97                       fontsize = opt.LabelSC;
98                   else
```

```
 99                              fontsize = 12;
100                          end
101                          text(opt.sdSC(:,1), opt.sdSC(:,2), ...
102                              [blanks(nsd)' int2str((1:nsd)')], ...
103                              'Color','r', 'FontSize',fontsize);
104                     end
105                 end
106         end
107         if isfield(opt, 'PlotNode') && opt.PlotNode
108             % ...  showing nodes by plotting a circle
109             plot(coord(:,1), coord(:,2), 'ko', ...
110                 'MarkerSize', markersize, 'LineWidth', LineWidth);
111         end
112         if isfield(opt, 'LabelNode') && ~isempty(opt.LabelNode) ...
113                 && opt.LabelNode
114             nNode = length(coord);
115             if opt.LabelNode > 1
116                 fontsize = opt.LabelNode;
117             else
118                 fontsize = 12;
119             end
120             % ...  showing nodes by plotting a circle
121             plot(coord(:,1), coord(:,2), 'ko', ...
122                 'MarkerSize', markersize, 'LineWidth', LineWidth);
123             % ...  label nodes with nodal number
124             text(coord(:,1),coord(:,2), ...
125                 [blanks(nNode)' int2str((1:nNode)')], ...
126                 'FontSize',fontsize);
127         end
128         if isfield(opt, 'BC_Disp') && ~isempty(opt.BC_Disp)
129             % ...  show fixed DOFs by a marker at the nodes
130             Node = opt.BC_Disp(:,1);
131             plot(coord(Node,1),coord(Node,2),'b>', ...
132                 'MarkerSize',8, 'LineWidth', LineWidth);
133         end
134         if isfield(opt, 'BC_Frc') && ~isempty(opt.BC_Frc)
135             % ...  show DOFs carrying external forces by a marker at the nodes
136             Node = opt.BC_Frc(:,1);
137             plot(coord(Node,1),coord(Node,2),'m^', ...
138                 'MarkerSize',8, 'LineWidth', LineWidth);
139         end
140     end
141     axis equal;
142 end
```

The deformed shape of the mesh is plotted by the function `PlotDeformedMesh.m` in the Code List 3.7. The nodal displacements are stored in the argument d. A magnification factor, which is the ratio of the maximum displacement in absolute value to the largest dimension of the problem domain, is specified by field `MagnFct` of the structure variable `opt`. The default value is 0.1. The nodal coordinates are augmented by the magnified nodal displacements. The plotting is performed by the function `PlotS-BFEMesh.m`. The undeformed mesh will also be plotted if the field `Undeformed` of `opt` is not empty.

Code List 3.7: Plot deformed mesh of polygon subdomains

```
1  function varargout = PlotDeformedMesh(d, coord, sdConn, opt)
2  %Plot deformed mesh
3  %
4  %Inputs:
5  %   coord(i,:)    - coordinates of node i
6  %   sdConn{isd,:}(ie,:)  - S-element connectivity. The nodes of
7  %                          line element ie in S-element isd.
8  %   d             - nodal displacements
9  %
10 %   options:
11 %   opt=struct('MagnFct', 0.1, 'Undeformed',':b');
12 %   where the options are:
13 %      opt.MagnFct    : magnification factor of deformed mesh
14 %      opt.Undeformed : style of undeformed mesh
15
16 if nargin > 3 && isfield(opt, 'MagnFct')
17     magnFct = opt.MagnFct;
18 else
19     magnFct = 0.1;  % ....................... default magnification factor
20 end
21
22 Umax = max(abs(d));  % ........................ maximum displacement
23 Lmax = max(max(coord)-min(coord));  %  maximum dimension of domain
24 fct = magnFct*Lmax/Umax;  % ......... factor to magnify the displacement
25 %  ... augment nodal coordinates
26 deformed = coord + fct*(reshape(d,2,[]))';
27 hold on
28 %  ... plot undeformed mesh
29 if nargin > 3 && isfield(opt, 'Undeformed') && ...
30           ~isempty(opt.Undeformed)
31     %  ... plotting option of undeformed mesh
32     undeformedopt = struct('LineSpec',opt.Undeformed);
33     PlotSBFEMesh(coord, sdConn, undeformedopt);
34 end
35 title('DEFORMED MESH')
36 %  ... plot deformed mesh
37 deformedopt = struct('LineSpec','-k');  % ........... plotting option
38 PlotSBFEMesh(deformed, sdConn, deformedopt);
39 varargout={deformed, fct};%ltx outputs
```

To reduce the repetitions of codes used for analysing the examples, the following script has been written. This script is called to perform a static analysis after the scaled boundary finite element models are defined, for example, by other MATLAB scripts.

Code List 3.8: 2D static analysis

```
1  %  ...  Solution of S-elements and assemblage of global stiffness
2  %  ...  and mass matrices
3  [sdSln, K, M] = SBFEMAssembly(coord, sdConn, sdSC, mat);
4
5  %  ...  Static solution of nodal displacements and forces
6  [d, F] = SolverStatics(K, BC_Disp, F);
7
8  %  ...  Plot deformed mesh
9  figure
10 opt = struct('MagnFct', 0.1, 'Undeformed','--k');
11 PlotDeformedMesh(d, coord, sdConn, opt)
```

3.6   Examples of Static Analysis Using Platypus

■ **Example 3.4   Uniaxial Tension of a Rectangular Body**

Consider the problem shown in Figure 3.3, Example 3.3 on page 90. A vertical force $F = 1000\,\text{KN/m}$ is applied at Nodes 3 and 8. Determine the nodal displacements and the support reactions.

Consider a unit thickness of the rectangular body, the exact solution of the uniaxial tension problem is expressed as

$$\sigma_x = 0$$

$$\sigma_y = \frac{F}{b}$$

$$\tau_{xy} = 0$$

$$u_x = -\nu\frac{F}{E}\frac{x-b}{b}$$

$$u_y = \frac{F}{E}\frac{y}{b}$$

The scaled boundary finite element analysis is performed by the MATLAB script listed below. This script imports the problem definition (mesh, material constants and boundary conditions) from MATLAB script `Exmpl3SElements.m` on page 90.

```
%A rectangular body under uniaxial tension
close all; clearvars; dbstop error;

% input problem definition
Exmpl3SElements;

% SBFEM analysis
SBFEM2DStaticAnalysisScript;
```

```
% output nodal displacements and forces
dofs = (1:numel(coord))';
format long g;
[ dofs d(dofs) F(dofs)]
```

The above MATLAB script can be obtained by scanning the QR-code to the right.

The deformed shape is plotted in Figure 3.5. The output of nodal displacements and forces at all DOFs are as follows:

DOF	U	F
1	2.50000000000001e-05	0
2	0	-500.000000000001
3	2.50000000000001e-05	0
4	0.0001	1.13686837721616e-13
5	2.50000000000005e-05	-8.5265128291212e-14
6	0.000300000000000001	1000
7	0	1.4210854715202e-13
8	0	-1000
9	7.11742644584424e-20	-2.8421709430404e-14
10	0.0001	0
11	-2.5e-05	0
12	0	-500

Figure 3.5 Deformed shape of a rectangular body under uniaxial tension.

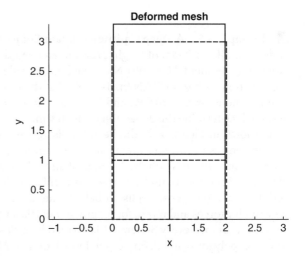

```
13       -2.49999999999999e-05          9.76996261670138e-15
14                      0.0001           5.6843418860808e-14
15       -2.49999999999996e-05         -5.6843418860808e-14
16                      0.0003                          1000
```

The nodal displacements are non-dimensionalized by F/E

Node	$u_x/(F/E)$	$u_y/(F/E)$
1	0.250000000000001	0
2	0.249999999999999	0.999999999999999
3	0.249999999999997	3
4	0	0
5	-6×10^{-16}	1
6	-0.25	0
7	-0.250000000000001	1
8	-0.250000000000003	3

The L^2 norm of error of the displacement solution ($\| \{u_{\text{exact}}\} - \{u\} \|/\| \{u_{\text{exact}}\} \|$) is equal to 1×10^{-15}.

The nodal forces are non-dimensionalized by F

Node	F_x/F	F_y/F
1	0	-0.5
2	-6×10^{-17}	6×10^{-17}
3	1×10^{-16}	1
4	-5×10^{-16}	-1
5	5×10^{-17}	-2×10^{-16}
6	-2×10^{-17}	-0.500000000000001
7	-7×10^{-17}	6×10^{-17}
8	6×10^{-17}	1

The displacement and force results are considered to be accurate up to the machine accuracy.

Example 3.5   A Deep Cantilever Beam Subject to Bending

A deep cantilever beam of height $H = 1$ m and length $L = 2$ m is shown in Figure 3.6a. A bending moment $M = 100$ kNm, which is equivalent to the linearly distributed surface traction with $p = 600$ kN/m, is applied at the free end. The material properties are Young's modulus $E = 10$ GPa and Poisson's ratio $v = 0.2$. Plane stress conditions are assumed. Determine the deflection of the beam.

As shown in Figure 3.6b, the beam is discretized using 7 polygon S-elements. The scaling centres of the S-elements are indicated by the symbol '$\oplus$' with the S-element numbers next to the symbols. An edge of an S-element is modelled by one 2-node line elements. There are 16 nodes in the mesh. This problem is defined using the MATLAB script ExmplDeepBeam.m listed below. The units are selected to be consistent with Newton for forces and metre for length and displacement.

In this example, every S-element is formed by a closed loop of 2-node line elements, a set of polygons are defined (see Line 9 of the MATLAB script) as an array of

(a)

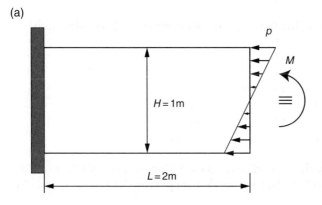

(b)

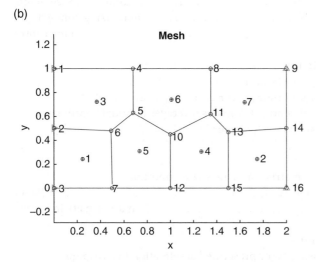

Figure 3.6 A deep cantilever beam subject to a force couple. (a) Geometry. (b) Mesh.

nodes to simplified the input. The sequence of the nodes in the array follows the counter-clockwise direction. The connectivity of line elements of the S-elements is formulated from the nodes of the polygon (see Line 21). The scaling centre of an S-element is chosen as the geometric centre of the set of nodes of the S-element. The coordinates are equal to the averages of the nodal coordinates (see Line 24). The mesh is plotted in Figure 3.6b.

MATLAB script: Cantilever beam under bending

```
1  % ... Mesh
2  % ... nodal coordinates. One node per row [x y]
3  coord = [    0      1;    0     0.5;    0          0;
4            0.68      1; 0.68    0.63;  0.49     0.48;
5             0.5      0; 1.35       1;     2          1;
6               1   0.45; 1.35    0.62;     1          0;
7             1.5   0.47;    2     0.5;   1.5          0;
```

```
8                 2         0  ];
9  % ... nodes of a polygon. The sequence follows counter-clockwise direction.
10 polygon = { [ 3      7      6      2];
11             [15     16     14     13];
12             [ 2      6      5      4      1];
13             [12     15     13     11     10];
14             [ 7     12     10      5      6];
15             [ 4      5     10     11      8];
16             [ 8     11     13     14      9] };
```

Input S-element connectivity as a cell array (One S-element per cell). In a cell, the connectivity of line elements is given by one element per row [Node-1 Node-2].

```
17 nsd = length(polygon); % ......................... number of S-elements
18 sdConn = cell(nsd,1); % ......................... initializing connectivity
19 sdSC = zeros(nsd,2); % ................................. scaling centre
20 for isub =1:nsd
21     % ... build connectivity
22     sdConn{isub}=[polygon{isub}; ...
23                   polygon{isub}(2:end) polygon{isub}(1)]';
24     % ... scaling centre at centroid of nodes (averages of nodal coordinates)
25     sdSC(isub,:) = mean(coord(polygon{isub},:));
26 end
27
28 % ... Materials: elasticity matrix for plane stress condition
29 mat.D = IsoElasMtrx(10E9, 0.2); % ......................... (E in Pa, v)
30 mat.den = 2000; % ................................. mass density in kg/m³
31
32 % ... Boundary conditions
33 % ... displacement constraints (or prescribed acceleration in a response
34 % ... history analysis. One constraint per row: [Node Dir Disp]
35 BC_Disp = [1 1 0; 1 2 0; 2 1 0; 2 2 0; 3 1 0; 3 2 0];
36
37 % ... assemble nodal forces
38 ndn = 2; % ............................................. 2 DOFs per node
39 NDof = ndn*size(coord,1); % ........................... number of DOFs
40 F = zeros(NDof,1); % .. initializing right-hand side of equation [K]{u} = {F}
41 edge = [16 14; 14 9]; % ....... edges subject to tractions, one row per edge
42 trac = [6E5 0 0 0; 0  0 -6E5 0]'; % ..... tractions, one column per edge
43 F = addSurfTraction(coord, edge, trac, F);
44 % ... plot mesh
45 figure
46 opt=struct('sdSC',sdSC, 'LabelSC', 14,'LineSpec','-k', ...
47            'PlotNode',1, 'LabelNode', 14,...
48            'BC_Disp',BC_Disp);
49 PlotSBFEMesh(coord, sdConn, opt);
50 title('MESH');
```

The analysis is performed using the following MATLAB script. This script imports the problem definition in MATLAB script ExmplDeepBeam.m on page 105.

```
%Deep Cantilever Beam Under Bending
close all; clearvars; dbstop error;

% input problem definition
ExmplDeepBeam

% SBFEM analysis
SBFEM2DStaticAnalysisScript

% output nodal displacements at lower edge
dofs = 2 * [3 7 12 15 16]';
[ dofs d(dofs)]
```

The above MATLAB script can be obtained by scanning the QR-code to the right.

The deformed shape of the beam is shown in Figure 3.7. The maximum magnitude of displacement is magnified to 10% of the length of the beam.

The vertical displacements of the nodes at the lower side of the beam are printed out:

```
DOF          U  (m)
  6          0
 14          0.0000152
 24          0.0000559
 30          0.0001266
 32          0.0002257
```

The vertical displacement at Node 16 (DOF 32) is equal to 0.0002257 m, which is normalized (considering a unit thickness) as $u_y/(M/(EH)) = 22.57$.

This example is only intended to illustrate the use of Platypus. The convergence study will be discussed in Chapter 4.

◨

◧ **Example 3.6   An Edge-cracked Rectangular Body Subject to Tension**
An edge-cracked rectangular body is shown in Figure 3.8a with the length $b = 0.1$ m. The base of the body is fixed and a uniform tension $p = 1$ MPa is applied at the top.

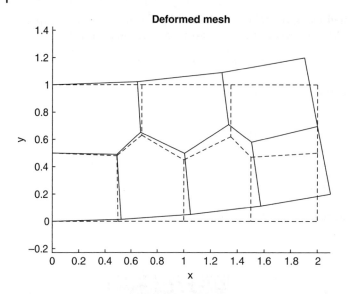

Figure 3.7 Deformed shape of deep cantilever beam subject to a force couple.

(a)   (b)

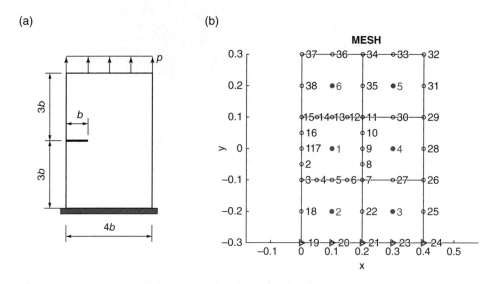

Figure 3.8 An edge-cracked rectangular body subject to tension. (a) Geometry. (b) Mesh.

The material properties are Young's modulus $E = 10\,\text{GPa}$ and Poisson's ratio $v = 0.25$. Plane stress conditions are assumed. Determine the crack opening displacements (CODs).

The scaled boundary finite element mesh is shown in Figure 3.8b. The problem domain is divided into 6 S-elements. The S-element numbers follow the '⊕' symbols indicating the locations of the scaling centres. S-element 1 encloses the crack tip. The crack tip is chosen as the scaling centre. Each of the four sides of S-element 1 is divided into four 2-node line elements. On the cracked side, two independent nodes

(Nodes 1 and 17 in Figure 3.8b) having the same coordinates, are introduced at the crack mouth. The lower and upper crack faces are generated by scaling these two nodes towards the scaling centre. No discretization is needed to represent the two crack faces. The line elements form an open loop (see Item 3a on cracked S-element and Figure 2.6). Except for the edges common with the first S-element, each edge of the other S-elements is divided into two 2-node line elements. As an edge can only be shared by two S-elements, any discretization of the edges is possible without causing mesh incompatibility.

A MATLAB script `ExmplEdgeCrackedRectangularPlate.m` listed below is written to define this problem. It also plots the mesh shown in Figure 3.8b.

MATLAB script: An edge-cracked rectangular body subject to tension

```
 1  clearvars; close all; dbstop if error
 2
 3  % ... Mesh
 4  % ... nodal coordinates in mm. One node per row [x y]
 5  b = 0.1;
 6  coord = b*[0 0; 0 -0.5; 0 -1; 0.5 -1; 1 -1; 1.5 -1; 2 -1;
 7      2 -0.5; 2 0; 2 0.5; 2 1; 1.5 1; 1 1; 0.5 1; 0 1; 0 0.5;
 8      0 1E-14; 0 -2; 0 -3; 1 -3; 2 -3; 2 -2; 3 -3; 4 -3;
 9      4 -2; 4 -1; 3 -1; 4 0; 4 1; 3 1; 4 2; 4 3; 3 3;
10      2 3; 2 2; 1 3; 0 3; 0 2];
```

Input S-element connectivity as a cell array (One S-element per cell). In a cell, the connectivity of line elements is given by one element per row [`Node-1 Node-2`].

```
11  sdConn = { [1:16; 2:17]'; % .............................. S-element 1
12      [3 18; 18 19; 19 20; 20 21; 21 22; ...
13      22 7; 4  3; 5  4; 6  5; 7  6]; % ................. S-element 2
14      [21 23; 23 24; 24 25; 25 26; 26 27; ...
15      27  7; 22 21; 7 22]; % ........................... S-element 3
16      [8  7; 9  8; 27 26; 7 27; 26 28; ...
17      28 29; 29 30; 30 11; 10  9; 11 10]; % ............ S-element 4
18      [30 29; 11 30; 29 31; 31 32; 32 33; ...
19      33 34; 34 35; 35 11]; % .......................... S-element 5
20      [12 11; 13 12; 14 13; 15 14; 35 34; ...
21      11 35; 34 36; 36 37; 37 38; 38 15]}; % ........... S-element 6
22  % ... coordinates of scaling centres of S-elements, one S-element per row
23  sdSC = b*[ 1  0; 1 -2; 3 -2; 3 0; 3 2; 1 2];
24
25  % ... Materials
26  % ... E: Young's modulus; p: Poisson's ratio
27  ElasMtrx = @(E, p) E/(1-p^2)*[1 p 0;p 1 0;0 0 (1-p)/2];
28  mat.D = ElasMtrx(10E9, 0.25); % ................................. E in Pa
29  mat.den = 2000; % ............................. mass density in kg/m³
30
31  % ... Boundary conditions
32  % ... displacement constraints. One constraint per row: [Node Dir Disp]
```

```
33  BC_Disp = [19 1 0; 19 2 0; 20 1 0; 20 2 0; ...
34        21 1 0; 21 2 0; 23 1 0; 23 2 0; 24 1 0; 24 2 0];
35  % ...  assemble load vector
36  ndn = 2; % ............................................. 2 DOFs per node
37  NDof = ndn*size(coord,1); % ........................... number of DOFs
38  F = zeros(NDof,1); % .. initializing right-hand side of equation [K]{u} = {F}
39  edge = [ 32 33; 33 34; 34 36; 36 37]; % ....... edges subject to traction
40  trac = [ 0 1E6 0 1E6]'; % ......... all edges have the same traction (in Pa),
41  F = addSurfTraction(coord, edge, trac, F);
42
43  % ...  Plot mesh
44  figure
45  opt=struct('sdSC',sdSC,'LineSpec','-k', ...
46        'PlotNode',1, 'LabelNode', 1,...
47        'BC_Disp',BC_Disp);
48  PlotSBFEMesh(coord, sdConn, opt);
49  title('MESH');
```

The following script performs the analysis to compute the nodal displacements.

```
%Edge Cracked Rectangular Body under Tension
close all; clearvars; dbstop error;

% input problem definition
ExmplEdgeCrackedRectangularPlate

% SBFEM analysis
SBFEM2DStaticAnalysisScript

% output crack opening displacements of Nodes 17 and 1
NodalDisp = (reshape(d, 2, []))';
disp(' Crack opening displacement');
COD = NodalDisp(17,:) - NodalDisp(1,:)
```

The above MATLAB script can be obtained by scanning the QR-code to the right.

The deformed shape of the mesh is shown in Figure 3.9.

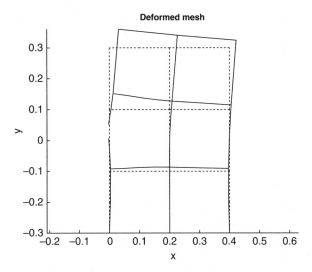

Figure 3.9 Deformation of edge-cracked rectangular body subject to tension.

The crack opening displacements are obtained as $\Delta u_x = -4.6 \times 10^{-7}$ m and $\Delta u_y = -7.83 \times 10^{-5}$ m. The results are expressed in dimensionless form as $\Delta u_x/(pb/E) = -0.046$ and $\Delta u_y/(pb/E) = 7.83$.

3.7 Evaluation of Internal Displacements and Stresses of an S-element

3.7.1 Integration Constants and Internal Displacements

The nodal displacements $\{d\} = \{u(\xi = 1)\}$ of an S-element (Figure 2.19) are extracted from the global solution based on the connectivity of the S-element. To determine the integration constants in the solution of the nodal displacement functions of the S-element, Eq. (3.21a) is formulated at $\xi = 1$ as

$$\{d\} = [\Phi_b^{(u)}]\{c\} \tag{3.55}$$

This leads to the integration constants

$$\{c\} = [\Phi_b^{(u)}]^{-1}\{u_b\} \tag{3.56}$$

The nodal displacement functions of the S-element, which describe the displacements along the radial lines connecting the scaling centre and a node on the boundary, follow from Eq. (3.18a) or Eq. (3.21a) in matrix form. The displacement field can be evaluated by interpolating the displacement functions using the shape function of the line elements (Eq. (2.79)).

The solution of the nodal displacement functions is a series of power function ξ^{λ_i} and the eigenvalues λ_i are sorted in descending order of their real parts. Except for the last two eigenvalues $\lambda_{n-1} = \lambda_n = 0$ corresponding to the two modes of translational rigid motion, all the eigenvalues λ_i ($i = 1, 2, \ldots, n - 2$) have positive real parts and the corresponding power functions ξ^{λ_i} decays as $\xi \to 0$. The rates of decay are proportional to the values of their real parts. Therefore, the contribution of the terms with a large real part of the eigenvalue to the displacement field is limited to a region adjacent to the boundary. At the scaling centre, $\xi = 0$, only the two modes of translational rigid modes remain. The displacements are equal to ($\xi^0 = 1$)

$$\{u(\xi = 0)\} = \sum_{i=n-1}^{n} c_i \{\phi_i^{(u)}\} \tag{3.57}$$

3.7.2 Strain/Stress Modes and Strain/Stress Fields

The solution of the displacement field is semi-analytical. In the radial direction ξ, the solution is expressed analytically by the nodal displacement functions (Eq. (3.18a)). In the circumferential direction η, the solution is given in discretized form and interpolated by the shape functions of the line elements (Eq. (2.79)).

To obtain the strain field of an S-element defined in the scaled boundary coordinates (Section 2.4), the partial derivative of nodal displacement functions in Eq. (3.18a) with respect to the ξ is taken analytically, resulting in

$$\{u(\xi)\}_{,\xi} = \sum_{i=1}^{n} c_i \lambda_i \xi^{\lambda_i - 1} \{\phi_i^{(u)}\} \tag{3.58}$$

Along the circumferential direction, the strain field is evaluated in a manner similar to the stress recovery in the standard finite element method. The partial derivative of the displacement field is evaluated piecewise for each line element independently. The strains are evaluated within an individual sector formed by scaling the line element and given by Eq. (2.83), which is reproduced below for convenience of reference

$$\{\varepsilon\} = [B_1]\{u^e(\xi)\}_{,\xi} + \frac{1}{\xi}[B_2]\{u^e(\xi)\} \tag{3.59}$$

where $\{u^e(\xi)\}$ are the displacement functions related to the nodes of the line element. They are extracted from the S-element solution (Eq. (3.18a)) using element connectivity and expressed as

$$\{u^e(\xi)\} = \sum_{i=1}^{n} c_i \xi^{\lambda_i} \{\phi_i^{(u)e}\} \tag{3.60}$$

where, in the same way as $\{u^e(\xi)\}$, the displacement modes of the line element $\{\phi_i^{(u)e}\}$ ($i = 1, 2, \ldots, n$) are extracted from the displacement modes $\{\phi_i^{(u)}\}$ of the S-element. The strains are obtained by substituting Eq. (3.60) into Eq. (3.59) as

$$\{\varepsilon\} = \sum_{i=1}^{n-2} c_i \xi^{\lambda_i - 1} \left(\lambda_i [B_1] + [B_2]\right) \{\phi_i^{(u)e}\} \tag{3.61}$$

The last two terms ($i = n - 1$, n) corresponding to the translational rigid body motions are excluded as they do not produce strains.

In a numerical implementation, the strains are evaluated at a few sampling values of the circumferential coordinate η. Typically, the values at Gauss points of a lower-order rule than that for full integration are selected. For a 2-node line element, the midpoint ($\eta = 0$) is used. The strain modes of an S-element, corresponding to the displacement modes, are constructed by concatenating the values of all the line elements

$$\{\phi_i^{(\varepsilon)}\} = \text{concat}\left[\left(\lambda_i[B_1] + [B_2]\right)\{\phi_i^{(u)e}\}\right] \tag{3.62}$$

where concat [•] stands for array concatenation. The strains (Eq. (3.61)) of the S-element are expressed as

$$\{\varepsilon\} = \sum_{i=1}^{n-2} c_i \xi^{\lambda_i - 1} \{\phi_i^{(\varepsilon)}\} \tag{3.63}$$

The solution of the strain field is also semi-analytical (i.e. analytical in the radial direction ξ and numerical in the circumferential direction defined by the natural coordinate η of the line elements). It can be regarded as the superposition of the strain modes. The values of the strains are along the radial lines passing through the scaled centre and the sampling points of the line elements.

The strains at the scaling centre ($\xi = 0$) are considered. When eigenvalues whose real parts are between 0 and 1, i.e. $0 < \text{Re}(\lambda_i) < 1$, exist, the corresponding power function $\xi^{\lambda_i - 1}$ in the strains (Eq. (3.63)) tends to infinity as $\xi \to 0$, leading to strain and stress singularities. Figure 2.6 illustrates such a case where the scaling centre is placed at a crack tip of a cracked S-element and the line elements form an open loop.

For the cases where the scaling centre is placed inside the S-element and the line elements form a closed loop, four eigenvalues $\lambda_i = 1$ ($i = n - 5$, $n - 4$, $n - 3$, $n - 2$) exist in the S-element solution (see Section 6.11). As $\xi^{\lambda_i - 1} = \xi^0 = 1$, they are the constant strain modes. The real part of all the other eigenvalues λ_i ($i = 1, 2, \ldots, n - 6$) is greater than 1. Their contributions vanish as the power function $\xi^{\lambda_i - 1} = 0$ at $\xi = 0$. The strains at the scaling centre are thus equal to

$$\{\varepsilon\} = \sum_{i=n-5}^{n-2} c_i \{\phi_i^{(\varepsilon)}\} \tag{3.64}$$

The results obtained from all sectors, each corresponding to a line element at the boundary, are numerically identical.

The above solutions for the displacement and strain fields in a sector defined by a line element can be expressed in matrix form. The displacement solution (Eq. 3.60) is expressed as (see Eq. (3.21a))

$$\{u^e(\xi)\} = [\Phi_b^{(u)e}]\xi^{\langle \lambda_b \rangle}\{c\} \tag{3.65}$$

where $[\Phi_b^{(u)e}]$ are the rows in $[\Phi_b^{(u)}]$ that are connected to the DOFs of the specified line element. The strain modes in Eq. (3.62) are assembled column-wise as

$$[\Phi_b^{(\varepsilon)}] = \left[\{\phi_1^{(\varepsilon)}\}, \{\phi_2^{(\varepsilon)}\}, \ldots, \{\phi_{n-2}^{(\varepsilon)}\}, 0, 0\right]$$

$$= [B_1][\Phi_b^{(u)e}]\langle \lambda_b \rangle + [B_2][\Phi_b^{(u)e}] \tag{3.66}$$

The strain field in Eq. (3.63) is rewritten as

$$\{\varepsilon\} = [\Phi_b^{(\varepsilon)}]\xi^{-1}\text{diag}(c_1\xi^{\lambda_1}, c_2\xi^{\lambda_2}, \ldots, c_{n-2}\xi^{\lambda_{n-2}}, 0, 0) \tag{3.67}$$

The stresses are obtained as (Eq. (3.63))

$$\{\sigma\} = \sum_{i=1}^{n-2} c_i \xi^{\lambda_i - 1} \{\phi_i^{(\sigma)}\} \tag{3.68}$$

where the stress modes are obtained using Eqs. (A.11) and (3.62) as

$$\{\phi_i^{(\sigma)}\} = \text{concat}\left[[D]\left(\lambda_i[B_1] + [B_2]\right)\{\phi_i^{(u)e}\} \right] \qquad (i = 1, 2, \ldots, n - 2) \tag{3.69}$$

with the elasticity matrix $[D]$ of the material. The stress solution in Eq. (3.68) is expressed in matrix form as

$$\{\sigma\} = [\Phi_b^{(\sigma)}]\xi^{-1}\text{diag}(c_1\xi^{\lambda_1}, c_2\xi^{\lambda_2}, \ldots, c_{n-2}\xi^{\lambda_{n-2}}, 0, 0) \tag{3.70}$$

where the matrix of stress modes are equal to

$$[\Phi_b^{(\sigma)}] = [D][\Phi_b^{(\varepsilon)}] = \left[\{\phi_1^{(\sigma)}\}, \{\phi_2^{(\sigma)}\}, \ldots, \{\phi_{n-2}^{(\sigma)}\}, \{0\}, \{0\} \right] \tag{3.71}$$

3.7.3 Shape Functions of Polygon Elements Modelled as S-elements

Various formulations for polygon elements have been proposed (Nguyen-Xuan (2017); Bishop (2014), to name a few). The key aspect in devising a polygon element is to construct the shape functions for interpolating over the polygon. The scaled boundary finite element method provides a systematic approach to this task when the polygon elements satisfy the scaling requirement.

Substituting Eq. (3.56) into Eq. (3.65) and using Eq. (2.79) leads to the displacement field in a sector defined by scaling one line element to the scaling centre

$$\{u\} = \underbrace{[N_u][\Phi_b^{(u)e}]\xi^{\langle\lambda_b\rangle}[\Phi_b^{(u)}]^{-1}}_{[N_V^e]}\{d\} \tag{3.72}$$

It relates the displacement field to the nodal displacements at the boundary of the polygon. The shape functions of the polygon elements is given sector-by-sector as

$$[N_V^e] = [N_u][\Phi_b^{(u)e}]\xi^{\langle\lambda_b\rangle}[\Phi_b^{(u)}]^{-1} \tag{3.73}$$

Using the shape functions, standard finite element procedures can be directly applied to handle dynamic, nonlinear and other types of analysis (Behnke et al., 2014; Ooi et al., 2014). The mass matrix of a (polygon) S-element is determined in Section 3.10.1.

3.8 Computer Program Platypus: Internal Displacements and Strains

MATLAB functions are provided to compute the displacements, strains and stresses inside S-elements. The boundary of an S-element is discretized by 2-node line elements.

The functions `SElementIntgConst.m` in Code List 3.9 determines the integration constants `sdIntgConst` of the S-elements using the nodal displacements d of the global system and the solution of the S-elements `sdSln` obtained with the function in Code List 3.5 on page 97.

Code List 3.9: Post-processing: Integration constants

```
1  function [ sdIntgConst ] = SElementIntgConst( d, sdSln )
2  %Integration constants of S-elements
3  %
4  % Inputs:
5  %    d              - nodal displacements
6  %    sdSln          - solutions for S-elements
7  % Outputs:
8  %    sdIntgConst - vector of intergration constants
9
10 Nsd = length(sdSln);  % .................... total number of S-elements
11 sdIntgConst = cell(Nsd,1);  % ......... initialization of output argument
12 for isd = 1:Nsd  % .............................. loop over S-elements
13
14     % ...  Integration constants
15     % ...  global DOFs of nodes in an S-element
16     dof = reshape([2*sdSln{isd}.node-1;2*sdSln{isd}.node], [], 1);
17     dsp = d(dof);  % ....... nodal displacements at boundary of S-element
18     sdIntgConst{isd} = (sdSln{isd}.v)\dsp;  % . integration constants,
   see Eq. (3.56)
19
20 end
```

The function `SElementStrainMode2NodeEle.m` is given in Code List 3.10 on page 115. This function evaluates the strain modes at the middle point ($\eta = 0$) of the element. The strain-displacement matrices in Eq. (2.117) at $\eta = 0$ become

$$[B_1] = [b_1][N^u] = \frac{1}{2|J_b|} \left[[C_1]\ [C_1] \right] \tag{3.74a}$$

$$[B_2] = [b_2][N^u]_{,\eta} = \frac{1}{2|J_b|} \left[-[C_2]\ [C_2] \right] \tag{3.74b}$$

where the matrices $[C_1]$ and $[C_2]$ are give in Eq. (2.114) and the Jacobian matrix at the boundary in Eq. (2.58). The output argument is a cell array `sdStrnMode` storing the strain modes. The function is documented in its list.

Code List 3.10: Post-processing: Strain modes

```
1  function [ sdStrnMode ] = SElementStrainMode2NodeEle( sdSln )
2  %Internal displacements and strain modes of S-elements
3  %
4  % Inputs:
5  %    sdSln          - solutions for S-elements
6  % Outputs:
7  %    sdStrnMode  - strain modes of S-elements
8
9  Nsd = length(sdSln);  % ........................ number of S-elements
10 sdStrnMode = cell(Nsd,1);  % ......... initialization of output argument
```

```
11   for isd = 1:Nsd % .............................. loop over S-elements
12
13       % ... number of DOFs of boundary nodes
14       nd = 2*length(sdSln{isd}.node);
15       v = sdSln{isd}.v; % ........................ displacement modes
16
17       % ... Strain modes. The last 2 columns equal 0 (rigid body motions)
18       % ... number of DOFs excluding 2 translational rigid-body motions
19       nd2 = nd-2;
20       d = sdSln{isd}.d(1:nd2); % ......................... eigenvalues
```

See Eq. (3.66). $[\Phi_b^{(u)}] \langle \lambda_b \rangle$ is computed. The element values are extracted based on element connectivity in Line 45

```
21       vb = v(:,1:nd2) .* (d(1:nd2,ones(1,nd)))';
22
23       n1 = sdSln{isd}.conn(:,1); % ..... first node of all 2-node elements
24       n2 = sdSln{isd}.conn(:,2); % .. second node of all 2-node elements
25       % ... LDof(i,:): Local DOFs of nodes of element i in an S-element
26       LDof = [ n1+n1-1 n1+n1 n2+n2-1 n2+n2 ];
27       xy = sdSln{isd}.xy; % nodal coordinates with origin at scaling centre
28       % ... dxy(i,:):[Δx, Δy] of i-th element, Eq. (2.50)
29       dxy = xy(n2,:) - xy(n1,:);
30       % ... mxy(i,:):[x̄, ȳ] of i-th element, Eq. (2.51)
31       mxy = (xy(n2,:) + xy(n1,:))/2;
32       % ... a(i):2|Jb| of i-th element, Eq. (2.57)
33       a = mxy(:,1).*dxy(:,2) - mxy(:,2).*dxy(:,1);
34
35       ne = length(n1); % ....................... numer of line elements
36       mode = zeros(3*ne, nd); % ............... initializing strain modes
37       for ie = 1:ne % .................... loop over elements at boundary
38           C1 = 0.5*[ dxy(ie,2) 0; 0 -dxy(ie,1); ...
39                   -dxy(ie,1) dxy(ie,2)]; % .................... Eq. (2.114a)
40           C2 = [-mxy(ie,2) 0; 0 mxy(ie,1); ...
41                   mxy(ie,1) -mxy(ie,2)]; % .................... Eq. (2.114b)
42           B1 = 1/a(ie)*[ C1 C1]; % ........................ Eq. (3.74a)
43           B2 = 1/a(ie)*[-C2 C2]; % ........................ Eq. (3.74b)
44           mode(3*(ie-1)+1:3*ie,1:nd2) = B1*vb(LDof(ie,:),:) ...
45                   + B2*v(LDof(ie,:),1:nd2); % ..... strain modes, Eq (3.66)
46       end
47
```

Store the ouput in cell array `sdPstP`. The number of S-element `isd` is the index of the array.
`GPxy(ie,:)`: the coordinates of the Gauss Point of element `ie` (middle point of 2-node element).
`strnMode(:,ie)`: the strain modes at the Gauss Point of element `ie`.

```
48       sdStrnMode{isd}= struct('xy',mxy, 'value',mode);
49
50   end
```

The function SElementInDispStrain.m in Code List 3.11 uses the integration constants and strain modes to evaluate the nodal displacement functions (Figure 2.18b) and strains on a coordinate line with a specified constant radial coordinate ξ. As illustrated in Figure 2.13, the geometry of the scaled boundary is defined in the same manner as the boundary by points and lines resulting from scaling the boundary. For convenience, the terms 'nodes' and 'elements' are still used to refer to these points and lines although they do not physically exist on the scaled boundary. The input arguments consist of the radial coordinate xi, the the solution of the S-element sdSln, the strain modes in sdStrnMode and the integration constants sdIntgConst. The output arguments include the nodal coordinates nodexy, nodal displacements dsp, nodal strains strnNode. The coordinates (GPxy) of the Gauss integration points (i.e. the middle points for the 2-node elements) of the elements and the strains strnEle at the Gauss points are also output. The nodal strains are obtained by averaging the Gauss point strains of the elements connected to it. The stresses can be evaluated from the strains using Eq. (A.11).

Code List 3.11: Post-processing: Displacements and strains

```
1  function [nodexy, dsp, strnNode, GPxy, strnEle] = ...
2      SElementInDispStrain(xi, sdSln, sdStrnMode, sdIntgConst)
3  %Displacements and strains at specified radial coordinate
4  % Inputs:
5  %    xi           - radial coordinate
6  %    sdSln        - solutions for S-element
7  %    sdStrnMode   - strain modes of S-element
8  %    IntgConst    - integration constants
9  %
10 % Outputs (All valus are on the scaled boundary, i.e. coodinate
11 %              line, at specified xi):
12 %    nodexy(i,:)   - coordinates of node i
13 %    dsp(i,:)      - nodal displacement funcitons of node i
14 %    strnNode(:,i) - strains on the radial line of node i
15 %    GPxy(ie,:)    - coordinates of middle point of element ie
16 %    strnEle(:,ie) - strains at middle point of element ie
17
```

Transform local coordinates (origin at scaling centre) at scaled boundary to global coordinates.

GPxy(ie,:) - coordinates of Gauss point (middle of 2-node element) of element ie after scaling.

nodexy(i,:) - coordinates of node i after scaling.

```
18 GPxy   = bsxfun(@plus, xi*sdStrnMode.xy, sdSln.sc);
19 nodexy = bsxfun(@plus, xi*sdSln.xy, sdSln.sc);
20
21 if xi >1.d-16 % ........ outside of a tiny region around the scaling centre
22     fxi = (xi.^sdSln.d).*sdIntgConst; % .................... ξ⟨λ_b⟩{c}
23     dsp = sdSln.v*fxi; % ............................................. Eq. (3.21a)
24     strnEle = sdStrnMode.value(:,1:end-2) ...
25                  *fxi(1:end-2)/xi; % ....................... Eq. (3.67)
26 else % ............................................... at scaling centre
```

```
27    dsp = sdSln.v(:,end-1:end)* ...
28        sdIntgConst(end-1:end);  % ..................... Eq. (3.57)
29    if(min(real(sdSln.d(1:end-2)))>0.999)
30        strnEle = sdStrnMode.value(:,end-5:end-2) ...
31            *sdIntgConst(end-5:end-2);  % ................ Eq. (3.64)
32    else  % ......................... stress singularity at scaling centre
33        strnEle = NaN(length(sdStrnMode.value),1);
34    end
35 end
36 % ...  remove possible tiny imaginary part due to numerical error
37 dsp = real(dsp);
38 strnEle = real(strnEle);
39
40 % ...  strnEle(1:3,ie) is the strains at centre of element ie after reshap-
ing.
41 strnEle = reshape(strnEle, 3, []);
42
43 % ...  nodal stresses by averaging element stresses
44 nNode = length(sdSln.node);  % ..................... number of nodes
45 strnNode = zeros(3,nNode);  % .......................... initialization
46 % ...  counters of number of elements connected to a node
47 count = zeros(1,nNode);
48 n = sdSln.conn(:,1); % vector of first node of elements
49 % ...  add element stresses to first node of elements
50 strnNode(:,n) = strnNode(:,n) + strnEle;
51 % ...  increment counters
52 count(n) = count(n) + 1;
53
54 n = sdSln.conn(:,2); % vector of second node of elements
55 % ...  add element stresses to second node of elements
56 strnNode(:,n) = strnNode(:,n) + strnEle;
57 % ...  increment counters
58 count(n) = count(n) + 1;
59 strnNode = bsxfun(@rdivide, strnNode, count);  % ........ averaging
60
61 end
```

■ **Example 3.7   A Rectangular Body Under Uniaxial Tension: Internal Displacements and Stresses**

The nodal displacements of the problem shown in Figure 3.3, Example 3.3 have been obtained in Example 3.4. Compute the displacements, strains and stresses of S-element 1 at the radial coordinate $\xi = 0.5$.

After running the script on page 102 to obtain the solutions of the 3 S-elements and the nodal displacements, the following script is used to compute the solution inside an S-element:

```
% strain modes of S-elements
sdStrnMode = SElementStrainMode2NodeEle( sdSln );

% integration constants
```

```
sdIntgConst = SElementIntgConst( d, sdSln );

isd = 1; % S-element number

% display integration constants
[(1:length(sdIntgConst{isd}))' sdIntgConst{isd}]
disp('strain modes')
sdStrnMode{isd}.value

xi = 0.5; % radial coordinate
% displacements and strains at specified raidal coordinate
[nodexy, dsp, strnNode, GPxy, strnEle] = ...
    SElementInDispStrain(xi, sdSln{isd}, ...
                    sdStrnMode{isd}, sdIntgConst{isd});

disp(['  x        y        ux        uy']);
[nodexy, (reshape(dsp,2,[]))']

disp('strains of Elements 1 and 2')
strnEle(:,1:2)
disp('stresses of Elements 1 and 2')
mat.D*strnEle(:,1:2)
```

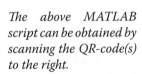

The above MATLAB script can be obtained by scanning the QR-code(s) to the right.

The command line [(1:length(sdIntgConst{isd}))' sdIntg-Const{isd}] displays the integration constants

i	c(i)
1	2.2197e-20
2	1.1825e-19
3	3.3305e-05
4	2.3174e-05
5	5.7693e-05
6	1.7784e-05
7	1.25e-05
8	5e-05

Note that the integrations constants are not uniquely defined as the eigenvectors are not (see Eq. (3.56)). Complex numbers may also occur.

For this uniaxial tension problem, the analytical solution of displacements (Example 3.4) are linear functions in coordinates x, y, and thus in ξ (see Eq. (2.5)).

Table 3.2 Strains modes of S-element 1 of a rectangular body under uniaxial tension.

		Strain mode					
		1	2	3	4	5	6
Ele. 1	ε_x	2	−0.2257	0.2179	0.8608	−0.5874	−1.0299
	ε_y	−0.9091	0.1026	0.9997	−0.1722	1.0444	0.5872
	γ_{xy}	8×10^{-16}	−1.0605	−1.8485	0.7291	0.6886	0.2777
Ele. 2	ε_x	7×10^{-16}	−0.8837	0.2179	0.8608	−0.5874	−1.0299
	ε_y	6×10^{-17}	1.9442	0.9997	−0.1722	1.0444	0.5872
	γ_{xy}	−1.0909	0.1231	−1.8485	0.7291	0.6886	0.2777
Ele. 3	ε_x	−2	0.2257	0.2179	0.8608	−0.5874	−1.0299
	ε_y	0.9091	−0.1026	0.9997	−0.1722	1.0444	0.5872
	γ_{xy}	4×10^{-16}	1.0605	−1.8485	0.7291	0.6886	0.2777
Ele. 4	ε_x	2×10^{-15}	0.8837	0.2179	0.8608	−0.5874	−1.0299
	ε_y	-4×10^{-16}	−1.9442	0.9997	−0.1722	1.0444	0.5872
	γ_{xy}	1.0909	−0.1231	−1.8485	0.7291	0.6886	0.2777

The scaled boundary finite element solution of the displacements is expressed in Eq. (3.18a) with the eigenvalues and eigenvectors for this problem given in Example 3.1 on page 80. The first two terms of the solution with the power function $\lambda^{1.915}$ vanish as the integration constants shown above are equal to zero numerically. The third to the sixth terms are linear in ξ and the last two terms are constant. Therefore, the scaled boundary finite element solution is consistent with the analytical solution.

The field value of the cell variable sdStrnMode is a matrix storing the strain modes at the middle point of the elements in columns. It is shown within the box in the Table 3.2 with annotations. The two strains modes (Modes 7 and 8) corresponding to the translational rigid-body motions are equal to 0 and omitted from the table. Each strain mode is a vector containing the three strain components at the middle points of the four elements. Modes 3-6 correspond to linear variations of displacement. As expected, in each of these four modes, the strains of all the four elements are identical representing a constant strain field.

The displacement output is as follows:

```
   x        y            ux                        uy
 0.25     0.25    1.87500000000001e-05      2.50000000000001e-05
 0.25     0.75    1.87500000000001e-05      7.50000000000002e-05
 0.75     0.25    6.25000000000002e-06      2.50000000000001e-05
 0.75     0.75    6.25000000000006e-06      7.50000000000001e-05
```

Compared with the analytical solution on page 102, it is observed that the results are accurate to machine precision.

The strains at the middle points of the elements on the scaled boundary are stored in the matrix strnEle. The 3 strain components at one point are stored as one column. The strains of Elements 1 and 2 are displayed and shown below with annotations

```
                    Element 1                     Element 2
Strain_xx               -2.5e-05        -2.50000000000001e-05
Strain_yy   9.99999999999999e-05       9.99999999999999e-05
Strain_xy   -5.9292306307801e-20       -1.09267250195805e-19
```

The stresses are computed (see Eq. (A.11)) and displayed for Elements 1 and 2

```
                    Element 1                     Element 2
Stress_xx   -5.24319355695487e-13       7.54469710410415e-15
Stress_yy                    1000          999.999999999999
Stress_xy   2.16840434497101e-13        3.25260651745651e-13
```

The nodal strains and stresses are stored and computed in the same way as the element strains and stresses are. For this simple case of uniaxial tension, the nodal values are equal to the element values and are not shown.

⬛

⬛ **Example 3.8   Patch Test on Distorted Polygon S-elements**
A patch test on a unit square shown in Figure 3.10a is performed. The analytical solutions for the three constant stress states are:

1) Constant σ_x:

$$\sigma_x = p; \qquad \sigma_y = 0; \qquad \tau_{xy} = 0; \qquad u_x = \frac{p}{E}x; \qquad u_y = -v\frac{p}{E}y$$

2) Constant σ_y:

$$\sigma_x = 0; \qquad \sigma_y = p; \qquad \tau_{xy} = 0; \qquad u_x = -v\frac{p}{E}x; \qquad u_y = \frac{p}{E}y$$

3) Constant τ_{xy}:

$$\sigma_x = 0; \qquad \sigma_y = 0; \qquad \tau_{xy} = p; \qquad u_x = \frac{p}{G}y; \qquad u_y = 0$$

For simplicity in presenting results, the Young's modulus and the constant stresses are chosen as a unit. It is assumed that a set of consistent units are adopted and the results are scalable for this linear analysis by the external loads and Young's modulus of the material. The Poisson's ratio is 0.25.

The patch of two distorted polygon S-elements in Figure 3.10a is considered. The dark shaped S-element is a 6-side concave polygon. Its interior angle at the vertex (x_1, y_1) is greater than 180°. Two locations of the node (x_1, y_1) are considered. One is at the centre of the square $x_1 = y_1 = 0.5$. The mesh is depicted in Figure 3.10b. The other is at $x_1 = 0.05$ and $y_1 = 0.95$, close to the upper-left corner of the square. As shown in Figure 3.10c, the polygon S-elements are extremely distorted.

The MATLAB script is listed below. After the input of nodal coordinates and connectivity, the locations of the scaling centres are specified (see Line 10). The scaling centre of S-element 1 is located at the middle of the line connecting Nodes 1 and 2. The scaling centre of S-element 2 is chosen on the extension of the line connecting Nodes 1 and 2. It is located in the middle of Node 2 and the intersecting point with the left or top edge

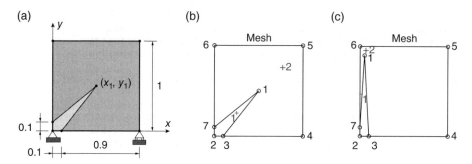

Figure 3.10 Patch of two irregular polygon S-elements. (a) Geometry. (b) Mesh of patch with $x_1 = y_1 = 0.5$. (c) Mesh of patch with $x_1 = 0.05$ and $y_1 = 0.95$.

of the square. The boundaries of both S-elements are directly visible from their scaling centres.

The boundary conditions are specified starting from Line 20. The displacement boundary conditions constrain the rigid body modes only. The surface tractions of the constant stress states are defined.

The MATLAB script SBFEM2DStaticAnalysisScript.m is called to obtain the nodal displacements. The stresses at the boundary of S-element 2 are computed (from Line 55 of the MATLAB script).

Patch test: A square modelled with two irregular polygon S-elements

```
1  close all; clearvars; dbstop error;
2
3  % ...  Mesh
4  % ...  nodal coordinates. One node per row [x y]
5  x1 = 0.5; y1 = 0.5; % Figure b
6  % x1 = 0.05; y1 = 0.95; % Figure c
7  coord = [ x1 y1; 0 0; 0.1  0; 1 0; 1 1; 0 1; 0 0.1];
```

Input S-element connectivity as a cell array (One S-element per cell). In a cell, the connectivity of line elements is given by one element per row [Node-1 Node-2].

```
8  sdConn = { [1 7; 7 2; 2 3; 3 1]; % ........................ S-element 1
9            [1 3; 3 4; 4 5; 5 6; 6 7; 7 1]}; % .................... S-element 2
10 % ...  coordinates of scaling centres of S-elements.
11 if x1 > y1 % ........ extension of line 21 intersecting right edge of the square
12     sdSC = [ x1/2 y1/2 ; (1+x1)/2 y1*(1+x1)/(2*x1)];
13 else % ................ extension of line 21 intersecting top edge of the square
14     sdSC = [ x1/2 y1/2 ;   x1*(1+y1)/(2*y1) (1+y1)/2];
15 end
16 % ...  Materials
17 mat.D = IsoElasMtrx(1, 0.25); % ........................ elasticity matrix
18 mat.den = 2; % ......................................... mass density
```

```
19
20  % ...  Boundary conditions
21  % ...  displacement constrains. One constrain per row: [Node Dir Disp]
22  BC_Disp = [2 1 0; 2 2 0; 4 2 0];
23  % ...  assemble load vector
24  ndn = 2; % ........................................ 2 DOFs per node
25  NDof = ndn*size(coord,1); % ........................... number of DOFs
26  F = zeros(NDof,1); % .. initializing right-hand side of equation [K]{u} = {F}
27  % %ltx  horizontal tension
28  % edge = [ 4 5; 6 7; 7 2];
29  % trac = [1 0 1 0; -1 0 -1 0; -1 0 -1 0]';
30  % %ltx vertical tension
31  % edge = [2 3; 3 4; 5 6];
32  % trac = [0 -1 0 -1; 0 -1 0 -1; 0 1  0 1]';
33  % ...  pure shear
34  edge = [2 3; 3 4; 4 5; 5 6; 6 7;
35         7 2]; % ................ edges subject to tractions, one row per edge
36  trac = [-1 0 -1 0; -1 0 -1 0; 0 1  0 1; 1 0 1 0;
37         0 -1 0 -1; 0 -1 0 -1]'; % ....... tractions, one column per edge
38  F = addSurfTraction(coord, edge, trac, F);
39
40  % ...  Plot mesh
41  figure
42  opt=struct('LineSpec','-k', 'sdSC',sdSC, ...
43      'PlotNode',1, 'LabelNode', 1); % ................. plotting options
44  PlotSBFEMesh(coord, sdConn, opt);
45  title('MESH');
46
47  % ...  Static solution
48  % ...  nodal displacements
49  SBFEM2DStaticAnalysisScript
50  disp('Nodal displacements')
51  for ii = 1:length(coord)
52      fprintf('%5d %25.15e %25.15d\n',ii, d(2*ii-1:2*ii))
53  end
54
55  % ...  Stresses
56  % ...  strain modes of S-elements
57  sdStrnMode = SElementStrainMode2NodeEle( sdSln );
58  % ...  integration constants
59  sdIntgConst = SElementIntgConst( d, sdSln );
60  % ...  displacements and strains at specified radial coordinate
61  isd = 2; % ................................................ S-element number
62  xi = 1; % ................................................ radial coordinate
63  [nodexy, dsp, strnNode, GPxy, strnEle] = ...
64          SElementInDispStrain(xi, sdSln{isd},  ...
65                              sdStrnMode{isd}, sdIntgConst{isd});
```

```
66  disp('Stresses of Elements 1 and 2')
67  mat.D*strnEle(:,1:2)
```

The scaled boundary finite element method passes the test on the S-element patches. Similar accuracy of the results are observed for all three stress states. Only the results for the constant shear stress ($\tau_{xy} = 1$) state are presented in this example, where the exact solution of displacements are

$$u_x = 2.5y; \qquad u_y = 0$$

Using the above MATLAB script, the results can be obtained for the other two stress states by selecting the corresponding nodal force boundary conditions.

The patch shown in Figure 3.10b is addressed (use x1 = 0.5; y1 = 0.5;). The nodal displacements are obtained as:

```
node            ux                        uy
 1     1.249999999999994e+00     -2.620873192519945e-15
 2     0.000000000000000e+00      000000000000000
 3    -4.676422590493927e-15     -3.386728472849379e-15
 4    -2.196395029118848e-15      000000000000000
 5     2.499999999999992e+00     -3.505108455514845e-16
 6     2.499999999999995e+00     -3.822042574728605e-15
 7     2.499999999999947e-01     -3.401327016801653e-15
```

The maximum error of displacements is about 5×10^{-15}.

The stresses at the middle of the line elements at the boundary ($\xi = 1$) of the S-element are computed. The stresses at the midpoints of Elements 1 (edge 1-3) and 2 (edge 3-4) are:

```
                    Element 1                  Element 2
Stress_xx    1.17203365265958e-13      3.79599866114004e-15
Stress_yy   -5.54602201134714e-14      8.74825963544821e-16
Stress_xy    1.00000000000001                     1
```

The maximum error of stresses is 1×10^{-13}. The stresses of other elements have the same order of accuracy.

The patch of the extremely-distorted S-elements (Figure 3.10c) is tested by selecting the coordinates $x_1 = 0.05$ and $y_1 = 0.95$ in the MATLAB script. The nodal displacements are obtained as:

```
node            ux                        uy
 1     2.374999999999539e+00     -1.003012244444069e-13
 2     0.000000000000000e+00      000000000000000
 3    -1.061243224900644e-13      2.285611248412104e-14
 4    -2.198284517713762e-13      000000000000000
 5     2.499999999999653e+00      1.427064972822626e-13
 6     2.499999999999519e+00     -1.207943951391976e-13
 7     2.499999999998503e-01     -1.478038453171263e-14
```

with a maximum error of 5×10^{-13}.

The stresses at the midpoints of Elements 1 and 2 are shown below:

```
               Element 1                Element 2
Stress_xx  -3.40606739229384e-13   -1.22250441648157e-13
Stress_yy  -2.18766106423843e-13    1.60000897584464e-14
Stress_xy   0.999999999999805       0.999999999999904
```

The maximum error is 4×10^{-13}.

Example 3.9 Edge-cracked Circular Body

A circular body under uniform radial tension p is shown in Figure 3.11a. A crack of length a exists at the left edge. The material constants are Young's modulus E, and Poisson's ratio $v = 0.25$. Assume plane stress condition. Determine the crack opening displacement (COD) Δu_y at left edge.

Figure 3.11 Edge-cracked circular body under uniform radial tension. (a) Geometry and boundary conditions. (b) Mesh. The crack faces are modelled by scaling the two nodes at the crack mouth and not discretized.

(a)

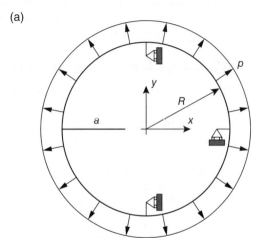

(b)

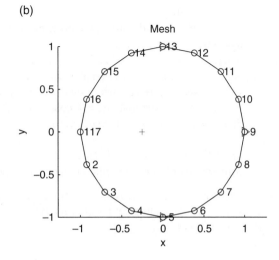

The analysis is performed using the following MATLAB script. The input data are shown at the beginning of the script. The crack opening displacement will be presented as a dimensionless value $\Delta u_y E/pR$. Any choices of Young's modulus E, surface traction p and radius R can be used and they will lead to the same dimensionless result.

The crack length $a = 0.75R$ is considered. For this simple geometry, a uniform mesh is generated automatically on the circle by specifying the number of elements to be used on a quarter of the circle. A mesh with four elements per quarter, i.e. 16 elements on the circle, is shown in Figure 3.11b as an example. There are two independent nodes (Nodes 1 and 17 in Figure 3.11b) at the crack mouth, one on each crack face. The scaling centre is selected at the crack tip. As illustrated in Figure 2.6, the crack faces are generated by scaling these two nodes at the crack mouth towards the scaling centre. No discretization is needed to represent the crack. In this case, the elements form an open loop. Statements to plot the mesh as shown in Figure 3.11b are included.

The radial tension applied at the boundary is in self-equilibrium. Supports are only provided to eliminate the rigid body motions so that the displacement solution is unique. The resulting support reactions are numerically equal to zero.

MATLAB script: Edge-cracked circular body under uniform radial tension

```
1  clearvars; close all; dbstop if error
2
3  % ...  Input
4  R = 1;  % ........................................ radius of circular body
5  p = 1000;  % .................................... radial surface traction (KPa)
6  E = 10E6;  % ............................................. E in KPa
7  nu = 0.25;  % ............................................. Poisson's ratio
8  den = 2;  % ........................................ mass density in Mg/m³
9
10  a = 0.75*R;  % ............................................. crack length
11  nq = 4;  % ............................... number of elements on one quadrant
12
13  % ...  Mesh
14  % ...  nodal coordinates. One node per row [x y]
15  n = 4*nq;  % ............................... number of element on boundary
16  dangle = 2*pi/n;  % .................... angular increment of an element Δθ
17  angle = -pi:dangle:pi+dangle/5;  % ........... angular coordinates of nodes
18  % ...  Note: there are two nodes at crack mouth with the same coordinates
19  % ...  nodal coordinates x = R cos(θ), y = R sin(θ)
20  coord = R*[cos(angle);sin(angle)]';
```

Input S-element connectivity as a cell array (one S-element per cell). In a cell, the connectivity of line elements is given by one element per row [Node-1 Node-2].

```
21  sdConn = { [1:n; 2:n+1]' };  % .......... Note: elements form an open loop
22  % ...  select scaling centre at crack tip
23  sdSC = [a-R 0];
24
```

```
25  % ...  Materials
26  % ...  E: Young's modulus; p: Poisson's ratio
27  ElasMtrx = @(E, p) E/(1-p^2)*[1 p 0;p 1 0;0 0 (1-p)/2];
28  mat.D = ElasMtrx(E, nu);
29  mat.den = den;
30
31  % ...  Boundary conditions
32  % ...  displacement constraints. One constraint per row: [Node Dir Disp]
33  BC_Disp = [  nq+1 1   0; 2*nq+1 2   0; ...
34                3*nq+1 1   0];  % ................. constrain rigid-body motion
35  eleF = R*dangle*p;  % ............. resultant radial force on an element pRΔθ
36  % ...  assemble radial nodal forces {Fr}
37  nodalF = [eleF/2, eleF*ones(1,n-1), eleF/2];
38  % ...  nodal forces. One node per row: [Node Dir F]
39     % ...  Fx = Fr cos(θ), Fy = Fr sin(θ)
40  BC_Frc = [1:n+1 1:n+1; ones(1,n+1) 2*ones(1,n+1); ...
41                nodalF.*cos(angle) nodalF.*sin(angle)]';
42
43  % ...  Plot mesh
44  h1 =   figure;
45  opt=struct('LineSpec','-k', 'sdSC',sdSC, ...
46       'PlotNode',1, 'LabelNode', 1,...
47        'BC_Disp',BC_Disp);  % ............................ plotting options
48  PlotSBFEMesh(coord, sdConn, opt);
49  title('MESH');
50
51  % ...  solution of S-elements and global stiffness and mass matrices
52  [sdSln, K, M] = SBFEMAssembly(coord, sdConn, sdSC, mat);
53
54  % ...  Assemblage external forces
55  ndn = 2;  % ............................................. 2 DOFs per node
56  NDof = ndn*size(coord,1);  % ........................... number of DOFs
57  F = zeros(NDof,1);  % ... initializing right-hand side of equation [K]{u} = {F}
58  F = AddNodalForces(BC_Frc, F);  % .............. add prescribed nodal forces
59
60  % ...  Static solution
61  [U, F] = SolverStatics(K, BC_Disp, F);
62
63  CODnorm = (U(end)-U(2))*E/(p*R);
64  disp(['Normalised crack openning displacement = ', num2str(CODnorm)])
65
66  % ...  plot deformed shape
67  Umax = max(abs(U));  % ........................... maximum displacement
68  fct = 0.2/Umax;  % .............. factor to magnify the displacement to 0.2 m
69  % ...  argument nodal coordinates
70  deformed = coord + fct*(reshape(U,2,[]))';
71  hold on
```

```
72   % ... plotting options
73   opt = struct('LineSpec','-ro', 'LabelNode', 1);
74   PlotSBFEMesh(deformed, sdConn, opt);
75   title('DEFORMED MESH');
76
77   % ... Internal displacements and stresses
78
79   % ... strain modes of S-elements
80   sdStrnMode =  SElementStrainMode2NodeEle( sdSln );
81   % ... integration constants of S-element
82   sdIntgConst = SElementIntgConst( U, sdSln );
83
84   isd = 1;  % ........................................... S-element number
85   xi = (1:-0.01:0).^2;  % ............................... radial coordinates
86   % ... initialization of variables for plotting
87   X = zeros(length(xi), length(sdSln{isd}.node));
88   Y = X; C = X;
89   % ... displacements and strains at the specified radial coordinate
90   for ii= 1:length(xi)
91       [nodexy, dsp, strnNode, GPxy, strnEle] = ...
92           SElementInDispStrain(xi(ii), sdSln{isd},  ...
93                               sdStrnMode{isd}, sdIntgConst{isd});
94       deformed = nodexy + fct*(reshape(dsp,2,[]))';
95       % ... coordinates of grid points
96       X(ii,:) =  deformed(:,1)';
97       Y(ii,:) =  deformed(:,2)';
98       strsNode = mat.D*strnNode;  % ......................... nodal stresses
99       C(ii,:) =  strsNode(2,:);  % ..................... store $\sigma_{yy}$ for plotting
100  end
101
102  % ... plot stress contour
103  h2 = figure('Color','white');
104  contourf(X,Y,C, (-1000:1000:10000), 'LineStyle','none');
105  hold on
106  axis off; axis equal;
107  xlabel('x'); ylabel('x');
108  colormap(jet);colorbar;
```

The crack opening displacement is obtained as the relative displacement between the two nodes at the crack mouth. Owing to symmetry, the relative horizontal displacement is zero. Only the vertical component Δu_y is presented below.

The analysis is performed using four meshes with increasing number of elements at the circular boundary. The dimensionless crack opening displacements ($\Delta u_y E/pR$) are shown in the following table. It is observed that the result converges as the mesh becomes finer. The difference between the results obtained with the last two meshes is less than 1%.

Number of elements	$\Delta u_y E / pR$	$a = 0.75R$ $\lambda_1^{(s)}$	$\lambda_2^{(s)}$
16	11.223	0.49867	0.49732
32	12.279	0.49966	0.49933
64	12.566	0.49992	0.49983
128	12.643	0.49998	0.49996

It is well known that the stresses are infinite (singular) at the crack tip according to linear elasticity theory. The stress field around the crack tip is described by the William's asymptotic expansion (Eq. (10.2) in Section 10.2.1), which is a power series of the distance r to the crack tip. There are two singular stress terms where the stresses are proportional to $r^{-0.5}$ close to the crack tip. Correspondingly, the displacements are proportional to $r^{0.5}$. The scaled boundary finite element solution in Eq. (3.18a) is a power series of the radial coordinate ξ, which is proportional to r (Eq. (2.6a) on page 34. The two eigenvalues of the singular stress terms, denoted as $\lambda_1^{(s)}$ and $\lambda_2^{(s)}$ (superscript 's' for singular), are also shown in the above table. The results converge to the exact value as the number of element increases.

The deformed shape of the boundary of the cracked circular plate is shown in Figure 3.12, where the boundary is divided evenly into 16 elements. The two nodes at the crack mouth separate from each other.

The computation of displacements and stresses at a specified array of radial coordinate ξ starts from Line 77 of the MATLAB script. To describe the rapid variation of stresses at the crack tip, the distribution of ξ is biased towards the scaling centre ($\xi = 0$). A small region within a radial coordinate $\xi < 1 \times 10^{-4}$ is excluded to avoid singular values. The contour of stress σ_{yy} is plotted on the deformed shape of the cracked body in Figure 3.13a. An enlarged view around the crack tip is shown in Figure 3.13b. It is observed that the crack opening close to the crack tip is dominated by the $\sqrt{r}$ term.

Cracks of varying length can easily be represented by moving the scaling centre with the crack tip in the scaled boundary finite element method. For the edge crack in Figure 3.11 but with a length $a = R$, the crack tip and scaling centre are at (0, 0) in the Cartesian coordinate system. It is sufficient to change Line 10 in the above MATLAB script to a = R. The results for the dimensionless crack opening displacement and the two eigenvalues of the singular stress modes are shown in the table below

Figure 3.12 Deformed mesh of edge-cracked circular plate under uniform radial tension.

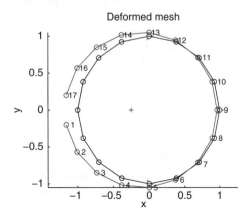

Deformed mesh

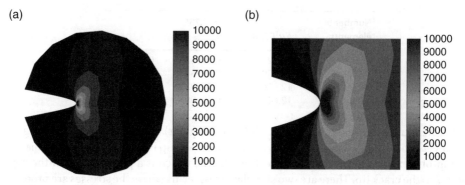

Figure 3.13 (a) Contour of stress σ_{yy} of edge-cracked circular plate under uniform radial tension. (b) Enlarged view around the crack tip.

Number of elements	$\Delta u_y E/pR$	$a = R$ $\lambda_1^{(s)}$	$\lambda_2^{(s)}$
16	25.159	0.49960	0.49930
32	26.837	0.49991	0.49983
64	27.282	0.49998	0.49996
128	27.399	0.49999	0.49999

◼ **Example 3.10   Shape Functions of a Pentagon Element**

The use of the scaled boundary finite element method to construct shape functions, as described by Eqs. (3.72) and (3.73), is demonstrated by considering the pentagon element shown in Figure 3.1, on page 82. A unit displacement $u_y = 1$ is prescribed at Node 4 and all other DOFs are fixed. The displacement field, which is the shape function related to this DOF, is computed and plotted using the following MATLAB script. The following script is written for this task. It uses the functions presented previously.

```
1  % ...  Shape function of a pentagon S-Element
2  clearvars; close all;
3
4  % ...  Mesh
5  % ...  nodal coordinates
6  coord = [cosd(-126:72:180);sind(-126:72:180)]';
7  % ...  connectivity
8  sdConn = { [1:5 ; 2:5 1]' };
9  sdSC = [ 0 0];  % .................................... one S-Element per row
10
11 % ...  Materials
12 % ...  E: Young's modulus; p: Poisson's ratio
13 ElasMtrx = @(E, p) E/(1-p^2)*[1 p 0;p 1 0;0 0 (1-p)/2];
14 mat.D = ElasMtrx(10E6, 0.25);  % .............................. E in KPa
```

```
15  mat.den = 2;  % .................................. mass density in Mg/m³
16
17  % ...  S-Element solution
18  [sdSln, K, M] = SBFEMAssembly(coord, sdConn, sdSC, mat);
19
20  % ...  Shape function
21  % ...  prescribed unit displacement
22  U = zeros(size(K,1),1);
23  U(2*4) = 1;
24
25  % ...  strain modes of S-element
26  sdStrnMode = SElementStrainMode2NodeEle( sdSln );
27
28  % ...  integration constants
29  sdIntgConst = SElementIntgConst( U, sdSln );
30
31  isd = 1;  % ............................................. S-element number
32  xi = 1:-0.01:0;  % ..................................... radial coordinates
33  % ...  initialization of variables for plotting
34  X = zeros(length(xi), length(sdSln{isd}.node)+1);
35  Y = X; Z = X;
36  % ...  displacements and strains at the specified radial coordinate
37  for ii= 1:length(xi)
38      [nodexy, dsp, strnNode, GPxy, strnEle] = ...
39          SElementInDispStrain(xi(ii), sdSln{isd},  ...
40                              sdStrnMode{isd}, sdIntgConst{isd});
41      % ...  coordinates of grid points forming a close loop
42      X(ii,:) =  [nodexy(:,1)' nodexy(1,1)];
43      Y(ii,:) =  [nodexy(:,2)' nodexy(1,2)];
44      Z(ii,:) =  [dsp(2:2:end)' dsp(2)];  % ......... store u_y for plotting
45  end
46
47  % ...  plot the shape function as a surface
48  figure('Color','white')
49  surf(X,Y,Z,'FaceColor','interp', 'EdgeColor','none',  ...
50          'FaceLighting','phong');
51  view(-110, 15);  % ................................. set direction of viewing
52  hold on
53  text(1.1*(coord(:,1)-0.02), 1.1*coord(:,2), ...
54      Z(1,1:end-1)'+0.05,int2str((1:5)'));  % .......... label the nodes
55  axis equal, axis off;
56  xlabel('x'); ylabel('y'); zlabel('N');  % ............... label the axes
57  colormap(jet)
58  plot3(X(1,:), Y(1,:), Z(1,:), '-b');  % ..................... plot edges
59
60  % ...  contour of the shape function
```

```
61  h = figure('Color','white');
62  contourf(X,Y,Z,10, 'LineStyle','none'); % ............. 10 contour lines
63  hold on
64  text(1.05*(coord(:,1)-0.02), 1.05*coord(:,2),int2str((1:5)'));
65  axis equal; axis off;
66  % ...  show a colourbar indicating the value of the shape function
67  colormap(jet)
68  caxis([0 1]); colorbar;
69  plot(X(1,:), Y(1,:), '-b'); % ............................... plot edges
```

The surface and contour plots of the shape function are shown in Figure 3.14. It is C^0 continuous along the circumferential direction and smooth in the radial direction.

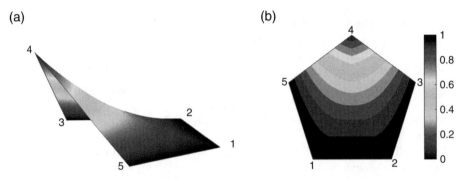

Figure 3.14 Shape function of a regular pentagon element. (a) Surface plot. The locations of the nodal numbers are changed from the MATLAB output for clearer view. (b) Contour plot.

3.9 Body Loads

Body loads have been addressed by Song and Wolf (1999); Deeks (2004); Song (2006). The scaled boundary finite element equation considering body loads is formulated as a system of non-homogeneous differential equations. A close-form solution is obtained. The formulation is able to consider complex distributions of body loads within an S-element. It is especially advantageous in modelling the singular stress field near the crack tips and other singular points. A simple alternative procedure using the shape functions of the S-element is presented in this section to handle simple body load distributions within an S-element. It postulates that the presence of body forces does not alter the basis functions (approximated by shape functions) of deformation, as in the finite element method. This is reasonably accurate when the size of an S-element is small. For complex body loads, subdivision of an S-element into smaller ones is necessary.

The principle of virtual work statement for an S-element in Eq. (2.93) is modified to include the body force $\{b\}$

$$\underbrace{\int_V \{\delta\varepsilon\}^T \{\sigma\} \mathrm{d}V}_{u_\varepsilon} = \{\delta d\}^T \{F\} + \underbrace{\int_V \{\delta u\}^T \{b\} \mathrm{d}V}_{W_B} \qquad (3.75)$$

The virtual increment W_B of the work done by the body force $\{b\}$ is addressed. It is evaluated sector-by-sector

$$W_B = \sum_e \int_{V^e} \{\delta u\}^T \{b\} \mathrm{d}V \tag{3.76}$$

The virtual displacement field in the S-element is obtained by interpolating the nodal values at the boundary (Eq. (3.72)) element-by-element

$$\{\delta u\} = [N_V^e]\{\delta d^e\} \tag{3.77}$$

Substituting Eq. (3.77) into Eq. (3.76) leads to

$$W_B = \sum_e \{\delta d\}_e^T \int_{V^e} [N_V^e]^T \{b\} \mathrm{d}V \tag{3.78}$$

Replacing the summation in Eq. (3.78) with the finite element assembly results in

$$W_B = \{\delta d\}^T \{F_B\} \tag{3.79}$$

where the body load vector, which consists of the nodal forces statically equivalent to the distributed body force, are expressed as

$$\{F_B\} = \sum_e \int_{V^e} [N_V^e]^T \{b\} \mathrm{d}V \tag{3.80}$$

Transposing the shape functions of the S-element in Eq. (3.73) is written as

$$[N_V^e]^T = [\Phi_b^{(u)}]^{-T} \xi^{\langle \lambda_b \rangle} [\Phi_b^{(u)e}]^T [N_u]^T \tag{3.81}$$

Substituting Eq. (3.81) and $\mathrm{d}V$ (Eq. (2.47)) into Eq. (3.80) leads to

$$\{F_B\} = [\Phi_b^{(u)}]^{-T} \int_0^1 \sum_e \int_{-1}^{+1} \xi^{\langle \lambda_b \rangle} [\Phi_b^{(u)e}]^T [N^u]^T \{b\} \xi |J_b| \mathrm{d}\xi \mathrm{d}\eta \tag{3.82}$$

When the spatial distribution of the body force is described by a polynomial function in the Cartesian coordinates x, y, the integration can be performed analytically (Song, 2006). In this section, only the constant distribution of body force is considered. The integrations in Eq. (3.82) with respect to the radial coordinates ξ and circumferential coordinates η can be separated. The body force vector is rewritten as

$$\{F_B\} = [\Phi_b^{(u)}]^{-T} \int_0^1 \xi^{\langle \lambda_b \rangle} \xi \mathrm{d}\xi [\Phi_b^{(u)}]^T \sum_e \int_{-1}^{+1} [N_u]^T |J_b| \mathrm{d}\eta \{b\} \tag{3.83}$$

where $[\Phi_b^{(u)}] = \sum_e [\Phi_b^{(u)e}]$ has been used. Performing the integration with respect to ξ analytically results in

$$\{F_B\} = [\Phi_b^{(u)}]^{-T} \left\langle \frac{1}{\lambda_b + 2} \right\rangle [\Phi_b^{(u)}]^T \{F_{B0}\} \tag{3.84}$$

with the coefficient vector of body force

$$\{F_{B0}\} = \sum_{e} \int_{-1}^{+1} [N_u]^T |J_b| d\eta \{b\} \qquad (3.85)$$

$\{F_{B0}\}$ is evaluated element-by-element at the boundary and assembled for the S-element. This operation is similar to the calculation of equivalent nodal forces to surface tractions in the finite element method.

An alternative way to consider a body load is to treat it as an inertial force caused by a time-invariant acceleration. The equivalent nodal forces can be obtained as the product of the mass matrix (obtained in Section 3.10.1) with the nodal accelerations.

As the shape functions are obtained as the solutions of the scaled boundary finite element equation, they satisfy the equation of equilibrium in Eq. (2.106). The virtual increment of the strain energy U_ε is given in Eq. (3.26). The virtual increments of the work done by the nodal forces are given in Eq. (3.79). Substituting them into Eq. (3.75), the principle of virtual work is expressed as

$$\{\delta d\}^T [K]\{d\} = \{\delta d\}^T \{F\} + \{\delta d\}^T \{F_B\} \qquad (3.86)$$

Considering that the virtual displacements are arbitrary, the standard finite element equation is obtained for the S-element

$$[K]\{d\} = \{F\} + \{F_B\} \qquad (3.87)$$

The assembly of the S-element equations (Eq. (3.87)) follows the procedure in Section 3.4.1. The body load vector is accumulated to load vector $\{F_G\}$ and the global system of equations in Eq. (3.38) remains the same.

■ **Example 3.11**   Consider the 2-node element shown in Figure 2.17 in Example 2.1. Evaluate the coefficient vector of body force (Eq. (3.85)).

Using Eq. (2.80), the shape functions $[N_u]$ of a 2-node element are expressed as

$$[N_u] = \begin{bmatrix} N_1 & 0 & N_2 & 0 \\ 0 & N_1 & 0 & N_2 \end{bmatrix} \qquad (3.88)$$

with (Eq. (2.16))

$$N_1 = \frac{1}{2}(1 - \eta) \qquad (3.89a)$$

$$N_2 = \frac{1}{2}(1 + \eta) \qquad (3.89b)$$

The Jacobian matrix $|J_b|$ (Eq. (2.58)) is independent of η for the 2-node element

$$|J_b| = \frac{1}{2}(x_1 y_2 - x_2 y_1) \qquad (3.90)$$

Considering the integrations

$$\int_{-1}^{+1} N_1 d\eta = \frac{1}{2} \int_{-1}^{+1} (1 - \eta) d\eta = 1 \qquad (3.91a)$$

$$\int_{-1}^{+1} N_2 d\eta = \frac{1}{2} \int_{-1}^{+1} (1 + \eta) d\eta = 1 \qquad (3.91b)$$

The vector $\{F_{B0}\}$ of the 2-node element for a body force $\{b\} = [b_x, b_y]^T$ is obtained as

$$\{F_{B0}\} = |J_b| \begin{bmatrix} b_x & 0 \\ 0 & b_y \\ b_x & 0 \\ 0 & b_y \end{bmatrix} \tag{3.92}$$

3.10 Dynamics and Vibration Analysis

Using the shape functions of S-elements in Eq. (3.72), standard finite element procedures are directly applicable to perform dynamic analysis. The aim of this section is to present the derivation of the mass matrix of an S-element and the same equation of motion as in the finite element method (Section 3.10.1). To illustrate the application of the scaled boundary finite element formulation, two MATLAB functions are provided. One function determines the natural frequencies and mode shapes (Section 3.10.2) and the other performs the time-history analysis using the Newmark method (Section 3.10.3).

3.10.1 Mass Matrix and Equation of Motion

The scaled boundary finite element method has also been derived for elastodynamics in Song and Wolf (1997). A solution procedure based on a continued fraction solution of the dynamic stiffness matrix is developed in Birk et al. (2014). This procedure can model high-frequency responses by increasing the order of continued fractions without meshing the interior of an S-element. In this section, a simple alternative based on standard finite element technique is presented. It is only valid when the S-element size is a fraction of a wavelength, i.e. at low frequencies. To model high-frequency responses of a problem, the mesh of S-elements has to be fine enough to represent the wave pattern.

The principle of virtual work statement in Eq. (2.93) is augmented to include the inertial forces $\rho\{\ddot{u}\}$ (see Eq. (A.10))

$$\underbrace{\int_V \{\delta\varepsilon\}^T \{\sigma\} dV}_{u_\varepsilon} + \underbrace{\int_V \{\delta u\}^T \rho\{\ddot{u}\} dV}_{u_\rho} = \{\delta d\}^T \{F\} \tag{3.93}$$

The increment of the virtual work of the inertial force U_ρ in Eq. (3.75) is determined sector-by-sector

$$U_\rho = \sum \int_{V^e} \{\delta u\}^T \rho\{\ddot{u}\} dV \tag{3.94}$$

The acceleration $\{\ddot{u}\}$ in the S-element is interpolated using the same shape functions as the displacement interpolation (Eq. (3.72))

$$\{\ddot{u}\} = [N_V^e]\{\ddot{d}\} \tag{3.95}$$

Substituting Eqs. (3.95) and (3.81) into Eq. (3.94) leads to

$$U_\rho = \sum_e \{\delta d^e\}^T \int_{V^e} [N_V^e]^T \rho [N_V^e] \mathrm{d}V \{\ddot{d}^e\} \tag{3.96}$$

Assembling the contributions from all the sectors, Eq. (3.96) is expressed as

$$U_\rho = \{\delta d\}^T [M] \{\ddot{d}\} \tag{3.97}$$

where the mass matrix

$$[M] = \sum_e \int_{V^e} [N_V^e]^T \rho [N_V^e] \mathrm{d}V \tag{3.98}$$

is introduced. Substituting Eqs. (3.26) and (3.97) into Eq. (3.93), the principle of virtual work is expressed as

$$\{\delta d\}^T [K] \{d\} + \{\delta d\}^T [M] \{\ddot{d}\} = \{\delta d\}^T \{F\} \tag{3.99}$$

Considering that the virtual displacements are arbitrary, Eq. (3.99) leads to the equation of motion of an S-element

$$[K]\{d\} + [M]\{\ddot{d}\} = \{F\} \tag{3.100}$$

The mass matrix of an S-element is determined by substituting $[N_V^e]^T$ in Eq. (3.81), $[N_V^e]$ in Eq. (3.73) and $\mathrm{d}V$ in Eq. (2.47) into Eq. (3.98) leading to

$$[M] = [\Phi_b^{(u)}]^{-T} \int_0^1 \xi^{\langle \lambda_b \rangle} \sum_e [\Phi_b^{(u)e}]^T \int_{-1}^{+1} [N_u]^T \rho [N_u] \xi |J_b|$$

$$\times \mathrm{d}\eta [\Phi_b^{(u)e}] \xi^{\langle \lambda_b \rangle} \mathrm{d}\xi [\Phi_b^{(u)}]^{-1}$$

$$= [\Phi_b^{(u)}]^{-T} \int_0^1 \xi^{\langle \lambda_b \rangle} [\Phi_b^{(u)}]^T [M_0][\Phi_b^{(u)}] \xi^{\langle \lambda_b \rangle} \xi \mathrm{d}\xi [\Phi_b^{(u)}]^{-1} \tag{3.101}$$

where the coefficient matrix $[M_0]$ is introduced

$$[M_0] = \sum_e [M_0^e]; \qquad \text{with} \quad [M_0^e] = \int_{-1}^{+1} [N_u]^T \rho [N_u] |J_b| \mathrm{d}\eta \tag{3.102}$$

The element coefficient matrix $[M_0^e]$ is symmetric and positive definite. The coefficient matrix $[M_0]$ of the S-element is obtained by assembling the element coefficient matrices $[M_0^e]$ according to the element connnectivity. To be able to perform the integration with respect to ξ in Eq. (3.101) analytically, the abbreviation is introduced

$$[m_0] = [\Phi_b^{(u)}]^T [M_0][\Phi_b^{(u)}] \tag{3.103}$$

The integration is rewritten as

$$[m] = \int_0^1 \xi^{\langle \lambda_b \rangle} [m_0] \xi^{\langle \lambda_b \rangle} \xi \mathrm{d}\xi \tag{3.104}$$

and Eq. (3.101) is expressed as

$$[M] = [\Phi_b^{(u)}]^{-T} [m][\Phi_b^{(u)}]^{-1} \tag{3.105}$$

Each entry of the matrix $[m]$ (Eq. (3.104)) is evaluated analytically, resulting in

$$
m_{ij} = \int_0^1 \xi^{\lambda_{bi}} m_{0ij} \xi^{\lambda_{bj}} \xi \mathrm{d}\xi
$$

$$
= \frac{m_{0ij}}{\lambda_{bi} + \lambda_{bj} + 2}
\tag{3.106}
$$

As the real parts of all the eigenvalues are non-negative, the magnitude of denominator $|\lambda_{bi} + \lambda_{bj} + 2| \geq 2$ applies. Like $[M_0]$, $[m]$ is symmetric and positive definite. After determining the matrix $[m]$, the mass matrix $[M]$ follows from Eq. (3.105) and is also symmetric and positive definite.

The equation of motion of the global system is constructed by assembling the equations of motion of the S-elements (Eq. (3.100)) as in the standard finite element procedure

$$
[K_G]\{d_G\} + [M_G]\{\ddot{d}_G\} = \{F_G\}
\tag{3.107}
$$

where $[M_G]$ is the global mass matrix obtained by assembling the mass matrices of the S-elements. All external forces are converted to equivalent nodal forces and stored in the external force vector $\{F_G\}$.

The dynamic response of the system under an external excitation is determined by solving the global equation of motion with prescribed boundary conditions. For simplicity in notation, the subscript G in Eq. (3.107) is omitted in the following. To enforce the prescribed nodal displacements, the nodal displacement vector is partitioned into $\{d_1\}$ containing unknown nodal displacements and $\{d_2\}$ containing the prescribed nodal displacements. Partitioning the stiffness matrix $[K]$, the mass matrix $[M]$ and the external force vector $\{F\}$ conformably, Eq. (3.107) is expressed as

$$
\begin{bmatrix} [K_{11}] & [K_{12}] \\ [K_{21}] & [K_{22}] \end{bmatrix} \left\{ \begin{array}{c} \{d_1\} \\ \{d_2\} \end{array} \right\} + \begin{bmatrix} [M_{11}] & [M_{12}] \\ [M_{21}] & [M_{22}] \end{bmatrix} \left\{ \begin{array}{c} \{\ddot{d}_1\} \\ \{\ddot{d}_2\} \end{array} \right\} = \left\{ \begin{array}{c} \{F_1\} \\ \{F_2\} \end{array} \right\}
\tag{3.108}
$$

Moving the terms with the known displacements $\{u_2\}$ to the right-hand side provides

$$
\begin{bmatrix} [K_{11}] \\ [K_{21}] \end{bmatrix} \{d_1\} + \begin{bmatrix} [M_{11}] \\ [M_{21}] \end{bmatrix} \{\ddot{d}_1\} = \left\{ \begin{array}{c} \{F_1\} \\ \{F_2\} \end{array} \right\} - \begin{bmatrix} [K_{12}] \\ [K_{22}] \end{bmatrix} \{d_2\} - \begin{bmatrix} [M_{12}] \\ [M_{22}] \end{bmatrix} \{\ddot{d}_2\}
\tag{3.109}
$$

The unknown displacements $\{u_1\}$ can be obtained by solving the equations in the first row block

$$
[K_{11}]\{d_1\} + [M_{11}]\{\ddot{d}_1\} = \{F_1\} - [K_{12}]\{d_2\} - [M_{12}]\{\ddot{d}_2\}
\tag{3.110}
$$

which is a system of 2nd order nonhomogeneous ordinary differential equations.

■ **Example 3.12** Consider the 2-node element shown in Figure 2.17 in Example 2.1. Evaluate the element coefficient matrix $[M_0^e]$ in Eq. (3.102).

Assuming that the mass density is constant, the element coefficient matrix is expressed as

$$
[M_0^e] = \rho|J_b| \int_{-1}^{+1} [N_u]^T [N_u] \mathrm{d}\eta
\tag{3.111}
$$

with $|J_b|$ given in Eq. (3.90). The integration $\int_{-1}^{+1}[N_u]^T[N_u]d\eta$ has been performed in Eqs. (3.47) and (3.48). The coefficient matrix of mass is obtained as

$$[M_0^e] = \frac{1}{3}\rho|J_b| \begin{bmatrix} 2 & 0 & 1 & 0 \\ 0 & 2 & 0 & 1 \\ 1 & 0 & 2 & 0 \\ 0 & 1 & 0 & 2 \end{bmatrix} \tag{3.112}$$

It is easy to find out that the coefficient matrix $[M_0^e]$ in Eq. (3.112) is symmetric and positive definite. The element coefficient matrix $[M_0^e]$ is determined by Line 31 in the function EleCoeff2NodeEle.m (Code List 2.1 on page 64).

Example 3.13 A square S-element is shown in Figure 3.15. Its boundary is divided into five 2-node line elements. The dimensions (unit: metre) are given in Figure 3.15. Assume that the mass density is $\rho = 2000 \text{ kg/m}^3$. Determine the mass matrix.

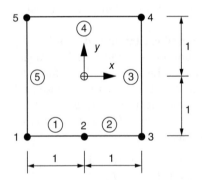

Figure 3.15 Five-node square S-element.

The mass matrix is computed using the following script. The input of the nodal coordinates and the connectivity of the 2-node line elements at the boundary follows the nodal and element numbers shown in Figure 3.15.

```
%% Compute mass matrix of a 5-node square S-element
close all; clearvars; dbstop error;

xy = [-1 -1; 0 -1; 1 -1; 1 1; -1 1];
conn = [1:5 ; 2:5 1]';
% elascity matrix (plane stress).
% E: Young's modulus; p: Poisson's ratio
ElasMtrx = @(E, p) E/(1-p^2)*[1 p 0; p 1 0; 0 0 (1-p)/2];
mat.D = ElasMtrx(10, 0); % E in GPa
mat.den = 2000; % mass density in kg per cubic metre
% coefficient matrices
[E0, E1, E2, M0 ] = SElementCoeffMtx(xy, conn, mat);
% S-element solution by eigenvalue method
[ ~, ~, ~, M] = SElementSlnEigenMethod(E0, E1, E2, M0);
```

```
sum(sum(M(1:2:end,1:2:end)))
sum(sum(M(2:2:end,2:2:end)))
```

The above MATLAB script can be obtained by scanning the QR-code to the right.

This script calls the function `EleCoeff2NodeEle.m` to calculate the element coefficient matrices and assemble them to form the coefficient matrices of the S-element. For example, the mass coefficient matrix $[M_0^e]$ is

666.67	0	333.33	0
0	666.67	0	333.33
333.33	0	666.67	0
0	333.33	0	666.67

for Element 1, and

1333.3	0	666.67	0
0	1333.3	0	666.67
666.67	0	1333.3	0
0	666.67	0	1333.3

for Element 3.

Assembling the element matrices, the mass coefficient matrix $[M_0]$ of the S-element is obtained as

$$[M_0] = \frac{1000}{3}\begin{bmatrix} 6 & 0 & 1 & 0 & 0 & 0 & 0 & 0 & 2 & 0 \\ 0 & 6 & 0 & 1 & 0 & 0 & 0 & 0 & 0 & 2 \\ 1 & 0 & 4 & 0 & 1 & 0 & 0 & 0 & 0 & 0 \\ 0 & 1 & 0 & 4 & 0 & 1 & 0 & 0 & 0 & 0 \\ 0 & 0 & 1 & 0 & 6 & 0 & 2 & 0 & 0 & 0 \\ 0 & 0 & 0 & 1 & 0 & 6 & 0 & 2 & 0 & 0 \\ 0 & 0 & 0 & 0 & 2 & 0 & 8 & 0 & 2 & 0 \\ 0 & 0 & 0 & 0 & 0 & 2 & 0 & 8 & 0 & 2 \\ 2 & 0 & 0 & 0 & 0 & 0 & 2 & 0 & 8 & 0 \\ 0 & 2 & 0 & 0 & 0 & 0 & 0 & 2 & 0 & 8 \end{bmatrix}$$

The mass matrix is computed by the function $\texttt{SElementSlnEigenMethod.m}$

$$[M] = \begin{bmatrix}
610 & 77 & 211 & 20 & 153 & -11 & 182 & 71 & 391 & 15 \\
77 & 505 & -0 & 254 & 11 & 48 & 63 & 122 & 24 & 331 \\
211 & -0 & 296 & -0 & 211 & 0 & 94 & 32 & 94 & -32 \\
20 & 254 & -0 & 544 & -20 & 254 & 47 & 214 & -47 & 214 \\
153 & 11 & 211 & -20 & 610 & -77 & 391 & -15 & 182 & -71 \\
-11 & 48 & 0 & 254 & -77 & 505 & -24 & 331 & -63 & 122 \\
182 & 63 & 94 & 47 & 391 & -24 & 896 & 86 & 438 & -0 \\
71 & 122 & 32 & 214 & -15 & 331 & 86 & 896 & 0 & 437 \\
391 & 24 & 94 & -47 & 182 & -63 & 438 & 0 & 896 & -86 \\
15 & 331 & -32 & 214 & -71 & 122 & -0 & 437 & -86 & 896
\end{bmatrix}$$

Summing up the entries of $[M]$ in the x–direction ($\texttt{sum(sum(M(1:2:end,1:2:end))))}$) and y–direction ($\texttt{sum(sum(M(2:2:end,2:2:end))))}$), respectively, leads to the mass of the square S-element 8000 kg/m (i.e. 8000 kg per unit thickness).

■

3.10.2 Natural Frequencies and Mode Shapes

The system of homogeneous ordinary differential equations associated with the equation of motion (Eq. (3.110)) is addressed

$$[K]\{d\} + [M]\{\ddot{d}\} = 0 \tag{3.113}$$

The subscripts are omitted for simplicity. Only unconstrained (free) degrees of freedom are retained in this equation.

The harmonic vibration is considered

$$\{d\} = \{D\} \sin \omega t \tag{3.114a}$$
$$\{\ddot{d}\} = -\omega^2 \{D\} \sin \omega t \tag{3.114b}$$

where $\{D\}$ denotes the amplitudes of the nodal displacements. ω is the circular frequency of the vibration. Substituting Eq. (3.114) into Eq. (3.113) yields a generalized eigenvalue problem

$$([K] - \lambda[M])\{D\} = 0 \tag{3.115}$$

where the eigenvalues λ leads to the natural frequencies

$$\omega = \sqrt{\lambda} \tag{3.116}$$

Only the positive roots are selected. The eigenvector $\{D\}$ represents the corresponding mode shapes. The period of free vibration is equal to $T = 2\pi/\omega$.

A MATLAB function $\texttt{SolverMode.m}$ written to demonstrate the computation of the natural frequencies and mode shapes is given in Code List 3.12. The input arguments are the number of modes ($\texttt{NMode}$) required for outputs, the stiffness matrix ($\texttt{K}$), the mass matrix ($\texttt{M}$) and the prescribed displacement boundary condition ($\texttt{BC\_Disp}$). When only a small portion (arbitrarily chosen as $< 1/20$) of the modes of large systems (arbitrarily chosen as the number of DOFs larger than 500) is required, the MATLAB function $\texttt{eigs}$ is called to compute the smallest NMode eigenvalues and the corresponding eigenvectors. The natural frequencies are sorted in ascending order. The output arguments are the first NMode smallest natural frequencies and the corresponding mode shapes.

Code List 3.12: Natural frequencies and mode shapes

```
1  function [freq, modeShape] = SolverMode(NMode, K, M, BC_Disp)
2  %Compute natural frequencies and mode shapes
3
4  ndn = 2; % ................................. number of DOFs per node
5
6  NDof = size(K,1); % ................................. number of DOFs
7  % ... find unconstrained (free) DOFs from displacement boundary condition
8  % ... initialization of free DOFs with unknown displacements
9  FDofs = 1:NDof;
10 if ~isempty(BC_Disp)
11     % ... constrained DOFs with prescribed displacements
12     CDofs = (BC_Disp(:,1)-1)*ndn +BC_Disp(:,2);
13     FDofs(CDofs) = []; % .................... remove constrained DOFs
14 end
15
16 % ... solve eigenproblem considering unconstrained DOFs only, see
   Eq. (3.115)
17 if NDof > 20*NMode && NDof > 500 % . compute the lowest NMode modes
18     [modeShape, freq] = eigs(K(FDofs,FDofs),...
19                              M(FDofs,FDofs), NMode, 'sm');
20 else % ............................................. compute all modes
21     [modeShape, freq] = eig(full(K(FDofs,FDofs)),...
22                              full(M(FDofs,FDofs)));
23 end
24
25 % ... sort eigenfrequencies (square-root of eigenvalues) in ascending order
26 [freq, idx] = sort(sqrt(diag(freq)),'ascend'); % ....... Eq. (3.116)
27 % ... arrange eigenvectors in the same order
28 mtmp = modeShape(:,idx(1:NMode));
29 % ... initialize mode shapes to include constrained DOFs
30 modeShape = zeros(NDof,NMode);
31 modeShape(FDofs,:) = mtmp; % .............. fill in unconstrained DOFs
32 freq = freq(1:NMode); % ............. output selected natural frequencies
```

⬛ **Example 3.14** Determine the first two natural frequencies and mode shapes of the deep beam shown in Figure 3.6 on page 105, Example 3.5. Assume that the mass density is $\rho = 2000 \, \text{kg/m}^3$.

The analysis is performed by the following script. The problem definition is imported by running the script ExmplDeepBeam.m. The stiffness and mass matrices are computed by calling the function SBFEMAssembly.m. The function Solver-Mode.m finds the first two modes. Their natural frequencies are equal to 512 rad/s and 1779 rad/s, respectively. The corresponding periods are equal to 0.01227 s and 0.00353 s. The two mode shapes are shown in Figure 3.16. The first mode is characterized by bending deformation and the second one by axial deformation.

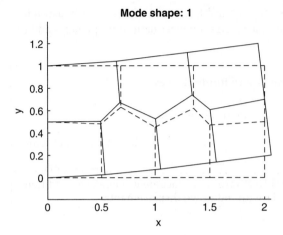

Mode shape: 1

Figure 3.16 Mode shapes of deep beam.

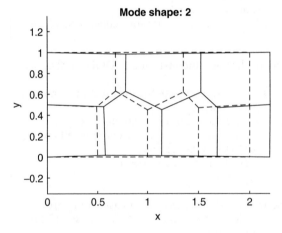

Mode shape: 2

```
%Modal analysis of cantilever beam
close all; clearvars; dbstop error;

% input problem definition
ExmplDeepBeam

% solution of S-elements and assembly of global stiffness
% and mass matrices
[sdSln, K, M] = SBFEMAssembly(coord, sdConn, sdSC, mat);

% natural frequencies and mode shapes
NMode = 4;
[freqAngular, modeShape] = SolverMode(NMode, K, M, BC_Disp);

fprintf('%6s %15s %15s\n', ' Mode', 'Frequency', 'Period')
for imode = 1:NMode %ltx mode number
    figure
    fprintf('%6d %15.7d %15.7d\n', imode, ...
```

```
        freqAngular(imode), 2*pi./freqAngular(imode));
    opt = struct('MagnFct', 0.1, 'Undeformed','--k');
    d = modeShape(:,imode); % use mode shape as displacements
    PlotDeformedMesh(d, coord, sdConn, opt)
    title(['MODE SHAPE: ',int2str(imode)]);
end
```

The above MATLAB script can be obtained by scanning the QR-codes to the right.

3.10.3   Response History Analysis Using the Newmark Method

The dynamic response history of a system is obtained by direct integration of Eq. (3.110) in time. The static stiffness matrix obtained by the scaled boundary finite element method is symmetric and semi-positive definite and the mass matrix is symmetric and positive definite. In principle, any scheme of time integration in the finite element method can be applied. A simple implementation of the Newmark method is given in the function `SolverNewmark.m` in the Code List 3.13 below.

The predictor-corrector scheme of the Newmark time integrator with the parameters γ and β is adopted. Denoting the size of the time step as Δ_t, Eq. (3.110) at the nth time step $t_n = n\Delta_t$ is expressed as

$$[K_{11}]\{d_1\}_n + [M_{11}]\{\ddot{d}_1\}_n = \{F_1\}_n - [K_{12}]\{d_2\}_n - [M_{12}]\{\ddot{d}_2\}_n \qquad (3.117)$$

The system is assumed to be initially at rest, i.e., $\{d\}_0 = \{\dot{d}\}_0 = 0$.

The displacements $\{d\}_n$ and velocity $\{\dot{d}\}_n$ are given by the correctors

$$\{d\}_n = \{\tilde{d}\}_n + \beta\Delta_t^2\{\ddot{d}\}_n \qquad (3.118a)$$

$$\{\dot{d}\}_n = \{\tilde{\dot{d}}\}_n + \gamma\Delta_t\{\ddot{d}\}_n \qquad (3.118b)$$

where $\{\tilde{d}\}_n$ and $\{\tilde{\dot{d}}\}_n$ are the predictors of the displacements and velocities, respectively

$$\{\tilde{d}\}_n = \{d\}_{n-1} + \Delta_t\{\dot{d}\}_{n-1} + (0.5 - \beta)\Delta_t^2\{\ddot{d}\}_{n-1} \qquad (3.119a)$$

$$\{\tilde{\dot{d}}\}_n = \{\dot{d}\}_{n-1} + (1 - \gamma)\Delta_t\{\ddot{d}\}_{n-1} \qquad (3.119b)$$

which are evaluated from the responses at the previous time station $(n - 1)$. The parameters γ and β control the stability of the step-by-step integration and the amount of numerical damping introduced into the system by the method. When $\gamma = 1/2$ and $\beta = 1/4$, the predictor-corrector scheme in Eqs. (3.118) and (3.119) is equivalent to the constant average acceleration method.

Substituting Eq. (3.118) into Eq. (3.117) and moving the known terms at time station n to the right-hand side provides a system of algebraic equations

$$\left(\beta\Delta_t^2[K_{11}] + [M_{11}]\right)\{\ddot{d}_1\}_n = \{F_1\}_n - [K_{11}]\{\tilde{d}_1\}_n - [K_{12}]\{d_2\}_n - [M_{12}]\{\ddot{d}_2\}_n$$

(3.120)

which is solved for the unknown accelerations $\{\ddot{d}_1\}_n$. The displacements $\{d_1\}_n$ and velocities $\{\dot{d}\}_n$ are then obtained from Eq. (3.118).

Initially (at time station $n = 0$), if external forces $\{F_1\}_0$ or prescribed accelerations $\{\ddot{d}_2\}_0$ exist, the initial accelerations $\{\ddot{d}_1\}_0$ are obtained by solving

$$[M_{11}]\{\ddot{d}_1\}_0 = \{F_1\}_0 - [M_{12}]\{\ddot{d}_2\}_0$$

(3.121)

The above algorithm is implemented in the function `SolverNewmark.m` in the Code List 3.13. The documentation of the code is included in the list. The input argument `TIMEPara` is an array. The first two entries are the number of time steps to be integrated and the size of the time step. The last two entries are optional. They change the default values of the parameters $\gamma = 0.5$ and $\beta = 0.25$ if specified. The argument `TIMEPara` is an array containing discrete points of a time-dependent factor to be multiplied to the external loadings. At a specified time, the time-dependent factor is obtained by linearly interpolating the data points in `forceHistory`. The displacement boundary condition is specified as nodal accelerations instead of nodal displacements. The rest of the inputs (stiffness matrix K, mass matrix M and external forces F) are the same as those of the function `SolverStatics.m` for static analysis. The outputs consist of the times (argument `t`), displacement responses (argument `dsp`), velocity responses (argument `vel`), and acceleration responses (argument `accl`).

Code List 3.13: Response history analysis using the Newmark method

```
1   function [t, dsp, vel, accl] = SolverNewmark(TIMEPara, ...
2                               forceHistory, K, M, BC_Accl, F)
3   %Newmark time integrator
4   %
5   %Inputs:
6   %    TIMEPara      : array containing parameters for Newmark method
7   %                    TIMEPara(1): number of time steps
8   %                    TIMEPara(2): size of time step
9   %                    TIMEPara(3): gamma, optional (default is 0.5)
10  %                    TIMEPara(4): beta, optional (default is 0.25)
11  %    forceHistory : external action at a given time is multiplied
12  %                    with a factor obtained by interpolating the array
13  %                    [0 f0; t1 f1; t2 f2, t3 f3, ...]
14  %    K             : stiffness matrix
15  %    M             : mass matrix
16  %    BC_Accl       : prescribed nodal accelerations in rows of
17  %                    [Node, Dir. (=1 for x; =2 for y), Acceleration]
18  %    F             : load vector
19  %
20  %Outputs:
21  %    t             : time
22  %    dsp           : response of nodal displacements
```

```
23  %     vel           : response of nodal velocity
24  %     accl          : response of nodal acceleration
25
26  ns = TIMEPara(1);  % ............................. number of time steps
27  dt = TIMEPara(2);  % ................................ size of time step
28  if length(TIMEPara)>2  % ........................... parameters γ and β
29      gamma = TIMEPara(3);
30      beta  = TIMEPara(4);
31  else  % ..................................................... default values
32      gamma = 0.5;
33      beta  = 0.25;
34  end
35
36  ndn = 2;  % ........................................... 2 DOFs per node
37  NDof = size(K,1);  % ................................... number of DOFs
38
39  % ...   displacement boundary condition
40  CDofs = [];  % ........................... initializing constrained DOFs
41  if ~isempty(BC_Accl)
42      % ...   constrained DOFs with prescribed accelerations
43      CDofs = (BC_Accl(:,1)-1)*ndn + BC_Accl(:,2);
44  end
45  FDofs    = (1:NDof);  % ............. initializing unconstrained (free) DOFs
46  FDofs(CDofs) = [];  % ......................... remove constrained DOFs
47
48  t = (0:ns)*dt;  % ..................... time at all time station 0, $\Delta_t$, $2\Delta_t$, ...
49  % ...   load factor at all time station by linear inter- or extra-polation
50  ft = interp1(forceHistory(:,1), forceHistory(:,2), t,...
51               'linear', 'extrap' );
52
53  % ...   initializing variables storing response history for output
54  dsp = zeros(NDof,ns+1);  % ................................ displacements
55  vel = dsp;  % ................................................. velocities
56  accl = dsp;  % .............................................. accelerations
57
58  % ...   initializing variables
59  dn = zeros(NDof,1);  % ................................... displacements
60  vn = dn;  % ................................................... velocities
61  an = dn;  % ................................................ accelerations
62
63  % ...   initial time step (t = 0)
64  if abs(ft(1)) > 1.d-20
65      % ...   balance initial forces with inertial forces
66      f = ft(1)*F;  % ......................... external nodal forces at $n = 0$
67      an(CDofs) = ft(1)*BC_Accl(:,3);  % ......... prescribed accelerations
68      if ~isempty(CDofs)
69          % ...   enforcing prescribed accelerations, Eq. (3.121)
70          f = f - M(:,CDofs)*an(CDofs);
71      end
72      % ...   solution of accelerations at unconstrained DOFs, Eq. (3.121)
73      an(FDofs) = (M(FDofs,FDofs)\f(FDofs));
```

```
74   end
75
76   % ... working variables, Eqs. (3.118) and (3.119)
77   fg1 = gamma*dt;
78   fg2 = (1.d0-gamma)*dt;
79   fb1 = beta*dt*dt;
80   fb2 = (0.5d0-beta)*dt*dt;
81
82   % ... dynamic-stiffness matrix of unconstrained DOFs
83   % ... βΔ²ₜ[K₁₁] + [M₁₁], Eq (3.120)
84   dstf = M(FDofs,FDofs) + fb1*K(FDofs,FDofs);
85   % ... Cholesky factorization of dynamic stiffness matrix
86   [Ldstf, ~, s] = chol(dstf,'lower','vector');
87
88   tmp = zeros(length(FDofs),1);
89   for it = 2:ns+1
90       dn = dn + dt*vn + fb2*an;  % ..... displacement predictor, Eq. (3.119a)
91       vn = vn + fg2*an;  % ................... velocity predictor, Eq. (3.119a)
92       f = ft(it)*F;  % ............................. external nodal forces
93       if ~isempty(CDofs)
94           an(CDofs) = ft(it)*BC_Accl(:,3);  % .... prescribed accelerations
95           % ... correctors for constrained DOFs, Eq. (3.119)
96           vn(CDofs) = vn(CDofs) + fg1*an(CDofs);  % ........... velocities
97           dn(CDofs) = dn(CDofs) + fb1*an(CDofs);  % ....... displacements
98           % ... add −[M₁₂]{ü₂}ₙ to right-hand side, Eq. (3.120)
99           f = f - M(:,CDofs)*an(CDofs);
100      end
101      % ... add −[K₁₁]{ũ₁}ₙ − [K₁₂]{u₂}ₙ to right-hand side, Eq. (3.120)
102      f = f - K*dn;
103      f = f(FDofs);  % ....................... right-hand side of Eq. (3.120)
104
105      % ... solve Eq. (3.120) for unknown accelerations
106      tmp(s) = Ldstf'\(Ldstf\(f(s)));
107      an(FDofs) = tmp;  % .... update acceleration vector containing all DOFs
108      % ... correctors, Eq. (3.118)
109      vn(FDofs) = vn(FDofs) + fg1*an(FDofs);  % ........... displacements
110      dn(FDofs) = dn(FDofs) + fb1*an(FDofs);  % ............... velocities
111
112      % ... store responses for output
113      dsp(:,it) = dn;
114      vel(:,it) = vn;
115      accl(:,it) = an;
116
117  end
```

Example 3.15 Compute the response of the deep beam shown in Figure 3.6 on page 105, Example 3.5. The time history of the external force couple is given by $F(t) = f(t) \times 100\text{kN/m}$, where the time variation factor $f(t)$ is depicted in Figure 3.17.

The analysis is performed using the MATLAB script below. The units are metres for length, Newton per unit thickness for force and seconds for time. The problem definition

Figure 3.17 Time variation factor $f(t)$ of force couple applied on deep beam.

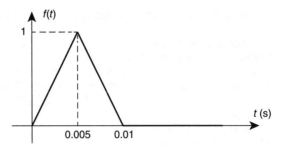

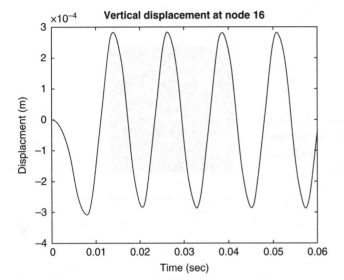

Figure 3.18 Response history of deep beam: vertical displacement at Node 16.

is imported by the MATLAB script ExmplDeepBeam.m on page 105. The computation of the stiffness and mass matrices of the S-elements and the assembly of the global system of equations are performed by the function SBFEMAssembly.m in Code List 3.2 on page 88.

The time variation factor plotted in Figure 3.17 is specified by the turning points stored in the variable forceHistory. The size of the time step is chosen as 0.0002 s. The Newmark time integration is performed over 300 time steps. The response of the vertical displacement of Node 16 is plotted in Figure 3.18.

```
%Response history analysis of cantilever beam
clearvars; close all; dbstop if error

% input problem definition
ExmplDeepBeam

% solution of S-elements and assembly of global stiffness
% and mass matrices
```

```
[sdSln, K, M] = SBFEMAssembly(coord, sdConn, sdSC, mat);

% response history
TIMEPara = [300 0.0002];
forceHistory = [0 0; 0.005 1; 0.01 0; 100 0];
[tn, dsp, vel, accl] = SolverNewmark(TIMEPara,forceHistory, ...
                        K, M, BC_Disp, F);
figure
plot(tn,accl(32,:),'-k')
xlabel('Time (sec)');
ylabel('Displacment (m)');
title('VERTICAL DISPLACEMENT AT NODE 16');
```

The above MATLAB script can be obtained by scanning the QR-code to the right.

4

Automatic Polygon Mesh Generation for Scaled Boundary Finite Element Analysis

4.1 Introduction

In the finite element method, a problem domain is discretized into a mesh of finite elements of simple geometries. In two dimensions, triangular and quadrilateral elements are commonly used. Generating a finite element mesh manually can be time-consuming, tedious and error-prone. Various methods have been developed to automate the task of mesh generation. A wealth of software packages for mesh generation are available in the public domain and commercially. Nowadays, a mesh generator often forms an integral part of a commercial finite element package.

Because the shapes of standard 2D finite elements are limited to triangles and quadrilaterals, almost all available finite element mesh generators produce meshes of only triangular and/or quadrilateral elements in two dimensions. A triangular mesh is the easiest type of mesh to be generated automatically and robustly. A polygon mesh (with number of sides more than 4) is closely related to a triangular mesh, for example, by a Voronoi diagram. As the polygon mesh is not commonly used in the finite element analysis, the information of a polygon mesh suitable for analysis can seldom be obtained directly from an available mesh generator.

In the 2D scaled boundary finite element method, S-elements can be polygons of any number of sides as long as the scaling requirement is satisfied. To automate the process of a scaled boundary finite element analysis, the generation of polygon meshes is addressed in this chapter. Its sole purpose is to generate meshes for use with the MATLAB code Platypus accompanying this book. It is not intended to develop a general, robust and efficient mesh generator.

This chapter starts with some background knowledge on the selected mesh generation algorithm and the implementation as MATLAB functions to generate polygon meshes suitable for a scaled boundary finite element analysis. Numerical examples are then presented to demonstrate the use of the MATLAB functions, which are parts of the accompanying computer program Platypus. The results of all the examples are produced with the provided MATLAB scripts, which can be obtained by scanning the QR-codes found after the scripts.

For readers who are not primarily interested in or familiar with automatic mesh generation, it would probably be the best to start this chapter from the numerical examples in Section 4.4 by following the execution of the MATLAB scripts using the debug toolbar of the MATLAB development environment.

The Scaled Boundary Finite Element Method: Introduction to Theory and Implementation,
First Edition. Chongmin Song.
© 2018 John Wiley & Sons Ltd. Published 2018 by John Wiley & Sons Ltd.
Companion website: www.wiley.com/go/author/scaledboundaryfiniteelementmethod

4.2 Basics of Geometrical Representation by Signed Distance Functions

In the triangular mesh generator DistMesh (Persson and Strang, 2004) and the polygon mesh generator PolyMesher (Talischi et al., 2012), the geometry of a problem domain is represented implicitly by signed distance functions. The basics of signed distance functions, which are necessary for understanding the problem definition functions in Section 4.4, are introduced in this section.

The definition of a signed distance function of a domain V is illustrated in Figure 4.1. The boundary of the domain is denoted as S. A given point P (x, y) is considered. Its distance to a boundary point B (x_b, y_b) is equal to

$$d(P, B) = \sqrt{(x_b - x)^2 + (y_b - y)^2} \tag{4.1}$$

The minimum distance from P to boundary S, i.e. the distance to the closest boundary point, is expressed as $\min_{B \in S}(d(P, B))$. The signed distance function at point P for the domain V is defined as

$$d_V(P) = \begin{cases} -\min_{B \in S}(d(P, B)) & \text{if } P \text{ is inside domain } V \\ +\min_{B \in S}(d(P, B)) & \text{if } P \text{ is outside of domain } V \end{cases} \tag{4.2}$$

The signed distance function $d_V(P)$ is chosen to be negative when P is inside the domain. If P is a point at the boundary, signed distance function $d_V(P) = 0$ applies.

The signed distance functions for some simple geometries can easily be obtained. The following ones used in DistMesh and PolyMesher are given as examples:

1) Straight lines

A straight line passing through two points $P_1(x_1, y_1)$ and $P_2(x_2, y_2)$ is shown in Figure 4.2. The signed distance function of the point $P(x, y)$ is chosen to be negative when P is located at the left of the line, i.e. the direction of the line follows the counter-clockwise direction around the point P.

The vectors **a** and **b** in Figure 4.2 are expressed as

$$\mathbf{a} = a_x\mathbf{i} + a_y\mathbf{j} = (x_2 - x_1)\mathbf{i} + (y_2 - y_1)\mathbf{j} \tag{4.3a}$$

$$\mathbf{b} = b_x\mathbf{i} + b_y\mathbf{j} = (x - x_1)\mathbf{i} + (y - y_1)\mathbf{j} \tag{4.3b}$$

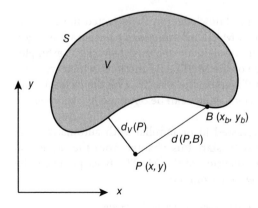

Figure 4.1 Signed distance function. The distance from a given point P to a point B at the boundary S of the domain V is denoted as $d(P, B)$. When the point P is outside of the domain V, the signed distance function d_V is the distance to the closest boundary point. When the point P is inside the domain V, the signed distance function $d_V(P)$ is the minus distance to the closest boundary point. When the point P is at the boundary, the signed distance function is equal to zero.

Figure 4.2 Signed distance function of a straight line passing through two points.

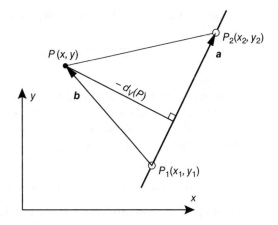

The magnitude of the cross-product of the two vectors is equal to twice the area of the triangle formed by the three points P, P_1 and P_2

$$\|\mathbf{a} \times \mathbf{b}\| = \begin{vmatrix} \mathbf{i} & \mathbf{j} & \mathbf{k} \\ a_x & a_y & 0 \\ b_x & b_y & 0 \end{vmatrix} = a_x b_y - a_y b_x \tag{4.4}$$

The signed distance function is minus the height of the triangle and equal to

$$d_V(P) = -\frac{a_x b_y - a_y b_x}{\|\mathbf{a}\|} \tag{4.5}$$

where $\|\mathbf{a}\| = \sqrt{a_x^2 + a_y^2}$ is the length of the vector $\mathbf{a}$.

A polygon can be constructed from straight lines using set operations. A MATLAB function `dpoly` is provided with DistMesh (Persson and Strang, 2004). MATLAB has also a built-in function `inpolygon` that determines whether given points are located inside or on the edge of a polygonal region.

2) Circles

A circle of radius R is shown in Figure 4.3. The centre O of the circle is located at (x_o, y_o). The minimum distance from the point $P(x, y)$ to the circle is on the radial

Figure 4.3 Signed distance function of a circle.

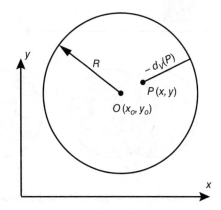

line passing through P. The signed distance function is negative inside the circle and equal to

$$d_V(P) = \sqrt{(x_o - x)^2 + (y_o - y)^2} - R \tag{4.6}$$

3) Rectangles

A rectangle oriented with its edges parallel to the Cartesian coordinates is shown in Figure 4.4. It is defined by its lower left corner $P_1(x_1, y_1)$ and upper right corner $P_2(x_2, y_2)$. The signed distance function (negative inside the rectangle) is minus the smallest distance to the four edges

$$d_V(P) = -\min(x - x_1, x_2 - x, y - y_1, y_2 - y)$$
$$= +\max(x_1 - x, x - x_2, y_1 - y, y - y_2) \tag{4.7}$$

This simple function produces the correct signs, but not the correct distances for the exterior regions (shaded in grey in Figure 4.4) from where the closest boundary points are the corners.

These simple geometries can be combined by set operations to create more complex shapes. Figure 4.5 illustrates the union, difference and intersection operations on domains A and B. The signed distance functions of the resulting geometries are expressed as

$$d_{A \cup B}(P) = \min(d_A(P), d_B(P)) \tag{4.8a}$$
$$d_{A \setminus B}(P) = \max(d_A(P), -d_B(P)) \tag{4.8b}$$
$$d_{A \cap B}(P) = \max(d_A(P), d_B(P)) \tag{4.8c}$$

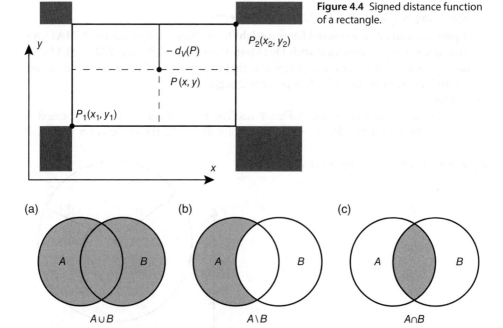

Figure 4.4 Signed distance function of a rectangle.

Figure 4.5 Set operations. (a) Union. (b) Difference of A from B. (c) Intersection.

■ Example 4.1   Signed Distance Functions

Figure 4.6 shows a circle (domain A) and a rectangle (domain B) in Cartesian coordinates x and y. The centre of the circle is located at $(4, 4)$ and the radius is equal to 3. The edges of the rectangle are parallel to the coordinate axes. The lower left corner and upper right corner of the rectangle are located at $(4, 1)$ and $(10, 5)$, respectively.

The signed distance functions of the circle and the rectangle are depicted in Figure 4.7 as surface plots in the xy-plane (left column) and contour plots (right column) inside the domains. The boundaries of the domains are shown by the contour lines labelled 0. The functions are negative inside the domains. Their values increase as a point approaches the boundary and are positive outside of the domains.

The resulting domains obtained by set operations on the circular and rectangular domains are shown in Figure 4.8. The signed distance functions are negative values inside the resulting domains and the boundaries are the contour lines labelled 0. ■

Figure 4.6 Geometry of a circle (domain A) and a rectangle (domain B).

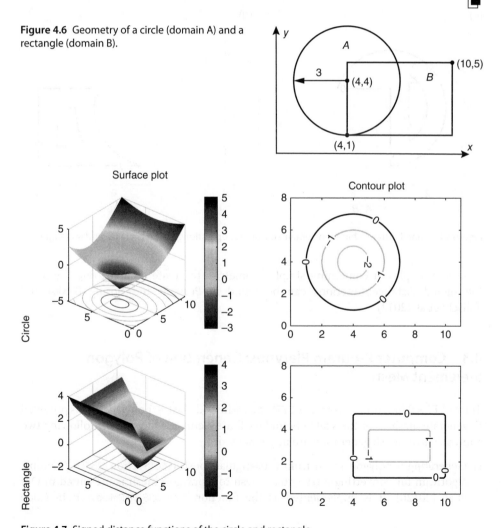

Figure 4.7 Signed distance functions of the circle and rectangle.

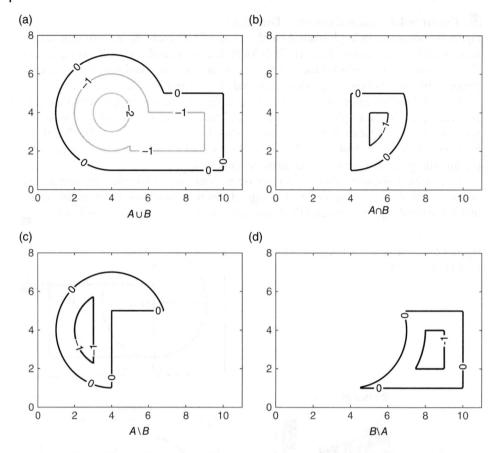

Figure 4.8 Signed distance functions obtained from set operations on the circle (A) and rectangle (B).

More examples of combining simple geometries to create complex ones by using the signed distance functions can be found in Persson and Strang (2004) and Talischi et al. (2012).

4.3 Computer Program Platypus: Generation of Polygon S-element Mesh

The MATLAB function `createSBFEMesh.m` in Code List 4.1 on page 156 is provided to generate meshes for the scaled boundary finite element analysis. The following two approaches for mesh generation are implemented:

1) Generating a polygon mesh from a triangular mesh (see Section 4.3.3). A simple algorithm for converting a triangular mesh to a polygon mesh is presented in Ooi et al. (2012b). It is implemented as the function `triToSBFEMesh.m` in Code

List 4.5. The data of a triangular mesh (see Section 4.3.1 for the data structure of triangular mesh) can be input manually or directly from a triangular mesh generator. In the function `createSBFEMesh.m`, DistMesh (a MATLAB code by Persson and Strang (2004) and available in public domain) is employed to generate the triangular mesh.

2) Converting a polygon mesh to a mesh of polygon S-elements (see Section 4.3.4). The function `polygonToSBFEMesh.m` in Code List 4.6 is written for this purpose. A general-purpose polygon mesh generator PolyMesher by Talischi et al. (2012) is linked with `polygonToSBFEMesh.m`.

By default, each edge of the polygons is modelled with one 2-node line element. Optionally, one edge of a polygon S-element can be divided into multiple elements (see Line 66) by calling the function `subdivideEdge.m` given in Section 4.3.5 on page 172.

The input arguments of the function `createSBFEMesh.m` are:

1) `probdef`: the handle to the function defining the problem to be analysed. It is adopted from the domain definition in PolyMesher. This function handle provides access to the function defining the given problem to be modelled, including the geometry of the problem domain, the boundary conditions, material constants, etc. The use of problem definition function will be discussed in Section 4.4 with examples.

2) `mesher`: control variable specifying the mesh generation approach. The options are:
   - `mesher=1` (Line 26): The geometry of the problem domain is defined by signed distance functions in a problem definition function. A triangular mesh is generated by DistMesh and converted to a mesh of polygon S-elements.
   - `mesher=2` (Line 37): A polygon mesh is generated by PolyMesher and converted to a mesh of polygon S-elements.
   - `mesher=3` (Line 44): A mesh of polygon S-elements is generated from the data of a triangular mesh provided in the problem definition function.
   - `mesher=4` (Line 53): A mesh of polygon S-elements is generated by converting the data of a polygon mesh provided in the problem definition function.
   - `mesher='keyword'` (Line 58): A mesh of polygon S-elements is included in the problem definition function and can be accessed using the `'keyword'` for analysis.

3) `para`: structure variable controlling the mesh size. The field "h0" is the initial edge length of DistMesh. For PolyMesher, the field "nPolygon" specifies the number of polygons. The field "nDiv" specifies the number of 2-node line elements on one edge of a polygon (This option will later on be used in Chapter 5.)

The output arguments of the function `createSBFEMesh.m` are:

1) `coord`: global nodal coordinates of the mesh.
2) `sdConn`: the connectivity of the line elements within individual S-elements.
3) `sdSC`: the coordinates of scaling centres of the polygon S-elements.

Examples of using the function `createSBFEMesh.m` can be found in Section 4.4.

Code List 4.1: Generate polygon S-element mesh

```
1   function [ coord, sdConn, sdSC ] = createSBFEMesh(probdef, ...
2                                                  mesher, para)
3   %Generate polygon S-element mesh
4   %
5   %Inputs:
6   %  probdef - handle of function defining the problem
7   %  mesher    = 1, DistMesh
8   %            = 2, PolyMesher
9   %            = 3, direct input of triangular mesh
10  %            = 4, direct input of polygon mesh
11  %            = 'keyword', access user-defined function to input
12  %                  or generate S-Element mesh
13  %  para      - parameters of element size
14  %              para.h0: Initial edge length (DistMesh only)
15  %              para.nPolygon: number of polygons (PolyMesher only)
16  %              nDiv:  number of 2-node line elements per edge
17  %
18  %Outputs:
19  %  coord(i,:)  -  coordinates of node i
20  %  sdConn{isd,:}(ie,:)  - S-element connectivity. The nodes of
21  %                         line element ie in S-element isd.
22  %  sdSC(isd,:)  - coordinates of scaling centre of S-element isd
23
24  switch(mesher)
25
26      case (1)  % .......................................... use DistMesh
27          fd=@(p) probdef('Dist_DistMesh',p); % .... distance function
28          fh=@(p) probdef('fh',p); % ................ scaled edge length
29          h0  = para.h0; % ......................... initial edge length
30          BdBox = probdef('BdBox'); % .................... bounding box
31          BdBox = [BdBox(1), BdBox(3); BdBox(2) BdBox(4)];
32          pfix = probdef('pfix'); % ....................... hard points
33          [p, t] = distmesh2d(fd,fh, h0, BdBox, pfix);
34          % ... convert a triangular mesh to a polygon mesh
35          [ coord,  sdConn, sdSC ] = triToSBFEMesh( p, t );
36
37      case (2)  % ........................................ use PolyMesher
38          % ... creat polygon mesh
39          nPolygon = para.nPolygon; % ..... number of polygons in mesh
40          [coord, polygon] = PolyMesher(probdef,nPolygon,100);
41          % ... convert polygons into S-elements
42          [sdConn, sdSC] = polygonToSBFEMesh(coord, polygon);
43
44      case (3)  % ............. use triangular mesh in problem definition file
45          ArgOut = probdef('TriMesh');
46          p = ArgOut{1}; t = ArgOut{2};
47          figure
48          PlotTriFEMesh( p, t )
49
50          % ... convert a triangular mesh to a polygon mesh
```

```
51        [ coord,   sdConn, sdSC ] = triToSBFEMesh( p, t );

52

53     case (4)  % ......... use polygon mesh in problem definition function
54          ArgOut = probdef('PolygonMesh');
55          coord = ArgOut{1}; polygon = ArgOut{2};
56          % ...  convert polygons into S-elements
57          [sdConn, sdSC] = polygonToSBFEMesh(coord, polygon);
58     otherwise  % .......... obtain mesh from problem definition function
59          [ glbMesh ] = probdef(mesher, para);
60          coord = glbMesh.coord;
61          sdConn = glbMesh.sdConn;
62          sdSC = glbMesh.sdSC;

63

64  end

65

66  % ...  subdivide one edge into multiple line elements
67  if nargin > 2 && isfield(para,'nDiv') && para.nDiv > 1
68       [ meshEdge, sdEdge ] = meshConnectivity( sdConn );
69       % ...  cell array edgeDivXii contains ξ of edge i
70       % ...  uniform subdivision points in local parametric coordinate ξ (0, 1)
71       edgeXi(1:length(meshEdge)) = {( 1:1:para.nDiv-1)'/para.nDiv };
72       [coord, sdConn ] = subdivideEdge(edgeXi,...
73            coord, meshEdge, sdEdge );
74  end

75

76  end
```

4.3.1 Mesh Data Structure

The function `createSBFEMesh.m` can create a mesh of polygon S-elements for the scaled boundary finite element analysis from the input of either a triangular mesh (e.g. generated by DistMesh) or a polygon mesh (e.g. generated by PolyMesher). A triangular mesh and a polygon mesh on a square domain of 1×1 are shown in Figure 4.9. The data

(a)   (b)

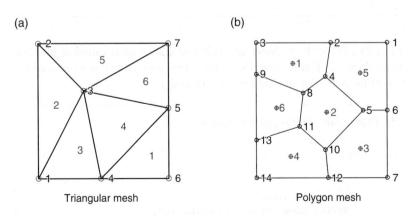

Triangular mesh   Polygon mesh

Figure 4.9 Example of triangular and polygon meshes. The dimension of the square domain is 1×1.

structures of the triangular and polygon meshes, as well as additional data necessary for generating scaled boundary polygon mesh and enforcing boundary conditions, are described below using this example.

The data of a triangular mesh consist of the nodal coordinates and the element connectivity. As an example, the following MATLAB script stores the nodal coordinates and triangular elements of the mesh in Figure 4.9a in the variables p and t, respectively. The nodal coordinates are stored in a matrix of two columns. The two entries of a row of the matrix are the x and y coordinates of a node. The triangular elements are stored in a matrix of three columns with one element per row. The three nodes of an element follow the counter-clockwise direction and the choice of the starting node is arbitrary.

```matlab
% Nodal coordinates
p = [ 0.00 0.00; 0.00 1.00; 0.35 0.65;
0.48 0.00; 1.00 0.52; 1.00 0.00;
1.00 1.00];

% Triangular elements
t = [ 5 4 6; 3 2 1; 1 4 3;
4 5 3; 2 3 7; 7 3 5];

% Plot mesh of triangular elements
opt =struct('LabelEle', 12, 'LabelNode', 10); % plotting options
PlotTriFEMesh( p, t, opt )
```

The above MATLAB script can be obtained by scanning the QR-code to the right.

The function `PlotTriFEMesh.m` shown in Code List 4.2 below is called to plot the triangular mesh as shown in Figure 4.9a. The built-in function `patch` is employed for plotting. The plot options are documented within the code.

Code List 4.2: Plot mesh of triangular elements

```matlab
1  function  PlotTriFEMesh( p, t, opt)
2  %Plot mesh of triangular elements
3  %
4  %Inputs
5  %     p(i,:)   - coordinates of node i
6  %     t(i,:)   - nodal numbers of triangle i
```

```
 7  %  plot options:
 8  %     opt=struct('LabelEle', 10, 'LabelNode', 8, ...
 9  %              'LineStyle','-','EdgeColor','k','LineWidth', 1)
10  %  where the options are:
11  %     opt.LabelEle  : = 0, do not label element;
12  %                       otherwise, font size of element number
13  %     opt.LabelNode : = 0, do not label nodes;
14  %                       otherwise, font size of element number
15  %     opt.LineStyle, opt.EdgeColor, opt.LineWidth: options of
16  %                       'patch' function with the same name
17
18  % ...  default options
19  LineStyle = '-';   EdgeColor = 'k';   LineWidth = 1;
20  % ...  use specified options if present
21  if nargin > 2
22      if isfield(opt, 'LineStyle') && ~isempty(opt.LineStyle)
23          LineStyle = opt.LineStyle;
24      end
25      if isfield(opt, 'EdgeColor') && ~isempty(opt.EdgeColor)
26          EdgeColor = opt.EdgeColor;
27      end
28      if isfield(opt, 'LineWidth') && ~isempty(opt.LineWidth)
29          LineWidth = opt.LineWidth;
30      end
31  end
32
33  % ...  plot triangular mesh
34  patch('Faces',t,'Vertices',p,'FaceColor','w',...
35      'LineStyle',LineStyle,'EdgeColor',EdgeColor, ...
36      'LineWidth',LineWidth);
37  axis equal; axis on; grid off;
38  hold on;
39
40  % ...  apply plot options
41  if nargin > 2
42      if isfield(opt, 'LabelNode') && opt.LabelNode  % ....... nodes
43          if opt.LabelNode <= 2  % ........... font size of nodal number
44              fontsize = 14;
45          else
46              fontsize = opt.LabelNode;
47          end
48          % ...  plot nodes
49          plot(p(:,1),p(:,2),'ko','LineWidth',LineWidth);
50          np = length(p);  % ......................... number of nodes
51          text(p(:,1), p(:,2), [blanks(np)' int2str((1:np)')], ...
52              'FontSize',fontsize );  % ................... label nodes
53      end
54      if isfield(opt, 'LabelEle') && opt.LabelEle  % ....... elements
55          if opt.LabelEle <= 2  % .......... font size of element number
56              fontsize = 14;
57          else
```

```
58              fontsize = opt.LabelEle;
59          end
60          % ... centroids
61          triCnt = (p(t(:,1),:)+p(t(:,2),:)+p(t(:,3),:))/3;
62          nTri = length(t); % ........... number of triangular elements
63          text(triCnt(:,1),triCnt(:,2), ...
64              [ int2str((1:nTri)')],...
65              'FontSize',fontsize, 'Color','r'); % .... label elements
66      end
67  end
```

A polygon mesh is similar to a triangular mesh except that the elements may have a different number of nodes. The polygon mesh in Figure 4.9b is described in the script below. The data of nodal coordinates are stored in the same way as in the case of the triangular mesh. To accommodate the difference in the number of nodes among the polygons, the polygons are stored in a cell array. Each cell contains the nodes of one polygon element in a row array. Again, the nodes are ordered in the counter-clockwise direction. The polygon mesh is converted to the S-element mesh using the function polygonToSBFEMesh.m given later in the Code List 4.6. Figure 4.9b is then plotted using the function PlotSBFEMesh.m provided in Code List 3.6 on page 98.

```
% Nodal coordinates
coord = [ 1.00 1.00; 0.57 1.00; 0.00 1.00;
          0.53 0.75; 0.82 0.50; 1.00 0.50;
          1.00 0.00; 0.36 0.63; 0.00 0.77;
          0.53 0.21; 0.33 0.38; 0.55 0.00;
          0.00 0.28; 0.00 0.00];
% Polygons
polygon = { [ 9 8 4  2   3]; [ 11 10 5 4   8];
            [ 10 12 7 6   5]; [ 14 12 10 11 13];
            [ 4  5  6 1   2]; [ 13 11 8 9]    };

% Plot mesh of polygon elements
[sdConn, sdSC] = polygonToSBFEMesh(coord, polygon) ;
opt=struct('LineSpec','-k', 'sdSC', sdSC, 'LabelSC', 14, ...
'PlotNode', 1, 'LabelNode', 14,'MarkerSize', 6);
PlotSBFEMesh(coord, sdConn, opt);
axis off
```

The above MATLAB script can be obtained by scanning the QR-codes to the right.

To convert a triangular mesh to an S-element mesh and, for later use, to enforce the prescribed boundary conditions, additional data on mesh connectivity are constructed by the function `meshConnectivity.m` in Code List 4.3. The input to this function, `sdConn`, is the element connectivity of a triangular, polygon or S-element mesh. The function differentiates the types of mesh and interprets the data according to the data structure of `sdConn`. The outputs of mesh connectivities include:

1) `meshEdge`: edges of elements in the mesh. It is a matrix of two columns with each row containing the two nodes at the ends of one edge. The nodes are sorted into ascending order and duplicated edges are removed.

   The edge numbers of the triangular mesh in Figure 4.9a are shown in the figure below with the nodes of edges 1, 3, 5, 7, 9 and 11 listed.

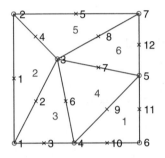

meshEdge (1:2:11,:)

1	2
1	4
2	7
3	5
4	5
5	6

2) `sdEdge`: edges of an element/S-element. It is a cell array. Each cell contains the edges of one element/S-element in a column. For the triangular mesh in Figure 4.9a, the edges of all the six elements are given column-wise as

```
[sdEdge{:}]
      9      4      3      9      4      8
     10      1      6      7      8      7
     11      2      2      6      5     12
```

3) `edge2sd`: elements/S-elements connected to each edge. This is a cell array. Each cell contains the elements/S-elements connected to one edge. An edge is at the boundary when there is only one element/S-element connected to it. For the triangular mesh in Figure 4.9a, the elements connected to edges 1 and 2 are displayed by

```
edge2sd{1} %S-elements connected to edge 1
      2

edge2sd{2} %S-elements connected to edge 2
      2
      3
```

Edge 1 is connected to one element only (element 1) and is at the boundary.

4) `node2Edge`: edges connected to each node. This is a cell array. Each cell contains the edges connected to one node. For the triangular mesh in Figure 4.9a, the edges connected to nodes 1 and 3 are displayed by

```
node2Edge{1} %edges connected to node 1
     1     2     3

node2Edge{3} %edges connected to node 3
     2     4     6     7     8
```

5) `node2sd`: elements/S-elements connected to each node. This is a cell array with each cell containing the elements/S-elements connected to one node.

```
node2sd{1} %elements/S-elements connected to node 1
     2
     3


node2sd{4} %elements/S-elements connected to node 4
     1
     3
     4
```

Detailed documentation of this function is included in the code.

Code List 4.3: Construct mesh connectivity data

```
1  function [ meshEdge, sdEdge, edge2sd, node2Edge, node2sd] = ...
2      meshConnectivity( sdConn )
3  %Construct mesh connectivity data
4  %
5  %Input
6  %    sdConn: S-element/element connectivity
7  %        when sdConn is a matrix,
8  %            sdConn(i,:) are the nodes of element i
9  %        when sdConn is a cell array of a 1D array,
10 %            sdConn{i} are the nodes of polygon i
11 %        when sdConn is a cell array of a matrix,
12 %            sdConn{i} are the nodes of line elements of S-element i
13 %
14 %Output
15 %    meshEdge(i,1:2)   - the 2 nodes on line i in a mesh.
16 %                        The node number of the first node is
17 %                        larger than that of the 2nd one
18 %    sdEdge{i}         - lines forming S-element i,
19 %                        >0 when a line follows anti-clockwise
20 %                            direction around the scaling centre.
21 %                        <0 otherwise
22 %    edge2sd{i}        - S-elements connected to edge i
23 %    node2Edge{i}      - edges connected to node i
24 %    node2sd{i}        - S-elements connected to node i
25
26 nsd = length(sdConn); % ................ number of S-elements/elements
27
28 % ...    construct connectivity of edges to S-elements/elements
29 % ...    sdConn is intepreted based on its data type
30 if ~iscell(sdConn)  % ............ sdConn is a matrix (triangular elements)
31     % ...  the following loop collects the edges of all elements
32     sdEdge = cell(nsd,1);  % ............................... initialization
33     i1 = 1;  % ........................................... counter of edges
```

```
34      meshEdge = cell(nsd,1);  % ........................... initialization
35      for ii = 1:nsd  % .............................. loop over elements
36          eNode = sdConn(ii,:);  % ................... nodes of an element
37          i2 = i1 + length(eNode) - 1;  % ............... count the edges
38          % ...  element edges. Each edge is defined by two nodes on a column
39          meshEdge{ii} = [eNode; eNode(2:end) eNode(1)];
40          sdEdge{ii} = i1:i2;  % .................... edges of an element
41          % ...  store the edges to be reversed in sorting as negative values
42          idx = find( meshEdge{ii}(1,:) > meshEdge{ii}(2,:) );
43          sdEdge{ii}(idx) = - sdEdge{ii}(idx);
44          i1 = i2 + 1;  % ........... update the counter for the next element
45      end
46      % ...  combine edges of all elements. The two nodes of an edge are sorted
47      % ...  in ascending order. Each edge is stored as one row.
48      meshEdge = (sort([meshEdge{:}]))';
49  else  % ................................ input of sdConn is a cell array
50      sdEdge = cell(nsd,1);  % ............................. initialization
51      i1 = 1;
52      if size(sdConn{1},1) > 1
53          % ...  S-elements (a cell has multiple rows)
54          % ...  all the edges of S-elements are numbered in this loop
55          for ii = 1:nsd  % ......................... loop over S-elements
56              i2 = i1 + length(sdConn{ii}) - 1;  % ...... count the edges
57              sdEdge{ii} = i1:i2;  % ................ edges of an S-element
58              % ...  store the edges to be reversed in sorting as negative values
59              % ...  each element edge is defined by two nodes as a row
60              idx = find( sdConn{ii}(:,1) > sdConn{ii}(:,2) );
61              sdEdge{ii}(idx) = - sdEdge{ii}(idx);
62              i1 = i2 + 1;
63          end
64          % ...  combine edges of all S-elements. The two nodes of an edge are
65          % ...  sorted in ascending order. Each edge is stored as one row.
66          meshEdge = sort(vertcat(sdConn{:}),2);
67      else
68          % ...  polygon element (closed loop specified by vertices)
69          % ...  the following loop collects the edges of all polygons
70          meshEdge = cell(nsd,1);  % ........................ initialization
71          for ii = 1:nsd  % ........................... loop over polygons
72              eNode = sdConn{ii};  % ................... nodes of a polygon
73              i2 = i1 + length(eNode) - 1;  % ............ count the edges
74              % ...  each element edge is defined by two nodes as a column
75              meshEdge{ii} = [eNode; eNode(2:end) eNode(1)];
76              sdEdge{ii} = i1:i2;  % .................... edges of a polygon
77              idx = find( meshEdge{ii}(1,:) > meshEdge{ii}(2,:) );
78              % ...  edge to be reversed
79              sdEdge{ii}(idx) = -sdEdge{ii}(idx);
80              i1 = i2 + 1;  % ....... update the counter for the next element
81          end
82          % ...  combine all edges of all elements. The two nodes of an edge are
83          % ...  sorted in ascending order. Each edge is stored as one row.
84          meshEdge = (sort([meshEdge{:}]))';
```

```
85        end
86    end
87
88    % ... remove duplicated entries of edges
89    [meshEdge, ~, ic] = unique(meshEdge,'rows');
90    for ii = 1:nsd  % ........................ loop over S-elements/elements
91        % ... update edge numbers
92        sdEdge{ii} = sign(sdEdge{ii}(:)).*ic(abs(sdEdge{ii}));
93    end
94
95    if nargout < 3
96        return;
97    end
98
99    % ... find S-elements/elements connected to an edge
100   a = abs(cell2mat(sdEdge));  % .......... edges of all S-elements/elements
101   % ... the following loop matchs S-element/element numbers to edges
102   asd = zeros(length(a),1);  % .............................. initialization
103   ib = 1;  % ................ pointer to the first edge of an S-element/element
104   for ii = 1:nsd  % ........................ loop over S-elements/elements
105       ie = ib + length(sdEdge{ii}) - 1;
106       asd(ib:ie) = ii;  % . edge a(i) is connected to S-element/element asd(i)
107       ib = ie + 1;  %update the pointer
108   end
109   % ... sort S-element numbers according to edge number
110   [c, indx] = sort(a); asd = asd(indx);
111   % ... the following loop collects the S-elements connected to nodes
112   ib = 1;  % ....... pointer to the 1st S-element/element connected to an edge
113   nMeshedges = length(meshEdge);  % .............. number of edges in mesh
114   edge2sd = cell(nMeshedges,1);  % ......................... initialization
115   for ii = 1:nMeshedges-1  % ........ loop over edges (except for the last one)
116       if c(ib+1) == ii  % .... two S-elements/elements connected to an edge
117           edge2sd{ii} = asd(ib:ib+1);  % ... store the S-elements/elements
118           ib = ib + 2;  % ............................. update the pointer
119       else  % ................... one S-element/element connected to an edge
120           edge2sd{ii} = asd(ib);  % .......... store the S-element/element
121           ib = ib + 1;  % ............................. update the pointer
122       end
123   end
124   % ... the S-elements/elements connected to the last edges
125   edge2sd{ii+1} = asd(ib:end);
126
127   if nargout < 4
128       return;
129   end
130
131   % ... find edges connected to a node
132   a = reshape(meshEdge',1,[]);  % ........................ nodes on edges
133   % ... edge numbers correponding to nodes in a
134   edgei = reshape([1:size(meshEdge,1); 1:size(meshEdge,1)],1,[]);
135   % ... sort edge number according to node number
```

```
136  [c, indx] = sort(a); edgei = edgei(indx);
137  ib = 1;  % ..................... pointer to the 1st edge connected to a node
138  nNode = c(end);  % ..................................... number of nodes
139  node2Edge = cell(nNode,1);  % ........................... initialization
140  for ii = 1:nNode-1  % ............ loop over nodes (except for the last one)
141      % ...  pointer to the last edge connected to a node
142      ie = ib-2+find(c(ib:end)~=ii, 1, 'first');
143      node2Edge{ii} = edgei(ib:ie);  % ...... store edges connected to node
144      ib = ie + 1;  % ................................ update the pointer
145  end
146  % ...  store edges connected to the last node
147  node2Edge{nNode} = edgei(ib:end);
148
149  % ...  find S-elements connected to a node
150  node2sd = cell(nNode,1);  % ............................. initialization
151  for ii = 1:nNode  % ..................................... loop over nodes
152      node2sd{ii} = unique(vertcat(edge2sd{node2Edge{ii}}));
153  end
154
155  end
```

4.3.2   Centroid of a Polygon

The centroids of polygons are required in the generation of an S-element mesh. The function `polygonCentroid.m` in Code List 4.4 is written to compute the coordinates of the centroid of a polygon from given coordinates of the vertices (ordered in counter-clockwise direction). The polygon is decomposed into a series of triangles formed by an edge of the polygon and the origin of the coordinate system. The three vertices of a triangle are denoted as (x_i, y_i), (x_{i+1}, y_{i+1}) and $(0, 0)$, where the direction of the line segment from (x_i, y_i) to (x_{i+1}, y_{i+1}) follows the direction of the edge of the polygon. Twice the signed area of the triangle is equal to (see Eq. 2.65)

$$2A_i = x_i y_{i+1} - x_{i+1} y_i \tag{4.9}$$

When the coordinate origin is outside of the polygon, the signed areas of some triangles will be negative. The centroid (x_{ci}, y_{ci}) of the triangle is equal to

$$x_{ci} = \frac{1}{3}(x_i + x_{i+1}) \tag{4.10a}$$

$$y_{ci} = \frac{1}{3}(y_i + y_{i+1}) \tag{4.10b}$$

The centroid of the polygon is given by

$$x = \frac{\sum_i 2A_i x_{ci}}{\sum_i 2A_i}; \tag{4.11a}$$

$$y = \frac{\sum_i 2A_i y_{ci}}{\sum_i 2A_i} \tag{4.11b}$$

Code List 4.4: Controid of a polygon

```
1   function [ cnt ] = polygonCentroid( xy )
2   %Centroid of a polygon
3   %Input
4   %    xy(i,:) - coordinates of vertices in counter-clockwise direction
5   %Output
6   %    cnt(:) - centroid of polygon
7
8   xy = [xy; xy(1,:)];  % .......... appending the 1st vertex after the last one
9   % ...  array of all triangles formed by the coordinate origin and an edge
10  % ...  of the polygon, Eq. (4.9)
11  area2 = xy(1:end-1,1).*xy(2:end,2)-xy(2:end,1).*xy(1:end-1,2);
12  % ...  array of centroids of triangles, Eq. (4.10)
13  c = [xy(1:end-1,1)+xy(2:end,1) xy(1:end-1,2)+xy(2:end,2)]/3;
14  % ...  centroid of polygon, Eq. (4.11)
15  cnt = sum([area2 area2].*c)/sum(area2);
16  end
```

4.3.3   Converting a Triangular Mesh to an S-element Mesh

An algorithm to convert a triangular mesh to a polygon mesh is presented by Ooi et al. (2012b). It is implemented in the function `triToSBFEMesh.m` provided in Code List 4.5 below to generate simple polygon S-element meshes for use with the accompanying computer program Platypus.

The algorithm implemented in `triToSBFEMesh.m` is illustrated in Figure 4.10. A triangular mesh is shown in Figure 4.10a. It also serves as the geometry description of the problem domain to generate a polygon mesh. To avoid confusion with the nodes of the polygon mesh to be generated, the nodes in the triangular mesh are referred to as points in this section.

In generating the polygon mesh, the centroids of the triangles and the midpoints of the boundary edges of the triangular mesh are determined. They will be used as nodes of the polygon mesh. The centroids of the triangles are equal to the averages of the coordinates of the vertices (Line 18). They are shown in Figure 4.10b by the symbol ∗. The edges of the triangles and the connectivity between edges, points and triangles are found by calling the function `meshConnectivity.m` (Line 20). The boundary edges are connected to only one triangle while the interior edges are connected to two. The midpoints of the boundary edges are shown in Figure 4.10b by the symbol ×. The boundary points that are connected to the boundary edges are identified. They will also be included as nodes of the polygon mesh. The numbering of the nodes of the polygon mesh is shown in Figure 4.10c.

One polygon S-element is generated around one point of the triangular mesh (see the part of Code List 4.5 starting from Line 45).

1) For an interior point (starting from Line 49 of the function `triToSBFEMesh.m`). A polygon is formed by connecting the centroids of all the triangles that this point is directly linked to, as shown by S-element 9 in Figure 4.10d. The interior point is

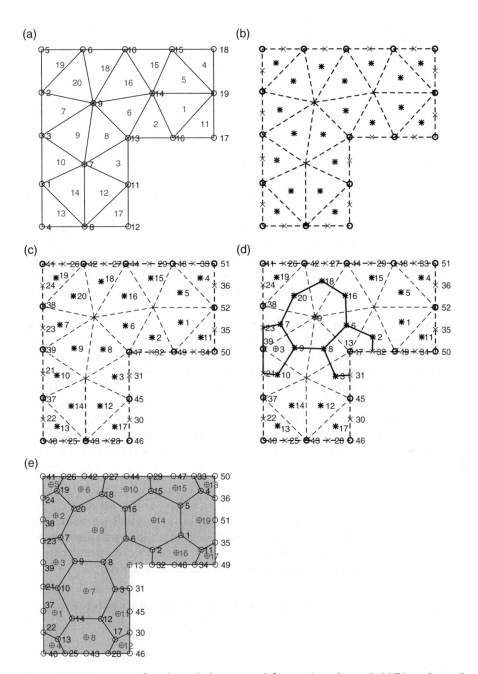

Figure 4.10 Generation of a polygon S-element mesh from a triangular mesh. (a) Triangular mesh. (b) The centroids "∗" of triangular elements, boundary nodes "○" of the triangular mesh and the midpoint "×" of edges at the boundary are identified. (c) Initial numbering of nodes of polygon S-elements. (d) Construction of three types of polygon S-element: Interior (S-element 9), Boundary (S-element 3) and Concave (S-element 13). The scaling centres are indicated by the symbol ⊕. (e) Final S-element mesh after removing unconnected nodes.

selected as the scaling centre. The nodes are sorted in the counter-clockwise direction around the scaling centre by the local function `sortNodes` (Line 108).

2) For a boundary point (starting from Line 57 of the function `triToSBFEMesh.m`), the midpoints of the boundary edges and the centroids of all the triangles this point directly connects to are considered (Figure 4.10d with S-elements 3 and 13 as examples). They are sorted in the counter-clockwise direction around the boundary point and ordered in such a way that the boundary nodes are at the beginning of the list. The internal angle at the boundary point is determined. Two types of polygons are generated, depending on whether the internal angle is smaller or greater than a specified angle at which the possible strain/stress singularity at this boundary point is expected to become important. This angle is rather arbitrarily chosen as 220° here. An example of each case is illustrated in Figure 4.10d by S-elements 3 and 13, respectively, .

   a) When the internal angle at the boundary point is less than 220°, the polygon is either convex or considered to be slightly concave. If a strain/stress singularity occurs, it is expected to be weak. As illustrated by S-element 3 in Figure 4.10d, the polygon is formed by a closed loop of line elements for such a case. The boundary point is included as a node (Node 39). The centroid of the polygon is chosen as the scaling centre.

   b) When the internal angle is not less than 220°, the polygon is concave and the strain/stress singularity at this boundary point may be potentially strong. Such a case occurs at the boundary point denoted as Node 47 in Figure 4.10c. Around this boundary point, S-element 13 in Figure 4.10d is constructed. The scaling centre is placed at the boundary point. The line elements form an open loop. This boundary point (Node 47) is not connected to any line elements. The edge connecting Nodes 47 and 31 and the edge connecting Nodes 47 and 32 in Figure 4.10c are represented by scaling Nodes 31 and 32 in Figure 4.10d, respectively. This choice of scaling centre provides a simple and effective way to model strain/stress singularities (see, e.g., Section 4.4.3).

After all the polygon S-elements are generated, the nodes that are not connected to any elements (for example, Node 47 in Figure 4.10c) are removed. This operation is performed by the part of code starting from Line 97. The remaining nodes are renumbered and the connectivity of the S-elements is updated. The final polygon S-element mesh is shown in Figure 4.10e. As illustrated by the shaded area, the problem domain is covered by scaling the edges of the polygons with respect to their scaling centres.

Code List 4.5: Convert a triangular mesh to a scaled boundary polygon mesh

```
1   function [ coord, sdConn, sdSC ] = triToSBFEMesh( p, t )
2   %Convert a triangular mesh to an S-element mesh
3   %
4   %Inputs
5   %  p(i,:)    - coordinates of node i
6   %  t(i,:)    - nodal numbers of triangle i
7   %
8   %Outputs:
9   %  coord(i,:)    - coordinates of node i
10  %  sdConn{isd,:}(ie,:)   - an S-element connectivity. The nodes of
```

```
11  %                             line element ie in S-element isd.
12  %  sdSC(isd,:)  - coordinates of scaling centre of S-element isd
13
14  np = length(p);  % ..................................... number of points
15  nTri = length(t);  % ............................... number of triangles
16  % ... centroids of triangles will be nodes of S-elements.
17  % ... triangular element numbers will be the nodal numbers of S-elements.
18  triCnt = (p(t(:,1),:)+p(t(:,2),:)+p(t(:,3),:))/3;  % ........... centroids
19
20  % ... construct data on mesh connectivity
21  [ meshEdge, ~, edge2sd, node2Edge, node2sd] = ...
22                             meshConnectivity( t );
23
24  % ... number of S-elements connected to an edge
25  edgeNsd = cellfun(@length,edge2sd);
26  % ... list of boundary edges (connected to 1 S-element only)
27  bEdge = find(edgeNsd==1);
28  % ... midpoints of boundary edges
29  bEdgeCentre = (p(meshEdge(bEdge,1),:)+p(meshEdge(bEdge,2),:))/2;
30  % ... list of points at boundary
31  bp = unique(meshEdge(bEdge,:));
32
33  % ... include the points in the middle of boundary edges as nodes of S-elements
34  nbEdge = length(bEdge);
35  bEdgeNode = nTri + (1:nbEdge)';  % ...................... nodal number
36  bEdgeIdx(bEdge(:)) = (1:nbEdge)';  % ............. index from edge number
37
38  % ... include the points at boundary as nodes of S-element mesh
39  nbp = length(bp);
40  bNode = nTri + nbEdge + (1:nbp)';  % ...................... nodal number
41  bpIdx(bp) = 1:nbp;  % ........................... index from point number
42  % ... nodal coordinates
43  coord = [triCnt; bEdgeCentre; p(bp,:)];
44
45  % ... construct polygon S-elements
46  sdConn = cell(np,1);  % ........................... initializing connectivity
47  sdSC   = zeros(np,2);  % ........................ initializing scaling centre
48  for ii = 1:np
49      if bpIdx(ii) == 0  % ................................... interior point
50          node = node2sd{ii};  % ........... S-elements connected to node ii
51          % ... sort nodes in counter-clockwise direction
52          [ node ] = sortNodes(node, coord(node,:), p(ii,:));
53          % ... scaling centre at current point
54          sdSC(ii,:) = p(ii,:);
55          % ... line element connectivity in an S-element
56          sdConn{ii}=[node; node(2:end) node(1)]';
57      else  % .... boundary point, which can become a node or a scaling centre
58          be = bEdgeIdx(node2Edge{ii});  % ........ edges connected to node
59          nodee = bEdgeNode( be(be~=0) );  % ...... nodes on boundary edges
60          % ... sort the nodes, except for the one at the current point
61          node = [ node2sd{ii}; nodee ];
62          [ node ] = sortNodes(node, coord(node,:), p(ii,:) );
```

```
63      % ... find the two boundary nodes in the node list
64      idx1 = find(node==nodee(1));
65      idx2 = find(node==nodee(2));
66      % ... maintain counter-clockwise direction and rearrange the nodes
        as:
67      % ... [boundary node 1, current point (node), boundary node 2,
        others]
68      if abs(idx1-idx2) > 1
69          % ... the two boundary nodes are the first and last in the list
70          node = [node(end) bNode(bpIdx(ii)) node(1:end-1)];
71      else
72          % ... the two boundary nodes are consecutive on the list
73          idx = min(idx1,idx2);
74          node = [node(idx) bNode(bpIdx(ii)) ...
75              node(idx+1:end) node(1:idx-1) ];
76      end
77      % ... internal angle between two boundary edges
78      dxy = diff(coord(node(1:3),:));  % ......... Δ_x, Δ_y of the 1st 2 edge
79      dl = sqrt(sum(dxy.^2,2));  % ............................. length
80      dxyn = dxy./[dl dl];  % ........................... direction cosin
81      % ... angle between 2 boundary edges
82      alpha = real(acosd(sum(dxyn(1,:).*dxyn(2,:))));
83      beta = 180 - sign(det(dxyn))*alpha;  % ............ internal angle
84      if beta < 220  % .................. include current point as a node
85          % ... line element connectivity in an S-element
86          sdConn{ii}=[node; node(2:end) node(1)]';
87          % ... select centroid as scaling centre
88          sdSC(ii,:) = polygonCentroid( coord(node,:) );
89      else  % ........ use current point (concave corner) as a scaling centre
90          sdSC(ii,:) = p(ii,:);
91          % ... line element connectivity in an S-element
92          sdConn{ii}=[node(3:end); node(4:end) node(1)]';
93      end
94    end
95 end
96
97 % ... remove unconnected nodes
98 a = reshape(vertcat(sdConn{:}),[],1);  % ...................... all nodes
99 [c, ~, ~] = unique(a);  % ................................. unique nodes
100 i(c) = 1:length(c);  % ............. new nodal numbers of the unique nodes
101 coord = coord(c,:);  % ............ update the nodal coordinates accordingly
102 % ... update line element connectivity in each S-element
103 for ii = 1:length(sdConn)
104     sdConn{ii} = reshape(i(sdConn{ii}(:)),[],2);
105 end
106 end
107
108 % ... function: sortNodes
109 function [ node ] = sortNodes(node, xy, c)
110 % ... sort nodes in counterclock direction around point c
111 xy = bsxfun(@minus, xy, c);
```

```
112  ang = atan2(xy(:,2), xy(:,1));  % ..................... angular coordinates
113  [~, ic] = sort(ang);  % .............. sort to increasing angular coordinates
114  node = (node(ic))';  % ................................ rearrange nodes
115  end
```

4.3.4 Use of Polygon Meshes Generated by PolyMesher in a Scaled Boundary Finite Element Analysis

A polygon mesh generated by PolyMesher (Talischi et al., 2012) is described by the coordinates of the vertices and the connectivity of the polygons. A polygon mesh of a cantilever beam generated by PolyMesher is shown in Figure 3.6 in Example 3.5 on page 104. The mesh is described in the MATLAB script ExmplDeepBeam.m. The coordinates of the vertices are listed after Line 2. It is assumed that the edges of a polygon form a closed loop. The connectivity of a polygon is defined by an array of vertices following the counter-clockwise direction (Line 9). In the computer program Platypus, a different data structure as explained in Section 2.6.2 on page 67 is adopted. The line elements at the boundary of a polygon S-element can form an open loop, which leads to an efficient and accurate way to model the stress singularity at a crack tip as demonstrated in Example 3.9 on page 125. In Example 3.5, a similar piece of code is given as a part of the MATLAB script on page 105 (see Line 21).

To convert a polygon mesh generated by PolyMesher to an S-element mesh for use with Platypus, the function polygonToSBFEMesh.m in the Code List 4.6 is written. The conversion of connectivity is illustrated using the polygon mesh in Figure 3.6. In function polygonToSBFEMesh.m, the scaling centre of a polygon S-element is placed at its centroid (see function polygonCentroid.m in Code List 4.4 on page 166).

Code List 4.6: Converting a polygon mesh (PolyMesher)

```
1   function [sdConn, sdSC] = polygonToSBFEMesh(coord, polygon)
2   %Convert a polygon mesh to an S-element mesh
3   %
4   %Inputs:
5   %   coord(i,:)    - coordinates of node i
6   %   polygon{i}    - array of vertices of polygon i.
7   %
8   %Outputs:
9   %   sdConn{isd,:}(ie,:)   - S-element connectivity. The nodes of
10  %                           line element ie in S-element isd.
11  %   sdSC(isd,:)   - coordinates of scaling centre of S-element isd
12
```

sdConn: S-element connectivity stored as a cell array (one S-element per cell). In a cell, the connectivity of line elements is given by one element per row [Node-1 Node-2].

```
13  nsd = length(polygon);  % ....................... number of S-elements
14  sdConn = cell(nsd,1);  % ....................... initializing connectivity
15  sdSC = zeros(nsd,2);  % ............................... scaling centre
```

```
16   for isub =1:nsd
17       % ...  build connectivity
18       sdConn{isub}=[polygon{isub}; ...
19                    polygon{isub}(2:end) polygon{isub}(1)]';
20       % ...  scaling centre at centroid of polygon
21       sdSC(isub,:) = polygonCentroid(coord(polygon{isub},:));
22   end
23
24   end
```

4.3.5   Dividing Edges of Polygons into Multiple Elements

In the scaled boundary finite element method, the geometry of an S-element only needs
to satisfy the scaling requirement. For a polygon S-element, the number of edges is not
limited. One edge of a polygon can also be subdivided into more than one line element.
As an example, Figure 4.11 depicts an S-element mesh with each edge subdivided into
2 line elements.

The function `subdivideEdge.m` (Code List 4.7 on page 173) modifies a mesh by
subdividing the edges of polygon S-elements into short segments. Each of the shorter
segment can be regarded as an edge of a new polygon mesh and modelled by one 2-node
line element.

The original S-element mesh is regarded as a geometrical model consisting of a
set of polygons. The input argument `coord` stores the coordinates of the vertices of
the polygons. The input argument `meshEdge` defines the lines that form the edges
of the polygons. Two adjacent polygons are connected by their common edge. The
input argument `sdEdge` is a cell array. Each cell contains the lines that form one

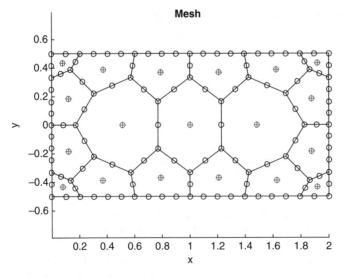

Figure 4.11 Example of polygon mesh (see Figure 4.12 on page 179) with each edge subdivided into
2 line elements.

polygon. The inputs `meshEdge` and `sdEdge` can be constructed from the S-element connectivity using the function `meshConnectivity.m` in Code List 4.3.

The coordinates of the intermediate nodes to be created on a line are calculated by interpolating the coordinates of the two ends of the line. The subdivision is controlled by the input argument `edgeXi`:

1) When `edgeXi` is an array of integers, it stores the numbers of subdivisions (equal to the numbers of 2-node line elements) of each line in `meshEdge`. The numbers of division can be different from one line to the other. The lines are subdivided uniformly.
2) When the input variable `edgeXi` is a cell array. The locations of the new nodes are specified by a local parametric coordinate defined in the range (0, 1). Each cell of `edgeXi` is a column vector and stores the values of the local parametric coordinates of the intermediate nodes to be created on a line.
3) If a cell array is null, the corresponding line is not subdivided.

A new mesh is created from the updated nodes and list of edges. When constructing the connectivity of an S-element, the directions of the elements on all edges have to follow the counter-clockwise direction around its scaling centre. Therefore, between two S-elements sharing one edge, the directions of the line elements on the common edge are opposite. Detailed documentation is included in the code list.

Code List 4.7: Dividing edges of polygons into multiple 2-node line elements

```
1  function [coord, sdConn ] = subdivideEdge(edgeXi, ...
2                        coord,  meshEdge, sdEdge )
3  %Subdivide selected edges of S-elements into multiple elements
4  %
5  %Inputs
6  %    edgeXi(i)          - When it is an array: number of subdivision
7  %                             of edge i
8  %                       - When it is a cell array: column vector of
9  %                             local parametric coordinate (between 0
10 %                             and 1) of subdivision of edge i
11 %    coord(i,:)         - coordinates of node i
12 %    meshEdge(i,1:2)    - the 2 nodes on line i in a mesh.
13 %                         The node number of the first node is
14 %                         larger than that of the 2nd one
15 %    sdEdge{i}          - lines forming S-element i,
16 %                           >0 when a line follows anti-clockwise
17 %                               direction around the scaling centre.
18 %                           <0 clockwise direction
19 %
20 %Outputs
21 %    coord  - nodal coordinates.  New nodes are padded to the end
22 %    sdConn{isd,:}(ie,:)  - S-element conncetivity. The nodes of
23 %                         line element ie in S-element isd.
24
25 nEdge0 = length(meshEdge);  % ......... number of edges in original mesh
26 nExtraPnt = 0;  % ............... number of extra points after subdivision
27 %  ... local parametric coordinate ξ (between 0 and 1) of subdivision
```

```
28  if ~iscell(edgeXi)
29      edgeDiv = edgeXi;  % .... save array of number of subdivision of edges
30      edgeXi = cell(nEdge0,1);
31      for ii = 1:nEdge0
32          if edgeDiv(ii) > 1
33              nDiv = edgeDiv(ii);  % ............. number of subdivision
34              edgeXi{ii} = (1:1:nDiv-1)'/nDiv;  % uniform subdivision
35              nExtraPnt = nExtraPnt + nDiv-1;
36          end
37      end
38  else
39      nExtraPnt = sum( cellfun(@length, edgeXi) );
40  end
41
42  % ... create new nodes and mesh edges
43  newEdges = cell(nEdge0,1);  % .............. to store edges after division
44  ib = length(coord) + 1;  % ........................ index for new nodes
45  coord = [coord; zeros(nExtraPnt, 2)];  % ........ including new nodes
46  for ii = 1:nEdge0  % ..................... dividing edges in orginal mesh
47      xi = edgeXi{ii};  % .............. local parametric coordinate ξ (0, 1)
48      if ~isempty(xi)
49          ie = ib + length(xi) - 1;  % ...... index for the last new node
50          xyb = coord(meshEdge(ii,1),:);  % .. coordinates of start point
51          xye = coord(meshEdge(ii,2),:);  % ... coordinates of end point
52          coord(ib:ie,:) = bsxfun(@times,xyb,1-xi) + ...
53              bsxfun(@times,xye,xi);  % ..... new nodes by interpolation
54          newEdges{ii} = [meshEdge(ii,1), ib:ie; ...
55              ib:ie meshEdge(ii,2)]';  % ....... new edges on an old one
56          ib = ie + 1;
57      else
58          newEdges{ii} = meshEdge(ii,:);  % ............... no division
59      end
60  end
61
62  % ... construct S-element connectivity
63  nsd = length(sdEdge);  % ........................ number of S-elements
64  sdConn = cell(nsd,1);  % ....................... initializing connectivity
65  for isd = 1:nsd
66      sdNewEdges = newEdges(abs(sdEdge{isd}));  % . update connectivity
67      % ... find edges that follow clockwise direction
68      idx = find(sdEdge{isd}(:) < 0);
69      for ii = idx'
70          % ... reverse the element orientation
71          sdNewEdges{ii} = [sdNewEdges{ii}(end:-1:1,2), ...
72              sdNewEdges{ii}(end:-1:1,1)];
73      end
74      % ... new element connectivity of an S-element
75      sdConn{isd} = vertcat(sdNewEdges{:});
76  end
77
78  end
```

4.4   Examples of Scaled Boundary Finite Element Analysis Using Platypus

Several examples commonly found in the literature to evaluate the accuracy of numerical methods are presented in this section. The aim is to illustrate the process of a typical scaled boundary finite element analysis. The accuracy of the scaled boundary finite element formulation for the simple 2-node line element developed in Chapters 2 and 3 is also evaluated. The examples are not selected to illustrate the advantages of the scaled boundary polygons. In Chapter 5, the high flexibility of the scaled boundary finite element method in meshing is shown using Platypus.

The MATLAB codes to compute the examples are organized with the following considerations: 1) Simplicity for readers to use as templates to model their own problems, 2) Flexibility for use to compute all examples in this chapter, and 3) Minimum duplications in the lists of codes. For each example, a problem definition file and a script driving the analysis are needed. The computations in all the examples are performed by the function SBFEPoly2NSolver.m given in Code List 4.8.

A problem definition file contains one main and several local functions. They store and provide access to data on the geometry, material properties, boundary conditions and other information on the problem. The coding of problem definition files will be explained in the descriptions of the examples (see, for example, ProbDefDeepBeam.m on page 181 for the problem definition file of a beam under pure bending).

The main file driving the analysis is a MATLAB script (see, for example, the script on page 185 for the analysis of a beam under pure bending). It saves the handler of the problem definition function that will be used to retrieve model data. This script generates the mesh by calling the function createSBFEMesh.m with the specified parameters. The mesh data and the handler of the problem definition function are passed to the function SBFEPoly2NSolver.m in Code List 4.8 to perform the scaled boundary finite element analysis.

The input arguments of function SBFEPoly2NSolver.m consist of the handler of the problem definition function prodef and the mesh data stored in coord (nodal coordinates), sdConn (connectivity of the S-elements), sdSC (coordinates of the scaling centres of the S-elements). This function is divided into the following blocks:

1) Input of material data (Line 4).
   The material data is accessed through the function handler prodef with the string keyword 'MAT'.
2) S-element solution and global stiffness and mass matrices (Line 7).
   The solutions of the S-elements are obtained and the global stiffness and mass matrices are assembled from the S-element solutions by executing the function SBFE-MAssembly.m in MATLAB Code List 3.2 on page 88.
3) Static analysis (from Line 15)
   The boundary conditions are accessed using the function handler prodef with the string keyword 'BCond'. The input data are grouped as one structure variable. The outputs are stored in a cell array. The passing of inputs and outputs should be coded consistently with the problem definition functions. The unknown nodal displacements are obtained by executing the functions SolverStatics.m (Code List 3.5) with the prescribed boundary conditions.

In the post-processing of results (from Line 22), the deformed mesh is plotted. The displacements required in the problem definition function are output. If the exact solution of displacements is provided in the problem definition function, the following relative error norm in displacement is computed

$$e_u = \frac{\|\{u_{ex}\} - \{u\}\|}{\|\{u_{ex}\}\|} \tag{4.12}$$

where $\{u_{ex}\}$ is the vector of exact solution of nodal displacements and $\{u\}$ the vector of numerical solution.

4) Modal analysis (from Line 47)

The boundary conditions for modal analysis are accessed using the string keyword 'BCondModal' in the same way as in the static analysis. The requested number of modes is specified by the variable opt.modalPara (field modalPara of the variable opt). The natural frequencies and mode shapes are obtained by calling the function SolverMode.m (Code List 3.12). The natural frequencies and period are printed, and the mode shapes are plotted.

5) Response history analysis (from Line 64)

The string keyword to access the boundary conditions is 'BCondTime'. The number of time steps and the size of the time step are specified by the array opt.TIMEPara (field TIMEPara of the variable opt). The response history is obtained by executing the function SolverNewmark.m (Code List 3.13). The string keyword 'OutputTime' accesses the degrees of freedom specified in the problem definition function for output. The response history of these degrees of freedom are plotted.

Code List 4.8: Scaled boundary finite element analysis using polygon S-elements

```
1  function varargout = SBFEPoly2NSolver(probdef, ...
2                          coord, sdConn, sdSC, opt)
```

Scaled boundary finite element analysis using polygon S-elements discretized with 2-node line elements at boundary

Inputs:

> probdef   -   handle of function defining the problem
> coord(i,:)   -   coordinates of node i
> sdConn{isd,:}(ie,:)   -   S-element connectivity. The nodes of line element ie in S-element isd.
> sdSC(isd,:)   -   coordinates of scaling centre of S-element isd
> opt   -   options of analysis. The fields are:
>> type   -   type of analysis, ='STATICS', 'MODAL' or 'TIME'
>> modalPara   -   number of modes
>> TIMEPara   -   an array with [number of time steps, size of time step]

Outputs:

1) *For static analysis*

> U   -   nodal displacements
> sdSln   -   solutions for S-element

2) *For modal analysis*

     `freqAngular`   -   angular frequencies

     `modeShape`   -   modal shapes

3) *For response history analysis*

     `tn`   -   time stations

     `dsp`   -   nodal displacements

     `vel`   -   nodal velocities

     `accl`   -   nodal accelerations

```
3
4    % ...  Material: elasticity matrix and mass density
5    mat = probdef('MAT');  % .......................... get material constants
6
7    % ...  S-element solution and global stiffness and mass matrix
8    [sdSln, K, M] = SBFEMAssembly(coord, sdConn, sdSC, mat);
9
10   if nargin < 5
11       opt.type = 'STATICS';
12   end
13
14   switch opt.type
15       case 'STATICS';  % .................................. Static analysis
16           % ...  Boundary conditions
17           % ...  outArg{1}: displacement constraints. outArg{2}: nodal forces
18           outArg = probdef('BCond', {coord, sdConn});
19           % ...  solution of nodal displacements and reactions
20           [U, ReactFrc] = SolverStatics(K, outArg{1}, outArg{2});
21
22           % ...  Post-processing
23           % ...  plot deformed mesh
24           figure
25           opt = struct('MagnFct', 0.1, 'Undeformed','--k');
26           PlotDeformedMesh(U, coord, sdConn, opt);
27           title('DEFORMED MESH');
28           % ...  output displacements at selected DOFs
29           output = probdef('Output', {coord, sdConn});
30           if ~isempty(output)
31               disp('    DOF          Displacement');
32               fprintf('%6d  %25.15e\n',[output.DispDOF ...
33                                         U(output.DispDOF)]');
34           end
35           % ...  compute error norm
36           Uex = probdef('EXACT',coord);
37           if ~isempty(Uex)
38               % ...  reshape the exact solution to a column vector
39               Uex = reshape(Uex(:,1:2)',[],1);
40               err = Uex-U;  % .......................... displacement error
41               eNorm = norm(err)/norm(Uex);
42               disp([' Displacement error norm = ', num2str(eNorm)]);
43           end
44           % ...  output
```

```
45        varargout = {U, sdSln};

46

47    case 'MODAL'  % .................................. Modal analysis
48        NMode = opt.modalPara;  % ..................... number of modes
49        outArg = probdef('BCondModal', {coord, sdConn});
50        [freqAngular, modeShape] = SolverMode(NMode, K, M, ...
51                outArg{1});  % ........ natural frequencies and mode shapes
52        % ... output natural frequency and period
53        fprintf('%6s %17s %17s\n', 'Mode', 'Frequency', 'Period')
54        for imode = 1:NMode;  % .......................... mode number
55            fprintf('%6d   %15.7d   %15.7d\n', imode, ...
56                freqAngular(imode), 2*pi./freqAngular(imode));
57            figure  % ................................. plot mode shape
58            opt = struct('MagnFct', 0.1, 'Undeformed','--k');
59            PlotDeformedMesh(modeShape(:,imode),coord,sdConn,opt);
60            title(['MODE SHAPE: ',int2str(imode)]);
61        end
62        varargout = {freqAngular, modeShape};

63

64    case 'TIME'  % .......................... Response history analysis
65        TIMEPara = opt.TIMEPara;
66        outArg = probdef('BCondTime', {coord, sdConn});
67        [tn, dsp, vel, accl] = SolverNewmark(TIMEPara, ...
68                outArg{3}, K, M, outArg{1}, outArg{2});
69        % ... plot displacements at selected DOFs
70        output = probdef('OutputTime', {coord, sdConn});
71        if ~isempty(output)
72            figure
73            plot(tn,dsp(output.DispDOF,:),'-k')
74            xlabel('Time (sec)');
75            ylabel('Displacment (m)');
76            title('DISPLACEMENT RESPONSE');
77        end
78        varargout = {tn, dsp, vel, accl};
79    otherwise

80

81  end
```

4.4.1 A Deep Beam

A deep beam with length $L = 2$ m and height $H = 1$ m is shown in Figure 4.12. The Young's modulus is $E = 10 \times 10^6$ kPa and the Poisson's ratio is 0.2. The mass density is 2 Mg/m^3. The plane stress state is assumed. The following types of analyses are performed:

1) Static analysis of the beam subject to pure bending (Figure 4.12a). The moment at the ends of the beam is $= 100$ kNm. The exact solution of displacements and stresses is

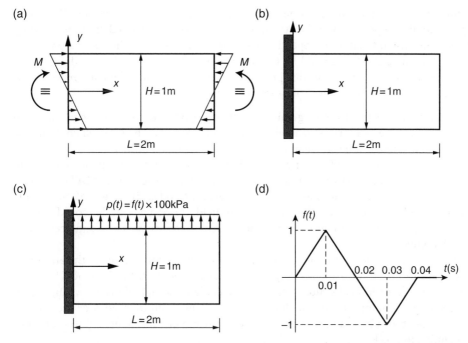

Figure 4.12 Geometry and boundary conditions of a deep beam. (a) Static analysis with surface tractions caused by pure bending. (b) Modal analysis of the beam as a cantilever. (c) Response history analysis of the beam as a cantilever subject to a uniform distributed load. (d) Time variation factor $f(t)$ of the surface traction.

expressed as

$$u = -\frac{M}{EI}xy; \qquad v = 0.5\frac{M}{EI}\left(x^2 + v\left(y^2 - \frac{H^2}{4}\right)\right)$$

$$\sigma_x = -\frac{M}{I}y; \qquad \sigma_y = 0; \qquad \tau_{xy} = 0$$

(4.13)

The surface transactions equivalent to the bending moments are shown in Figure 4.12. The exact solution of displacements in Eq. (4.13) is specified as the boundary conditions at the left side of the beam. The surface tractions are prescribed at the right side of the beam. The top and bottom are traction free.

2) Modal analysis of the beam as a cantilever with the left side of the beam being fixed (Figure 4.12b).
3) Response history analysis with a uniform surface traction $p(t) = f(t) \times 100\,\text{kPa}$ applied to the top of the beam. The time variation factor $f(t)$ is depicted in Figure 4.12d.

The MATLAB file `ProbDefDeepBeam.m` containing the main problem definition function and the local functions are listed on page 181. The main problem definition function provides access to the model data stored within itself and the local functions. The input argument Demand is a string keyword (string of characters). It specifies the data requested and, if applicable, the local functions to call. The input argument `Arg` contains the arguments to be passed to the local functions. The actual type of variable depends on the local function to call. In the problem definition function, the dimensions and material properties of the beam are stored in the structure variable `para`. A rectangular bounding box (`BdBox`) required by the meshers is defined and also used in other instances. In this problem definition function, data are provided for the following keywords of `Demand` (see Line 16):

- `'TriMesh'`: Data of a triangular mesh that are pre-stored in the local function `TriMesh` (Line 155) are returned as an array of two cells. The first cell contains the nodal coordinates and the second the triangular elements. Both are in a row format.
- `'RectMesh'`: A grid of rectangular S-elements is generated by the local function `RectMesh` according to a desired size of the S-elements. The built-in MATLAB function `meshgrid` is called to generate square grids of nodes. The mesh of S-elements is returned as a structure variable with the fields `'coord'` (nodal coordinates) , `'sdConn'` (S-element connectivity) and `'sdSC'` (coordinates of scaling centres of S-elements).
- `'Dist'`: The values of the signed distance function at all the points in the argument `Arg` are returned for use with PolyMesher (Talischi et al., 2012). The local function `DistFuc` (Line 51) calls `dRectangle` to calculate the signed distance function of a rectangle. The output also includes the distances to all four sides of the rectangle. `dRectangle` is a part of PolyMesher available in Talischi et al. (2012) and not included in this book.
- `'BC'`: A cell of two empty arrays is return. It is used in PolyMesher to control the plotting of boundary conditions and is not used here.
- `'Dist_DistMesh'`: The values of the signed distance function at all the points in the argument `Arg` are returned for use in DistMesh (Persson and Strang, 2004). To avoid defining the same geometry separately for the two meshers, the function `dRectangle` of PolyMesher is used and only the last column, which contains the minimum distance to the whole rectangle, is returned.
- `'fh'`: The scaled edge lengths at the points in the argument `Arg` are returned for use in DistMesh. In the example, the length of edges of the triangular mesh are uniform.
- `'pfix'`: The coordinates of hard points, where nodes of the triangular elements have to be created, are returned for use in DistMesh. The four corners of the beam are specified as hard points.
- `'BdBox'`: The data of a rectangular box bounding the problem domain is returned.
- `'MAT'`: The elasticity matrix and mass density of the material are returned.
- `'BCond'`, `'BCondModal'` and `'BCondTime'`: These keywords refer to the local functions on Lines 56, 92 and 106 for the definitions of boundary conditions in the static, modal and response history analyses, respectively.

The nodal coordinates and the connectivity of the polygon S-element mesh are passed to the local functions as fields of the input argument `Arg`. In the static analysis, the nodes at the left side of the beam are identified and the exact solution of displacements is enforced. To prescribe surface tractions, the edges of the polygons are identified and those at the right side of the beam are selected. The equivalent nodal forces are computed and added to the global nodal force vector `F` (see the function `addSurfTraction.m` in Code List 3.4 on page 95). In the modal and response history analyses, all the nodes at the left side of the beam are fixed. In the response history analysis, the time variation factor $f(t)$ (Figure 4.12d) is stored in the variable forceHistory and returned as the last output argument.

- 'Output' and 'OutputTime': The specifications of the outputs are returned as a structure variable. In the static analysis, the output is printed on screen. In the response history analysis, the output is plotted in a figure. For this example, both keywords refer to the same function. The vertical displacement at the upper-right corner is requested.

- 'EXACT': The exact solutions of displacements and stresses (Eq. (4.13 on page 179)) at specified coordinates xy are returned.

```
1   function [x] = ProbDefDeepBeam(Demand,Arg)
2   %Problem definition function of a deep beam
3   %
4   %inputs
5   %    Demand     - option to access data in local functions
6   %    Arg        - argument to pass to local functions
7   %output
8   %    x          - cell array of output.
9
10  % ...  L: Length; H: Height;
11  % ...  E: Young's modulus; pos: Poisson's ratio; den: mass density
12  para = struct('L',2, 'H',1, 'M', 100, ...
13      'E', 10E6, 'pos',0.2, 'den',2);
14  % ...  bounding box [x1,x2,y1,y2]
15  BdBox = [0 para.L -para.H/2 para.H/2];
16  switch(Demand)  % .............. access local functions according to Demand
17      case('para');     x = para;  % ...... parameters of problem definitions
18
19      case('TriMesh'); x = TriMesh;  % ....... direct input triangular mesh
20      case('RectMesh'); x = RectMesh(Arg, para);
21
22      % ...  for use with polygon mesh generator PolyMesher
23      case('Dist');     x = DistFnc(Arg,BdBox);
24      case('BC');       x = {[], []};
25
26      % ...  for use with triangular mesh generator DistMesh
27      case('Dist_DistMesh');   x = DistFnc(Arg,BdBox);
28          x = x(:,end);  % ...................... only use the last column
29      case('fh');       x = huniform(Arg);
30      case('pfix');     x = [BdBox([1 3]); BdBox([1 4]); ...
31              BdBox([2 3]); BdBox([2 4])];
32
33      case('BdBox');    x = BdBox;  % . for use with PolyMesher and DistMesh
34
35      case('MAT');      x = struct('D',IsoElasMtrx(para.E, ...
36              para.pos), 'den', para.den);  % ........ material constants
37
```

```
38      case('BCond');    x = BoundaryCond(Arg, para, BdBox);
39      case('BCondModal');   x = BoundaryCondModal(Arg, para, BdBox);
40      case('BCondTime');    x = BoundaryCondTime(Arg, para, BdBox);
41
42      case('Output');   x = OutputRequests(Arg, BdBox);
43      case('OutputTime');   x = OutputRequests(Arg, BdBox);
44
45      case('EXACT');    x = ExactSln(Arg, para);   % ....... exact solution
46      otherwise
47          warning('Unexpected keyword in ProbDefDeepBeam.')
48  end
49  end
50
51  % ...  Signed distance function
52  function Dist = DistFnc(P,BdBox)
53  Dist = dRectangle(P,BdBox(1),BdBox(2),BdBox(3),BdBox(4));
54  end
55
56  % ...  Boundary conditions
57  function [x] = BoundaryCond(Arg, beam, BdBox)
58  coord = Arg{1};  % ..................................... nodal coordinates
59  sdConn = Arg{2};  % .............................. element connectivity
60
61  eps = 1.d-4*sqrt(beam.L*beam.H/size(coord,1));
62  % ...  Displacement constraints (prescribed acceleration in a response history
63  % ...  analysis). One constraint per row: [Node Dir]
64  lNodes = find(abs(coord(:,1)-BdBox(1))<eps);  % ..... nodes at left edge
65  llNode = find(abs(coord(lNodes,1)-BdBox(1))<eps & ...
66      abs(coord(lNodes,2)-BdBox(3))<eps);  % ............ lower left corner
67  n1 = length(lNodes);
68  ex = ExactSln(coord(lNodes,:), beam);  % ...... exact solution at left side
69  BC_Disp = [ [lNodes; lNodes(llNode)], [ones(n1,1); 2], ...
70      [ex(:,1); ex(llNode,2)]];  % ....... displacement boundary condition
71
72  % ...  Surface traction at right side of beam
73  % ...  edges of polygon S-elements
74  MeshEdges = meshConnectivity( sdConn );
75  % ...  centres of edges of S-elements
76  centres = (coord(MeshEdges(:,1),:)+coord(MeshEdges(:,2),:))/2;
77  % ...  find edges at the right side of the beam
78  rEdges = MeshEdges(abs(centres(:,1)-BdBox(2))<eps,:);
79  % ...  surface traction at nodes
80  ex = ExactSln(coord([rEdges(:,1); rEdges(:,2)],:), beam);
81  n2  = length(rEdges);
82  trac = [ex(1:n2,3)'; zeros(1,n2); ex(n2+1:end,3)'; zeros(1,n2)];
83  % ...  nodal forces
84  F = zeros(2*size(coord,1),1);  % ................. initializing force vector
```

```
85  % ...  add nodal forces equivalent to surface traction
86  F = addSurfTraction(coord, rEdges, trac, F);
87
88  % ...  output
89  x = {BC_Disp, F};
90  end
91
92  % ...  Boundary conditions for modal analysis
93  function [x] = BoundaryCondModal(Arg, beam, BdBox)
94  coord = Arg{1}; % ................................... nodal coordinates
95
96  eps = 1.d-4*sqrt(beam.L*beam.H/size(coord,1));
97  lNodes = find(abs(coord(:,1)-BdBox(1))<eps); % ..... nodes at left edge
98  n1 = length(lNodes);
99  BC_Disp = [ [lNodes; lNodes], [ones(n1,1); 2*ones(n1,1)], ...
100      zeros(2*n1,1)]; % ................... displacement boundary condition
101
102 % ...  output
103 x = {BC_Disp};
104 end
105
106 % ...  Boundary conditions for response history analysis
107 function [x] = BoundaryCondTime(Arg, beam, BdBox)
108 coord = Arg{1}; % ................................... nodal coordinates
109 sdConn = Arg{2}; % ................................ element connectivity
110
111 eps = 1.d-4*sqrt(beam.L*beam.H/size(coord,1));
112 lNodes = find(abs(coord(:,1)-BdBox(1))<eps); % ..... nodes at left edge
113 n1 = length(lNodes);
114 BC_Disp = [ [lNodes; lNodes], [ones(n1,1); 2*ones(n1,1)], ...
115      zeros(2*n1,1)]; % ................... displacement boundary condition
116
117 % ...  surface traction at the top of the beam
118 % ...  edges of polygon S-elements
119 MeshEdges = meshConnectivity( sdConn );
120 % ...  centres of edges of S-elements
121 centres = (coord(MeshEdges(:,1),:)+coord(MeshEdges(:,2),:))/2;
122 % ...  find edges at the top of the beam
123 tEdges = MeshEdges(abs(centres(:,1)-BdBox(2))<eps,:);
124 F = zeros(2*size(coord,1),1); % ................. initializing force vector
125 trac = [ 0; 100; 0; 100]; % ..................... uniform surface traction
126 F = addSurfTraction(coord, tEdges, trac, F); % equivalent nodal forces
127 forceHistory = [0 0; 0.01 1; 0.03 -1; 0.04 0; 100 0];
128
129 % ...  output
130 x = {BC_Disp, F, forceHistory};
131 end
```

```
132
133 % ...  Output Requests
134 function [outputs] =OutputRequests(Arg, BdBox)
135 coord = Arg{1}; %nodal coordinates
136 % ...  find the node at upper-right corner
137 inode = find( abs(coord(:,1)-BdBox(2)) <1.d-5 & ...
138     abs(coord(:,2)-BdBox(4)) <1.d-5 );
139 outputs = struct('DispDOF', 2*inode);  % .. request vertical displacement
140 end
141
142 % ...  Exact solution (see Eq. (4.13))
143 function [x] = ExactSln(xy, beam)
144
145 x = xy(:,1);   y = xy(:,2);  % ................................ coordinates
146 M = beam.M;  % ......................................... bending moment
147 I = beam.H^3/12;  % .................................. moment of inertia
148 E = beam.E; pos = beam.pos;
149 ux = -M/(E*I).*x.*y;  % .................................... displacements
150 uy = 0.5*M/(E*I).*( x.^2 + pos*(y.^2-(beam.H/2)^2) );
151 sx = -M*y/I; sy = zeros(length(x),1); sxy = sy;  % ........... stresses
152 x  = [ux, uy, sx, sy, sxy];  % .................................. output
153 end
154
155 % ...  A pre-stored triangular mesh
156 function [x] = TriMesh()
157 % ...  a triangular mesh
158
159 % ...  nodal coordinates
160 p = [ 0.00  -0.16;   0.00  0.16;   0.00 -0.50;   0.00   0.50;
161       0.41  -0.50;   0.41  0.50;   0.51 -0.00;   0.80  -0.50;
162       0.80   0.50;   1.00  0.00;   1.20  0.50;   1.20  -0.50;
163       1.49   0.00;   1.59  0.50;   1.59 -0.50;   2.00  -0.50;
164       2.00   0.50;   2.00  0.16;   2.00 -0.16 ];
165
166 % ...  triangular elements
167 t = [5    1    3;   15   16   19;    7    2    1;    1    5    7;
168      5    8    7;   18   17   14;    4    2    6;    2    7    6;
169     10    8   12;   10    7    8;   12   15   13;   13   10   12;
170     13   15   19;   19   18   13;   13   18   14;    9    6    7;
171      7   10    9;   11    9   10;   11   13   14;   10   13   11];
172 % ...  output
173 x = {p, t};
174 end
175
176 % ...  Rectangular mesh
177 function [x] = RectMesh(para, beam)
178 % ...  a rectangular mesh
```

```
179
180  h0 = para.h0; % ........................ desired size of square S-elements
181  L = beam.L; H = beam.H; % ................................ dimensions
182  nx = ceil(L/h0); ny = ceil(H/h0); % ............. number of S-elements
183  dx = L/nx; dy = H/ny; % ........................ actual size of S-elements
184  [x, y] = meshgrid((0:nx)*dx, (0:ny)*dy-H/2); % ......... grid of nodes
185  coord = [x(:), y(:)]; % ............................. nodal coordinates
186
187  % ...  construct S-elements
188  [ic, ir] = meshgrid(0:nx-1, 0:ny-1); % .............. grid of S-elements
189  i0 = (ny+1)*(ic(:))+ir(:)+1; % ... node at low-left corner of an S-element
190  sdConn = cell(nx*ny,1); % ...................... initializing connectivity
191  for ii = 1:nx*ny % ................................. loop over S-elements
192      i1 = i0(ii);  i2 = i1+ny+1; % lower left and right nodes
193      sdConn{ii}=[i1 i2; i2 i2+1; i2+1 i1+1; i1+1 i1];
194  end
195  % ...  scaling centre
196  [x, y] = meshgrid((0.5:nx)*dx, (0.5:ny)*dy-H/2); % .... grid of points
197  sdSC = [x(:), y(:)]; % ..................... coordinates of scaling centres
198
199  % ...  output
200  x = struct('coord',coord , 'sdConn',{sdConn}, 'sdSC',sdSC );
201  end
```

The following MATLAB script drives the analysis of this example. The handler of the problem definition function is saved in the variable `probdef`. The approach for mesh generation is specified by the variable `mesher` and size of the polygon S-elements by the structure variable `para` (see the documentation on the inputs of function `createS-BFEMesh.m` on page 155).

```
%A deep beam
close all; clearvars; dbstop error;

% handler of problem definition function
probdef = @ProbDefDeepBeam;

%% Mesh generation
mesher = 3; % = 1: DistMesh; = 3: pre-stored triangular mesh
para.h0 = 0.1; % element size when using DistMesh
[ coord, sdConn, sdSC ] = createSBFEMesh(probdef, mesher, para);

figure;
PlotSBFEMesh(coord, sdConn);
title('MESH');


%% static analysis (pure bending)
slnOpt.type='STATICS'; % 'STATICS'; 'MODAL'; 'TIME';
```

```
[U] = SBFEPoly2NSolver(probdef, coord, sdConn, sdSC, slnOpt);

%% modal analysis (cantilever beam)
slnOpt.type='MODAL';
slnOpt.modalPara = 4;
[f] = SBFEPoly2NSolver(probdef, coord, sdConn, sdSC, slnOpt);

%% response history analysis (cantilever beam)
slnOpt.type='TIME';
slnOpt.TIMEPara = [500 0.0002];
[t, dsp] = SBFEPoly2NSolver(probdef, coord, sdConn, sdSC, slnOpt);
```

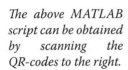

The above MATLAB script can be obtained by scanning the QR-codes to the right.

4.4.1.1 Static Analysis

A scaled boundary polygon mesh is generated from the pre-stored triangular mesh by specifying `mesher=3`. The mesh is plotted in Figure 4.13a. The solution for the nodal displacements is obtained by the function `SBFEPoly2NSolver.m`. The printout of result on screen is

```
DOF           Displacement
100        2.330109796700603e-04
Displacement error norm = 0.027896
```

Therefore, the vertical displacement at the upper-right corner is approximately 2.33×10^{-4} m. The error norm in displacement is about 3%. The deformed mesh under the prescribed bending moment is shown in Figure 4.13b.

The convergence behaviour of the scaled boundary finite element results with increasing number of polygon S-elements and nodes is investigated. The S-element mesh is generated by calling the function `createSBFEMesh.m`. The access to the geometric data of the problem domain is provided by the function handler `prodef`. When the S-element meshes are generated from the triangular meshes created by DistMesh (i.e. `mesher=1` in MATLAB script on page 185), a series of analyses are performed by reducing the element size specified by the parameter `para.h0`. The results of relative error norm in displacement (Eq. (4.12)) as the mesh is refined are shown in Table 4.1.

A similar convergence study is also performed by generating the S-element meshes via PolyMesher. In this case, a mesh is refined by increasing the number of polygons (`para.nPolygon` in the MATLAB script on the preceding page). The results of the relative error norm in displacement are given in Table 4.2.

Figure 4.13 A beam subject to pure bending. (a) Mesh generated from the pre-stored triangular mesh in the problem definition function `ProbDefDeepBeam.m` on page 181 (see Line 155). (b) Deformed mesh.

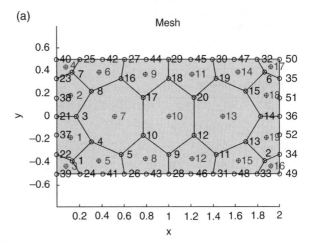

(a)

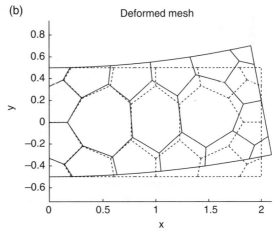

(b)

Table 4.1 Relative error norm in displacement of a beam subject to pure bending. The polygon S-element meshes are converted from triangular meshes generated by DistMesh.

Element size (m)	S-elements	Nodes	Error e_u
0.2	65	159	6.2×10^{-3}
0.1	248	557	1.4×10^{-3}
0.05	974	2073	3.3×10^{-4}
0.025	3786	7822	8.3×10^{-5}

Table 4.2 Relative error norm in displacement of a beam subject to pure bending. The polygon S-element meshes are generated by PolyMesher.

S-elements	Nodes	Error e_u
64	127	7.4×10^{-3}
256	510	1.6×10^{-3}
1024	2037	3.7×10^{-4}
4096	8119	1.0×10^{-4}

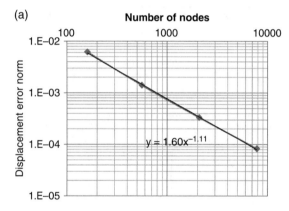

(a)

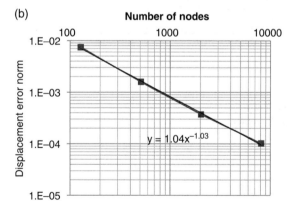

(b)

Figure 4.14 Convergence of displacements of a beam subject to pure bending. (a) The S-element meshes are generated via triangular meshes of DistMesh. (b) The S-element meshes are generated via PolyMesher.

It is observed that the errors in Tables 4.1 and 4.2 decrease as the mesh is refined. For a similar number of nodes, the errors resulting from both types of mesh generation are very close. The error norms are plotted versus the number of nodes for the two mesh generation approaches in Figure 4.14a and Figure 4.14b, respectively. The optimal rate of convergence (= 1) is achieved in both cases.

4.4.1.2 Modal Analysis

The modes of free vibration of the deep beam are computed in the MATLAB script on page 185. The first 4 modes are requested by setting the option slnOpt. modalPara=4. With the polygon mesh in Figure 4.13a (generated from the pre-stored triangular mesh by specifying mesher=3), the natural frequencies (radians per second) and periods (seconds) are obtained as

Mode	Frequency	Period
1	4.9468067e+02	1.2701497e-02
2	1.7652880e+03	3.5592976e-03
3	1.9268996e+03	3.2607747e-03
4	4.1193427e+03	1.5252883e-03

The mode shapes of the first four modes are plotted in Figure 4.15.

A convergence study of the natural frequencies of the deep beam is performed. The S-element meshes are generated from the triangular meshes created by DistMesh (i.e. mesher=1 in MATLAB script on page 185). The parameter para.h0, which specifies the element size, is chosen as 0.2 and is halved subsequently in each analysis. As can be observed from Table 4.3, the four natural frequencies converge.

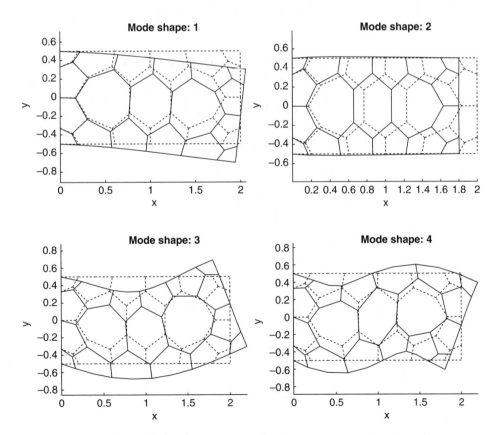

Figure 4.15 Mode shapes of a deep beam.

Table 4.3 Convergence of the first four natural frequencies (rad/s) of a deep beam.

Element size (m)	Mode			
	1	2	3	4
0.2	490.14	1761.0	1897.5	4002.5
0.1	488.94	1760.0	1890.7	3976.2
0.05	488.64	1759.7	1889.0	3969.8
0.025	488.56	1759.7	1888.6	3968.2

4.4.1.3 Response History Analysis

The last section of the MATLAB script on page 185 performs the response history analysis. The boundary condition and external load are shown in Figures 4.12c and 4.12d. The polygon mesh is generated with the options mesher=1 and para.h=0.1, which leads to accurate natural frequencies. The time integration is executed for 500 steps with the size of the time step 0.0002 s. The number of time steps and the size of time step are specified in the structure variable slnOpt.TIMEPara = [500 0.0002]. The response history of the vertical displacement at the upper-right corner is plotted in Figure 4.16.

4.4.1.4 Pure Bending of a Beam: 2 Line Elements on an Edge of Polygons

The problem of pure bending of a beam shown in Figure 4.12, Section 4.4.1 is modelled with every edge of the polygon mesh being divided into two 2-node line elements. The same geometry, material properties and boundary conditions are selected. The problem definition file ProbDefDeepBeam.m listed on page 181 is used to obtain the model data. The analysis is driven by the following script

```
%Pure bending of beam with subdivision of edges
close all; clearvars; dbstop error;

probdef = @ProbDefPureBendingBeam; % function handler

mesher = 1; % = 1: DistMesh
para.h0 = 0.2; % element size
para.nDiv = 2; % number of 2-node line elements per edge
[ coord, sdConn, sdSC ] = createSBFEMesh(probdef, mesher, para);

figure; % plot mesh
opt=struct('sdSC', sdSC, 'LabelSC',1, 'PlotNode',1, ...
    'MarkerSize', 4);
PlotSBFEMesh(coord, sdConn, opt);
title('MESH');

% analysis
[U, sdSln] = SBFEPoly2NSolver(probdef, coord, sdConn, sdSC);
```

The above MATLAB script can be obtained by scanning the QR-code to the right.

The S-element mesh is converted by the function `createSBFEMesh.m` from a triangular mesh generated by DistMesh (option `mesher = 1`) with an element size of 0.2 m (option `para.h0 = 0.2`). Every edge of the polygons is subdivided into 2 line elements (option `para.nDiv = 2`). The final S-element mesh is shown in Figure 4.17a. The analysis is performed by calling the function `SBFE-Poly2NSolver.m`. The deformed mesh is plotted in Figure 4.17b.

A convergence study is performed. The element size h (`para.h0`) of the initial triangular mesh generated by DistMesh is chosen as 0.4 m. The value of h (`para.h0`) is halved successively in each mesh refinement. The division of each edge of a polygon into 2 line elements is maintained during the mesh refinement. The results of relative error norm in displacement and the mesh information are listed in the table on the left side of Figure 4.18. The error decreases as the mesh is refined. The relative error norm in displacement is plotted in Figure 4.18 as a function of the number of nodes in the

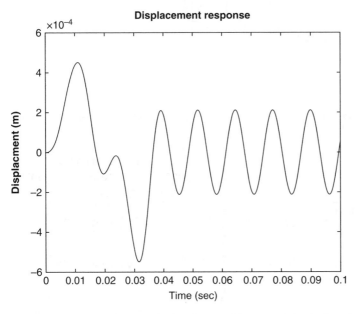

Figure 4.16 Response history of a deep beam subject to a triangular impulse of uniform surface traction.

(a)

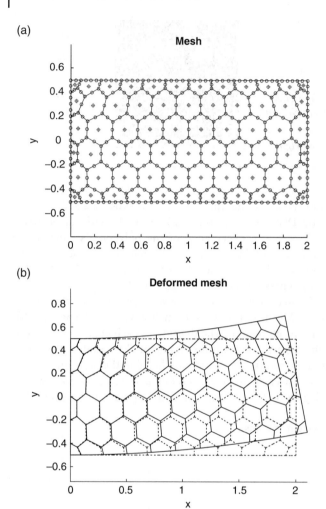

(b)

Figure 4.17 Beam subject to pure bending. (a) Mesh. Every edge of the polygons is discretized with 2line elements. (b) Deformed mesh.

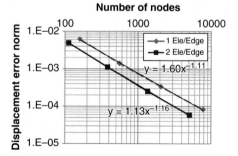

h (m)	Polygons	Nodes	Error e_u
0.4	19	112	4.9×10^{-3}
0.2	65	382	1.1×10^{-3}
0.1	248	1361	2.5×10^{-3}
0.05	974	5119	5.9×10^{-3}

Figure 4.18 Convergence of displacement of a beam subject to pure bending. The results are obtained by discretizing every edge of the polygon mesh with 2 line elements (see Figure 4.17).

mesh. The line is labelled with '2 Ele/Edge'. Higher accuracy than using one line element per edge (labelled with '1 Ele/Edge') is observed at the same number of nodes.

4.4.2 A Circular Hole in an Infinite Plane Under Remote Uniaxial Tension

A circular hole of radius a in an infinite plane under remote uniaxial tension p is shown in Figure 4.19.

The analytical solution of this problem is expressed in the polar coordinates (r, θ) as

$$\sigma_x = \frac{p}{2}\left(2 - \frac{a^2}{r^2}\left(3\cos 2\theta + \left(2 - 3\frac{a^2}{r^2}\right)\cos 4\theta\right)\right) \tag{4.14a}$$

$$\sigma_y = -\frac{pa^2}{2r^2}\left(\cos 2\theta - \left(2 - 3\frac{a^2}{r^2}\right)\cos 4\theta\right) \tag{4.14b}$$

$$\tau_{xy} = -\frac{pa^2}{2r^2}\left(\sin 2\theta + \left(2 - 3\frac{a^2}{r^2}\right)\sin 4\theta\right) \tag{4.14c}$$

for stresses, and

$$u_x = \frac{pa}{8G}\left(\frac{r}{a}(1+\kappa)\cos\theta + \frac{2a}{r}((1+\kappa)\cos\theta + \cos 3\theta) - \frac{2a^3}{r^3}\cos 3\theta\right) \tag{4.15a}$$

$$u_y = \frac{pa}{8G}\left(\frac{r}{a}(\kappa - 3)\sin\theta + \frac{2a}{r}((1-\kappa)\sin\theta + \sin 3\theta) - 2\frac{a^3}{r^3}\sin 3\theta\right) \tag{4.15b}$$

for displacements with the shear modulus G, Poisson's ratio ν and Kolosov constant

$$\kappa = \begin{cases} \dfrac{3-\nu}{1+\nu} & \text{for plane stress} \\[2mm] 3 - 4\nu & \text{for plane strain} \end{cases} \tag{4.16}$$

This problem is modelled by replacing the infinite plane with a square body of dimension $L \times L$ around the circular hole. The exact solution of displacement (Eq. (4.15)) is prescribed as the boundary condition on the four edges of the square.

The problem definite function `ProbDefInfPlateWithHole.m` is listed below. The main function and the local functions are structured in the same way as those of the problem in Section 4.4.1. Plane stress state is considered in this example. The

Figure 4.19 A circular hole in an infinite plane under remote uniaxial tension.

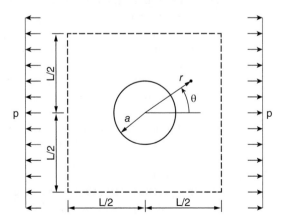

following parameters are used: length of the square plate $L = 2\,\text{m}$, radius of the circular hole $a = 0.4\,\text{m}$, Young's modulus $E = 10^3\,\text{kPa}$, Poisson's ratio is $v = 0.25$, mass density $\rho = 2\,\text{Mg/m}^3$ and remote tension $p = 1\,\text{kPa}$. They are stored in the structure variable `para`.

The geometry of the problem domain is defined by the local function `'Dist'` (Line 49) using the signed distance function. It is modelled as the difference between a solid square of dimension $L \times L$ and a circle of radius a.

```
1   function [x] = ProbDefInfPlateWithHole(Demand,Arg)
2   %Problem definition function of a circular hole in
3   % an infinite plane under remote uniaxial tension
4   %
5   %inputs
6   %    Demand    - option to access data in local functions
7   %    Arg       - argument to pass to local functions
8   %output
9   %    x         - cell array of output.

11  % ...  L: Length; D: Height; R: Radius of hole;
12  % ...  E: Young's modulus; pos: Poisson's ratio; den: mass density
13  para = struct('L',2, 'a', 0.4, ...
14      'E', 1E3, 'pos',0.25, 'den', 2, 'trac', 1);
15  BdBox = [-para.L/2 para.L/2 -para.L/2 para.L/2];  % ..... [x_1,x_2,y_1,y_2]
16  switch(Demand)
17      case('para');    x = para;  % ...... parameters of problem definitions
18          % ...  direct input triangular mesh
19      case('TriMesh'); x = [];

21      % ...  for use with polygon mesh generator PolyMesher
22      case('Dist');    x = DistFnc(Arg,para,BdBox);
23      case('BC');      x = {[],[]};

25      % ...  for use with triangular mesh generator DistMesh
26      case('Dist_DistMesh');   x = DistFnc(Arg,para,BdBox);
27          x = x(:,end);  % ....................... only use the last column
28      case('fh');
29          x = huniform(Arg);  % ........................... element size
30      case('pfix');
31          x = [BdBox(1) BdBox(3); BdBox(1) BdBox(4); ...
32              BdBox(2) BdBox(3); BdBox(2) BdBox(4)];

34      % ...  for use with PolyMesher and DistMesh
35      case('BdBox');   x = BdBox;

37      case('MAT');
38          x = struct('D',IsoElasMtrx(para.E, para.pos), ...
39              'den', para.den);  % .................... material constants
```

In lines 13–14, the subscripts shown are $[x_1,x_2,y_1,y_2]$.

```
40
41      case('BCond');   x = BoundryConditions(Arg, para, BdBox);
42
43      case('Output');  x = OutputRequests(Arg, BdBox);
44
45      case('EXACT');   x =  ExactSln(Arg, para);
46    end
47    end
48
```

49 % ... **Signed distance function**
```
50    function Dist = DistFnc(P,plate,BdBox)
51    d1 = dRectangle(P,BdBox(1),BdBox(2),BdBox(3),BdBox(4));
```
52   d2 = dCircle(P, 0, 0, plate.a); % ..... Centre and radius of circle x_c, y_c, r
```
53    Dist = dDiff(d1,d2);
54    end
55
```

56 % ... **Boundary conditions**
```
57    function [x] = BoundryConditions(Arg, para, BdBox)
```
58   coord = Arg{1}; % .... nodal coordinates
```
59    eps = 0.01*sqrt((BdBox(2)-BdBox(1))*...
60        (BdBox(4)-BdBox(3))/size(coord,1));  % ................... tolerance
```
61   % ... displacement constrains (prescribed acceleration in a response
62   % ... history analysis). One constraint per row: [Node Dir]
```
63    d = dRectangle(coord,BdBox(1),BdBox(2), ...
```
64   BdBox(3),BdBox(4)); % .... distance from nodes to the square boundary
65   bNodes = find(abs(d(:,end))<eps); % ...... nodes on the square boundary
66   ex = ExactSln(coord(bNodes,:), para); % .... exact solution
```
67    n1 = length(bNodes);
68    BC_Disp = [ [bNodes; bNodes], [ones(n1,1); 2*ones(n1,1)], ...
```
69   [ex(:,1); ex(:,2)]]; % .... displacement boundary condition
```
70
```

71 % ... nodal forces
72   F = zeros(2*size(coord,1),1); % .... initializing force vector
```
73
```

74 % ... output
```
75    x = {BC_Disp, F};
76    end
77
```

78 % ... **Output Requests**
```
79    function [outputs] = OutputRequests(Arg, BdBox)
80    outputs = {};
81    end
82
```

83 % ... **Exact solution**
```
84    function x = ExactSln(xy, para)
```
85   p = para.trac; % .... remote tension
86   a = para.a; % .... radius of hole

```
87  pos = para.pos; G = para.E/(2*(1+pos));  %.......... material constants
88  x = xy(:,1);  y = xy(:,2);
89  ar = a./sqrt(x.^2 + y.^2); c = atan2(y,x);  %....... polar coordinates
90  % ... displacements (see Eq. (4.15))
91  ka = (3-pos)/(1+pos);  %................. Kolosov constant for plane stress
92  ux = p*a/8/G *( 1./ar*(1+ka).*cos(c) + ...
93       2*ar.*((1+ka)*cos(c)+cos(3*c)) - 2*ar.^3.*cos(3*c) );
94  uy = p*a/8/G *( 1./ar*(ka-3).*sin(c) + ...
95       2*ar.*((1-ka)*sin(c)+sin(3*c)) - 2*ar.^3.*sin(3*c) );
96  % ... stresses (see Eq. (4.14))
97  sx  = p/2*( 2 - ar.^2.*(3*cos(2*c) + (2-3*ar.^2).*cos(4*c)) );
98  sy  = -p*ar.^2/2.*(cos(2*c) - (2-3*ar.^2).*cos(4*c));
99  sxy = -p*ar.^2/2.*(sin(2*c) + (2-3*ar.^2.).*sin(4*c));
100 x = [ux, uy, sx, sy, sxy];  %................................... output
101 end
```

The analysis is performed using the following MATLAB script. Two options (`mesher=1` or `=2`) for automatic generation of scaled boundary polygon meshes are considered.

```
%A circular hole in an infinite plane under remote uniaxial tension
close all; clearvars; dbstop error;

% handler of problem definition function
probdef = @ProbDefInfPlateWithHole;
% Mesh generation
mesher = 2; % = 1: DistMesh ; = 2 PolyMesher
para.h0 = 0.2; % element size when using DistMesh
para.nPolygon = 128; % number of polygons when using PolyMesher
[ coord, sdConn, sdSC ] = createSBFEMesh(probdef, mesher, para);

figure;
opt = struct('sdSC', sdSC);
PlotSBFEMesh(coord, sdConn, opt);
title('MESH');

% analysis
[U, sdSln] = SBFEPoly2NSolver(probdef, coord, sdConn, sdSC);
```

The above MATLAB script can be obtained by scanning the QR-code to the right.

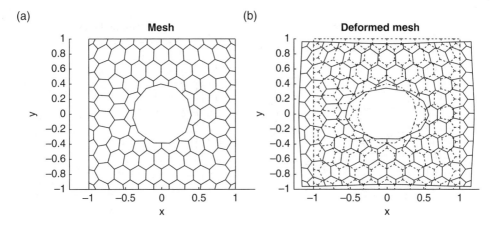

Figure 4.20 (a) Polygon S-element mesh of a square with a circular hole converted from a triangular mesh generated by DistMesh with the element size $h = 0.2$ m. (b) Deformed mesh simulating a circular hole in an infinite plane under uniform remote tension.

To generate an S-element mesh via the triangular mesh produced by DistMesh, the option mesher=1 is specified. The elements are uniformly distributed (Line 29 in ProbDefInfPlateWithHole.m). The element size h is controlled by para.h0 (field h0 of the structure variable para). The S-element mesh generated with the element size $h = 0.2$ m (para.h0=0.2) is shown in Figure 4.20a. The mesh consists of 113 polygon S-elements and 280 nodes. The deformed mesh obtained with the boundary condition simulating an infinite plane is plotted by SBFEPoly2N.m and shown in Figure 4.20b. The relative error norm in displacement defined in Eq. (4.12) is 1.0×10^{-2}.

PolyMesher is selected to generate S-element meshes by specifying mesher=2. The number of polygon is controlled by para.nPolygon (field nPolygon of the structure variable para). A mesh of 128 polygons is shown in Figure 4.21a. The mesh has 262 nodes, which is similar to the number of nodes in the mesh converted from the triangular mesh (Figure 4.20a). It can be observed that the geometry of the circular hole is represented better by PolyMesher than by the converted mesh. The deformed mesh is plotted in Figure 4.21b. The relative error norm in displacement is 4.2×10^{-3}.

A convergence study of the displacement solution is performed. When using DistMesh, the element size is halved repeatedly. When using the PolyMesher, the number of polygons is quadrupled. The relative error norms in displacement (Eq. (4.12)) are listed in Table 4.4. They are plotted against the number of nodes in Figure 4.22. The results converge at the optimal rate ($= 1$) for both types of meshes. The meshes generated via PolyMesher lead to higher accuracy than the meshes converted from triangular meshes as a result of better representations of the geometry of the circular hole.

4.4.3 An L-shaped Panel

An L-shaped panel is shown in Figure 4.23a. The dimensions are indicated in the figure with $b = 1$ m. The material properties are Young's modulus $E = 10 \times 10^6$ kPa, Poisson's

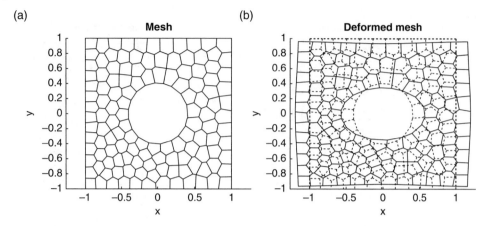

Figure 4.21 (a) Polygon S-element mesh of a square with a circular hole generated by PolyMesher with 128 polygons. (b) Deformed mesh simulating a circular hole in an infinite plane under uniform remote tension.

Table 4.4 Displacement error norm (Eq. (4.12)) of an infinite plate under remote uniaxial tension. (a) The polygon S-element meshes (see Fig 4.20a for an example) are converted from triangular meshes generated via DistMesh. (b) The polygon S-element meshes (see Fig 4.21a for an example) are converted from polygon meshes generated via PolyMesher.

	(a)				(b)		
h (m)	Polygons	Nodes	Error e_u		Polygons	Nodes	Error e_u
0.2	113	280	1.0×10^{-2}		128	262	4.2×10^{-3}
0.1	435	982	2.2×10^{-3}		512	1019	1.1×10^{-3}
0.05	1674	3570	5.5×10^{-4}		2048	4058	2.6×10^{-4}
0.025	6556	13556	1.4×10^{-4}		8192	16254	5.0×10^{-5}

ratio $v = 1/3$ and mass density $\rho = 2\,\mathrm{Mg/m^3}$. The horizontal displacements at the right side and the vertical displacements at the bottom are constrained. Plane stress state is considered. The following types of analyses are performed:

1) Static analysis with a uniform surface traction $p = 100\,\mathrm{kPa}$ applied to the top of the panel (Section 4.4.3.1).
2) Modal analysis.
3) Response history analysis with a uniform surface traction $p(t) = f(t) \times 100\,\mathrm{kPa}$ applied to the top of the panel. The time variation factor $f(t)$ is depicted in Figure 4.23b.

The problem definition function of the L-shaped panel is listed below. The structures of the main function and the local functions are the same as those in the problem definition function of the beam under pure bending on page 181. The geometry of the L-shaped panel is given by the signed distance function (starting from Line 48), which is the difference of the signed distance functions of two squares. The same displacement boundary conditions are applied in all three types of analyses. The time variation factor

Figure 4.22 Convergence of displacements of an infinite plate under remote uniaxial tension. (a) The polygon S-element meshes are converted from the triangular meshes generated by DistMesh. (b) The polygon S-element meshes are generated by PolyMesher.

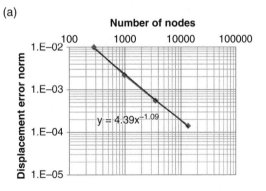

(a)

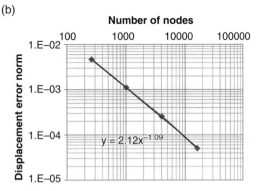

(b)

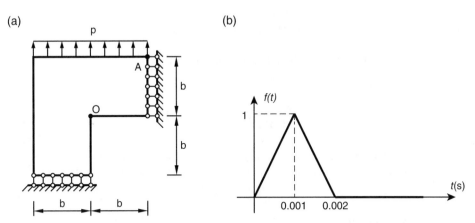

Figure 4.23 An L-shaped panel. (a) Geometry. (b) Time variation factor $f(t)$ of surface traction in response history analysis.

(Figure 4.23b) is given on Line 84. The vertical displacement at the upper-right corner is requested as output in the static and response history analyses.

```
1  function [x] = ProbDefLShapedPlate(Demand,Arg)
2  %Problem definition function of an L-shaped panel
3  %
```

```
 4  %inputs
 5  %    Demand     - option to access data in local functions
 6  %    Arg        - argument to pass to local functions
 7  %output
 8  %    x          - cell array of output.
 9
10  % ...  L1: Height of panel; L2: Height of opening;
11  % ...  E: Young's modulus; pos: Poisson's ratio; den: mass density
12  para = struct('L1',2, 'L2',1, 'E',10E6, 'pos',1/3, 'den',2);
13  % ...  bounding box [x₁,x₂,y₁,y₂]
14  BdBox = [0 para.L1 0 para.L1];
15  switch(Demand)  % .............. access local functions according to Demand
16      % ...  direct input triangular mesh
17      case('TriMesh'); x = TriMesh;
18
19      % ...  options for mesh generators
20      % ...  for use with PolyMesher
21      case('Dist');      x = DistFnc(Arg,para);
22      case('BC');        x = {[],[]};
23
24      % ...  for use with DistMesh
25      case('Dist_DistMesh');  x = DistFnc(Arg,para);
26          x = x(:,end);  % ...................... only use the last column
27      case('fh');        x = huniform(Arg);
28      case('pfix');   x = [0 para.L1; 0 0; para.L2 para.L2; ...
29                  para.L2 0; para.L1 para.L2; para.L1 para.L1];
30
31      % ...  for use with PolyMesher and DistMesh
32      case('BdBox');  x = BdBox;
33
34      case('MAT');    x = struct('D',IsoElasMtrx(para.E, ...
35                  para.pos), 'den', para.den);  % ........ material constants
36
37      case('BCond');  x = BoundryConditions(Arg,  BdBox);
38      case('BCondModal');   x = BoundryConditions(Arg,  BdBox);
39      case('BCondTime');   x = BoundryConditions(Arg,  BdBox);
40
41      case('Output'); x = OutputRequests(Arg, BdBox);
42      case('OutputTime'); x = OutputRequests(Arg, BdBox);
43
44      case('EXACT'); x = [];
45  end
46  end
47
48  % ...  Compute distance functions
49  function Dist = DistFnc(P,plate)
50  d1 = dRectangle(P,0,plate.L1,0,plate.L1); %(P,x1,x2,y1,y2)
```

```
51  d2 = dRectangle(P,plate.L2,plate.L1,0,plate.L2);
52  Dist = dDiff(d1,d2);
53  end
54
55  %  ...  Prescribe boundary conditions
56  function [x] = BoundryConditions(Arg,BdBox)
57  coord = Arg{1};  % ..................................... nodal coordinates
58  sdConn = Arg{2}; % .............................. element connectivity
59
60  %  ...  displacement constraints (prescribed acceleration in a response
61  %  ...  history analysis). One constraint per row: [Node Dir]
62  eps = 0.001*sqrt((BdBox(2)-BdBox(1))*...
63      (BdBox(4)-BdBox(3))/size(coord,1));
64  rightNodes = find(abs(coord(:,1)-BdBox(2))<eps);
65  n1 = length(rightNodes);
66  bottomNodes = find(abs(coord(:,2)-BdBox(3))<eps);
67  n2 = length(bottomNodes);
68  BC_Disp = [ [rightNodes; bottomNodes] ...
69      [ones(n1,1); 2*ones(n2,1)] zeros(n1+n2,1)];
70
71  %  ...  surface traction
72  %  ...  edges of polygon S-elements
73  MeshEdges = meshConnectivity( sdConn );
74  %  ...  centres of edges of S-elements
75  centres = (coord(MeshEdges(:,1),:)+coord(MeshEdges(:,2),:))/2;
76  %  ...  edges at the right side of the square
77  edge = MeshEdges(abs(centres(:,2)-BdBox(4))<eps,:);
78  trac = [ 0; 100; 0; 100];  % .................... uniform surface traction
79  trac = trac(:, ones(1,length(edge)));  % ... surface traction on all edges
80  %  ...  nodal forces
81  F = zeros(2*size(coord,1),1);  % ................ initializing force vector
82  %  ...  add nodal forces equivalent to surface traction
83  F = addSurfTraction(coord, edge, trac, F);
84  forceHistory = [0 0; 0.001 1; 0.002  0; 200 0];  % ... see Figure 4.23b
85
86  %  ...  output
87  x = {BC_Disp, F, forceHistory};
88  end
89
90  %  ...  Output Requests
91  function [outputs] =OutputRequests(Arg, BdBox)
92  coord = Arg{1};
93  %  ...  find the node at upper-right corner
94  inode = find( abs(coord(:,1)-BdBox(2)) <1.d-5 & ...
95          abs(coord(:,2)-BdBox(4)) <1.d-5 );
96  outputs = struct('DispDOF', 2*inode);  % .. request vertical displacement
97  end
```

```
 98
 99   % ...  Triangular mesh
100   function [x] = TriMesh()
101   % ...  nodal coordinates
102   p = [    -0.0000      0.4854;    -0.0000      1.5174;
103       -0.0000      1.0308;     0.0000      0.0000;
104        0.0000      2.0000;     0.4828      2.0000;
105        0.4960      0.7059;     0.4980      0.0000;
106        0.5994      1.3949;     0.9679      2.0000;
107        1.0000      0.4734;     1.0000      0.0000;
108        1.0000      1.0000;     1.2864      1.5042;
109        1.5142      2.0000;     1.5267      1.0000;
110        2.0000      1.0000;     2.0000      2.0000;
111        2.0000      1.5021 ];
112   % ...  triangles
113   t = [ 19       14       16 ;      16       14       13 ;
114        13        7       11 ;      19       18       15 ;
115        15       14       19 ;      13       14        9 ;
116         2        3        9 ;       9        7       13 ;
117         9        3        7 ;       7        3        1 ;
118        19       16       17 ;       8       11        7 ;
119         8        1        4 ;       7        1        8 ;
120        14       15       10 ;      10        9       14 ;
121        11        8       12 ;       9       10        6 ;
122         6        5        2 ;       2        9        6 ];
123   % ...  output
124   x = {p, t};
125   end
```

All the three types of analyses are performed using the following MATLAB script.

```
%An L-shaped plate
close all; clearvars; dbstop error;

probdef = @ProbDefLShapedPlate;% function handler

mesher = 1; % = 1: DistMesh
para.h0 = 0.1; % element size when using DistMesh
[ coord, sdConn, sdSC ] = createSBFEMesh(probdef, mesher, para);

figure;
opt = struct('sdSC', sdSC, 'fill', [0.9 0.9 0.9]);
PlotSBFEMesh(coord, sdConn, opt);
title('MESH');

% static analysis
[U, sdSln] = SBFEPoly2NSolver(probdef, coord, sdConn, sdSC);
```

```
% modal analysis
slnOpt.type='MODAL';
slnOpt.modalPara = 4;
[f, mShp] = SBFEPoly2NSolver(probdef, coord, sdConn, sdSC, slnOpt);

% response history analysis
slnOpt.type='TIME';
slnOpt.TIMEPara = [1000 0.00005];
[t, dsp] = SBFEPoly2NSolver(probdef, coord, sdConn, sdSC, slnOpt);
```

The above MATLAB script can be obtained by scanning the QR-codes to the right.

4.4.3.1 Static Analysis

In the static analysis, a convergence study on the vertical displacement at the upper-right corner (point A in Figure 4.23a) is performed. The S-element meshes are generated via triangular meshes created by DistMesh (option `mesher=1`). The mesh density is controlled by the element size `para.h0`. The mesh generated with `para.h0=0.1` is shown in Figure 4.24a. Note that the re-entrant corner is the scaling centre of an open-loop S-element surrounding it (see Item 2b on page 168 and Figure 4.10 on page 167). This choice allows the scaled boundary finite element method to more accurately model the stress singularity at the re-entrant corner .

A convergence study is also performed using a commercial finite element package Ansys. The vertical displacement at point A (Figure 4.23a) converged up to the first 5 significant digits is obtained as $(v_A)_{\mathrm{ref}} = 1.0560 \times 10^{-4}$ m. It is used as a reference solution to evaluate the error of the present solution v_A

$$e_v = \frac{|(v_A)_{\mathrm{ref}} - v_A|}{(v_A)_{\mathrm{ref}}}$$

The scaled boundary finite element results obtained with a sequence of decreasing size of S-elements are listed in Table 4.5.

A similar study is also performed using Ansys. The L-shaped panel is modelled with a series of increasingly fine uniform meshes of bilinear 4-node elements (Plane 182). The finite element mesh with element size 0.025 m is shown in Figure 4.24b as an example. The results of vertical displacement are shown in Table 4.6. It is observed that the accuracy of the scaled boundary finite element result obtained using the mesh in Figure 4.24a with 835 nodes (2nd row of results in Table 4.5) is similar to that of the finite element result obtained using the mesh in Figure 4.24b with 4961 nodes (3rd row of results in Table 4.6). Both the scaled boundary finite element results and the finite element results

(a)

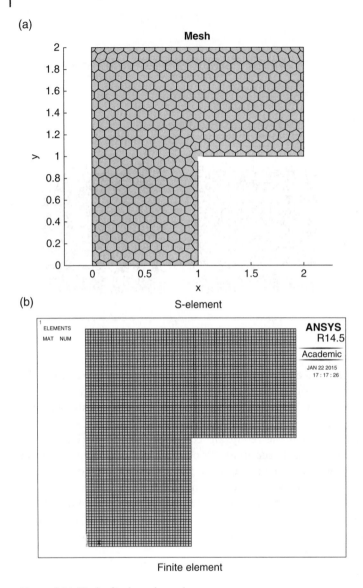

Figure 4.24 Mesh of L-shaped panel.

by Ansys are plotted in Figure 4.25. Based on the linear elastic theory, the stresses are singular at the re-entrant corner (point *O* in the figure). Both sets of results converge at sub-optimal rates owing to the existence of the stress singularity. The scaled boundary finite element results are considerably more accurate for the same number of nodes.

4.4.3.2 Modal Analysis

The MATLAB script on page 202 computes the natural frequencies and mode shapes of the first four modes (`slnOpt.modalPara=4`) of the L-shaped panel. The natural frequencies (radians per second) and periods (seconds) are obtained with the mesh in Figure 4.24a (`para.h0=0.1`) as

Table 4.5 Vertical displacement at upper-right corner of L-shaped panel obtained by the scaled boundary finite element method.

h (m)	S-elements	Nodes	v_A (10^{-4} m)	Error e_v
0.2	100	240	1.0503	5.4×10^{-3}
0.1	376	835	1.0542	1.8×10^{-3}
0.05	1428	3024	1.0553	7.0×10^{-4}
0.025	5611	11563	1.0558	2.6×10^{-4}

Table 4.6 Vertical displacement at upper-right corner of L-shaped panel obtained by the finite element method.

Element Size (m)	Elements	Nodes	v_A ($\times 10^{-4}$ m)	Error e_v
0.1	300	341	1.0434	1.2×10^{-2}
0.05	1200	1281	1.0514	4.3×10^{-3}
0.025	4800	4961	1.0542	1.7×10^{-3}
0.01	3000	30401	1.0555	5.4×10^{-4}
0.005	12000	120801	1.0558	2.4×10^{-4}

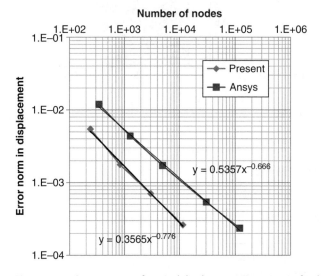

Figure 4.25 Convergence of vertical displacement at point *A* of L-shaped panel.

Mode	Frequency	Period
1	6.0000367e+02	1.0471911e-02
2	1.1480765e+03	5.4727934e-03
3	1.7098009e+03	3.6748052e-03
4	2.1396031e+03	2.9366125e-03

The four mode shapes are plotted in Figure 4.26. Modes 1 and 3 are anti-symmetric and Modes 2 and 4 are symmetric about the −45° diagonal, respectively.

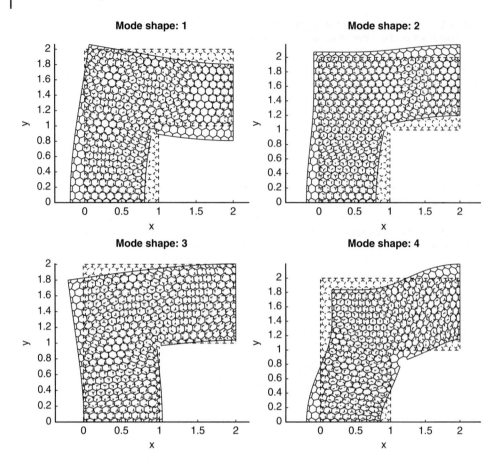

Figure 4.26 Mode shapes of L-shaped panel.

Table 4.7 Convergence of the first four natural frequencies (rad/s) of L-shaped panel.

Element Size (m)	Mode			
	1	2	3	4
0.2	601.33	1150.9	1710.9	2145.5
0.1	600.00	1148.1	1709.8	2139.6
0.05	599.69	1147.0	1709.5	2137.9
0.025	599.61	1146.3	1709.4	2137.3
0.0125	599.58	1146.1	1709.4	2137.1

Figure 4.27 Response history of L-shaped panel subject to a triangular impulse of uniform surface traction.

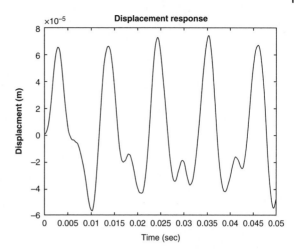

A convergence study of the natural frequencies is performed. The parameter para.h0, which specifies the element size, is chosen as 0.2 and is halved subsequently in each analysis. As can be observed from Table 4.7, the four natural frequencies converge.

4.4.3.3 Response History Analysis

The response history of the L-shaped panel subject to a triangular impulse of uniform surface traction (Figure 4.23) is analysed. The number of time steps and the size of time step are specified in the variable slnOpt.TIMEPara = [1000 0.00005]. The response of the vertical displacement at at point A indicated in Figure 4.23a is plotted in Figure 4.27.

5

Modelling Considerations in the Scaled Boundary Finite Element Analysis

In a numerical method, a mathematical model is approximated by a discrete numerical model. The quality of the discrete model affects the accuracy of the analysis. The scaled boundary finite element method has several distinctive features in comparison with the popular finite element method. The numerical model is obtained by dividing the problem domain into S-domains that satisfy the scaling requirement but are not limited by the number of faces and edges. The solution of an S-domain is sought semi-analytically. The boundary of the S-domain is discretized into line elements in the manner of the finite element method, resulting in an S-element. The solution inside the S-element is obtained analytically. Many of the modelling techniques of the finite element method are applicable to the scaled boundary finite element method. In this chapter, only the modelling considerations pertinent to the scaled boundary finite element method are discussed with examples to illustrate its salient features

1) In the S-element formulation, a scaling centre is selected from where the whole boundary is directly visible. The condition of the scaled boundary transformation is directly related to the visibility of the boundary. The effect of the location of the scaling centre on the accuracy of the solution is investigated in Section 5.1.
2) An S-element can have an arbitrary number of edges as long as the scaling requirement is met. Furthermore, an edge can be divided into any number of line elements. This feature offers a high degree of flexibility in mesh generation. The effect of the subdivision of the edges of S-elements is investigated in Section 4.3.5. The application to mesh transition in local mesh refinement is illustrated.
3) The high flexibility and robustness of S-elements provide a simple procedure to seamless joint domains discretized with non-matching meshes on the interfaces. The process does not require any additional operations or approximations. Examples on the treatment of non-matching meshes are presented in Section 5.3.
4) The semi-analytical solution of an S-element provides an effective approach to model stress singularities. The modelling considerations of this type of applications are discussed in Section 5.4.

5.1 Effect of Location of Scaling Centre on Accuracy

In the scaled boundary finite element method, the geometry of an S-domain has to satisfy the scaling requirement, i.e. there exists a region from which every point on the

The Scaled Boundary Finite Element Method: Introduction to Theory and Implementation,
First Edition. Chongmin Song.
© 2018 John Wiley & Sons Ltd. Published 2018 by John Wiley & Sons Ltd.
Companion website: www.wiley.com/go/author/scaledboundaryfiniteelementmethod

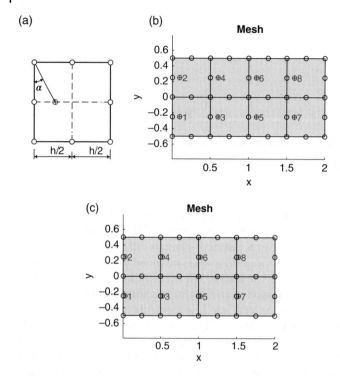

Figure 5.1 Variation of location of scaling centre of square S-elements for the modelling of beam under pure bending. (a) A square S-element is discretized with eight 2-node elements on boundary, where the smallest angle between the radial line and the boundary is denoted as α. (b) Mesh of square S-elements with length $h = 0.5$ m. The scaling centres are placed in locations such that $\tan \alpha = 0.4$ ($\alpha = 21.8°$). (c) $\tan \alpha = 0.1$ ($\alpha = 5.7°$).

boundary is directly visible (see Section 2.2.1). The scaling centre can be selected at any point within this region. An S-element is constructed by discretizing the boundary of the S-domain. The geometry of the S-element is then described in the scaled boundary coordinates that consist of a radial coordinate pointing away from the scaling centre and a circumferential coordinate that is tangential to the boundary (Section 2.2.3). Mathematically, the scaling requirement ensures that the scaled boundary coordinates are not collinear and the scaled boundary transformation is well defined. As explained in Section 2.2.1, when the angle formed by the two scaled coordinates is small (i.e., the visibility of the boundary from the scaling centre is low), the Jacobian matrix of the scaled boundary coordinate transformation may become ill-conditioned.

The visibility of the boundary of an S-element depends on the location of the scaling centre. In this section, the effect of the location of the scaling centre on the accuracy of a scaled boundary finite element analysis is investigated numerically. The problem of a beam under pure bending shown in Figure 4.12 on page 179, Section 4.4.1, is used for this purpose. The beam is discretized by the square S-element shown in Figure 5.1a. Each edge of the square S-element is divided into two elements (see Section 4.3.5). The optimal location of the scaling centre is at the centre of the square. When the scaling centre is shifted horizontally to the left, the smallest angle between a radial line emanating from the scaling centre and the edges is shown in the figure as α ($\alpha = 45°$ when the

Table 5.1 Relative error norm in displacement of a beam subject to pure bending. The meshes are grids of square S-elements with 2 line elements on each edge of a square (see Figure 5.1).

h (m)	Nodes	$\tan\alpha = 1$ ($\alpha = 45°$) Disp. v (10^{-4} m)	Error e_u (10^{-3})	$\tan\alpha = 0.4$ ($\alpha = 21.8°$) Disp. v (10^{-4} m)	Error e_u (10^{-3})	$\tan\alpha = 0.1$ ($\alpha = 5.7°$) Disp. v (10^{-4} m)	Error e_u (10^{-3})
0.5	37	2.3752	9.2	2.3664	12	2.3535	17
0.25	121	2.3935	2.3	2.3909	3.1	2.3875	4.3
0.125	433	2.3983	0.57	2.3976	0.79	2.3967	1.1
0.0625	1633	2.3996	0.14	2.3994	0.20	2.3991	0.27
0.03125	6337	2.3999	0.03	2.3998	0.05	2.3998	0.07

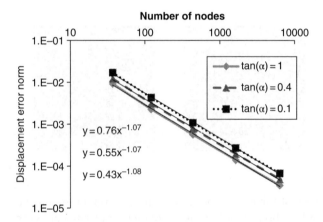

Figure 5.2 Effect of locations of scaling centres of square S-elements on the convergence of displacement of a beam under pure bending.

scaling centre is located at the centre of the square). The beam is discretized uniformly. A mesh with square S-elements of a length of $h = 0.5$ m is shown in Figure 5.1b and c. The scaling centres of the S-elements are chosen in such locations that $\tan\alpha = 0.4$ ($\alpha = 21.8°$) applies in Figure 5.1b and $\tan\alpha = 0.1$ ($\alpha = 5.7°$) in Figure 5.1c.

A convergence study is performed using the following MATLAB script for three cases of angle α: $\tan\alpha = 1$ ($\alpha = 45°$), $\tan\alpha = 0.4$ ($\alpha = 21.8°$) and $\tan\alpha = 0.1$ ($\alpha = 5.7°$). For each case, the length h of the square S-element, specified by the variable `para.h0`, is halved successively starting from $h = 0.5$ m. The vertical displacement at the upper-right corner of the beam and the relative error norm of the nodal displacement are listed in Table 5.1. The error norm is plotted in Figure 5.2.

```
%Pure bending of beam: shift scaling centres
close all; clearvars; dbstop error;

probdef = @ProbDefPureBendingBeam; % function handler

mesher = 'RectMesh'; % keyword to access square grid of S-elements
```

```
para.h0 = 0.5; % size of square S-elements
para.nDiv = 2; % number of line elements per edge
[ coord, sdConn, sdSC ] = createSBFEMesh(probdef, mesher, para);

a = 0.2; % tan(alpha)
sdSC(:,1) = sdSC(:,1) - (1-a)*para.h0/2; % shift scaling centre

h = figure;
opt=struct('sdSC', sdSC, 'LabelSC',12, ...
    'fill', [0.9 0.9 0.9], 'PlotNode',1, 'MarkerSize', 6);
PlotSBFEMesh(coord, sdConn,opt);
title('MESH');

% analysis
[U, sdSln] = SBFEPoly2NSolver(probdef, coord, sdConn, sdSC);
```

The above MATLAB script can be obtained by scanning the QR-codes to the right.

It is observed that the accuracy of the results decreases with the minimum angle α formed between a radial line and the boundary, but the decrease is not significant even at the angle $\alpha = 5.7°$. Furthermore, the rate of convergence is barely affected. From practical application point of view, an angle of $\alpha > 10°$ is satisfactory.

5.2   Mesh Transition

In the scaled boundary finite element method, an S-element may have any number of edges as long as the scaling requirement is satisfied. An edge can also be divided into multiple elements. The S-element provides higher flexibility in mesh transition than a finite element does.

5.2.1   Local Mesh Refinement

In a stress analysis, it is often necessary to employ gradations of element size in a mesh. For example, it is desirable to refine the mesh locally in the region where stress varies rapidly. The application of the scaled boundary finite element method to local mesh refinement is straightforward. It is demonstrated by modelling the L-shaped panel shown in Figure 4.23, Section 4.4.3 on page 197. The same dimensions, material properties and boundary conditions are used here. At the re-entrant corner, a stress singularity exists. It is expected that the stress variation is more rapid around the re-entrant corner than elsewhere.

The analysis is driven by the following script:

```
% L-shaped plate with mesh transition
close all; clearvars; dbstop error;

probdef = @ProbDefLShapedPlate;%problem definition function handler

mesher = 3; % = 3: stored triangular mesh
[ coord, sdConn, sdSC ] = createSBFEMesh(probdef, mesher);

[ meshEdge, sdEdge ] = meshConnectivity( sdConn );
nDiv = 1; % number of subdivisions of an edge
edgeDiv = nDiv*ones(length(meshEdge),1); % initialization
% number of subdivisions of edges connected to S-element 13
edgeDiv(abs(sdEdge{13})) = 2*nDiv;
[coord, sdConn ] = subdivideEdge(edgeDiv, coord, meshEdge, sdEdge);

h = figure;
opt=struct('sdSC', sdSC, 'LabelSC', 12, ...
    'fill', [0.9 0.9 0.9], 'PlotNode', 1);
PlotSBFEMesh(coord, sdConn, opt);
title('MESH');

% analysis
[U, sdSln] = SBFEPoly2NSolver(probdef, coord, sdConn, sdSC);
```

The above MATLAB script can be obtained by scanning the QR-codes to the right.

Two S-element meshes are show in Figure 5.3. They are converted from the same triangular mesh stored in the problem definition file (Line 99 of the problem definition file on page 199). The mesh connectivity in terms of the lines that form the edges of the polygons is constructed from the S-element connectivity using the function meshConnectivity.m in Code List 4.3. The re-entrant corner is the scaling centre of S-element 13 in the mesh shown in Figure 5.3. To improve the accuracy of the solution, the line elements at the edges connected to S-element 13 are refined locally. Globally, one edge of a polygon is divided into nDiv line elements. Locally within S-element 13, one edge is divided 2*nDiv line elements. The function subdivideEdge.m is called to subdivide the edges. The S-element meshes after subdivision with nDiv=1 and nDiv=2 are shown in Figures 5.3a and 5.3b, respectively.

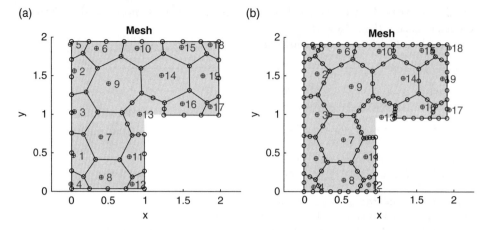

Figure 5.3 S-element mesh of L-shaped panel. The line elements at the edges of S-element 13 is refined locally. (1) Two line elements per edge for S-element 13 and one line element per edge elsewhere. (2) Four line elements per edge for S-element 13 and two line elements per edge elsewhere.

nDiv	Nodes	$v_A(10^{-4}\text{m})$	Error e_v
1	56	1.0367	1.8×10^{-2}
2	129	1.0525	3.3×10^{-3}
4	275	1.0552	8.0×10^{-4}
8	567	1.0559	1.6×10^{-4}
16	1151	1.0560	2.3×10^{-5}

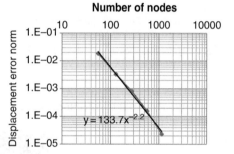

Figure 5.4 Convergence of vertical displacement at point A of L-shaped panel as the edges of polygon mesh (Figure 5.3) are divided into increasing number ($\texttt{nDiv}$) of line elements.

As only two S-elements can be connected to one given edge, and an S-element can have any number of elements on its edges, the effect of subdividing an edge is limited to the two S-elements directly connected to it. This leads to a high degree of flexibility and simplicity in local mesh refinement.

The convergence behaviour is investigated by increasing the number of subdivisions $\texttt{nDiv}$ of an edge while the polygons remain the same as in Figure 5.3. The analysis is performed by calling the function $\texttt{SBFEPoly2NSolver.m}$. The vertical displacement v_A at point A in Figure 4.23 and its relative error are shown in the table on the left side of Figure 5.4. The relative error is plotted as a function of the number of nodes in Figure 5.4. Rapid convergence with a rate larger than 2 is observed.

5.2.2 Rapid Mesh Transition

The effect of rapid mesh transition on the accuracy of the solution is investigated. The problem of beam subject to pure bending (Figure 4.12, Section 4.4.1) is considered. The dimensions, material properties and boundary conditions remain the same.

The analysis is driven by the script below. In the script, the problem definition function `ProbDefDeepBeam.m` with the keyword `mesher = 'RectMesh'` is used to access the local function `RectMesh` (page 184). It generates a grid of 8 square S-elements with the size specified by the variable `para.h0=0.5` (see Figure 5.5a). To illustrate the simplicity in mesh transition, the edges connected to S-elements 7 and 8 are further subdivided into multiple line elements (see Figure 5.5b). S-elements 7 and 8 could also be further divided in to smaller S-elements or connected to other S-elements of smaller sizes.

```
% Rectangular mesh of beam: Illustration of mesh transition
close all; clearvars; dbstop error;

probdef = @ProbDefPureBendingBeam; % function handler

mesher = 'RectMesh'; % keyword to access square grid of S-elements
para.h0 = 0.5; % size of a square S-element
[ coord, sdConn, sdSC ] = createSBFEMesh(probdef, mesher, para);

[ meshEdge, sdEdge ] = meshConnectivity( sdConn );
edgeDiv = ones(length(meshEdge),1); % initializaiton
% number of subdivisions of edges connected to S-elements 7 and 8
edgeDiv(abs(sdEdge{7})) = 5;
edgeDiv(abs(sdEdge{8})) = 11;
[coord, sdConn ] = subdivideEdge(edgeDiv, coord, meshEdge, sdEdge);

figure;
opt=struct('sdSC', sdSC, 'LabelSC',14, 'PlotNode',1);
PlotSBFEMesh(coord, sdConn,opt);
title('MESH');

% analysis
[U, sdSln] = SBFEPoly2NSolver(probdef, coord, sdConn, sdSC);
```

The above MATLAB script can be obtained by scanning the QR-codes to the right.

As a reference case, the mesh in Figure 5.5a, with every edge of the squares being modelled with 1 line element, is analysed. The error norm in displacement is obtained as 5.8%. The mesh in Figure 5.5b, where the edges connected to S-element 7 and 8 are subdivided into 5 and 9 line elements, shows abrupt change in element size. Using this mesh, the error norm in displacement is equal to 5.4%, which is slightly smaller than the

(a)

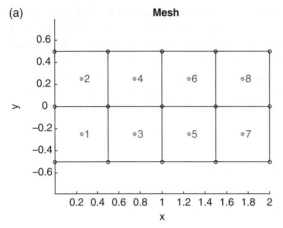

Figure 5.5 Illustration of mesh transition using a beam discretized with square S-elements. (a) Every edge of the squares is discretized with 1 line element. (b) Edges (except for the common edge with S-element 8) connected to S-element 7 are discretized with 5 line elements and edges connected to S-element 8 with 9 line elements.

(b)

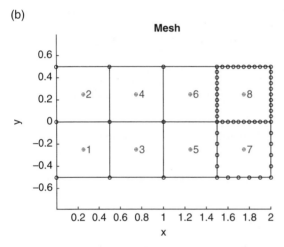

reference case. The major part of the error is caused by S-elements 1-6 that have large line elements on the boundary. Further analyses also confirm that this simple approach of mesh transition does not adversely affect the accuracy of the results even with rather abrupt changes in element size.

5.2.3 Effect of Nonuniformity of Line Element Length on the Boundary of S-elements

The effect of nonuniformity of line element length on the boundary of S-elements on the accuracy of results is examined. A discrete method allows the use of high nonuniform mesh without meaningful loss of accuracy and reduces the burden in mesh genera-tion. The nonuniformity of the length of line elements is simulated by the subdivision of S-element edges into shorter line elements of different lengths.

The L-shaped panel shown in Fig 4.23, Section 4.4.3 on page 197, is considered. The same dimensions, material properties and boundary conditions apply. The problem def-inition function is provided on page 199. The mesh shown in Figure 5.3a and its result in Figure 5.4 (nDiv=1) are used as the reference case.

In the following script, each line element in Figure 5.3a is divided into 2 line elements. The point of subdivision is specified by the parametric coordinate (xi), which ranges from 0 at the starting node to 1 at the ending node of the original element. It represents the length ratio between a new element to the original element. The mesh after subdivision with the length ratio 0.1 (xi=0.1), resulting in S-elements with strongly nonuniform line elements on the edges, is shown in Figure 5.6.

```
%An L-shaped plate with nonuniform subdivision of edges
close all; clearvars; dbstop error;

probdef = @ProbDefLShapedPlate; % function handler

mesher = 3; % = 3: stored triangular mesh
[ coord, sdConn, sdSC ] = createSBFEMesh(probdef, mesher);

[ meshEdge, sdEdge ] = meshConnectivity( sdConn );
% parametric coordinates of subdivision points (0, 1)
xi = [ 0.1]; % if xi = []; no subdivision
edgeDiv(1:length(meshEdge)) = { xi };
% on edges connected to S-Element 13
edgeDiv(abs(sdEdge{13})) = { [ xi 1 1+xi]'/2 };
[coord, sdConn ] = subdivideEdge(edgeDiv, coord, meshEdge, sdEdge);

figure;
opt=struct('sdSC', sdSC, 'LabelSC', 12, ...
    'fill', [0.9 0.9 0.9], 'PlotNode', 1, 'MarkerSize',4);
PlotSBFEMesh(coord, sdConn, opt);
title('MESH');

% analysis
[U, sdSln] = SBFEPoly2NSolver(probdef, coord, sdConn, sdSC);
```

The above MATLAB script can be obtained by scanning the QR-codes to the right.

The vertical displacement at point *A* obtained with the length ratios 0.1, 0.01 and 0.001 and the errors are shown in the table in Figure 5.6. The result of the reference case (mesh in Figure 5.3a without subdivision) is obtained by setting xi = [] and is also listed in the table (The same result is shown in the table on the left side of Figure 5.4 with nDiv=1). It can be observed that the accuracy decreases as the length ratio of the short element becomes smaller (i.e. as the nonuniformity of the subdivision increases). On the other

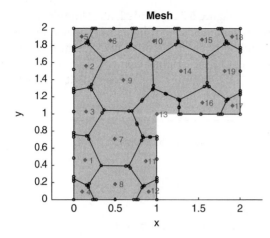

Length ratio	$v_A(10^{-4}\text{m})$	Error e_v
0.1	1.0447	1.1×10^{-2}
0.01	1.0387	1.6×10^{-2}
0.001	1.0372	1.8×10^{-2}
[]	1.0367	1.8×10^{-2}

Figure 5.6 Effect of nonuniform subdivision of edges on accuracy of vertical displacement at point A of L-shaped panel. The S-element mesh on the left is obtained by subdividing the elements in the S-element mesh shown in Figure 5.3a into 2 elements. The length ratio between the shorter element to the original element is 0.1.

hand, even if the length ratio is only 0.001, the accuracy is no less than that of the reference case.

This example shows that the accuracy of the result is not adversely affected when nonuniform subdivisions (or extra nodes) are introduced for the purpose of facilitating mesh transition.

5.3 Connecting Non-matching Meshes of Multiple Domains

The flexibility of the S-elements in mesh transition is demonstrated in Section 5.2. This feature of the S-elements provides a simple and robust approach to connect non-matching meshes as occurring in the domain decomposition method and contact mechanics.

This approach is illustrated in Figure 5.7. The two rectangular domains in Figure 5.7a are meshed independently. Any type of displacement-based elements, including standard finite elements, is possible although the S-elements are used here. The two domains are to be connected by shifting them vertically. At their interface, the nodes (referred to as interface nodes) of the two meshes exist at different locations and do not match. The two meshes are non-matching as the displacement continuity cannot be enforced at the nodes.

The meshes can easily be modified using S-elements so that they match at the interface. As depicted in Figure 5.7b, the non-matching interface nodes of one mesh are inserted to the other mesh. For example, the modification of the lower mesh is considered. All the interface nodes of the upper mesh that do not match any nodes of the lower match are identified. These nodes are inserted on the interface of the lower mesh as indicated by the square nodes and arrows pointing downwards. They subdivide the interfaces edge of the S-elements into multiple line elements as discussed in Section 4.3.5. The upper mesh is modified in the same way and the non-matching nodes at the interface

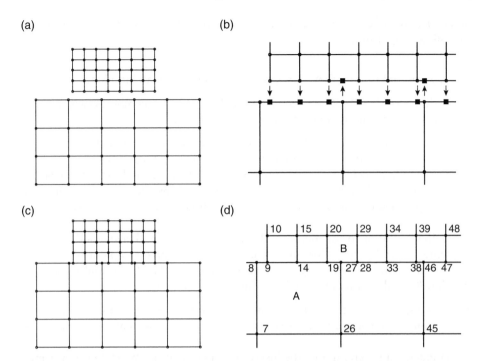

Figure 5.7 Connecting of non-matching meshes by S-elements. (a) Two non-matching meshes on two rectangular domains to be connected by shifting them vertically. (b) Close-up view of the S-elements adjacent to the left part of the interface. At the interface, an edge of an S-element of a mesh is subdivided at the locations marked by squares to match the nodes of the other mesh at the opposite side (indicated by arrows). (c) The combined mesh satisfies compatibility at the interface. (d) Close-up view of the S-elements adjacent to the left part of the interface after connecting the two meshes. The upper edge of S-element A is subdivided into 4 line elements and the lower edge of S-element B into 2 line elements.

of the lower mesh are inserted by subdividing the interface edges into multiple line elements (as indicated by the square nodes and upward arrows in Figure 5.7b).

After the modifications of the meshes at the interface, the two meshes become matching and can be combined by connecting the interface nodes. Since the S-elements can have any number of edges and tolerate highly nonuniform length of line elements, no further change of the mesh is required. The resulting mesh is shown in Figure 5.7c. The close-up view of the S-elements adjacent to the left part of the interface is shown in Figure 5.7d with the nodal numbers of the final mesh. For example, the upper edge of S-element A in the lower rectangle is subdivided into four 2-node line elements ($27 - 19$, $19 - 14$, $14 - 9$ and $9 - 8$). The upper edge of S-element B in the upper rectangle is subdivided into two 2-node line elements ($19 - 27$ and $27 - 28$). The displacement compatibility is satisfied at the interface.

It is worthwhile mentioning that the original meshes can also be triangular and quadrilateral finite elements. In this case, it is sufficient to replace these elements that have been modified with polygon S-elements.

Simple functions for combining non-conforming meshes are included in the accompanying computer program Platypus. It is only intended to demonstrate the above approach and be used for the examples in this chapter. These functions are presented in the following sections.

5.3.1 Computer Program Platypus: Combining Two Non-matching Meshes

The function `combineTwoMeshes.m` in Code List 5.1 below combines two independently meshed domains into one. The input arguments are:

1) `d`: A cell array storing the meshes of the two domains to be combined.
2) `onLine`: The handle of a function that finds the points on the interface from a given array of points specified by their coordinates and returns the indices of these points on the interface. The function is defined as a part of the problem definition function (see the problem definition function `ProbDefCombinedPlateWithHole.m` on page 226).
3) `res`: The resolution (or tolerance) of the nodal coordinates. To avoid nodes that are extremely close to each other and are duplicates from practical point of view, the coordinates of the nodes on the interface will be rounded up to the multiples of the resolution. Nodes with duplicated coordinates are merged.

The output arguments `coord`, `sdConn` and `sdSC` store, respectively, the nodal coordinates, S-element connectivity and the scaling centres of the S-elements of the combined mesh.

For simplicity in programming, the insertion of nodes on the interface is performed separately for the two meshes. This function consists of three sections. In the first section, the interface edges and interface nodes are identified (Line 3) using the function specified by the function handle `onLine` for both meshes. In the second section (Line 22), the interface nodes of one mesh are inserted into the other mesh as illustrated in Figure 5.7b. The locations of the interface nodes of a mesh on the interface edges of the opposite mesh are determined by the function `findPointsOnLine.m` (see Code List 5.2 on page 223 and the explanation preceding it). The output is provided in terms of a local parametric coordinate that is equal to 0 at the first point of the edge and 1 at the second point. The interface edges are subdivided into line elements by inserting nodes at these locations using the function `subdivideEdge.m` (Code List 4.7 on page 173). In the last section (Line 40), the two meshes are connected at the interface and the duplicate nodes are removed to generate the combined mesh.

Code List 5.1: Combining two non-matching meshes

```
1  function [coord, sdConn, sdSC] = combineTwoMeshes(d, onLine, res)
```

Combine two non-matching meshes on a line defined by the function onLine

Inputs:

```
      d{i}      -   mesh data of domain i
      onLine{i}  -   handler of function defining the interface between domain i and i+1
      res   -   resolution of coordinates. Coordinates of the nodes on the interface (except for the
               end nodes) will be rounded up to multiples of the resolution and nodes at the same coor-
               dinates will be merged.
```

Outputs:

> `coord(i,:)` - Coordinates of node i
> `sdConn{isd,:}(ie,:)` - S-element connectivity. The nodes of line element `ie` in S-element `isd`.
> `sdSC(isd,:)` - scaling centres of S-element `isd`.

```
 2
 3   % ...  find interface edges and nodes
 4   ctLine = cell(2,1);  % ....... cells store the data of the two meshes at their interface
 5   for ii = 1:2
 6       % ...  find interface edges (with both nodes on interface)
 7       ctLine{ii}.edgeID = intersect( ...
 8               onLine(d{ii}.coord(d{ii}.meshEdge(:,1),:), res), ...
 9               onLine(d{ii}.coord(d{ii}.meshEdge(:,2),:), res));
10       edge = d{ii}.meshEdge(ctLine{ii}.edgeID,:)';
11       % ...  round the coordinates of the nodes on interface to a specified tolerance
12       % ...  to avoid inserting nodes very close to an existing nodes
13       [node, ~, ic] = unique(edge);  % ................. nodes on interface
14       binCounter = histc(ic, 1:length(node));
15       endNode = node(binCounter==1);  % . nodes at the two ends of interface
16       midNode = setdiff(node, endNode);  % ..... middle nodes on interface
17       d{ii}.coord(midNode,:) = round(d{ii}.coord(midNode,:)/res)*res;
18       d{ii}.edgexy =  d{ii}.coord(edge,:);
19       % ...  coordinates of nodes on interface
20       ctLine{ii}.xy = d{ii}.coord(node,:);
21   end
22
```

`edgeXi{1}{i}` stores the parametric coordinates ($0 < \xi < 1$) of the points that will be inserted into the interface edge i of mesh 1 to match the interface nodes of mesh 2 (see Figure 5.7). Correspondingly, `edgeXi{2}{i}` is for the points to be inserted into the interface edges of mesh 2.

```
23   edgeXi{1} = findPointsOnLine(d{1}.edgexy, ctLine{2}.xy, res);
24   edgeXi{2} = findPointsOnLine(d{2}.edgexy, ctLine{1}.xy, res);
25
26   % ...  subdivide the interface edges by inserting points to match the two meshes
27   for ii = 1:2
28       edgeDiv = cell(length(d{ii}.meshEdge),1);  % ......... initialization
29       % ...  parametric coordinates of points of subdivision at interface edges
30       edgeDiv(ctLine{ii}.edgeID) = edgeXi{ii};
31       n1 = length(d{ii}.coord)+1;  % .. number of nodes before subdivision
32       % ...  subdivide edges and update the mesh of a domain
33       [d{ii}.coord, d{ii}.sdConn ] = subdivideEdge(edgeDiv, ...
34               d{ii}.coord, d{ii}.meshEdge, d{ii}.sdEdge);
35       % ...  round the coordinates of newly created nodes on interface
36       d{ii}.coord(n1+1:end,:) = round( ...
37               d{ii}.coord(n1+1:end,:)/res )*res;
38   end
```

```
39
40   % ... append the nodes of the two meshes
41   coord  = round([d{1}.coord; d{2}.coord]/(0.5*res));
42   [~, ia, ic] = unique(coord,'rows');  % .................... merge nodes
43   coord = coord(ia(:),:)*(0.5*res);
44   sdSC = [d{1}.sdSC; d{2}.sdSC];
45   % ... append the S-element connectivity of meshes
46   sdConn = [ d{1}.sdConn; d{2}.sdConn];
47   % ... shift nodal numbers in mesh 2 by the number of nodes in mesh 1
48   shft = [ zeros(1,length(d{1}.sdConn)) ...
49       length(d{1}.coord)*ones(1,length(d{2}.sdConn))];
50   % ... update S-element connectivity
51   for ii = 1:length(sdConn)
52       sdConn{ii} = ic(shft(ii)+sdConn{ii});  % ................. shifting
53   end
54
55   end
```

The function `findPointsOnLine.m` in Code List 5.2 on page finds the points on given lines from a list of points. A line is defined by the coordinates of its two ends in the argument `line`. The points are specified by their coordinates in the argument pnt. The tolerance of the coordinates is given by the argument `res`.

The algorithm is explained using Figure 5.8 where one line defined by Points 1 and 2 is shown. The origin of coordinates is shifted to Point 1. The length of the line from Point 1 to Point 2 is equal to $\sqrt{x_2^2 + y_2^2}$. In this coordinate system, a query point, e.g. Point A (x_q, y_q), is considered. Twice the area of the triangle formed by the three points A, 1 and 2 is equal to $|x_q y_2 - y_q x_2|$ (Line 11). When the height of the triangle is equal to zero (considering the tolerance in coordinates and round-off error), the point is located on the line (see Points B, C and D in Figure 5.8).

Only the points on the line are considered further. A local parametric coordinate ξ (Note that this symbol does not denote the radial coordinate in this section) is defined on the line. It is equal to 0 at Point 1 and 1 at Point 2. The parametric coordinates of the points on the line are evaluated (see Line 15). The parametric coordinates of the points located between Points 1 and 2, which are between 0 and 1, are sorted in ascending order and returned as outputs (argument `lineXi`) of the function.

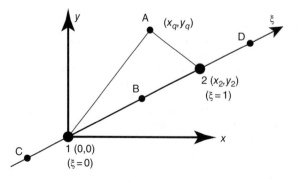

Figure 5.8 Find points on a line defined by points 1 and 2.

Code List 5.2: Find points on lines and determine their local parametric coordinates

```
1  function [ lineXi ] = findPointsOnLine(line, pnt, res)
```

From a given list of points, find those on a line and determine their parametric coordinates ($0 < \xi < 1$)

Inputs:

> `line(2*i-1:2*i,:)` - coordinates of 2 points defining line `i`
> `xyq(i,:)` - coordinates of point `i` on the list of points
> `res` - resolution (tolerance) of coordinates

Outputs:

> `lineXi{i}` - parametric coordinates ($0 < \xi < 1$) of points on edge `i`.

```
2
3  nline = length(line)/2;  % ........................... number of lines
4  lineXi = cell(nline,1);  % ........................... initialization
5  for ii = 1:nline  % ..................................... loop over lines
6      xy = line(2*ii-1:2*ii,:);  % . coordinates of the two ends of the line
7      % ...  shift coordinate orgin to the first point of the line
8      xyq = bsxfun(@minus, pnt, xy(1,:));
9      xy = xy(2,:)-xy(1,:);
10     leng = sqrt(xy(1)^2+xy(2)^2);
11     % ...  areas of triangles formed by points in the list and the line
12     a2 = abs( xyq(:,1)*xy(2) - xyq(:,2)*xy(1) );
13     % ...  points on line (the height of triangle is equal to zero)
14     xyq = xyq( a2/leng < res, :);
15     % ...  parametric coordinates of points on line
16     if abs(xy(2)) > abs(xy(1))  % use the coordinate with larger change
17         xi = xyq(:,2)/xy(2);
18     else
19         xi = xyq(:,1)/xy(1);
20     end
21     % ...  select the points with their parametric coordinates between 0 and 1
22     lineXi{ii} = sort(xi( (1d-5<xi)&(1-xi)>1d-5) );
23 end
24
25 end
```

5.3.2 Computer Program Platypus: Modelling of a Problem by Multiple Domains with Non-matching Meshes

In the domain decomposition method, a complex problem is split into smaller domains. Each domain is meshed independently. The compatibility and equilibrium are enforced during the iterative solution procedure. When the meshes of the domains do not match, special treatments are required.

A function `createMultiDomainMesh.m` is given in Code List 5.3 to illustrate the use of the scaled boundary finite element method in the context of domain decomposition. The problem domain is defined as an assembly of simple domains. This function automatically generates the S-element mesh of each domain independently. The global mesh is obtained by combining the non-matching meshes at their interface.

The input arguments of the function `createMultiDomainMesh.m` consist of the structure variable `dDefs` that stores the data defining the domains and the interfaces, and the cell array `para` that stores the meshing parameters of the individual domains for automatic mesh generation. The data of the combined global mesh are stored in the output arguments `coord` (nodal coordinates), `sdConn` (S-element connectivity) and `sdSC` (coordinates of scaling centres of S-elements). Detailed documentation on the input and output arguments is provided in the code list.

This function generates the meshes of all the domains (Line 6) and estimates the length of the shortest edge in all the meshes. The resolution (or tolerance) of the coordinates is set as a fraction, which is arbitrarily set at 10%, of the length of the shortest edge (Line 20). The meshes of all the domains are combined one by one by executing repeatedly (Line 25) the function `combineTwoMeshes.m` in Code List 5.1 on page 220.

Code List 5.3: Creating a mesh composed of multiple domains

```
1  function [coord,sdConn,sdSC] = createMultiDomainMesh(dDefs, para)
```

Create a mesh of multiple domains by combining meshes generated independently for each domain

Inputs:

> `dDefs` - structure variable storing the data of the problem domains and the interfaces in the following fields:
>> `prodef{i}`: handle of problem definition function of domain i
>> `cshft(i,:)`: amount of translation of coordinates of domain i
>> `onLine{i}`: handle of function defining the interface between domain i-1 and domain i
>
> `para{i}` - cell array of structure variables containing parameters for mesh generation of domain i. The fields are:
>> `mesher`: = 1,   DistMesh
>>   = 2,   PolyMesher
>>   = 3,   direct input of triangular mesh
>>   = 4,   direct input of polygon mesh
>>   = 'keyword',   access user-defined function to input or generate S-element mesh
>> `h0`:   initial edge length (DistMesh only)
>> `nPolygon`:   number of polygons (PolyMesher only)
>> `nDiv`:   number of 2-node line elements per edge

Outputs:

> `coord(i,:)` - Coordinates of node i
> `sdConn{isd,:}(ie,:)` - S-element connectivity. The nodes of line element ie in S-element isd.

```
         sdSC(isd,:)   -   scaling centers of S-element isd.

 2
 3  nDomain = length(dDefs.prodef); % ................. number of domains
 4  d = cell(nDomain,1); % ............. initialize cell variable storing domains
 5  minLeng = 1d20; % .......... initialize minimum length of edges in a mesh
 6  for id = 1:nDomain % .............................. loop over domain
 7      [ coord, sdConn, sdSC ] = createSBFEMesh( ...
 8          dDefs.prodef{id}, para{id}.mesher, para{id});
 9      [ meshEdge, sdEdge] = meshConnectivity( sdConn );
10      % ...  shift coordinates of nodes and scaling centres
11      coord = bsxfun(@plus, coord, dDefs.cshft(id,:));
12      sdSC = bsxfun(@plus, sdSC, dDefs.cshft(id,:));
13      d{id} = struct('coord',coord, 'sdConn',{sdConn}, 'sdSC',sdSC, ...
14          'meshEdge',{meshEdge}, 'sdEdge',{sdEdge});
15      % ...  length of shortest edge in mesh
16      dxy = coord(meshEdge(:,1),:) - coord(meshEdge(:,2),:);
17      minLeng = min( minLeng , sqrt(min(sum(dxy.*dxy,2))) );
18  end
19  % ...  resolution of coordinates (round to the first digit in 1, 2 or 5).
20  res = 0.1*minLeng; % ................ the factor is arbitrarily chosen as 0.1
21  expnt = floor(log10(res));
22  firstDigit = floor(res/10^expnt);
23  roundTo = [1 2 2 5 5 5 5 10 10];
24  res = 10^expnt*roundTo(firstDigit); % ............. round the resolution
25  % ...  Combining the domains one by one iteratively
26  da = d{1}; % ................................. including the first domain
27  for id = 2:nDomain; % .................. loop over the remaining domains
28      [ da.coord, da.sdConn, da.sdSC ] = combineTwoMeshes( ...
29          {d{id}, da}, dDefs.onLine{id-1}, res);
30      [ da.meshEdge, da.sdEdge] = meshConnectivity( da.sdConn );
31  end
32
33  coord = da.coord; sdConn = da.sdConn; sdSC = da.sdSC; % ...... output
34
35  end
```

5.3.3 Examples

Two examples are presented to demonstrate the modelling of problems consisting of multiple domains. The function createMultiDomainMesh.m is used to generate the meshes. For each example, a problem definition function is written. The execution is controlled by a MATLAB script.

■ Example 5.1   Rectangle with a Hole Composed of Two Domains

The problem of a circular hole in an infinite plane under remote uniaxial tension in Section 4.4.2 is considered. In this example, a part of the infinite plane consisting of two domains is considered. One is the square domain ($L = 2$ m) with the hole shown in Figure 4.19. The other is a rectangular domain of 2 m × 1 m adjacent to the lower edge of

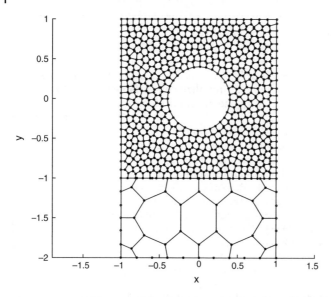

Figure 5.9 Mesh of a rectangular plate with a circular hole obtained by combining a mesh of the upper square domain with the hole and a mesh of the lower rectangular domain. The meshes of the two domains are generated independently.

the square domain. The global mesh obtained by combining those of the two domains is shown in Figure 5.9a. The two domains are meshed independently with high contrast in the sizes of the polygon S-elements. The interface between the two domains is easily identified visually.

The problem definition function `ProbDefCombinedPlateWithHole.m` is listed on page 226. Its structure is similar to those given previously for a single domain. A new local function `definitionOfDomains` (Line 20) defines the composition of the problem. It is accessed by the keyword `'Domains'`. The outputs consist of three variables. The cell array `prodef` contains the handlers of the individual problem definitions of the constituent domains, which are the lower rectangle defined in the function `ProbDefDeepBeam.m` (page 181, Section 4.4.1) and the upper square with a circular hole defined in the function `ProbDefInfPlateWithHole.m` (page 194, Section 4.4.2). A row of the matrix `cshft` specifies the amount of coordinate translation to move a domain to its desired position in the problem domain. The first domain (lower rectangle) is shifted by $(-1\text{ m}, -1.5\text{ m})$. The origin of coordinates remains at the centre of the hole and the exact solution given in the function `ProbDefInfPlateWithHole.m` can be used directly (see Line 71). The interface ($y = -1$ m) is defined by the anonymous function `onLine`.

Two sets of boundary conditions are considered. One set is accessible by the string keyword `'BCond'`. It is intended for a static analysis to simulate the problem of a circular hole in an infinite plane under remote uniaxial tension. The exact displacement solution of this problem is prescribed on the boundary nodes (Line 33). The other set is accessible by the string keyword `'BCondModal'`. It is intended for a modal analysis where the bottom of the rectangular plate is fixed (Line 33).

```
1   function [x] = ProbDefCombinedPlateWithHole(Demand,Arg)
2   %Plate with a hole composed of two domains
3   %
4   %inputs
5   %   Demand    - option to access data in local functions
6   %   Arg       - argument to pass to local functions
7   %output
8   %   x         - cell array of output.
9
10  switch(Demand)  % .............. access local functions according to Demand
11      case('Domains'); x = definitionOfDomains();
12      case('MAT');     x = ProbDefInfPlateWithHole('MAT');
13      case('BCond');   x = BoundryConditions(Arg);
14      case('BCondModal');   x = BoundryConditionsModal(Arg);
15      case('Output'); x = [];
16      case('EXACT');   x =  ExactSln(Arg);
17  end
18  end
19
20  % ... Definition of composition of domains
21  function [x] = definitionOfDomains ()
22
23  % ... handler of problem definition functions of constituent domains
24  prodef ={@ProbDefPureBendingBeam, @ProbDefInfPlateWithHole};
25  cshft = [ -1 -1.5;   0 0 ]; % ... amount of shifting of x- and y-coordinates
26
27  % ... Definition of the interface between the two domains
28  onLine = {@(xy, eps) find(abs(xy(:,2)+1) < eps)}; % ..... lines y = -1
29
30  x = struct('prodef',{prodef}, 'cshft',cshft, 'onLine',{onLine});
%   ................................................................ output
31  end
32
33  % ... Prescribe boundary conditions
34  function [x] = BoundryConditions(Arg)
35  coord = Arg{1}; % .................................... nodal coordinates
36
37  % ... displacement constraints (prescribed acceleration in a response
38  % ... history analysis). One constraint per row: [Node Dir]
39  eps = 1d-4*(max(coord(:,1))-min(coord(:,1)));
40  bNodes = find(abs(coord(:,1)-min(coord(:,1)))<eps | ...
41      abs(coord(:,1)-max(coord(:,1)))<eps | ...
42      abs(coord(:,2)-min(coord(:,2)))<eps | ...
43      abs(coord(:,2)-max(coord(:,2)))<eps );
44  % ... get exact solution at specified nodes
45  ex = ProbDefInfPlateWithHole('EXACT', coord(bNodes,:));
46  BC_Disp = [ [bNodes; bNodes ], [ones(length(bNodes),1);...
```

```
47        2*ones(length(bNodes),1)], [ex(:,1); ex(:,2)] ];
48
49  % ... nodal forces
50  F = zeros(2*size(coord,1),1); % ................ initializing force vector
51
52  % ... output
53  x = {BC_Disp, F};
54  end
55
56  % ... Prescribe boundary conditions
57  function [x] = BoundryConditionsModal(Arg)
58  coord = Arg{1}; % ..................................... nodal coordinates
59
60  % ... displacement constraints (prescribed acceleration in a response
61  % ... history analysis). One constraint per row: [Node Dir]
62  eps = 1d-4*(max(coord(:,1))-min(coord(:,1)));
63  bNodes = find(abs(coord(:,2)-min(coord(:,2)))<eps);
64  BC_Disp = [ [bNodes; bNodes ], [ones(length(bNodes),1);...
65        2*ones(length(bNodes),1)], zeros(2*length(bNodes),1) ];
66
67  % ... output
68  x = {BC_Disp};
69  end
70
71  function x = ExactSln(xy)  % ........................... exact solution
72  x = ProbDefInfPlateWithHole('EXACT', xy);
73  end
```

Both the static analysis and the modal analysis are driven by the following script. To make it easier for readers to modify the mesh and attempt the solution, the mesh parameters are specified in the cell array `para` in this script instead of in the problem definition function. An entry of the cell defines the parameters of one domain. The mesh of the lower rectangle is created from a triangular mesh generated by DistMesh with the element size h0=0.4. The mesh of the upper square is generated by PolyMesher with 512 polygons. The variables are documented in the function `createSBFEMesh.m` (Matlab Code List 4.1 on page 156) in Section 4.3. The mesh is generated by executing the function `createMultiDomainMesh.m`.

```
%A circular hole in a rectangle modelled by combining two domains
close all; clearvars; dbstop error;

probdef = @ProbDefCombinedPlateWithHole; % function handler

% mesh parameters
para = {struct('mesher',1, 'h0',0.4), ... % lower rectangle
        struct('mesher',2, 'nPolygon',512)}; % upper square with hole
dDefs = probdef('Domains'); % get domain definitions
```

```
[ coord, sdConn, sdSC ] = createMultiDomainMesh( dDefs, para );

figure; % plot mesh
opt=struct('sdSC', sdSC, 'PlotNode',1, 'MarkerSize',2);
PlotSBFEMesh(coord, sdConn, opt);
title('MESH');

% static analysis
slnOpt.type='STATICS'; % 'STATICS'; 'MODAL'; 'FREQUENCY':'TIME';
[U, sdSln] = SBFEPoly2NSolver(probdef, coord, sdConn, sdSC, slnOpt);

% modal analysis
slnOpt.type='MODAL';
slnOpt.modalPara = 4;
[f, mShp] = SBFEPoly2NSolver(probdef, coord, sdConn, sdSC, slnOpt);
```

The above MATLAB script can be obtained by scanning the QR-codes to the right.

The static analysis is performed by the function SBFEPoly2NSolver.m (Code List 4.8) with the option slnOpt.type='STATICS'. The deformed mesh is shown in Figure 5.10. It can be observed that the displacement field is continuous across the interface of the two domains. The relative displacement error norm (Eq. (4.12)) is obtained as 0.13%. It is slightly larger than the error shown in Table 4.4 with the same number (512) of polygons in the square domain.

Figure 5.10 Deformed mesh of a circular hole in an infinite plate under remote uniaxial tension.

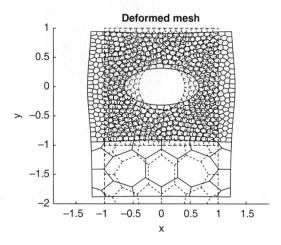

The modal analysis is performed by specifying the option `slnOpt.type='MODAL'`. The natural frequencies of the first 4 modes are:

```
Mode         Frequency              Period
   1       3.8940959e+00        1.6135158e+00
   2       1.1311491e+01        5.5546925e-01
   3       1.3107297e+01        4.7936544e-01
   4       1.8144105e+01        3.4629349e-01
```

The first two mode shapes are plotted in Figure 5.11. The mode shapes are continuous at the interface between the square and rectangular domains.

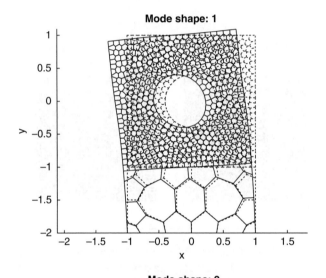

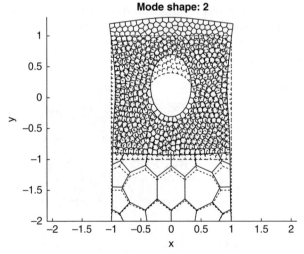

Figure 5.11 Mode shapes of the first two modes of a rectangular plate with a circular hole.

■ Example 5.2   A Mosaic Composed of Five Domains

To demonstrate the combination of multiple domains, a mosaic composed of five domains that are meshed independently, as shown in Figure 5.12a, is considered.

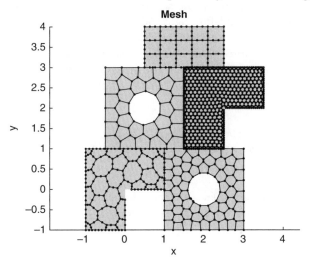

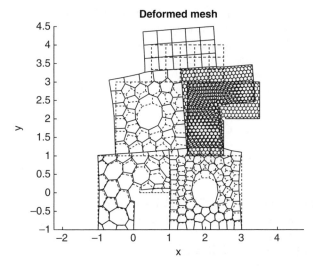

Figure 5.12 A mosaic composed of five domains meshed independently.

The problem definition function `ProbDefMosaicFiveDomains.m` is provided below. The local function `definitionOfDomains` (Line 19) defines the composition of the mosaic. It is accessible using the keyword `'Domains'`. The five domains include:

1) Two instances of the L-shaped panel (`ProbDefLShapedPlate.m`) on page 199, Section 4.4.3.
2) Two instances of the square plate with a circular hole (`ProbDefInfPlateWithHole.m`) on page 194, Section 4.4.2.
3) One rectangle (`ProbDefDeepBeam.m`) on page 181, Section 4.4.1.

The handlers of their definition functions are stored in the variable `prodef`. The domains are translated by the distances in matrix `cshft` to match each other at their edges. In the function `createMultiDomainMesh.m` (Code List 5.3 on page 224), the domains are combined one after the other in the sequence given in the variable `prodef`. In this example the sequence is from left to right and bottom to top. The interfaces are defined by following the same sequence. The boundary condition are specified by the local function `BoundryConditions` (Line 39). The bottom of the mosaic is fixed and the top is subject to a uniform tension of 1 kPa. The material properties given in the problem of a square with a circular hole (`ProbDefInfPlateWithHole.m` on page 194, Section 4.4.2) are used.

```matlab
 1  function [x] = ProbDefMosaicFiveDomains(Demand,Arg)
 2  %Problem definition function of a mosaic composed of five domains
 3  %
 4  %inputs
 5  %    Demand     - option to access data in local functions
 6  %    Arg        - argument to pass to local functions
 7  %output
 8  %    x          - cell array of output.
 9
10  switch(Demand)   % .............. access local functions according to Demand
11      case('Domains'); x = definitionOfDomains();
12      case('MAT'); x = ProbDefInfPlateWithHole('MAT');  % ..... material
13      case('BCond');   x = BoundryConditions(Arg);
14      case('Output'); x = [];
15      case('EXACT');   x = [];
16  end
17  end
18
19  % ...  Definition of composition of domains
20  function [x] = definitionOfDomains ()
21
22  prodef ={@ProbDefLShapedPlate, @ProbDefInfPlateWithHole, ...
23      @ProbDefInfPlateWithHole,  @ProbDefLShapedPlate, ...
24      @ProbDefPureBendingBeam};
25  % ...  amount of shifting of x- and y- coordinates
26  cshft = [-1 -1; 2 0;   0.5 2; 1.5 1; 0.5 3.5];
27
28  % ...  Definitions of the interfaces between the domains
29  onLine = {@(xy, eps) find(abs(xy(:,1)-1) < eps),  ... % ... line x = 1
30      @(xy, eps) find(abs(xy(:,2)-1) < eps),  ... % .......... line y = 1
31      @(xy, eps) find( (abs(xy(:,2)-1) < eps) | ...
32      (abs(xy(:,1)-1.5) < eps)),  ... % ............ lines y = 1 and x = 1.5
33      @(xy, eps) find(abs(xy(:,2)-3) < eps)};  ... % ......... line y = 3
34
35  x = struct('prodef',{prodef}, 'cshft',cshft, ...
```

```
36         'onLine',{onLine}); % ..................................... output
37    end
38
39    % ...  Prescribe boundary conditions
40    function [x] = BoundryConditions(Arg)
41    coord = Arg{1}; % ..................................... nodal coordinates
42    sdConn = Arg{2}; % ................................. element connectivity
43
44    % ...  displacement constraints (prescribed acceleration in a response
45    % ...  history analysis). One constraint per row: [Node Dir Disp]
46    eps = 0.001*(max(coord(:,1))-min(coord(:,1)));
47    bNodes = find(abs(coord(:,2)-min(coord(:,2)))<eps);
48    BC_Disp = [ [bNodes; bNodes], [ones(length(bNodes),1);...
49         2*ones(length(bNodes),1)], 0*[bNodes; bNodes] ];
50
51    % ...  surface traction
52    % ...  edges of polygon S-elements
53    [ MeshEdges ] = meshConnectivity( sdConn );
54    % ...  find edges at the top
55    ymax = max(coord(:,2));
56    tEdges = MeshEdges((abs(coord(MeshEdges(:,1),2)-ymax)+ ...
57         abs(coord(MeshEdges(:,2),2)-ymax))<eps,:);
58    % ...  nodal forces
59    F = zeros(2*size(coord,1),1); % ................. initializing force vector
60    % ...  add nodal forces equivalent to surface traction
61    F = addSurfTraction(coord, tEdges, [0; 1; 0; 1], F);
62
63    % ...  output
64    x = {BC_Disp, F};
65    end
```

The analysis is driven by the script below. The mesh parameters of the five domains are specified in the cell array `para`. For the purpose of demonstration, the meshes of the domains are generated with three options and differ noticeably (Figure 5.12a). Each edge of the polygon S-elements is divided into 2 line elements in the L-shaped panel at the bottom and 3 line elements in the rectangle at the top. The deformed mesh is shown in Figure 5.12b. The displacement field is continuous across the interfaces between the domains.

```
%A mosaic composed of five domains
close all; clearvars; dbstop error;

probdef = @ProbDefMosaicFiveDomains; % function handler

% mesh parameters
para = {struct('mesher',1, 'h0', 0.4, 'nDiv',2), ...
```

```
          struct('mesher',2, 'nPolygon',64), ...
          struct('mesher',2, 'nPolygon',32), ...
          struct('mesher',1, 'h0', 0.1), ...
          struct('mesher','RectMesh', 'h0',0.4, 'nDiv',3)};
dDefs = probdef('Domains', para); % get domain definitions
[ coord, sdConn, sdSC ] = createMultiDomainMesh( dDefs, para );

figure; % plot mesh
opt=struct('sdSC', sdSC, 'PlotNode',1, 'MarkerSize',2, ...
    'fill',[0.9 0.9 0.9]);
PlotSBFEMesh(coord, sdConn, opt);
title('MESH');

% analysis
slnOpt.type='STATICS'; % 'STATICS'; 'MODAL'; 'FREQUENCY':'TIME';
[U, sdSln] = SBFEPoly2NSolver(probdef, coord, sdConn, sdSC, slnOpt);
```

The above MATLAB script can be obtained by scanning the QR-codes to the right.

5.4 Modelling of Stress Singularities

The scaled boundary finite element method offers a simple and accurate way to model singularity problems. The modelling of the singular stress field around a crack tip is presented in Example 3.9. Detailed discussions and further examples of the modelling of stress singularities, including interface cracks and multi-material corners, can be found in Song and Wolf (2002); Song (2005) and Chapter 10.

The scaled boundary finite element solution is semi-analytical inside an S-element. Along the circumferential direction parallel to the boundary, the solution is given at discrete points and interpolated in between. Along the radial direction emanating from the scaling centre, the solution is analytical and more suitable than a numerical solution to represent singularities.

To further discuss the modelling considerations, the L-shaped panel analysed in Sections 4.4.3 and 5.2.1 is re-examined here. As depicted in Figure 5.13, the L-shaped panel is modelled as one S-element. The scaling centre is selected as the re-entrant corner. Except for the two edges intersecting at the scaling centre (the re-entrant corner), the boundary of the panel is divided into four 2-node line elements.

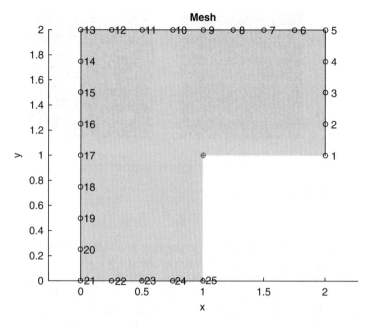

Figure 5.13 Modelling of L-shaped panel by one scaled boundary finite element S-element.

The analysis is performed with the following MATLAB script.

```
%An L-shaped panel modelled as one S-element
close all; clearvars; dbstop error;

probdef = @ProbDefLShapedPlate;% function handler
%nodal coordinates. One node per row
coord = [ 2 1; 2 1.25; 2 1.5; 2 1.75; ...
        2 2; 1.75 2; 1.5 2; 1.25 2; ...
        1 2; 0.75 2; 0.5 2; 0.25 2; ...
        0 2; 0 1.75; 0 1.5; 0 1.25; ...
        0 1; 0 0.75; 0 0.5; 0 0.25; ...
        0 0; 0.25 0; 0.5 0; 0.75 0; 1 0];
sdConn = { [1:24; 2:25]' }; %connectivity
sdSC = [1 1]; %scaling centre

figure;
opt=struct('sdSC', sdSC, 'LabelSC', 0, ...
    'fill', [0.9 0.9 0.9], 'LabelNode',12);
PlotSBFEMesh(coord, sdConn, opt);
title('MESH');

% analysis
[U, sdSln] = SBFEPoly2NSolver(probdef, coord, sdConn, sdSC);
```

The above MATLAB script can be obtained by scanning the QR-codes to the right.

The printout of the vertical displacement at point A is

```
Vertical displacement =    1.054015e-04
```

As given in Section 4.4.3, the result converged up to the first five significant digits is $(v_A)_{\mathrm{ref}} = 1.0560 \times 10^{-4}$ m. The error of the present result is about 1.9×10^{-3}.

Different discretization schemes are employed in Section 4.4.3, Section 5.2.1 and the present analysis. In all these cases, the boundaries of the S-elements are clearly visible from the scaling centres. In Section 4.4.3, the problem domain is divided into a mesh of polygon S-elements and one edge of an S-element is modelled by one 2-node line element. This discretization scheme is similar to that of the finite element method. In Section 5.2.1, only a few polygon S-elements are used. The edges of the S-elements are divided into more than one line element. In the present scheme, the whole panel is modelled by one S-element while each edge is divided into multiple element. With the three discretization schemes, the meshes in Figures 5.13, 5.3a and 5.13 lead to similar accuracy. It becomes evident that it is more efficient to increase the number of line elements on the boundary of the S-element surrounding the singular point than to increase the number of S-elements. This observation is conditioned on the high visibility of boundary from the scaling centre. If the visibility is poor, an S-element has to be subdivided into smaller ones to improve the visibility.

Part II

Theory and Applications of the Scaled Boundary Finite Element Method

In Part II, the basic concepts of the scaled boundary finite element method are presented for two-dimensional elasticity problems. In this method, a problem domain is divided into S-domains satisfying the scaling requirement, i.e. there exists a region from which every point on the boundary of an S-domain is directly visible. The geometry of such an S-domain can be described by scaling its boundary continuously with respect to the scaling centre selected to have direct visibility of the boundary. The scaling operation is performed with a radial coordinate emanating from the scaling centre. The major steps of a scaled boundary finite element analysis are summarized as follows:

1) To model practical problems, the boundary of an S-domain is divided into line elements. After discretization, the S-domain is referred to as an S-element. The circumferential direction of the S-element parallel to the boundary is represented in discretized form by the line elements and their natural coordinates. The radial coordinate for scaling and the natural coordinate of a line element are selected as the scaled boundary coordinates, which are employed to describe the geometry of the S-element.

2) Nodal displacement functions are defined along the radial lines connecting the nodes on the boundary of the S-element and the scaling centre. Along the circumferential direction, the displacements are interpolated using the shape functions of the line elements, leading to a semi-analytical representation of the displacement field of the S-element.

3) Applying the virtual work principle, the governing differential equations are expressed in weak form along the circumferential direction. This results in the scaled boundary finite element equation, which is expressed as a system of ordinary differential equations of the displacement functions with the radial coordinate as the independent variable.

4) The displacement functions and the stiffness matrix of the S-element are obtained by solving the scaled boundary finite element equation using an eigenvalue method. The assembly of the S-element equations to form the global equation and the solution of the global equation follow the standard finite element procedure.

This part of the book is built on the fundamentals presented in Part I. The theory of the scaled boundary finite element method is generalized. Additional numerical and

The Scaled Boundary Finite Element Method: Introduction to Theory and Implementation,
First Edition. Chongmin Song.
© 2018 John Wiley & Sons Ltd. Published 2018 by John Wiley & Sons Ltd.
Companion website: www.wiley.com/go/author/scaledboundaryfiniteelementmethod

analytical techniques are introduced. Since the assembly procedure has been illustrated in Part I, it will not be repeated. This part will focus on three-dimensional elasticity problems.

In Chapter 6, the derivation of the scaled boundary finite element method is extended to three dimensions. Useful properties of the scaled boundary finite element equation are investigated. In Chapter 7, a robust solution procedure based on the matrix exponential function is described. In Chapter 8, high-order elements are applied to the scaled boundary finite element method. In Chapter 9, a mesh generation technique that is highly complementary to the scaled boundary finite element method is introduced. This allows fully automatic integration of complex geometric models with stress analysis. In Chapter 10, the application to linear fracture mechanics, where the scaled boundary finite element method is advantageous owing to its semi-analytical solution, is described. Numerical examples are presented to illustrate the applications of the theory described in this part.

6

Derivation of the Scaled Boundary Finite Element Equation in Three Dimensions

6.1 Introduction

In this chapter, the scaled boundary finite element equation is derived for the linear elastic analysis of a three-dimensional (3D) S-domain. As described in Section 2.2.1 for the two-dimensional (2D) case, the geometry of the S-domain has to satisfy the scaling requirement. The scaling of the boundary is addressed in Section 6.2. A scaling centre is selected in a region from where every point on the boundary is directly visible. In Section 6.3, the boundary of the 3D S-domain is divided into surface elements, yielding a 3D S-element. In Section 6.4, the scaled boundary coordinates are defined, and the coordinate transformation is performed. In Section 6.5, the geometrical properties of the scaled boundary transformation are discussed. In Section 6.6, the governing partial differential equations of elastodynamics are formulated in the scaled boundary coordinates. In Section 6.7, the Galerkin's weighted residual technique is applied to the governing partial differential equations of elasticity. Satisfying the governing equations in weak form along the circumferential directions leads to the scaled boundary finite element equation in displacement. Formulations of the scaled boundary finite element method in 2D and 3D are unified in Section 6.8. For the numerical solution of the scaled boundary finite element equation, the order of differential equations is reduced from two to one in Section 6.9. The properties of the coefficient matrices of the equations are investigated in Section 6.10. The linear completeness of the solution is proven in Section 6.11. The formulation of the scaled boundary finite element equations in stiffness matrix is presented in Section 6.12.

6.2 Scaling of Boundary

The scaling requirement presented in Section 2.2.1 for the two-dimensional case is directly applied to the three-dimensional case. Figure 6.1 shows a three-dimensional S-domain V (i.e. a domain that satisfies the scaling requirement). Mathematically, this particular S-domain in Figure 6.1 is a star-shaped domain. The scaling centre is selected at a point from where every point on the boundary is directly visible.

The Scaled Boundary Finite Element Method: Introduction to Theory and Implementation,
First Edition. Chongmin Song.
© 2018 John Wiley & Sons Ltd. Published 2018 by John Wiley & Sons Ltd.
Companion website: www.wiley.com/go/author/scaledboundaryfiniteelementmethod

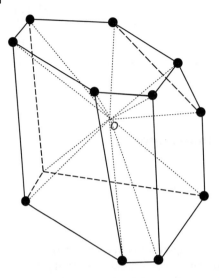

Figure 6.1 A three-dimensional S-domain.

A Cartesian coordinate system x, y, z is introduced. The origin of the coordinate system is chosen at the scaling centre. The position vector of a point (x, y, z) is expressed as

$$\mathbf{r} = x\mathbf{i} + y\mathbf{j} + z\mathbf{k} \tag{6.1}$$

where $\mathbf{i}, \mathbf{j}$ and $\mathbf{k}$ are the unit vectors along the x, y, z coordinates.

In a spherical coordinate system with its origin at the scaling centre, the radial coordinate r, azimuthal angle θ and polar angle ϕ are expressed by the Cartesian coordinates as

$$r = \sqrt{x^2 + y^2 + z^2} \tag{6.2a}$$

$$\theta = \tan^{-1}\frac{y}{x} \tag{6.2b}$$

$$\phi = \cos^{-1}\frac{z}{r} \tag{6.2c}$$

θ is defined as the angle in the xy-plane from the x-axis. Its principal value is in the range of $0 \leq \theta < 2\pi$. ϕ is the angle from the positive z-axis and has the principal value $0 \leq \phi \leq \pi$.

Denoting the coordinates on the boundary as x_b, y_b, z_b, a point on the boundary is described by its position vector

$$\mathbf{r}_b = x_b\mathbf{i} + y_b\mathbf{j} + z_b\mathbf{k} \tag{6.3}$$

For an S-domain satisfying the scaling requirement, a point on the boundary is uniquely defined by the azimuthal angle θ and polar angle ϕ. The boundary of the S-domain can be described by a single-valued function $r_b(\theta, \phi)$ or parametrically by three single-valued functions $x_b(\theta, \phi)$, $y_b(\theta, \phi)$ and $z_b(\theta, \phi)$

$$\begin{cases} x_b = x_b(\theta, \phi) \\ y_b = y_b(\theta, \phi) \\ z_b = z_b(\theta, \phi) \end{cases} \tag{6.4}$$

A dimensionless radial coordinate ξ, emanating from the scaling centre towards the boundary, is introduced. ξ is chosen as $\xi = 0$ at the scaling centre and as $\xi = 1$ on the boundary. In the direction of the radial vector $\mathbf{r}$, a point (x, y, z) in the domain is obtained by scaling the point (x_b, y_b, z_b) on the boundary with respect to the scaling centre. The scaling operation is expressed using the position vectors as

$$\mathbf{r} = \xi \mathbf{r}_b \tag{6.5}$$

In Cartesian coordinates x, y, z, Eq (6.5) is expressed as (Eqs. (6.1) and (6.3))

$$x = \xi x_b \tag{6.6a}$$
$$y = \xi y_b \tag{6.6b}$$
$$z = \xi z_b \tag{6.6c}$$

Similar to the two-dimensional case (Section 2.2.1), the S-domain is generated by applying the scaling operation to the whole boundary. The scaling requirement guarantees that every point in the domain is covered once and once only during the scaling process.

Choosing the origin of the Cartesian coordinates x, y, z at the scaling centre simplifies the equations in the scaled boundary finite element method. In the computer implementation (for example, in the MATLAB code Platypus accompanying this book), the coordinates are easily transformed from a global system to one defined at the scaling centre of an S-domain. This choice is, however, not essential. When the scaling centre is located as a point (x_0, y_0, z_0), the position vector of the scaling centre is equal to

$$\mathbf{r}_0 = x_0 \mathbf{i} + y_0 \mathbf{j} + z_0 \mathbf{k} \tag{6.7}$$

The radial vector from the scaling centre to a point (x_b, y_b, z_b) on the boundary is equal to $(\mathbf{r}_b - \mathbf{r}_0)$. The radial vector from the scaling centre to a point (x, y, z) in the domain $(\mathbf{r} - \mathbf{r}_0)$ is obtained by scaling $(\mathbf{r}_b - \mathbf{r}_0)$

$$\mathbf{r} - \mathbf{r}_0 = \xi(\mathbf{r}_b - \mathbf{r}_0) \tag{6.8}$$

The scaling is expressed in Cartesian coordinates as

$$x - x_0 = \xi(x_b - x_0) \tag{6.9a}$$
$$y - y_0 = \xi(y_b - y_0) \tag{6.9b}$$
$$z - z_0 = \xi(z_b - z_0) \tag{6.9c}$$

It is easily observed that a coordinate translation to the scaling centre is performed. Alternatively, Eqs. (6.8) and (6.9) are rewritten as

$$\mathbf{r} = \mathbf{r}_0 + \xi(\mathbf{r}_b - \mathbf{r}_0) \tag{6.10}$$

and

$$x = x_0 + \xi(x_b - x_0) \tag{6.11a}$$
$$y = y_0 + \xi(y_b - y_0) \tag{6.11b}$$
$$z = z_0 + \xi(z_b - z_0) \tag{6.11c}$$

respectively. This expression including the coordinates of the scaling centre is useful in studying the sensitivity of the S-element with respect to the location of the scaling centre. It is employed by Chowdhury et al. (2014) and Long et al. (2014) to evaluate the sensitivity of stress intensity factors to crack length and orientation.

6.3 Boundary Discretization of an S-domain

A 3D S-element (Figure 6.2) is created by dividing the boundary of an S-domain (Figure 6.1) into piecewise continuous surface elements, while the volume of the domain is not discretized. The surface mesh is much simpler than a volume mesh to generate.

Standard isoparametric finite element technology is directly applicable to the surface mesh of the S-element. Adopting the isoparametric formulation, the same set of shape functions is used to interpolate both the geometry and the displacements from the nodal values. As the S-element is obtained by scaling its boundary, the interpolation applies equally on any scaled boundary (i.e. at any radial coordinate ξ).

The shape functions of isoparametric elements are constructed on parent elements in the natural coordinates η, ζ (ξ has been used for the radial coordinate). The interpolation of geometry of a surface element is expressed formally as

$$x_b(\eta, \zeta) = \sum_i N_i(\eta, \zeta)x_i = [N(\eta, \zeta)]\{x\} \tag{6.12a}$$

$$y_b(\eta, \zeta) = \sum_i N_i(\eta, \zeta)y_i = [N(\eta, \zeta)]\{y\} \tag{6.12b}$$

$$z_b(\eta, \zeta) = \sum_i N_i(\eta, \zeta)z_i = [N(\eta, \zeta)]\{z\} \tag{6.12c}$$

where the subscript i indicates the nodal number. $N_i(\eta, \zeta)$ is the shape function related to node i. x_i, y_i, z_i are the nodal coordinates of node i. The shape functions and the nodal coordinates are written in matrix form, respectively, as

$$[N(\eta, \zeta)] = [N_1(\eta, \zeta) \quad N_2(\eta, \zeta) \quad \ldots] \tag{6.13}$$

and

$$\{x\} = [x_1 \quad x_2 \quad \ldots]^T \tag{6.14a}$$

$$\{y\} = [y_1 \quad y_2 \quad \ldots]^T \tag{6.14b}$$

$$\{z\} = [z_1 \quad z_2 \quad \ldots]^T \tag{6.14c}$$

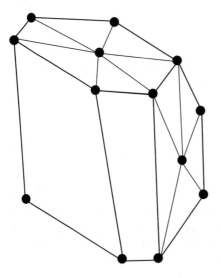

Figure 6.2 Surface discretization of an S-element. Faces with more than four vertices are divided into triangular elements. A node is inserted at the centroid of such a face (optional).

The same interpolation applies to the displacements $u_{bx}(\eta, \zeta)$, $u_{by}(\eta, \zeta)$, and $u_{bz}(\eta, \zeta)$ on a surface element

$$u_{bx}(\eta, \zeta) = \sum_i N_i(\eta, \zeta) u_{ix} = [N(\eta, \zeta)]\{u_x\} \tag{6.15a}$$

$$u_{by}(\eta, \zeta) = \sum_i N_i(\eta, \zeta) u_{iy} = [N(\eta, \zeta)]\{u_y\} \tag{6.15b}$$

$$u_{bz}(\eta, \zeta) = \sum_i N_i(\eta, \zeta) u_{iz} = [N(\eta, \zeta)]\{u_z\} \tag{6.15c}$$

with the nodal displacement vectors

$$\{u_x\} = [u_{1x} \quad u_{2x} \quad \ldots]^T \tag{6.16a}$$

$$\{u_y\} = [u_{1y} \quad u_{2y} \quad \ldots]^T \tag{6.16b}$$

$$\{u_z\} = [u_{1z} \quad u_{2z} \quad \ldots]^T \tag{6.16c}$$

The coordinates $x_b(\eta, \zeta)$, $y_b(\eta, \zeta)$, $z_b(\eta, \zeta)$ and displacements $u_{bx}(\eta, \zeta)$, $u_{by}(\eta, \zeta)$ and $u_{bz}(\eta, \zeta)$ are functions of the natural coordinates η, ζ. Equations (6.12) and (6.15) form a set of parametric equations (with the natural coordinates η, ζ as the parameter variables) to implicitly describe the displacement field on the surface element.

As in the two-dimensional case (Section 2.2.2), the shape functions should have the following properties:

- Kronecker delta functions

$$N_i(\eta_j, \zeta_j) = \delta_{ij} = \begin{cases} 1 & \text{when} \quad i = j \\ 0 & \text{when} \quad i \neq j \end{cases} \tag{6.17}$$

- Partition of unity

$$\sum_i N_i(\eta, \zeta) = 1 \tag{6.18}$$

The partial derivatives of this equation with respect to η, ζ lead to

$$\sum_i \frac{\partial N_i}{\partial \eta} = \sum_i \frac{\partial N_i}{\partial \xi} = 0 \tag{6.19}$$

The isoparametric technique allows the use of extremely simple geometries to construct the shape functions of an element. The formulations of two-dimensional parent elements are provided in Section 6.3.1 for quadrilateral elements and in Section 6.3.2 for triangular elements. The reader is referred to textbooks on the finite element method (Cook et al. (2002); Zienkiewicz et al. (2005); Reddy (2005), to name a few) for details. High-order quadrilateral spectral elements are described in Chapter 8. It is worthwhile recalling that the scaled boundary finite element method requires a boundary mesh only. The boundary of a three-dimensional S-element is decretized by surface elements that are obtained by mapping two-dimensional parent elements.

6.3.1 Isoparametric Quadrilateral Elements

6.3.1.1 Four-node Quadrilateral Element

A 4-node quadrilateral element is obtained by mapping the parent element shown in Figure 6.3. The parent element is a square with a side length of 2 units. The natural coordinates η and ζ of the nodes are listed Figure 6.3.

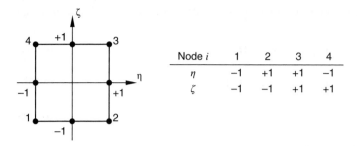

Figure 6.3 Four-node parent element in natural coordinates η, ζ.

The shape functions can be obtained by following the procedure in Section 2.2.2. The geometry and displacements over the element are interpolated by a bilinear function, for example,

$$x = a_0 + a_1\eta + a_2\zeta + a_3\eta\zeta \tag{6.20}$$

where the interpolation constants a_i ($i = 1, 2, 3, 4$) are determined by evaluating this expression at the nodes. The shape functions $N_i(\eta, \zeta)$ ($i = 1, 2, 3, 4$) follow after the terms are grouped according to the nodal values.

For the simple geometry of this parent element, it is simpler to determine the shape functions using their Kronecker delta property given in Eq. (6.17). The shape function $N_1(\eta, \zeta)$ is considered as an example (The arguments η and ζ will be omitted from the shape functions for conciseness in the following.) Since $N_1 = 0$ at nodes 2, 3 and 4 is required and the interpolation is linear along the vertical and horizontal directions, N_1 must vanish at the right and top edges of the element. Therefore, the bilinar shape function N_1 should be of the form

$$N_1 = c_1(1 - \eta)(1 - \zeta) \tag{6.21}$$

As $N_1 = 1$ at node 1, the constant c_1 is equal to $1/4$. Similarly, the other shape functions are determined, leading to

$$N_1 = \frac{1}{4}(1 - \eta)(1 - \zeta) \tag{6.22a}$$

$$N_2 = \frac{1}{4}(1 + \eta)(1 - \zeta) \tag{6.22b}$$

$$N_3 = \frac{1}{4}(1 + \eta)(1 + \zeta) \tag{6.22c}$$

$$N_4 = \frac{1}{4}(1 - \eta)(1 + \zeta) \tag{6.22d}$$

It is observed that the shape functions are the products of the 1D linear shape functions in Eq. (2.13) formulated in the η and ζ directions. The bilinear interpolation can be interpreted as applying the one-dimensional linear interpolation in Eq. (2.12) once to each direction. Note that the sum of the shape functions in Eq. (6.22) is equal to 1, i.e. the condition of partition of unity (Eq. (6.18)) is satisfied.

The shape functions can be written in a unified expression

$$N_i = \frac{1}{4}(1 + \eta_i\eta)(1 + \zeta_i\zeta) \qquad (i = 1, 2, 3, 4) \tag{6.23}$$

where η_i, ζ_i are the natural coordinates of node i (see the table in Figure (6.3)).

The derivatives of shape functions in Eq. (6.23) with respect to the natural coordinates η and ζ are equal to

$$\frac{\partial N_i}{\partial \eta} = \frac{1}{4}\eta_i(1 + \zeta_i\zeta)$$
$$\frac{\partial N_i}{\partial \zeta} = \frac{1}{4}\zeta_i(1 + \eta_i\eta)$$
$\qquad (i = 1, 2, 3, 4) \qquad\qquad\qquad$ (6.24)

It can be verified that the summation of the partial derivatives of the 4 shape functions with respect to a natural coordinate is equal to 0 (Eq. (6.19)).

6.3.1.2 Quadrilateral Element of Variable Number of Nodes

A 9-node parent element is depicted in Figure 6.4. The parent element is also a square with a side length of 2 units in the natural coordinates η and ζ. In addition to the four corner nodes, four midside nodes and an internal node exist. The natural coordinates of the nodes are listed in the figure. This parent element can be mapped to a doubly curved surface element, which is advantageous in fitting curved boundaries. The shape functions of this 9-node quadrilateral element can be obtained as the product of the shape functions (Eq. (2.19)) of the 3-node line element in Figure 2.11 formulated in the $\eta-$ and $\zeta-$directions. The resulting shape functions are complete second-degree polynomials in both directions. This element is referred to as a second-order or quadratic element.

Here, an alternative formulation that allows the use of quadrilateral elements of variable number of nodes is presented. The reader is referred to Cook et al. (2002) for details. With the standard numbering scheme of the nodes shown in Figure 6.4, the following shape functions apply

$$N_1 = \frac{1}{4}(1 - \eta)(1 - \zeta) - \frac{1}{2}(N_8 + N_5) - \frac{1}{4}N_9 \qquad (6.25a)$$

$$N_2 = \frac{1}{4}(1 + \eta)(1 - \zeta) - \frac{1}{2}(N_5 + N_6) - \frac{1}{4}N_9 \qquad (6.25b)$$

$$N_3 = \frac{1}{4}(1 + \eta)(1 + \zeta) - \frac{1}{2}(N_6 + N_7) - \frac{1}{4}N_9 \qquad (6.25c)$$

$$N_4 = \frac{1}{4}(1 - \eta)(1 + \zeta) - \frac{1}{2}(N_7 + N_8) - \frac{1}{4}N_9 \qquad (6.25d)$$

$$N_5 = \frac{1}{2}(1 - \eta^2)(1 - \zeta) - \frac{1}{2}N_9 \qquad (6.25e)$$

$$N_6 = \frac{1}{2}(1 + \eta)(1 - \zeta^2) - \frac{1}{2}N_9 \qquad (6.25f)$$

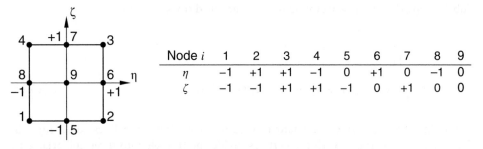

Node i	1	2	3	4	5	6	7	8	9
η	−1	+1	+1	−1	0	+1	0	−1	0
ζ	−1	−1	+1	+1	−1	0	+1	0	0

Figure 6.4 Nine-node parent element in natural coordinates η, ζ.

$$N_7 = \frac{1}{2}(1 - \eta^2)(1 + \zeta) - \frac{1}{2}N_9 \qquad (6.25\text{g})$$

$$N_8 = \frac{1}{2}(1 - \eta)(1 - \zeta^2) - \frac{1}{2}N_9 \qquad (6.25\text{h})$$

$$N_9 = (1 - \eta^2)(1 - \zeta^2) \qquad (6.25\text{i})$$

The shape functions are derived from their Kronecker delta property starting with node 9 (the internal node). The shape function N_9 (Eq. (6.25i)) is equal to 0 on all the four edges $\eta = \pm1$ and $\zeta = \pm1$ (i.e. at nodes $1 - 8$) and a unit at the centre $\eta = \zeta = 0$ (node 9).

The shape functions from N_5 to N_8 (Eqs. (6.25e) to (6.25h)) are obtained by blending two terms. At all the corner and midside nodes, the first terms of N_5 to N_8 are Kronecker delta functions and the second terms $-1/2N_9$ vanish. At node 9 ($\eta = \zeta = 0$), the first terms of N_5 to N_8 are all equal to $1/2$ and cancel with the second terms $-1/2N_9$. Therefore, N_5 to N_8 are Kronecker delta functions at the nodes.

Similarly, the shape functions from N_1 to N_4 (Eqs. (6.25a) to (6.25d)) are constructed by blending three terms. The first terms are the shape functions of the 4-node quadrilateral elements (Eq. (6.22)), which are Kronecker delta functions at the four corner nodes. The second terms cancel the values of the first terms at the midside nodes without any effect at the corner nodes. The third terms $(-1/2N_9)$ render the shape functions N_1 to N_4 equal to 0 at node 9 without affecting their values at all the other nodes.

This formulation of shape functions is directly applicable to a quadrilateral element having from four to nine nodes, i.e. four corner nodes and any of other five nodes. When any of the midside nodes or the internal node are missing from an element, the corresponding shape functions are omitted or set to 0. It can be verified that the remaining shape functions have the Kronecker delta property.

6.3.2 Isoparametric Triangular Elements

It is convenient to express the shape functions of triangular elements in area coordinates. The definition of the area coordinates is illustrated in Figure 6.5. The three vertices of the triangle are numbered 1, 2 and 3 following the counter-clockwise direction. The Cartesian coordinates of the vertices are denoted as (x_1, y_1), (x_2, y_2) and (x_3, y_3), respectively. The area of the triangle is equal to

$$A = \frac{1}{2} \begin{vmatrix} 1 & x_1 & y_1 \\ 1 & x_2 & y_2 \\ 1 & x_3 & y_3 \end{vmatrix} \qquad (6.26)$$

Substracting the first row from the second and third rows leads to

$$A = \frac{1}{2} \begin{vmatrix} 1 & x_1 & y_1 \\ 0 & x_2 - x_1 & y_2 - y_1 \\ 0 & x_3 - x_1 & y_3 - y_1 \end{vmatrix}$$

$$= \frac{1}{2}\left((x_2 - x_1)(y_3 - y_1) - (x_3 - x_1)(y_2 - y_1)\right) \qquad (6.27)$$

A point P (x, y) inside the triangle is considered. The three lines (dash lines in Figure 6.5) connecting P with the vertices divide the triangle into three sub-triangles.

Figure 6.5 Area coordinates L_1, L_2, L_3 in a triangle.

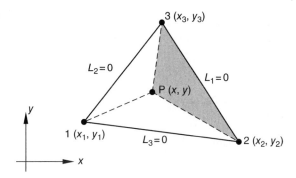

The area of the sub-triangle $P23$ (shaded area in Figure 6.5) is obtained from Eq. (6.27) by replacing the coordinates (x_1, y_1) of vertex 1 of the triangle with the coordinate (x, y) of point P

$$A_{P23} = \frac{1}{2}\left((x_2 - x)(y_3 - y) - (x_3 - x)(y_2 - y)\right) \tag{6.28a}$$

Similarly, the areas of the sub-triangles $P31$ and $P12$ are equal to

$$A_{P31} = \frac{1}{2}\left((x_3 - x)(y_1 - y) - (x_1 - x)(y_3 - y)\right) \tag{6.28b}$$

$$A_{P12} = \frac{1}{2}\left((x_1 - x)(y_2 - y) - (x_2 - x)(y_1 - y)\right) \tag{6.28c}$$

The area coordinates L_1, L_2, L_3 are defined as the ratios of the areas of the sub-triangles to the area of the whole triangle:

$$L_1 = \frac{A_{P23}}{A} \tag{6.29a}$$

$$L_2 = \frac{A_{P31}}{A} \tag{6.29b}$$

$$L_3 = \frac{A_{P12}}{A} \tag{6.29c}$$

Note that the three area coordinates are not independent. Since the area of the total triangle is equal to the sum of the areas of the three sub-triangles ($A = A_{P23} + A_{P31} + A_{P12}$), the area coordinates are related by

$$L_1 + L_2 + L_3 = 1 \tag{6.30}$$

The area coordinates are linear functions of the Cartesian coordinates x, y since the two terms of xy in each of Eqs. (6.28a)–(6.28c) cancel.

It can be observed from Figure 6.5 that an area coordinate L_i ($i = 1, 2, 3$) is equal to 1 at node i and to 0 on the opposite side of the triangle. For example, the area coordinate L_1 is equal to 1 at node 1 and to 0 on side $\overline{23}$.

6.3.2.1 Three-node Triangular Elements

The parent element of a 3-node triangular element is shown in Figure 6.6 in the natural coordinates η, ζ. It is identified that the area coordinates L_2 and L_3 are equal to

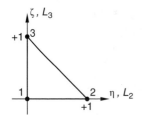

Node i	1	2	3
L_1	1	0	0
L_2, η	0	1	0
L_2, ζ	0	0	1

η and ζ, respectively,

$$L_2 = \eta \tag{6.31a}$$

$$L_3 = \zeta \tag{6.31b}$$

The other coordinate L_1 follows from Eq. (6.30) and Eq. (6.31) as

$$L_1 = 1 - \eta - \zeta \tag{6.31c}$$

The nodal values of the area coordinates are indicated in Figure 6.6. It is observed that the area coordinates have the Kronecker delta property. They can be chosen as the shape functions

$$N_1 = L_1 = 1 - \eta - \zeta \tag{6.32a}$$

$$N_2 = L_2 = \eta \tag{6.32b}$$

$$N_3 = L_3 = \zeta \tag{6.32c}$$

where Eq. (6.31) has been substituted into Eq. (6.32a). The sum of the three shape functions is equal to 1 (partition of unity).

6.3.2.2 Six-node Triangular Elements

The parent element of 6-node (quadratic) triangular elements is shown in Figure 6.7 in the natural coordinates η, ζ. The nodal values of the area coordinates and the natural coordinates are listed in Figure 6.7.

The shape functions are conveniently constructed using the area coordinates and based on their Kronecker delta property. Using Eq. (6.31), they are expressed as

$$N_1 = L_1(2L_1 - 1) = (1 - \eta - \zeta)(1 - 2\eta - 2\zeta) \tag{6.33a}$$

$$N_2 = L_2(2L_2 - 1) = \eta(2\eta - 1) \tag{6.33b}$$

$$N_3 = L_3(2L_3 - 1) = \zeta(2\zeta - 1) \tag{6.33c}$$

$$N_4 = 4L_1L_2 = 4(1 - \eta - \zeta)\eta \tag{6.33d}$$

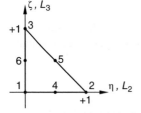

Node i	1	2	3	4	5	6
L_1	1	0	0	0.5	0	0.5
L_2, η	0	1	0	0.5	0.5	0
L_2, ζ	0	0	1	0	0.5	0.5

Figure 6.7 Six-node parent element in natural coordinates η, ζ.

$$N_5 = 4L_2L_3 = 4\eta\zeta \tag{6.33e}$$
$$N_6 = 4L_3L_1 = 4\zeta(-\eta - \zeta) \tag{6.33f}$$

Shape function N_1 is considered as an example of the corner nodes. It should be equal to 1 at node 1 and 0 at all the five other nodes (nodes 2–6). Three (nodes 1, 2 and 4) of the five nodes are located at its opposite side where the area coordinate $L_1 = 0$ applies. The area coordinates of the other two nodes (midside nodes 5 and 6) are $L_1 = 0.5$. To have the Kronecker delta property, the shape function N_1 must be the product of the factors L_1, $(2L_1 - 1)$

$$N_1 = c_1L_1(2L_1 - 1) \tag{6.34}$$

where the constant $c_1 = 1$ is determined by $N_1 = 1$ at node 1, leading to Eq. (6.33a).

Shape function N_4, as a example of the midside nodes, is considered. It should be equal to 1 at node 4 and 0 at all other five nodes. These five nodes are located at either the area coordinate $L_1 = 0$ (node 2, 3 and 5) or the area coordinate $L_2 = 0$ (node 1, 3 and 6). The shape function must have the factors L_1 and L_2

$$N_4 = c_4L_1L_2 \tag{6.35}$$

where the constant $c_4 = 4$ is obtained from $N_4 = 1$ at node 4 ($L_1 = L_2 = 0.5$), leading to Eq. (6.33d).

The partition of unity of the shape functions can be verified by considering the identity $L_1 + L_2 + L_3 = 1$.

6.4 Scaled Boundary Transformation of Geometry

The scaled boundary transformation of the geometry of an S-element is formulated by combining the scaling operation of the boundary in Section 6.2 and the boundary discretization in Section 6.3. The transformation of a specific 9-node surface element is illustrated in Figure 6.8. The numbering of the corner nodes has to follow the counter-clockwise direction around the outward normal vector of the boundary. The other nodes are numbered following the sequence adopted in the parent element.

Figure 6.8 Scaled boundary transformation of geometry defined on one surface element.

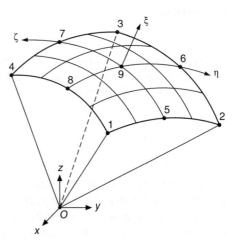

Substituting Eq. (6.12), which describes the geometry of the surface element, into Eq. (6.11), which describes an S-element by scaling its boundary, leads to

$$x(\xi, \eta, \zeta) = x_0 + \xi([N(\eta, \zeta)]\{x\} - x_0) \tag{6.36a}$$

$$y(\xi, \eta, \zeta) = y_0 + \xi([N(\eta, \zeta)]\{y\} - y_0) \tag{6.36b}$$

$$z(\xi, \eta, \zeta) = z_0 + \xi([N(\eta, \zeta)]\{z\} - z_0) \tag{6.36c}$$

ξ, η and ζ are called the *scaled boundary coordinates* in three dimensions and constitute a right-hand coordinate system. ξ is a (dimensionless) radial coordinate pointing from the scaling centre to the boundary. η and ζ are defined on the boundary or any surface parallel to the boundary. They are referred to as the circumferential coordinates. Equation (6.36) defines the scaled boundary transformation. The scaling requirement guarantees that the transformation is unique and that the radial coordinate ξ forms an acute angle with the outward normal of the boundary.

The governing differential equations (see Appendix A) are expressed in the Cartesian coordinates x, y and z. In the derivation, the field of a displacement component (the subscript indicating the direction is omitted for simplicity in notation) is expressed as

$$u = u(x, y, z) \tag{6.37}$$

The infinitesimal displacement increment is written as

$$du = \frac{\partial u}{\partial x}dx + \frac{\partial u}{\partial y}dy + \frac{\partial u}{\partial z}dz = [dx \quad dy \quad dz]\begin{Bmatrix} \dfrac{\partial u}{\partial x} \\[2mm] \dfrac{\partial u}{\partial y} \\[2mm] \dfrac{\partial u}{\partial z} \end{Bmatrix} \tag{6.38}$$

In the scaled boundary finite element method (similar to the isoparametric finite element formulation), the displacement field is described implicitly by parametric equations written in the scaled boundary coordinates ξ, η, ζ. The system of equations involves the expression of the displacement field (see Eq. (6.86) in Section 6.7) expressed as

$$u = u(\xi, \eta, \zeta) \tag{6.39}$$

To avoid proliferation of notation, the same symbol u is used for the displacement field in the Cartesian coordinates and in the scaled boundary coordinates. The geometry of the S-element is described by

$$x = x(\xi, \eta, \zeta) \tag{6.40a}$$

$$y = y(\xi, \eta, \zeta) \tag{6.40b}$$

$$z = z(\xi, \eta, \zeta) \tag{6.40c}$$

For a given point in the scaled boundary coordinates (ξ, η, ζ), the Cartesian coordinates (x, y, z) and displacement u at the point are evaluated from Eqs. (6.40) and (6.39), respectively. The function $u(x, y, z)$ and its partial derivatives with respect to the Cartesian coordinates x, y and z are not available explicitly.

To determine the partial derivatives from the parametric expressions, the infinitesimal displacement increment in Eq. (6.39) is expressed as

$$du = \frac{\partial u}{\partial \xi} d\xi + \frac{\partial u}{\partial \eta} d\eta + \frac{\partial u}{\partial \zeta} d\zeta = [d\xi \quad d\eta \quad d\zeta] \begin{Bmatrix} \dfrac{\partial u}{\partial \xi} \\[2mm] \dfrac{\partial u}{\partial \eta} \\[2mm] \dfrac{\partial u}{\partial \zeta} \end{Bmatrix} \tag{6.41}$$

Using Eq. (6.38) to replace du, the partial derivatives in the Cartesian and scaled boundary coordinate systems are related by

$$[dx \quad dy \quad dz] \begin{Bmatrix} \dfrac{\partial u}{\partial x} \\[2mm] \dfrac{\partial u}{\partial y} \\[2mm] \dfrac{\partial u}{\partial z} \end{Bmatrix} = [d\xi \quad d\eta \quad d\zeta] \begin{Bmatrix} \dfrac{\partial u}{\partial \xi} \\[2mm] \dfrac{\partial u}{\partial \eta} \\[2mm] \dfrac{\partial u}{\partial \zeta} \end{Bmatrix} \tag{6.42}$$

The infinitesimal increments of the Cartesian coordinates in Eq. (6.40) are equal to

$$dx = x_{,\xi} \, d\xi + x_{,\eta} \, d\eta + x_{,\zeta} \, d\zeta \tag{6.43a}$$
$$dy = y_{,\xi} \, d\xi + y_{,\eta} \, d\eta + y_{,\zeta} \, d\zeta \tag{6.43b}$$
$$dz = z_{,\xi} \, d\xi + z_{,\eta} \, d\eta + z_{,\zeta} \, d\zeta \tag{6.43c}$$

and expressed in matrix form as

$$[dx \quad dy \quad dz] = [d\xi \quad d\eta \quad d\zeta] \begin{bmatrix} x_{,\xi} & y_{,\xi} & z_{,\xi} \\ x_{,\eta} & y_{,\eta} & z_{,\eta} \\ x_{,\zeta} & y_{,\zeta} & z_{,\zeta} \end{bmatrix} \tag{6.44}$$

Introducing the Jacobian matrix

$$[J] = \begin{bmatrix} x_{,\xi} & y_{,\xi} & z_{,\xi} \\ x_{,\eta} & y_{,\eta} & z_{,\eta} \\ x_{,\zeta} & y_{,\zeta} & z_{,\zeta} \end{bmatrix} \tag{6.45}$$

Eq. (6.44) is inversed and rewritten as

$$[d\xi \quad d\eta \quad d\zeta] = [dx \quad dy \quad dz][J]^{-1} \tag{6.46}$$

Substituting Eq. (6.46) into Eq. (6.42) results in

$$\begin{Bmatrix} \dfrac{\partial u}{\partial x} \\[2mm] \dfrac{\partial u}{\partial y} \\[2mm] \dfrac{\partial u}{\partial z} \end{Bmatrix} = [J]^{-1} \begin{Bmatrix} \dfrac{\partial u}{\partial \xi} \\[2mm] \dfrac{\partial u}{\partial \eta} \\[2mm] \dfrac{\partial u}{\partial \zeta} \end{Bmatrix} \tag{6.47}$$

for arbitrary infinitesimal increments dx, dy and dz. The partial derivatives with respects to x, y, z are thus transformed to those with respect to ξ, η, ζ by the inverse of the Jacobian matrix.

The entries of the Jacobian matrix (Eq. 6.45) are determined from Eq. (6.11) as

$$x_{,\xi} = x_b - x_0 \qquad y_{,\xi} = y_b - y_0 \qquad z_{,\xi} = y_b - y_0 \tag{6.48a}$$

$$x_{,\eta} = \xi x_{b,\eta} \qquad y_{,\eta} = \xi y_{b,\eta} \qquad z_{,\eta} = \xi z_{b,\eta} \tag{6.48b}$$

$$x_{,\zeta} = \xi x_{b,\zeta} \qquad y_{,\zeta} = \xi y_{b,\zeta} \qquad z_{,\zeta} = \xi z_{b,\zeta} \tag{6.48c}$$

The arguments η and ζ are omitted for conciseness. Substituting Eq. (6.48) into Eq. (6.45), the Jacobian matrix is expressed as

$$[J] = \begin{bmatrix} 1 & 0 & 0 \\ 0 & \xi & 0 \\ 0 & 0 & \xi \end{bmatrix} \begin{bmatrix} x_b - x_0 & y_b - y_0 & y_b - y_0 \\ x_{b,\eta} & y_{b,\eta} & z_{b,\eta} \\ x_{b,\zeta} & y_{b,\zeta} & z_{b,\zeta} \end{bmatrix} \tag{6.49}$$

where the radial coordinate ξ is separated from the circumferential coordinates η and ζ that describe the boundary (x_b, y_b, z_b) in Eq. (6.12). Substituting Eq. (6.49) into Eq. (6.47), the partial differential operators (where the function u is omitted) with respect to the Cartesian coordinates x, y, z are written as

$$\left\{ \begin{array}{c} \dfrac{\partial}{\partial x} \\[2mm] \dfrac{\partial}{\partial y} \\[2mm] \dfrac{\partial}{\partial z} \end{array} \right\} = [J_b]^{-1} \left\{ \begin{array}{c} \dfrac{\partial}{\partial \xi} \\[2mm] \dfrac{1}{\xi} \dfrac{\partial}{\partial \eta} \\[2mm] \dfrac{1}{\xi} \dfrac{\partial}{\partial \zeta} \end{array} \right\} \tag{6.50}$$

with the Jacobian matrix on the boundary $([J_b] = [J(\xi = 1)])$

$$[J_b] = \begin{bmatrix} x_b - x_0 & y_b - y_0 & y_b - y_0 \\ x_{b,\eta} & y_{b,\eta} & z_{b,\eta} \\ x_{b,\zeta} & y_{b,\zeta} & z_{b,\zeta} \end{bmatrix} \tag{6.51}$$

The entries of the Jacobian matrix $[J_b]$ are obtained from Eq. (6.12) as

$$x_b - x_0 = \sum_i N_i x_i = [N]\{x\} \quad y_b - y_0 = [N]\{y\} \quad z_b - z_0 = [N]\{z\} \tag{6.52a}$$

$$x_{b,\eta} = \sum_i N_{i,\eta} x_i = [N_{,\eta}]\{x\} \quad y_{b,\eta} = [N_{,\eta}]\{y\} \quad z_{b,\eta} = [N_{,\eta}]\{z\} \tag{6.52b}$$

$$x_{b,\zeta} = \sum_i N_{i,\zeta} x_i = [N_{,\zeta}]\{x\} \quad y_{b,\zeta} = [N_{,\zeta}]\{y\} \quad z_{b,\zeta} = [N_{,\zeta}]\{z\} \tag{6.52c}$$

where the arguments η, ζ are omitted. Note that $[J_b]$ depends on the geometry of the boundary (shape functions and nodal coordinates) only.

The determinant of the matrix $[J_b]$ is equal to

$$|J_b| = (x_b - x_0)(y_{b,\eta} z_{b,\zeta} - z_{b,\eta} y_{b,\zeta}) + (y_b - y_0)(z_{b,\eta} x_{b,\zeta} - x_{b,\eta} z_{b,\zeta})$$
$$+ (z_b - z_0)(x_{b,\eta} y_{b,\zeta} - y_{b,\eta} x_{b,\zeta}) \tag{6.53}$$

The inverse of the 3×3 Jacobian matrix $[J_b]$ is obtained explicitly as

$$[J_b]^{-1} = \begin{bmatrix} j_{11} & j_{12} & j_{13} \\ j_{21} & j_{22} & j_{23} \\ j_{31} & j_{32} & j_{33} \end{bmatrix} \tag{6.54}$$

where the following abbreviations are introduced

$$j_{11} = \frac{1}{|J_b|}(y_{b,\eta} z_{b,\zeta} - z_{b,\eta} y_{b,\zeta}) \tag{6.55a}$$

$$j_{21} = \frac{1}{|J_b|}(z_{b,\eta} x_{b,\zeta} - x_{b,\eta} z_{b,\zeta}) \tag{6.55b}$$

$$j_{31} = \frac{1}{|J_b|}(x_{b,\eta} y_{b,\zeta} - y_{b,\eta} x_{b,\zeta}) \tag{6.55c}$$

$$j_{12} = \frac{1}{|J_b|}((z_b - z_0)y_{b,\zeta} - (y_b - y_0)z_{b,\zeta}) \tag{6.55d}$$

$$j_{22} = \frac{1}{|J_b|}((x_b - x_0)z_{b,\zeta} - (z_b - z_0)x_{b,\zeta}) \tag{6.55e}$$

$$j_{32} = \frac{1}{|J_b|}((y_b - y_0)x_{b,\zeta} - (x_b - x_0)y_{b,\zeta}) \tag{6.55f}$$

$$j_{13} = \frac{1}{|J_b|}((y_b - y_0)z_{b,\eta} - (z_b - z_0)y_{b,\eta}) \tag{6.55g}$$

$$j_{23} = \frac{1}{|J_b|}((z_b - z_0)x_{b,\eta} - (x_b - x_0)z_{b,\eta}) \tag{6.55h}$$

$$j_{33} = \frac{1}{|J_b|}((x_b - x_0)y_{b,\eta} - (y_b - y_0)x_{b,\eta}) \tag{6.55i}$$

Numerical inversion can be performed in a computer program.

Substituting this equation into Eq. (6.50) yields the following expression of the differential operator

$$\left\{ \begin{array}{c} \dfrac{\partial}{\partial x} \\ \dfrac{\partial}{\partial y} \\ \dfrac{\partial}{\partial z} \end{array} \right\} = \left\{ \begin{array}{c} j_{11} \\ j_{21} \\ j_{31} \end{array} \right\} \dfrac{\partial}{\partial \xi} + \left\{ \begin{array}{c} j_{12} \\ j_{22} \\ j_{32} \end{array} \right\} \dfrac{1}{\xi}\dfrac{\partial}{\partial \eta} + \left\{ \begin{array}{c} j_{13} \\ j_{23} \\ j_{33} \end{array} \right\} \dfrac{1}{\xi}\dfrac{\partial}{\partial \zeta} \tag{6.56}$$

For later use, it is noted that the entries of $[J_b]^{-1}$ satisfy the identities

$$(|J_b|j_{12})_{,\eta} + (|J_b|j_{13})_{,\zeta} = -2(|J_b|j_{11}) \tag{6.57a}$$

$$(|J_b|j_{22})_{,\eta} + (|J_b|j_{23})_{,\zeta} = -2(|J_b|j_{21}) \tag{6.57b}$$

$$(|J_b|j_{32})_{,\eta} + (|J_b|j_{33})_{,\zeta} = -2(|J_b|j_{31}) \tag{6.57c}$$

6.5 Geometrical Properties in Scaled Boundary Coordinates

The geometrical properties of an S-element in the scaled boundary coordinates are discussed. A part of the S-element obtained by scaling a surface element is illustrated in

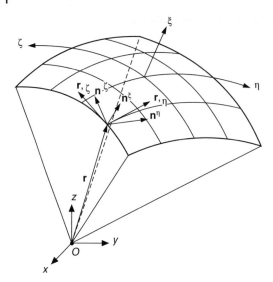

Figure 6.9 Geometrical properties in the scaled boundary coordinates of a part of an S-domain obtained by scaling a surface element.

Figure 6.9. The coordinate surfaces with constant values of one of the scaled boundary coordinates ξ, η, ζ are shown. The volume has a pyramid shape with a curved base. The scaling centre is the apex of the pyramid. The base has a constant radial coordinate ξ and is parallel to the surface element on the boundary. The side faces have either a constant circumferential coordinate η or ζ, as indicated by the coordinate axes. The geometrical properties of the three types of coordinate surfaces are considered, which will also be used in the transformation of an infinitesimal volume.

A coordinate surface with a constant radial coordinate ξ is denoted by S^ξ, which is also referred to as an (η, ζ) surface. The geometry around a point on the surface is described by two tangential vectors along the coordinate curves with constant η and ζ. Since any surface S^ξ is parallel to the boundary, the surface element on the boundary (i.e. $\mathbf{r} = \mathbf{r}_b$ and $\xi = 1$) is addressed first. The two tangential vectors (Figure 6.9 without subscript b) are obtained as the partial derivatives of the position vector $\mathbf{r}_b$ (Eq. (6.3)) with respect to the circumferential coordinates η and ζ

$$\mathbf{r}_{b,\eta} = x_{b,\eta}\,\mathbf{i} + y_{b,\eta}\,\mathbf{j} + z_{b,\eta}\,\mathbf{k} \tag{6.58a}$$

$$\mathbf{r}_{b,\zeta} = x_{b,\zeta}\,\mathbf{i} + y_{b,\zeta}\,\mathbf{j} + z_{b,\zeta}\,\mathbf{k} \tag{6.58b}$$

The outward normal vector to the surface element, which is described by the circumferential coordinates η and ζ, is equal to

$$\begin{aligned}
\mathbf{g}^\xi &= \mathbf{r}_{b,\eta} \times \mathbf{r}_{b,\zeta} \\
&= \begin{vmatrix} \mathbf{i} & \mathbf{j} & \mathbf{k} \\ x_{b,\eta} & y_{b,\eta} & z_{b,\eta} \\ x_{b,\zeta} & y_{b,\zeta} & z_{b,\zeta} \end{vmatrix} \\
&= (y_{b,\eta}\,z_{b,\zeta} - z_{b,\eta}\,y_{b,\zeta})\mathbf{i} + (z_{b,\eta}\,x_{b,\zeta} - x_{b,\eta}\,z_{b,\zeta})\mathbf{j} + (x_{b,\eta}\,y_{b,\zeta} - y_{b,\eta}\,x_{b,\zeta})\mathbf{k}
\end{aligned} \tag{6.59}$$

The unit outward normal vector is written as (Figure 6.9)

$$\mathbf{n}^\xi = n_x^\xi\,\mathbf{i} + n_y^\xi\,\mathbf{j} + n_z^\xi\,\mathbf{k} = \frac{\mathbf{g}^\xi}{|g^\xi|} \tag{6.60}$$

where $|g^\xi|$ denotes the magnitude of the vector $\mathbf{g}^\xi$. It is expressed in matrix form as

$$\{n^\xi\} = \begin{Bmatrix} n_x^\xi \\ n_y^\xi \\ n_z^\xi \end{Bmatrix} = \frac{1}{|g^\xi|} \begin{Bmatrix} y_{b,\eta}\, z_{b,\zeta} - z_{b,\eta}\, y_{b,\zeta} \\ z_{b,\eta}\, x_{b,\zeta} - x_{b,\eta}\, z_{b,\zeta} \\ x_{b,\eta}\, y_{b,\zeta} - y_{b,\eta}\, x_{b,\zeta} \end{Bmatrix} \tag{6.61}$$

The tangential vectors to a surface S^ξ at any constant value of ξ (Figure 6.9) are given by differentiating Eq. (6.10)

$$\mathbf{r},_\eta = \xi \mathbf{r}_{b,\eta} = \xi x_{b,\eta}\,\mathbf{i} + \xi y_{b,\eta}\,\mathbf{j} + \xi z_{b,\eta}\,\mathbf{k} \tag{6.62a}$$

$$\mathbf{r},_\zeta = \xi \mathbf{r}_{b,\zeta} = \xi x_{b,\zeta}\,\mathbf{i} + \xi y_{b,\zeta}\,\mathbf{j} + \xi z_{b,\zeta}\,\mathbf{k} \tag{6.62b}$$

Note that the tangential vectors are parallel to those on the boundary (Eq. 6.58). The unit outward normal vector to a surface S^ξ is independent of ξ and given by Eq. (6.60). An infinitesimal area on a surface S^ξ is equal to

$$dS^\xi = |\mathbf{r},_\eta \times \mathbf{r},_\zeta|\, d\eta d\zeta \tag{6.63}$$

Using Eqs. (6.62) and (6.59), it is obtained as

$$dS^\xi = |\xi \mathbf{r}_{b,\eta} \times \xi \mathbf{r}_{b,\zeta}|\, d\eta d\zeta = \xi^2 |g^\xi|\, d\eta d\zeta \tag{6.64}$$

The side faces are coordinate surfaces with one of the two circumferential coordinates η and ζ being constant. A coordinate surface with a constant η is denoted as S^η. It defines a (ζ, ξ) surface. Similarly, a coordinate surface with a constant ζ is denoted as S^ζ and is a (ξ, η) surface. To define the geometry around a point of the side faces, the tangential vector to the radial direction is obtained by differentiating the position vector (Eq. (6.1)) with respect to the radial coordinate ξ

$$\mathbf{r},_\xi = x,_\xi\,\mathbf{i} + y,_\xi\,\mathbf{j} + z,_\xi\,\mathbf{k} \tag{6.65}$$

Using Eqs. (6.10), (6.3) and (6.7), Eq. (6.65) is rewritten as

$$\mathbf{r},_\xi = \mathbf{r}_b - \mathbf{r}_0 = (x_b - x_0)\mathbf{i} + (y_b - y_0)\mathbf{j} + (z_b - z_0)\mathbf{k} \tag{6.66}$$

As the radial coordinate ξ is a straight line in the Cartesian coordinate, this tangential direction is the same as the radial direction originating from the scaling centre.

The outward normal vectors to the surfaces S^η and S^ζ are determined at $\xi = 1$ (i.e. on the intersection with the surface element). Using Eqs. (6.58) and (6.66), they are obtained as

$$\mathbf{g}^\eta = \mathbf{r}_{b,\zeta} \times \mathbf{r},_\xi = ((z_b - z_0)y_{b,\zeta} - (y_b - y_0)z_{b,\zeta})\mathbf{i} + ((x_b - x_0)z_{b,\zeta} - (z_b - z_0)x_{b,\zeta})\mathbf{j}$$
$$+ ((y_b - y_0)x_{b,\zeta} - (x_b - x_0)y_{b,\zeta})\mathbf{k} \tag{6.67a}$$

$$\mathbf{g}^\zeta = \mathbf{r},_\xi \times \mathbf{r}_{b,\eta} = ((y_b - y_0)z_{b,\eta} - (z_b - z_0)y_{b,\eta})\mathbf{i} + ((z_b - z_0)x_{b,\eta} - (x_b - x_0)z_{b,\eta})\mathbf{j}$$
$$+ ((x_b - x_0)y_{b,\eta} - (y_b - y_0)x_{b,\eta})\mathbf{k} \tag{6.67b}$$

The unit normal vectors of the surfaces S^η and S^ζ (Figure 6.9) follow from Eq. (6.67) as

$$\mathbf{n}^\eta = n_x^\eta\mathbf{i} + n_y^\eta\mathbf{j} + n_z^\eta\mathbf{k} = \frac{\mathbf{g}^\eta}{|g^\eta|} \tag{6.68a}$$

$$\mathbf{n}^\zeta = n_x^\zeta\mathbf{i} + n_y^\zeta\mathbf{j} + n_z^\zeta\mathbf{k} = \frac{\mathbf{g}^\zeta}{|g^\zeta|} \tag{6.68b}$$

They are written in matrix form as

$$\{n^{\eta}\} = \begin{Bmatrix} n_x^{\eta} \\ n_y^{\eta} \\ n_z^{\eta} \end{Bmatrix} = \frac{1}{|g^{\eta}|} \begin{Bmatrix} (z_b - z_0)y_{b,\zeta} - (y_b - y_0)z_{b,\zeta} \\ (x_b - x_0)z_{b,\zeta} - (z_b - z_0)x_{b,\zeta} \\ (y_b - y_0)x_{b,\zeta} - (x_b - x_0)y_{b,\zeta} \end{Bmatrix} \tag{6.69a}$$

$$\{n^{\zeta}\} = \begin{Bmatrix} n_x^{\zeta} \\ n_y^{\zeta} \\ n_z^{\zeta} \end{Bmatrix} = \frac{1}{|g^{\zeta}|} \begin{Bmatrix} (y_b - y_0)z_{b,\eta} - (z_b - z_0)y_{b,\eta} \\ (z_b - z_0)x_{b,\eta} - (x_b - x_0)z_{b,\eta} \\ (x_b - x_0)y_{b,\eta} - (y_b - y_0)x_{b,\eta} \end{Bmatrix} \tag{6.69b}$$

The infinitesimal areas on the surfaces S^{η} and S^{ζ} for any ξ are determined using Eqs. (6.62), (6.66) and (6.67) as

$$dS^{\eta} = |\mathbf{r}_{,\zeta} \times \mathbf{r}_{,\xi}| d\zeta d\xi = |\xi \mathbf{r}_{b,\zeta} \times \mathbf{r}_{,\xi}| d\zeta d\xi = \xi |g^{\eta}| d\zeta d\xi \tag{6.70a}$$

$$dS^{\zeta} = |\mathbf{r}_{,\xi} \times \mathbf{r}_{,\eta}| d\xi d\eta = |\mathbf{r}_{,\xi} \times \xi \mathbf{r}_{b,\eta}| d\xi d\eta = \xi |g^{\zeta}| d\xi d\eta \tag{6.70b}$$

The infinitesimal volume dV is determined using Eqs. (6.66) and (6.62), as

$$dV = \mathbf{r}_{,\xi} \cdot (\mathbf{r}_{,\eta} \times \mathbf{r}_{,\zeta}) d\xi d\eta d\zeta = \xi^2 (\mathbf{r}_b - \mathbf{r}_0) \cdot (\mathbf{r}_{b,\eta} \times \mathbf{r}_{b,\zeta}) d\xi d\eta d\zeta \tag{6.71}$$

Note that the vector triple product is simply equal to the determinant of the Jacobian matrix on the boundary $[J_b]$ in Eq. (6.51)

$$(\mathbf{r}_b - \mathbf{r}_0) \cdot (\mathbf{r}_{b,\eta} \times \mathbf{r}_{b,\zeta}) = \begin{vmatrix} x_b - x_0 & y_b - y_0 & z_b - z_0 \\ x_{b,\eta} & y_{b,\eta} & z_{b,\eta} \\ x_{b,\zeta} & y_{b,\zeta} & z_{b,\zeta} \end{vmatrix} = |J_b| \tag{6.72}$$

Using Eq. (6.72), Eq. (6.71) is expressed as

$$dV = \xi^2 |J_b| d\xi d\eta d\zeta \tag{6.73}$$

The outward normals of the coordinate surfaces are related to the inverse of the Jacobian matrix on the boundary. Comparing the right-hand side of Eq. (6.61) to Eqs. (6.55a)–(6.55c), Eq. (6.69a) to Eqs. (6.55d)–(6.55f) and Eq. (6.69b) to Eqs. (6.55g)–(6.55i), the following equations are identified

$$\begin{Bmatrix} j_{11} \\ j_{21} \\ j_{31} \end{Bmatrix} = \frac{|g^{\xi}|}{|J_b|} \begin{Bmatrix} n_x^{\xi} \\ n_y^{\xi} \\ n_z^{\xi} \end{Bmatrix} \tag{6.74a}$$

$$\begin{Bmatrix} j_{12} \\ j_{22} \\ j_{32} \end{Bmatrix} = \frac{|g^{\xi}|}{|J_b|} \begin{Bmatrix} n_x^{\eta} \\ n_y^{\eta} \\ n_z^{\eta} \end{Bmatrix} \tag{6.74b}$$

$$\begin{Bmatrix} j_{13} \\ j_{23} \\ j_{33} \end{Bmatrix} = \frac{|g^{\xi}|}{|J_b|} \begin{Bmatrix} n_x^{\zeta} \\ n_y^{\zeta} \\ n_z^{\zeta} \end{Bmatrix} \tag{6.74c}$$

The inverse of $[J_b]$ (Eq. (6.54)) is expressed as

$$[J_b]^{-1} = \frac{1}{|J_b|} \begin{bmatrix} |g^\xi| n_x^\xi & |g^\eta| n_x^\eta & |g^\zeta| n_x^\zeta \\ |g^\xi| n_y^\xi & |g^\eta| n_y^\eta & |g^\zeta| n_y^\zeta \\ |g^\xi| n_z^\xi & |g^\eta| n_z^\eta & |g^\zeta| n_z^\zeta \end{bmatrix} \tag{6.75}$$

Using Eq. (6.74), Eq. (6.56) is rewritten as

$$\left\{ \begin{array}{c} \dfrac{\partial}{\partial x} \\[2mm] \dfrac{\partial}{\partial y} \\[2mm] \dfrac{\partial}{\partial z} \end{array} \right\} = \frac{|g^\xi|}{|J_b|}\{n^\xi\}\frac{\partial}{\partial \xi} + \frac{1}{\xi}\left(\frac{|g^\eta|}{|J_b|}\{n^\eta\}\frac{\partial}{\partial \eta} + \frac{|g^\zeta|}{|J_b|}\{n^\zeta\}\frac{\partial}{\partial \zeta} \right) \tag{6.76}$$

The partial differential operators with respect to x, y, z are expressed as functions of those with respect to the scaled boundary coordinates ξ, η, ζ in terms of the geometrical properties of the coordinate surfaces.

Using Eq. (6.57), Eq. (6.61) multiplied with $|g^\xi|$, Eq. (6.69a) multiplied with $|g^\eta|$ and Eq. (6.69b) multiplied with $|g^\zeta|$, the following identity can be verified

$$\left(|g^\eta|\{n^\eta\} \right)_{,\eta} + \left(|g^\zeta|\{n^\zeta\} \right)_{,\zeta} = -2|g^\xi|\{n^\xi\} \tag{6.77}$$

6.6 Governing Equations of Elastodynamics with Geometry in Scaled Boundary Coordinates

In the scaled boundary finite element method, the scaled boundary transformation is applied to the geometry of the S-element. Only the spatial coordinates are affected. The components of the displacements, strains and stresses are still defined in the original Cartesian coordinate system x, y, z. This is analogous to the procedure of mapping parent elements to curvilinear elements in the finite element method.

For the governing partial differential equations of elastodynamics in Section A.1, only the differential operator $[L]$ in Eq. (A.6) is addressed. Using Eq. (6.56), Eq. (A.6) is expressed as

$$[L] = [b_1]\frac{\partial}{\partial \xi} + \frac{1}{\xi}\left([b_2]\frac{\partial}{\partial \eta} + [b_3]\frac{\partial}{\partial \zeta} \right) \tag{6.78}$$

with

$$[b_1] = \begin{bmatrix} j_{11} & 0 & 0 \\ 0 & j_{21} & 0 \\ 0 & 0 & j_{31} \\ 0 & j_{31} & j_{21} \\ j_{31} & 0 & j_{11} \\ j_{21} & j_{11} & 0 \end{bmatrix} \tag{6.79a}$$

$$[b_2] = \begin{bmatrix} j_{12} & 0 & 0 \\ 0 & j_{22} & 0 \\ 0 & 0 & j_{32} \\ 0 & j_{32} & j_{22} \\ j_{32} & 0 & j_{12} \\ j_{22} & j_{12} & 0 \end{bmatrix} \tag{6.79b}$$

$$[b_3] = \begin{bmatrix} j_{13} & 0 & 0 \\ 0 & j_{23} & 0 \\ 0 & 0 & j_{33} \\ 0 & j_{33} & j_{23} \\ j_{33} & 0 & j_{13} \\ j_{23} & j_{13} & 0 \end{bmatrix} \tag{6.79c}$$

where j_{kl} (k, l = 1, 2, 3) are entries of the inverse of the Jacobian matrix $[J_b]$ (Eq. (6.54)). The explicit expressions are given in Eq. (6.55). Note that $[b_1]$, $[b_2]$ and $[b_3]$ are independent of ξ. Using Eq. (6.57) or (6.77), it can be verified that Eq. (6.81) satisfies

$$(|J_b|[b_2])_{,\eta} + (|J_b|[b_3])_{,\zeta} = -2|J_b|[b_1] \tag{6.80}$$

For later use, the matrices in Eq. (6.79) are also expressed using the normal vectors of the coordinate surfaces. Substituting Eq. (6.74) into Eq. (6.79) yields

$$[b_1] = \frac{|g^\xi|}{|J_b|} \begin{bmatrix} n_x^\xi & 0 & 0 \\ 0 & n_y^\xi & 0 \\ 0 & 0 & n_z^\xi \\ 0 & n_z^\xi & n_y^\xi \\ n_z^\xi & 0 & n_x^\xi \\ n_y^\xi & n_x^\xi & 0 \end{bmatrix} \tag{6.81a}$$

$$[b_2] = \frac{|g^\eta|}{|J_b|} \begin{bmatrix} n_x^\eta & 0 & 0 \\ 0 & n_y^\eta & 0 \\ 0 & 0 & n_z^\eta \\ 0 & n_z^\eta & n_y^\eta \\ n_z^\eta & 0 & n_x^\eta \\ n_y^\eta & n_x^\eta & 0 \end{bmatrix} \tag{6.81b}$$

$$[b_3] = \frac{|g^\zeta|}{|J_b|} \begin{bmatrix} n_x^\zeta & 0 & 0 \\ 0 & n_y^\zeta & 0 \\ 0 & 0 & n_z^\zeta \\ 0 & n_z^\zeta & n_y^\zeta \\ n_z^\zeta & 0 & n_x^\zeta \\ n_y^\zeta & n_x^\zeta & 0 \end{bmatrix} \tag{6.81c}$$

Substituting the differential operator in the scaled boundary coordinates Eq. (6.78) into Eq. (A.5), the strains $\{\varepsilon\} = \{\varepsilon(\xi, \eta, \zeta)\}$ in Eq. (A.4) are expressed as

$$\{\varepsilon\} = [b_1]\frac{\partial\{u\}}{\partial\xi} + \frac{1}{\xi}\left([b_2]\frac{\partial\{u\}}{\partial\eta} + [b_3]\frac{\partial\{u\}}{\partial\zeta}\right) \tag{6.82}$$

The stress-strain relationship Eq. (A.11) and the elasticity matrix $[D]$ are not affected by the transformation of the spatial coordinates.

Using Eq. (6.78), the equations of motion given by Eq. (A.9) are expressed as

$$[b_1]^T \frac{\partial\{\sigma\}}{\partial\xi} + \frac{1}{\xi}\left([b_2]^T \frac{\partial\{\sigma\}}{\partial\eta} + [b_3]^T \frac{\partial\{\sigma\}}{\partial\zeta}\right) + \{p\} = \rho\{\ddot{u}\} \qquad (6.83)$$

The surface tractions $\{t\} = [\, t_x \ t_y \ t_z \,]^T$ at a point with the outward unit normal vector $\{n\} = \{n_x, \ n_y, \ n_z\}^T$ and the stresses $\{\sigma\}$ are given by Eq. (A.26). It is noted that the matrices in Eqs. (A.26) and (6.81) are arranged in the same pattern. On the coordinate surfaces (η, ζ), (ζ, ξ) and (ξ, η), the surface tractions are obtained from Eqs. (A.26) and Eq. (6.81) as

$$\{t^\xi\} = \frac{|J_b|}{|g^\xi|}[b_1]^T\{\sigma\} \qquad (6.84a)$$

$$\{t^\eta\} = \frac{|J_b|}{|g^\eta|}[b_2]^T\{\sigma\} \qquad (6.84b)$$

$$\{t^\zeta\} = \frac{|J_b|}{|g^\zeta|}[b_3]^T\{\sigma\} \qquad (6.84c)$$

6.7 Derivation of the Scaled Boundary Finite Element Equation by the Galerkin's Weighted Residual Technique

The scaled boundary finite element equation is derived by applying the virtual work principle in Section 2.5. In this section, the derivation based on the Galerkin's weighted residual method (Song and Wolf, 1997) is presented.

6.7.1 Displacement, Strain Fields and Nodal Force Functions in Scaled Boundary Coordinates

In the scaled boundary finite element method, the boundary of an S-domain is discretized with surface elements. The domain is modelled by an S-element obtained by scaling the surface elements. The derivation is performed on a pyramid sector of the S-element corresponding to one surface element, as shown in Figure 6.8 on page 249. The notation follows that in the derivation based on the virtual work principle in Section 2.5. For conciseness, the superscript (e) indicating one element on boundary is omitted.

The displacements are represented analytically in the radial direction and numerically in the circumferential directions. The radial lines $\overline{O1}, \overline{O2}, \ldots$ that pass through the scaling centre O and a node on the boundary are considered. Along these radial lines, the circumferential coordinates η, ζ are constant and the nodal displacement functions

$$\{u(\xi)\} = [u_{1x}(\xi) \quad u_{1y}(\xi) \quad u_{1z}(\xi) \quad u_{2x}(\xi) \quad u_{2y}(\xi) \quad u_{2z}(\xi) \quad \ldots]^T \qquad (6.85)$$

are introduced.

Isoparametric formulations of displacement-based elements are employed. The same shape functions are used to interpolate both the geometry (Section 6.3) and the displacement field piecewise. The interpolation in the circumferential directions is applied not only to the surface element on the boundary but also to any S^ξ surfaces (i.e. surfaces with a constant radial coordinate ξ and is thus defined by the circumferential coordinates η and ζ) obtained by scaling the surface element. The displacement field is thus described by

$$u_x(\xi, \eta, \zeta) = \sum_i N_i(\eta, \zeta) u_{ix}(\xi) = [N(\eta, \zeta)]\{u_x(\xi)\} \tag{6.86a}$$

$$u_y(\xi, \eta, \zeta) = \sum_i N_i(\eta, \zeta) u_{iy}(\xi) = [N(\eta, \zeta)]\{u_y(\xi)\} \tag{6.86b}$$

$$u_z(\xi, \eta, \zeta) = \sum_i N_i(\eta, \zeta) u_{iz}(\xi) = [N(\eta, \zeta)]\{u_z(\xi)\} \tag{6.86c}$$

The displacement field $\{u\} = \{u(\xi, \eta, \zeta)\} = [u_x(\xi, \eta, \zeta) \quad u_y(\xi, \eta, \zeta) \quad u_z(\xi, \eta, \zeta)]^T$ in Eq. (6.86) is expressed concisely in matrix form as

$$\{u\} = [N_u]\{u(\xi)\} \tag{6.87}$$

where the shape function matrix $[N_u] = [N_u(\eta, \zeta)]$ is written as

$$[N_u] = [N_1[I] \quad N_2[I] \quad \ldots] = \begin{bmatrix} N_1 & 0 & 0 & N_2 & 0 & 0 & \ldots \\ 0 & N_1 & 0 & 0 & N_2 & 0 & \ldots \\ 0 & 0 & N_1 & 0 & 0 & N_2 & \ldots \end{bmatrix} \tag{6.88}$$

with a 3×3 identity matrix $[I]$. The shape functions N_i $(i = 1, 2, \ldots)$ are given in Section 6.3 for standard triangular and quadrilateral surface elements and in Chapter 8 for high-order spectral elements.

The partial derivatives of the displacement field (Eq. (6.87)) with respect to the circumferential coordinates η and ζ are expressed as

$$\frac{\partial \{u\}}{\partial \eta} = [N_u]_{,\eta} \{u(\xi)\} \tag{6.89a}$$

$$\frac{\partial \{u\}}{\partial \zeta} = [N_u]_{,\zeta} \{u(\xi)\} \tag{6.89b}$$

Substituting Eq. (6.89) into Eq. (6.82), the strain field $\{\varepsilon\} = \{\varepsilon(\xi, \eta, \zeta)\}$ corresponding to the nodal displacement functions $\{u(\xi)\}$ is expressed in the scaled boundary coordinates as

$$\{\varepsilon\} = [b_1][N_u]\{u(\xi)\}_{,\xi} + \frac{1}{\xi}\left([b_2][N_u]_{,\eta} + [b_3][N_u]_{,\zeta}\right)\{u(\xi)\} \tag{6.90}$$

Introducing the matrices $[B_1] = [B_1(\eta, \zeta)]$ and $[B_2] = [B_2(\eta, \zeta)]$

$$[B_1] = [b_1][N_u] \tag{6.91a}$$

$$[B_2] = [b_2][N_u]_{,\eta} + [b_3][N_u]_{,\zeta} \tag{6.91b}$$

Eq. (6.90) is rewritten as

$$\{\varepsilon\} = [B_1]\{u(\xi)\}_{,\xi} + \frac{1}{\xi}[B_2]\{u(\xi)\} \tag{6.92}$$

$[B_1]$ and $[B_2]$ depend on the geometry of the surface element only.

The stress field is expressed as (Eq. (A.11))

$$\{\sigma\} = [D]\left([B_1]\{u(\xi)\}_{,\xi} + \frac{1}{\xi}[B_2]\{u(\xi)\}\right) + \{\sigma_0\} \tag{6.93}$$

where $[D]$ is the elasticity matrix and $\{\sigma_0\}$ the initial stresses.

To apply the Galerkin's weighted residual technique, the field of weighting function $\{w\} = \{w(\xi, \eta, \zeta)\}$ is constructed in the same way as the displacement field in Eq. (6.87)

$$\{w\} = [N_u]\{w(\xi)\} \tag{6.94}$$

where $\{w(\xi)\}$ is the vector of the weighting functions along the radial lines passing through the nodes on boundary.

Corresponding to the internal displacement functions $\{u(\xi)\}$, internal nodal force functions $\{q(\xi)\}$ are introduced along the radial lines passing through the nodes on the boundary

$$\{q(\xi)\} = [q_{1x}(\xi) \quad q_{1y}(\xi) \quad q_{1z}(\xi) \quad q_{2x}(\xi) \quad q_{2y}(\xi) \quad q_{2z}(\xi) \quad \dots]^T \tag{6.95}$$

On any surface S^ξ with a constant ξ, $\{q(\xi)\}$ are statically equivalent to the surface traction $\{t^\xi\}$

$$\{w(\xi)\}^T\{q(\xi)\} = \int_{S^\xi} \{w\}^T\{t^\xi\}dS^\xi \tag{6.96}$$

Substituting the weighting functions $\{w\}$ in Eq. (6.94) and the infinitesimal surface area dS^ξ in Eq. (6.64) yields for an arbitrary $\{w(\xi)\}$

$$\{q(\xi)\} = \int_{S^\xi} [N_u]^T\{t^\xi\}\xi^2|g^\xi|d\eta d\zeta \tag{6.97}$$

Substituting Eq. (6.84a) into Eq. (6.97) yields

$$\{q(\xi)\} = \xi^2 \int_{S^\xi} [N_u]^T[b^1]^T\{\sigma\}|J_b|d\eta d\zeta \tag{6.98}$$

Using Eq. (6.91a) leads to

$$\{q(\xi)\} = \xi^2 \int_{S^\xi} [B^1]^T\{\sigma\}|J_b|d\eta d\zeta \tag{6.99}$$

Substituting Eq. (6.93) into Eq. (6.99) yields

$$\{q(\xi)\} = \xi^2 \int_{S^\xi} [B^1]^T[D]\left([B^1]\{u(\xi)\},_\xi + \frac{1}{\xi}[B^2]\{u(\xi)\} + \{\sigma_0\}\right)|J_b|d\eta d\zeta \tag{6.100}$$

Introducing the coefficient matrices

$$[E^0] = \int_{S^\xi} [B^1]^T[D][B^1]|J_b|d\eta d\zeta \tag{6.101}$$

$$[E^1] = \int_{S^\xi} [B^2]^T[D][B^1]|J_b|d\eta d\zeta \tag{6.102}$$

results in the internal nodal forces

$$\{q(\xi)\} = [E^0]\xi^2\{u(\xi)\},_\xi + [E^1]^T\xi\{u(\xi)\} + \{q_0(\xi)\} \tag{6.103}$$

with the contribution of the initial stresses

$$\{q_0(\xi)\} = \xi^2 \int_{S^\xi} [B^1]^T\{\sigma_0\}|J_b|d\eta d\zeta \tag{6.104}$$

The equilibrium on the boundary ($\xi = 1$) with the external nodal forces $\{F\}$ leads to

$$\{F\} = \{q(\xi = 1)\} \tag{6.105}$$

6.7.2 The Scaled Boundary Finite Element Equation

The Galerkin's weighted-residual technique is applied to the governing partial differential equations. A domain V covered by scaling an element on the boundary in the range $\xi_1 \leq \xi \leq \xi_2$ is considered. Equation (6.83) is pre-multiplied by the weighting functions $\{w\}^T$ and integrated over the domain

$$\int_V \{w\}^T \left([b_1]^T \{\sigma\}_{,\xi} + \frac{1}{\xi} \left([b_2]^T \{\sigma\}_{,\eta} + [b_3]^T \{\sigma\}_{,\zeta} \right) + \{p\} - \rho\{\ddot{u}\} \right) dV = 0$$

(6.106)

Substituting the semi-analytical expressions of the weighting functions in Eq. (6.94) and displacement field in Eq. (6.87), Eq. (6.106) is separated into four terms W_I, W_{II}, W_{III} and W_{IV}

$$\underbrace{\int_V \{w(\xi)\}^T [B_1]^T \{\sigma\}_{,\xi} \, dV}_{W_I} + \underbrace{\int_V \{w(\xi)\}^T \frac{1}{\xi} [N_u]^T \left([b_2]^T \{\sigma\}_{,\eta} + [b_3]^T \{\sigma\}_{,\zeta} \right) dV}_{W_{II}}$$

$$+ \underbrace{\int_V \{w(\xi)\}^T [N_u]^T \{p\} dV}_{W_{III}} - \underbrace{\int_V \{w(\xi)\}^T [N_u]^T \rho [N_u]\{\ddot{u}(\xi)\} dV}_{W_{IV}} = 0 \qquad (6.107)$$

where $[B_1] = [b_1][N_u]$ (Eq. (6.91a)) has been substituted into the first term W_I. Equation (6.107) is addressed term-by-term. In all terms, the volume integration is expressed using Eq. (6.73) as

$$\int_V (\bullet) dV = \int_{\xi_1}^{\xi_2} \int_{S^\xi} (\bullet)\xi^2 |J_b| d\xi d\eta d\zeta = \int_{\xi_1}^{\xi_2} \xi^2 \int_{S^\xi} (\bullet) |J_b| d\eta d\zeta \, d\xi \qquad (6.108)$$

where $(\bullet)$ stands for the integrand, and S^ξ a coordinate surface with a constant ξ.

The first term W_I is expressed as

$$W_I = \int_{\xi_1}^{\xi_2} \{w(\xi)\}^T \xi^2 \int_{S^\xi} [B_1]^T \{\sigma\}_{,\xi} |J_b| d\eta d\zeta \, d\xi \qquad (6.109)$$

The second term of Eq. (6.107) is equal to

$$W_{II} = \int_{\xi_1}^{\xi_2} \{w(\xi)\}^T \xi \int_{S^\xi} [N_u]^T \left(|J_b|[b_2]^T \{\sigma\}_{,\eta} + |J_b|[b_3]^T \{\sigma\}_{,\zeta} \right) d\eta d\zeta \, d\xi \qquad (6.110)$$

Applying Green's theorem to the surface integral over S^ξ results in

$$W_{II} = \int_{\xi_1}^{\xi_2} \{w(\xi)\}^T \xi \left(\underbrace{\oint_{\Gamma^\xi} [N_u]^T \left(|J_b|[b_2]^T \{\sigma\} d\zeta + |J_b|[b_3]^T \{\sigma\} d\eta \right)}_{\{P_S(\xi)\}} \right.$$

$$\left. -\xi \underbrace{\int_{S^\xi} \left(\left([N_u]^T |J_b|[b_2]^T \right)_{,\eta} + \left([N_u]^T |J_b|[b_3]^T \right)_{,\zeta} \right) \{\sigma\} d\eta d\zeta}_{W_{IIB}} \right) d\xi \qquad (6.111)$$

where Γ^ξ denotes the edges of the surface S^ξ. Γ^ξ is formed by scaling the edges of the surface finite element. Using Eqs. (6.84b) and (6.84c), the contour integration $\{P_S(\xi)\}$ in the first term on the right side of Eq. (6.111) is rewritten as

$$\{P_S(\xi)\} = \xi \oint_{\Gamma^\xi} [N_u]^T \left(\{t^\eta\}|g^\eta|d\zeta + \{t^\zeta\}|g^\zeta|d\eta \right) \tag{6.112}$$

which represents simply the nodal force functions equivalent to the surface traction at ξ. The expression within the parentheses in the second term W_{IIB} on the right side of Eq. (6.111) is expanded as

$$W_{IIB} = [N_u]^T \left(\left(|J_b|[b_2]^T \right)_{,\eta} + \left(|J_b|[b_3]^T \right)_{,\zeta} \right) + [N_u]_{,\eta}^T |J_b|[b_2]^T + [N_u]_{,\zeta}^T |J_b|[b_3]^T \tag{6.113}$$

Using the identity in Eq. (6.80), it is expressed as

$$W_{IIB} = -2[N_u]^T |J_b|[b_1]^T + ([N_u]_{,\eta}^T [b_2]^T + [N_u]_{,\zeta}^T [b_3]^T)|J_b| \tag{6.114}$$

Introducing $[B_1]$ and $[B_2]$ defined in Eq. (6.91) leads to

$$W_{IIB} = (-2[B_1]^T + [B_2]^T)|J_b| \tag{6.115}$$

Using Eqs. (6.112) and (6.115), Eq. (6.111) is written as

$$W_{II} = \int_{\xi_1}^{\xi_2} \{w(\xi)\}^T \left(\{P_S(\xi)\} + \xi \int (2[B_1]^T - [B_2]^T)\{\sigma\}|J_b|d\eta d\zeta \right) d\xi \tag{6.116}$$

The third term of Eq. (6.107) represents the body force and is equal to

$$W_{III} = \int_{\xi_1}^{\xi_2} \{w(\xi)\}^T \xi^2 \int_{S^\xi} [N_u]^T \{p\}|J_b|d\eta d\zeta d\xi \tag{6.117}$$

Introducing the equivalent nodal force functions

$$\{P_B(\xi)\} = \xi^2 \int_{S^\xi} [N_u]^T \{p\}|J_b|d\eta d\zeta \tag{6.118}$$

Eq. (6.117) is rewritten as

$$W_{III} = \int_{\xi_1}^{\xi_2} \{w(\xi)\}^T \{P_B(\xi)\}d\xi \tag{6.119}$$

The fourth term of Eq. (6.107) is the contribution of the inertial forces. It is expressed as

$$W_{IV} = \int_{\xi_1}^{\xi_2} \{w(\xi)\}^T \xi^2 \int_{S^\xi} [N_u]^T \rho[N_u]|J_b|d\eta d\zeta \{\ddot{u}(\xi)\}d\xi \tag{6.120}$$

Introducing the mass coefficient matrix

$$[M_0] = \int_{S^\xi} [N_u]^T \rho[N_u]|J_b|d\eta d\zeta \tag{6.121}$$

Equation (6.120) is written as

$$W_{IV} = \int_0^1 \{w(\xi)\}^T \xi^2 [M_0]\{\ddot{u}(\xi)\}d\xi \tag{6.122}$$

Substituting W_I (Eq. (6.109)), W_{II} (Eq. (6.116)), W_{III} (Eq. (6.119)) and W_{IV} (Eq. (6.122)) into Eq. (6.107) yields, after simplifying,

$$\int_{\xi_1}^{\xi_2} \{w(\xi)\}^T \left(\int_{S^\xi} [B_1]^T (\xi^2\{\sigma\}_{,\xi} +2\xi\{\sigma\})|J_b|\mathrm{d}\eta\mathrm{d}\zeta - \xi \int_{S^\xi} [B_2]^T \{\sigma\}|J_b|\mathrm{d}\eta\mathrm{d}\zeta \right.$$
$$\left. + \{P_S(\xi)\} + \{P_B(\xi)\} - \xi^2[M_0]\{\ddot{u}(\xi)\} \right) \mathrm{d}\xi = 0 \tag{6.123}$$

Considering $\xi^2\{\sigma\}_{,\xi} +2\xi\{\sigma\} = (\xi^2\{\sigma\})_{,\xi}$, this equation is rewritten as

$$\int_{\xi_1}^{\xi_2} \{w(\xi)\}^T \left(\int_{S^\xi} [B_1]^T (\xi^2\{\sigma\})_{,\xi} |J_b|\mathrm{d}\eta\mathrm{d}\zeta - \xi \int_{S^\xi} [B_2]^T \{\sigma\}|J_b|\mathrm{d}\eta\mathrm{d}\zeta \right.$$
$$\left. + \{P_S(\xi)\} + \{P_B(\xi)\} - \xi^2[M_0]\{\ddot{u}(\xi)\} \right) \mathrm{d}\xi = 0 \tag{6.124}$$

Equation (6.124) is satisfied by setting the integrand of the integral over ξ equal to zero

$$\int_{S^\xi} [B_1]^T (\xi^2\{\sigma\})_{,\xi} |J_b|\mathrm{d}\eta\mathrm{d}\zeta - \xi \int_{S^\xi} [B_2]^T \{\sigma\}|J_b|\mathrm{d}\eta\mathrm{d}\zeta$$
$$+ \{P_S(\xi)\} + \{P_B(\xi)\} - \xi^2[M_0]\{\ddot{u}(\xi)\} = 0 \tag{6.125}$$

This corresponds to satisfying the equations of motion in the radial direction in strong form.

Using the internal nodal force functions $\{q(\xi)\}$ in Eq. (6.99), the surface integration in the first term of Eq. (6.125) is rewritten as

$$\int_{S^\xi} [B_1]^T (\xi^2\{\sigma\})_{,\xi} |J_b|\mathrm{d}\eta\mathrm{d}\zeta = \left(\xi^2 \int_{S^\xi} [B_1]^T \{\sigma\}|J_b|\mathrm{d}\eta\mathrm{d}\zeta \right)_{,\xi}$$
$$= \{q(\xi)\}_{,\xi} \tag{6.126}$$

Substituting Eq. (6.93) into the second term of Eq. (6.125) results in

$$\xi \int_{S^\xi} [B_2]\{\sigma\}|J_b|\mathrm{d}\eta\mathrm{d}\zeta$$
$$= \xi \int_{S^\xi} [B_2]^T \left([D][B^1]\{u(\xi)\}_{,\xi} +\frac{1}{\xi}[D][B^2]^T \{u(\xi)\} + \{\sigma_0\} \right) |J_b|\mathrm{d}\eta\mathrm{d}\zeta \tag{6.127}$$

Using the coefficient matrix $[E_1]$ defined in Eq. (6.102) and introducing the coefficient matrix

$$[E_2] = \int_{S^\xi} [B_2]^T [D][B_2]^T |J|\mathrm{d}\eta\mathrm{d}\zeta \tag{6.128}$$

Eq. (6.127) is expressed as

$$\xi \int_{S^\xi} [B_2]\{\sigma\}|J_b|\mathrm{d}\eta\mathrm{d}\zeta = [E_1]\xi\{u(\xi)\}_{,\xi} +[E_2]\{u(\xi)\} + \xi \int_{S^\xi} [B_2]^T \{\sigma_0\}|J_b|\mathrm{d}\eta\mathrm{d}\zeta \tag{6.129}$$

Using Eqs. (6.126) and (6.129), Eq. (6.125) is reformulated as

$$\{q(\xi)\}_{,\xi} -[E_1]\xi\{u(\xi)\}_{,\xi} -[E_2]\{u(\xi)\} - \xi \int_{S^\xi} [B_2]^T \{\sigma_0\}|J_b|\mathrm{d}\eta\mathrm{d}\zeta$$
$$+ \{P_S(\xi)\} + \{P_B(\xi)\} - \xi^2[M_0]\{\ddot{u}(\xi)\} = 0 \tag{6.130}$$

For convenience in the subsequent derivation, the contribution $\{P_0(\xi)\}$ from the initial stress $\{\sigma_0\}$ is introduced as

$$\{P_0(\xi)\} = \{q_0(\xi)\}_{,\xi} - \xi \int_{S^\xi} [B_2]^T \{\sigma_0\} |J_b| d\eta d\zeta \tag{6.131}$$

where $\{q_0(\xi)\}$ is given in Eq. (6.104). $\{P_0(\xi)\}$ is rewritten as

$$\{P_0(\xi)\} = \int_{S^\xi} \left([B^1]^T (\xi^2 \{\sigma_0\})_{,\xi} - [B_2]^T \xi \{\sigma_0\} \right) |J_b| d\eta d\zeta \tag{6.132}$$

It is evaluated from given initial stress field σ_0. Using Eq. (6.131) to eliminate the integral in Eq. (6.130) and simplifying yield

$$(\{q(\xi)\} - \{q_0(\xi)\})_{,\xi} - [E_1]\xi \{u(\xi)\}_{,\xi} - [E_2]\{u(\xi)\}$$
$$+ \{P_S(\xi)\} + \{P_B(\xi)\} + \{P_0(\xi)\} - \xi^2 [M_0]\{\ddot{u}(\xi)\} = 0 \tag{6.133}$$

Introducing the force functions

$$\{P(\xi)\} = \{P_S(\xi)\} + \{P_B(\xi)\} + \{P_0(\xi)\} \tag{6.134}$$

that consists of the contributions of the external loads inside the S-domain, Eq. (6.133) is rewritten as

$$(\{q(\xi)\} - \{q_0(\xi)\})_{,\xi} - [E_1]\xi \{u(\xi)\}_{,\xi} - [E_2]\{u(\xi)\} + \{P(\xi)\} - \xi^2 [M_0]\{\ddot{u}(\xi)\} = 0 \tag{6.135}$$

Equations (6.103) and (6.135) are a system of equations for the nodal displacement functions $\{u(\xi)\}$ and nodal force functions $\{q(\xi)\}$. To eliminate the nodal force functions $\{q(\xi)\}$, Eq. (6.103) is differentiated and simplified as

$$(\{q(\xi)\} - \{q_0(\xi)\})_{,\xi} = [E_0]\xi^2 \{u(\xi)\}_{,\xi\xi} + (2[E_0] + [E_1]^T)\xi \{u(\xi)\}_{,\xi} + [E_1]^T \{u(\xi)\} \tag{6.136}$$

Substituting Eq. (6.136) into Eq. (6.135) leads to the scaled boundary finite element equation in displacement

$$[E_0]\xi^2 \{u(\xi)\}_{,\xi\xi} + (2[E_0] + [E_1]^T - [E_1])\xi \{u(\xi)\}_{,\xi}$$
$$+ ([E_1]^T - [E_2])\{u(\xi)\} + \{P(\xi)\} - \xi^2 [M_0]\{\ddot{u}(\xi)\} = 0 \tag{6.137}$$

In the above derivation of the scaled boundary finite element equation, a pyramid sector is covered by scaling one surface element on the boundary. To model the total S-element, the equations of the individual sectors are assembled as in Section 2.5 for the two-dimensional case. To simplify the nomenclature, the same symbols for the element matrices and vectors are used in the chapter for the assembled matrices (such as the coefficient matrices $[E_0], [E_1], [E_2], [M_0]$ and the assembled nodal displacement functions $\{u(\xi)\}$ and force functions $\{q(\xi)\}$) of the S-element. In the assemblage process the contribution of the surface tractions on the side faces of the pyramid $\{P_S(\xi)\}$ in Eq. (6.134) will cancel at the common side of two adjacent pyramid sections. Only external loads applied to the S-element (for example, when the scaled boundary is not a closed surface) will remain.

For easy reference, the key equations of the scaled boundary finite element method derived in this section are summarized in the following:

1) The scaled boundary finite element equation
   a) In terms of nodal displacement functions $\{u(\xi)\}$ and nodal force functions $\{q(\xi)\}$ (Eqs. (6.103) and (6.135))

$$\{q(\xi)\} - \{q_0(\xi)\} - [E_0]\xi^2\{u(\xi)\}_{,\xi} - [E_1]^T\xi\{u(\xi)\} = 0 \tag{6.138a}$$

$$(\{q(\xi)\} - \{q_0(\xi)\})_{,\xi} - [E_1]\xi\{u(\xi)\}_{,\xi} - [E_2]\{u(\xi)\} + \{P(\xi)\} - \xi^2[M_0]\{\ddot{u}(\xi)\} = 0 \tag{6.138b}$$

   b) In terms of nodal displacement functions $\{u(\xi)\}$

$$[E_0]\xi^2\{u(\xi)\}_{,\xi\xi} + (2[E_0] + [E_1]^T - [E_1])\xi\{u(\xi)\}_{,\xi}$$
$$+ ([E_1]^T - [E_2])\{u(\xi)\} + \{P(\xi)\} - \xi^2[M_0]\{\ddot{u}(\xi)\} = 0 \tag{6.139}$$

2) The element coefficient matrices and force vectors
   a) The coefficient matrices (Eqs. (6.101), (6.102), (6.128) and (6.121))

$$[E_0] = \int_{S^\xi} [B_1]^T[D][B_1]|J_b|\mathrm{d}\eta\mathrm{d}\zeta \tag{6.140a}$$

$$[E_1] = \int_{S^\xi} [B_2]^T[D][B_1]|J_b|\mathrm{d}\eta\mathrm{d}\zeta \tag{6.140b}$$

$$[E_2] = \int_{S^\xi} [B_2]^T[D][B_2]|J_b|\mathrm{d}\eta\mathrm{d}\zeta \tag{6.140c}$$

$$[M_0] = \int_{S^\xi} [N_u]^T\rho[N_u]|J_b|\mathrm{d}\eta\mathrm{d}\zeta \tag{6.140d}$$

where $[B_1]$ and $[B_2]$ (Eq. (6.91)) are similar to the B–matrix of standard finite elements

$$[B_1] = [b_1][N_u] \tag{6.141a}$$
$$[B_2] = [b_2][N_u]_{,\eta} + [b_3][N_u]_{,\zeta} \tag{6.141b}$$

$[b_1]$, $[b_2]$ and $[b_3]$ (Eq. (6.81)) and $|J_b|$ (Eq. (6.53)) depend on the geometry of the surface element only.

Note that the integrations are limited to the surface elements. In other words, no volume integration is needed. Standard numerical integration techniques in the finite element method, such as Gauss quadrature, are directly applicable. The order of quadrature can also be chosen following the guidelines of standard isoparametric finite elements (for example, Cook et al., 2002).

   b) The load vector $\{P(\xi)\}$ consists of the contributions of the external loads inside the S-element

$$\{P(\xi)\} = \{P_S(\xi)\} + \{P_B(\xi)\} + \{P_0(\xi)\} \tag{6.142}$$

   • The surface tractions applied to the side faces (Eq. (6.112))

$$\{P_S(\xi)\} = \oint_{\Gamma^\xi} [N_u]^T \left(\{t^\eta\}|g^\eta|\mathrm{d}\zeta + \{t^\zeta\}|g^\zeta|\mathrm{d}\eta\right) \tag{6.143}$$

- The body force (Eq. (6.118))

$$\{P_B(\xi)\} = \int_{S^\xi} [N_u]^T \{p\} |J| \mathrm{d}\eta \mathrm{d}\zeta \tag{6.144}$$

- The initial stresses (Eq. (6.132))

$$\{P_0(\xi)\} = \int_{S^\xi} \left([B^1]^T (\xi^2 \{\sigma_0\})_{,\xi} - [B_2]^T \xi \{\sigma_0\} \right) |J_b| \mathrm{d}\eta \mathrm{d}\zeta \tag{6.145}$$

Note that the initial stresses also contribute to the internal nodal forces by $\{q_0(\xi)\}$ as appear in Eqs. (6.138a) and (6.138b). $\{q_0(\xi)\}$ is given in Eq. (6.104). It is not directly used in the solution for nodal displacement functions as can be observed from Eq. (6.139).

After computing and assembling the coefficient matrices $[E_0]$, $[E_1]$, $[E_2]$ and $[M_0]$ and, if applicable, the external load vector $\{P(\xi)\}$ (Eq. (6.142)) acting inside the S-element, the scaled boundary finite element equation (Eq. (6.215) or Eq. (6.216)) can be solved by the eigenvalue method in Chapter 3 or the block-diagonal Schur decomposition in Chapter 7.

6.8 Unified Formulations in Two and Three Dimensions

To unify the expressions of the scaled boundary finite element equations in two and three dimensions, a variable s is introduced to denote the spatial dimensions ($s = 2$ for two dimensions and $s = 3$ for three dimensions). In terms of the nodal displacement functions and $\{u(\xi)\}$ and force function $\{q(\xi)\}$, the scaled boundary finite element equations in two dimensions (Eq. (2.111) for statics without body forces) and in three dimensions (Eq. (6.138)) are written in two unified equations

$$\{q(\xi)\} - \{q_0(\xi)\} = \xi^{s-2} \left([E_0]\xi\{u(\xi)\}_{,\xi} + [E_1]^T \{u(\xi)\} \right) \tag{6.146a}$$

$$\xi^{3-s}(\{q(\xi)\} - \{q_0(\xi)\})_{,\xi} = [E_1]\xi\{u(\xi)\}_{,\xi} + [E_2]\{u(\xi)\} - \{P(\xi)\} + \xi^2 [M_0]\{\ddot{u}(\xi)\} \tag{6.146b}$$

In terms of the nodal displacement functions, the scaled boundary finite element equations in two dimensions (Eq. (2.112)) and in three dimensions (Eq. (6.139)) are expressed as

$$[E_0]\xi^2\{u(\xi)\}_{,\xi\xi} + ((s-1)[E_0] - [E_1] + [E_1]^T)\xi\{u(\xi)\}_{,\xi}$$
$$+ ((s-2)[E_1]^T - [E_2])\{u(\xi)\} + \{P(\xi)\} - \xi^2 [M_0]\{\ddot{u}(\xi)\} = 0 \tag{6.147}$$

Equation (6.147) can also be obtained by substituting Eq. (6.146a) into Eq. (6.146b) to eliminate $(\{q(\xi)\} - \{q_0(\xi)\})$.

The procedure presented in Chapter 3 can be directly applied to solve the above scaled boundary finite element equation. The static case is addressed first leading to the static stiffness matrix (Section 3.1) and the internal displacement field (Section 3.7). A numerically robust procedure based on the block-diagonal Schur decomposition is presented in Chapter 7. The body force (Section 3.9) and mass matrix (Section 3.10.1) are evaluated using the internal displacement field as shape functions of the S-domain.

Alternatively, body forces are considered by solving nonhomogeneous ordinary differential equations in Song and Wolf (1999); Song (2006). This procedure leads to accurate solutions without resorting to subdivision of an S-domain. It is especially useful when a stress singularity exists. The solution procedure for dynamics is presented in Song (2009); Birk et al. (2014). These advanced solution procedures will not be discussed in this book.

6.9 Formulation of the Scaled Boundary Finite Element Equation as a System of First-order Differential Equations

The scaled boundary finite element equation in displacement (Eq. 6.147) is a system of second-order ordinary differential equation in the radial coordinate ξ. In the solution procedure for displacement functions, it is convenient to reduce the second-order equations to first-order ones as in Section 3.1.

The scaled boundary finite element equation in statics is considered. Setting $\omega = 0$ (i.e. the inertial force vanishes), Eq. (6.215) in terms of the nodal displacement and force functions is rewritten as

$$\{q(\xi)\} - \{q_0(\xi)\} = \xi^{s-2} \left([E_0]\xi\{u(\xi)\}_{,\xi} +[E_1]^T\{u(\xi)\}\right) \tag{6.148a}$$

$$\xi^{3-s}(\{q(\xi)\} - \{q_0(\xi)\})_{,\xi} = [E_1]\xi\{u(\xi)\}_{,\xi} +[E_2]\{u(\xi)\} - \{P(\xi)\} \tag{6.148b}$$

Equation (6.147) in terms of the nodal displacement functions is expressed as

$$[E_0]\xi^2\{u(\xi)\}_{,\xi\xi} +((s-1)[E_0] - [E_1] + [E_1]^T)\xi\{u(\xi)\}_{,\xi}$$
$$+ ((s-2)[E_1]^T - [E_2])\{u(\xi)\} + \{P(\xi)\} = 0 \tag{6.149}$$

For an S-element with n degrees of freedom on the boundary, Eq. (6.148) consists of two sets of n ordinary differential equations with n nodal displacement functions $\{u(\xi)\}$ and n internal nodal forces $\{q(\xi)\}$ as the unknowns. Equation (6.148a) is rearranged to obtain

$$\xi\{u(\xi)\}_{,\xi} = \xi^{2-s}[E_0]^{-1}(\{q(\xi)\} - \{q_0(\xi)\}) - [E_0]^{-1}[E_1]^T\{u(\xi)\} \tag{6.150}$$

Substituting $\xi\{u(\xi)\}_{,\xi}$ in this equation into Eq. (6.148b) results in

$$\xi^{3-s}(\{q(\xi)\} - \{q_0(\xi)\})_{,\xi} -\xi^{2-s}[E_1][E_0]^{-1}(\{q(\xi)\} - \{q_0(\xi)\})$$
$$+ ([E_1][E_0]^{-1}[E_1]^T - [E_2])\{u(\xi)\} + \{P(\xi)\} = 0 \tag{6.151}$$

To simplify the dependence of Eqs. (6.150) and (6.151) on ξ, the change of variables

$$\{u(\xi)\} = \xi^{-0.5(s-2)}\{\bar{u}(\xi)\} \tag{6.152a}$$

$$\{q(\xi)\} - \{q_0(\xi)\} = \xi^{+0.5(s-2)}(\{\bar{q}(\xi)\} - \{\bar{q}_0(\xi)\}) \tag{6.152b}$$

is introduced. Differentiating Eq. (6.152) yields

$$\xi\{u(\xi)\}_{,\xi} = \xi^{-0.5(s-2)}(\xi\{\bar{u}(\xi)\}_{,\xi} -0.5(s-2)\{\bar{u}(\xi)\}) \tag{6.153a}$$

$$\xi(\{q(\xi)\} - \{q_0(\xi)\})_{,\xi} = \xi^{+0.5(s-2)}(\xi(\{\bar{q}(\xi)\} - \{\bar{q}_0(\xi)\})_{,\xi}$$
$$+ 0.5(s-2)(\{\bar{q}(\xi)\} - \{\bar{q}_0(\xi)\})) \tag{6.153b}$$

Substituting Eqs. (6.152b) and (6.153a) into Eqs. (6.150) results in

$$\xi\{\bar{u}(\xi)\}_{,\xi} = \xi^{2-s}[E_0]^{-1}(\{\bar{q}(\xi)\} - \{\bar{q}_0(\xi)\}) + (0.5(s-2)[I] - [E_0]^{-1}[E_1]^T)\{\bar{u}(\xi)\}$$

(6.154)

Substituting Eqs. (6.152) and (6.153b) into Eqs. (6.150), the following equation is obtained

$$\xi(\{\bar{q}(\xi)\} - \{\bar{q}_0(\xi)\})_{,\xi} + (-[E_1][E_0]^{-1} + 0.5(s-2)[I])(\{\bar{q}(\xi)\} - \{\bar{q}_0(\xi)\})$$
$$+ ([E_1][E_0]^{-1}[E_1]^T - [E_2])\{\bar{u}(\xi)\} = 0$$

(6.155)

Introducing the variable with $2n$ unknown functions

$$\{X(\xi)\} = \left\{ \begin{array}{c} \{\bar{u}(\xi)\} \\ \{\bar{q}(\xi)\} - \{\bar{q}_0(\xi)\} \end{array} \right\} = \left\{ \begin{array}{c} \xi^{+0.5(s-2)}\{u(\xi)\} \\ \xi^{-0.5(s-2)}(\{q(\xi)\} - \{q_0(\xi)\}) \end{array} \right\}$$

(6.156)

Eqs. (6.154) and (6.155) are reformulated as a system of nonhomogeneous first-order ordinary differential equations

$$\xi\{X(\xi)\}_{,\xi} = [Z_p]\{X(\xi)\} - \left\{ \begin{array}{c} 0 \\ \{P(\xi)\} \end{array} \right\}$$

(6.157)

with the coefficient matrix $[Z_p]$ specified in

$$[Z_p] = \left[\begin{array}{cc} -[E_0]^{-1}[E_1]^T + 0.5(s-2)[I] & [E_0]^{-1} \\ [E_2] - [E_1][E_0]^{-1}[E_1]^T & [E_1][E_0]^{-1} - 0.5(s-2)[I] \end{array} \right]$$

(6.158)

When the body forces including the external loads acting inside the S-domain and initial stresses vanish ($\{P(\xi)\} = 0$ and $q_0(\xi) = 0$), Eq. (6.148) is expressed as

$$\{q(\xi)\} = \xi^{s-2} \left([E_0]\xi\{u(\xi)\}_{,\xi} + [E_1]^T\{u(\xi)\} \right)$$ (6.159a)

$$\xi^{3-s}\{q(\xi)\}_{,\xi} = [E_1]\xi\{u(\xi)\}_{,\xi} + [E_2]\{u(\xi)\}$$ (6.159b)

and Eq. (6.149) as

$$[E_0]\xi^2\{u(\xi)\}_{,\xi\xi} + ((s-1)[E_0] - [E_1] + [E_1]^T)\xi\{u(\xi)\}_{,\xi}$$
$$+ ((s-2)[E_1]^T - [E_2])\{u(\xi)\} = 0$$

(6.160)

The system of nonhomogeneous equations in Eq. (6.157) becomes a system of homogeneous equations

$$\xi\{X(\xi)\}_{,\xi} = [Z_p]\{X(\xi)\}$$

(6.161)

which is a system of first-order Euler-Cauchy equations.

6.10 Properties of Coefficient Matrices

The coefficient matrices $[E_0]$, $[E_1]$, $[E_2]$ and $[M_0]$ are given in Eq. (6.140) for three-dimensional problems, and in Eqs. (2.110) and (3.102) for two-dimensional problems. It is easily observed that $[E_0]$, $[E_2]$ and $[M_0]$ are symmetric when the elasticity matrix $[D]$ is.

6.10.1 Coefficient Matrices $[E_0]$ and $[M_0]$

The coefficient matrix $[E_0]$ in three dimensions (Eq. (6.140a)) is rewritten, using Eq. (6.141a), as

$$[E_0] = \int_{S^\xi} [N_u]^T [b_1]^T [D][b_1][N_u]|J_b| d\eta d\zeta \qquad (6.162)$$

The matrix $[b_1]$ is given in Eq. (6.81a). For a properly constructed surface element, the determinant of the Jacobian matrix $|J_b|$ is positive. Since the three components of the unit outward normal will not vanish identical, the rank of $[b_1]$ is thus 3. The product $[b_1]^T [D][b_1]$ will be positive definite when the elasticity matrix $[D]$ is. Invoking the definition of positive definiteness of a matrix, the quadratic form of Eq. (6.162) for a nonzero displacement vector $\{u_b\}$ is expressed as

$$\{u_b\}^T [E_0]\{u_b\} = \int_{S^\xi} ([N_u]\{u_b\})^T [b_1]^T [D][b_1][N_u]\{u_b\}|J_b| d\eta d\zeta \qquad (6.163)$$

Since the displacement $[N_u]\{u_b\}$ does not vanish identically over the surface element,

$$\{u_b\}^T [E_0]\{u_b\} > 0 \qquad (6.164)$$

applies. Thus, $[E_0]$ is positive definite. The same proof applies to the coefficient matrix $[E_0]$ in two dimensions (2.110a).

Based on the same arguments, the coefficient matrix $[M_0]$ (Eq. (6.140d) in three dimensions and Eq. (3.102) in two dimensions) is also positive definite since the mass density ρ is positive.

6.10.2 Coefficient Matrix $[E_2]$

The coefficient matrix $[E_2]$ in three dimensions (Eq. (6.140a)) is addressed. A translational rigid-body motion of the S-element (i.e. the displacement field is uniform) is considered

$$u_x(x, y, z) = a_{x0}; \qquad u_y(x, y, z) = a_{y0}; \qquad u_z(x, y, z) = a_{z0} \qquad (6.165)$$

where the arbitrary constants a_{x0}, a_{y0} and a_{z0} are the amount of displacements in the $x-$, $y-$and $z-$direction, respectively. The corresponding nodal displacement vector on the boundary is denoted $\{u_{b0}\}$ as and equal to

$$\{u_{b0}\} = [a_{x0} \ \ a_{y0} \ \ a_{z0} \ \ a_{x0} \ \ a_{y0} \ \ a_{z0} \quad \cdots \quad a_{x0} \ \ a_{y0} \ \ a_{z0}]^T \qquad (6.166)$$

The product $[B_2]\{u_{b0}\}$, where $[B_2]$ is given in Eq. (6.91b), is examined first. Using Eqs. (6.88) and the property of the shape functions in Eq. (6.19), the partial derivatives of the displacements on the surface element (Eq. (6.89) at $\xi = 1$) vanish

$$[N_u]_{,\eta} \{u_{b0}\} = \begin{Bmatrix} a_{x0} \sum_i N_{i,\eta} \\ a_{y0} \sum_i N_{i,\eta} \\ a_{z0} \sum_i N_{i,\eta} \end{Bmatrix} = \begin{Bmatrix} 0 \\ 0 \\ 0 \end{Bmatrix} \qquad (6.167a)$$

$$[N_u]_{,\zeta} \{u_{b0}\} = \begin{Bmatrix} a_{x0} \sum_i N_{i,\zeta} \\ a_{y0} \sum_i N_{i,\zeta} \\ a_{z0} \sum_i N_{i,\zeta} \end{Bmatrix} = \begin{Bmatrix} 0 \\ 0 \\ 0 \end{Bmatrix} \qquad (6.167b)$$

Using Eqs. (6.91b) and (6.167) results in

$$[B_2]\{u_{b0}\} = ([b_2][N_u]_{,\eta} + [b_3][N_u]_{,\zeta})\{u_{b0}\} = 0 \tag{6.168}$$

Substituting Eq. (6.168) into the definition of $[E_2]$ in Eq. (6.140a) leads to

$$[E_2]\{u_{b0}\} = 0 \tag{6.169}$$

applies.

Following the steps in proving the positive definiteness of $[E_0]$ in Section 6.10.1 and considering Eq. (6.169), it is found that $[E_2]$ is semi-positive definite. It can be shown by considering Eq. (2.110b) with Eq. (2.82b) that Eqs. (6.168) and (6.169) also hold in two dimensions.

6.10.3 Matrix $[Z_p]$

As the matrices $[E_0]$ and $[E_2]$ are symmetric, the matrix $[Z_p]$ in Eq. (6.158) is a Hamiltonian matrix (Laub, 1979) satisfying

$$([J_{2n}][Z_p])^T = [J_{2n}][Z_p] \tag{6.170}$$

where the matrix $[J_{2n}]$ of order $2n$ is defined as ($[I]$ is a $n \times n$ identity matrix)

$$[J_{2n}] = \begin{bmatrix} 0 & [I] \\ -[I] & 0 \end{bmatrix} \tag{6.171}$$

The eigenvalues of the Hamiltonian matrix $[Z_p]$ in Eq. (3.5) occur in pairs of $(\lambda, -\lambda)$, i.e. if λ is an eigenvalue of a Hamiltonian matrix, $-\lambda$ is also an eigenvalue. This can be shown by examining the characteristic polynomial of $[Z_p]$ at an eigenvalue λ

$$\left| [Z_p] - \lambda[I]_{2n} \right| =$$
$$\begin{vmatrix} -[E_0]^{-1}[E_1]^T + 0.5(s-2)[I] - \lambda[I] & [E_0]^{-1} \\ [E_2] - [E_1][E_0]^{-1}[E_1]^T & [E_1][E_0]^{-1} - 0.5(s-2)[I] - \lambda[I] \end{vmatrix} = 0 \tag{6.172}$$

The transpose of the determinant leads to

$$\begin{vmatrix} -[E_1][E_0]^{-1} + 0.5(s-2)[I] - \lambda[I] & [E_2] - [E_1][E_0]^{-1}[E_1]^T \\ [E_0]^{-1} & [E_0]^{-1}[E_1]^T - 0.5(s-2)[I] - \lambda[I] \end{vmatrix} = 0 \tag{6.173}$$

Swapping the positions of the two row blocks

$$\begin{vmatrix} [E_0]^{-1} & [E_0]^{-1}[E_1]^T - 0.5(s-2)[I] - \lambda[I] \\ -[E_1][E_0]^{-1} + 0.5(s-2)[I] - \lambda[I] & [E_2] - [E_1][E_0]^{-1}[E_1]^T \end{vmatrix} = 0 \tag{6.174}$$

and followed by swapping the positions of the two columns results in

$$\begin{vmatrix} [E_0]^{-1}[E_1]^T - 0.5(s-2)[I] - \lambda[I] & [E_0]^{-1} \\ [E_2] - [E_1][E_0]^{-1}[E_1]^T & -[E_1][E_0]^{-1} + 0.5(s-2)[I] - \lambda[I] \end{vmatrix} = 0 \tag{6.175}$$

Multiplying the first row block and then the last column with $-[I]$ yields

$$\begin{vmatrix} -[E_0]^{-1}[E_1]^T + 0.5(s-2)[I] + \lambda[I] & -[E_0]^{-1} \\ [E_2] - [E_1][E_0]^{-1}[E_1]^T & [E_1][E_0]^{-1} - 0.5(s-2)[I] + \lambda[I] \end{vmatrix} = 0$$

(6.176)

Comparing Eq. (6.176) to Eq. (6.172) shows that $-\lambda$ is also an eigenvalue.

Although not used in this book, it is worthwhile mentioning that a large amount of literature studying the eigen-problem of a Hamiltonian matrix exists (Laub, 1979; Golub and van Loan, 1996).

6.11 Linear Completeness of the Scaled Boundary Finite Element Solution

In Example 3.8 on page 121, a patch test is performed. It is demonstrated that the scaled boundary finite element method can reproduce a linear displacement (i.e. constant stress) field up to machine accuracy. The linear completeness is shown analytically in the following by considering the scaled boundary finite element equation (Section 6.9). The constant terms and the linear terms of a displacement field are considered separately.

6.11.1 Constant Displacement Field

An arbitrary constant displacement field $\{u_{b0}\}$ (i.e. translational rigid-body motions) in three dimensions is given in Eq. (6.166). The two-dimensional counterpart is written by omitting the component a_{z0}. The nodal displacement functions along the radial lines are expressed as

$$\{u(\xi)\} = \{u_{b0}\}$$

(6.177)

Equation (6.159), the scaled boundary finite element equation in $\{u(\xi)\}$ and $\{q(\xi)\}$, is addressed. Substituting this equation into Eq. (6.159a), the nodal force functions are expressed as

$$\{q(\xi)\} = \xi^{s-2}[E_1]^T\{u_{b0}\}$$

(6.178)

since the nodal displacement functions are independent of the radial coordinate ξ. Using Eq. (6.140b) and considering Eq. (6.168),

$$[E_1]^T\{u_{b0}\} = \int_{S^\xi} [B_1]^T[D][B_2]\{u_{b0}\}|J_b|d\eta d\zeta = 0$$

(6.179)

applies. The same can be shown in two dimensions using Eq. (2.110b). Therefore, the nodal force functions vanishes for translational rigid-body motions (Eq. (6.178))

$$\{q(\xi)\} = 0$$

(6.180)

Using Eqs. (6.177), (6.180) and (6.169) it can be shown that Eq. (6.159b) holds.

Equation (6.160), the scaled boundary finite element equation in $\{u(\xi)\}$ is then considered. Using Eqs. (6.179) and (6.169), it is easy to show that the nodal displacement functions in Eq. (6.177) satisfies Eq. (6.160).

Lastly, the scaled boundary finite element equation formulated as first-order differential equation (6.161) is examined. This equation can be solved by the eigenvalue method in Section 3.1. Postulating the solution as

$$\{X(\xi)\} = \xi^\lambda \{\phi\} \tag{6.181}$$

Eq. (6.161) leads to an eigenvalue problem of the coefficient matrix $[Z_p]$ (Eq. (6.158))

$$[Z_p]\{\phi\} = \lambda\{\phi\} \tag{6.182}$$

The vector consisting of an arbitrary translational rigid-body motion $\{u_{b0}\}$ and the corresponding nodal forces $\{q(\xi)\} = 0$

$$\{\phi\} = \left\{ \begin{array}{c} \{u_{b0}\} \\ 0 \end{array} \right\} \tag{6.183}$$

is considered. Substituting $\{\phi\}$ into Eq. (6.182) with $[Z_p]$ given in Eq. (6.158) the eigenvalue

$$\lambda = 0.5(s - 2) \tag{6.184}$$

is obtained from the first row block, while the second row block is satisfied by Eqs. (6.179) and (6.169). In two and three dimensions, $\lambda = 0$ and $\lambda = 0.5$ applies, respectively. It can be seen by comparing Eq. (6.156) to the solution in Eq. (6.181) with Eqs. (6.183) and (6.184) that the translational rigid-body motion is a solution of Eq. (6.161).

Note that 0 is not an eigenvalue of $[Z_p]$ in three dimensions ($s = 3$).

6.11.2 Linear Displacement Field

The linear term of the displacement field is addressed. To avoid lengthy expressions, only the two-dimensional case is presented in details. Only key equations are given in three dimensions. An arbitrary linear displacement field is expressed as

$$\{u(x,y)\} = \left\{ \begin{array}{c} u_x(x,y) \\ u_y(x,y) \end{array} \right\} = \left\{ \begin{array}{c} a_{11}x + a_{12}y \\ a_{21}x + a_{22}y \end{array} \right\} \tag{6.185}$$

where a_{ij} $(i,j = 1,2)$ are arbitrary constants. For a p-th order element with $p + 1$ nodes, the nodal displacements on the boundary are expressed as

$$\{u_{b1}\} = \left\{ \begin{array}{c} u_x(x_1,y_1) \\ u_y(x_1,y_1) \\ u_x(x_2,y_2) \\ \vdots \\ u_y(x_p,y_p) \\ u_x(x_{p+1},y_{p+1}) \\ u_y(x_{p+1},y_{p+1}) \end{array} \right\} = \left\{ \begin{array}{c} a_{11}x_1 + a_{12}y_1 \\ a_{21}x_1 + a_{22}y_1 \\ a_{11}x_2 + a_{12}y_2 \\ \vdots \\ a_{21}x_p + a_{22}y_p \\ a_{11}x_{p+1} + a_{12}y_{p+1} \\ a_{21}x_{p+1} + a_{22}y_{p+1} \end{array} \right\} \tag{6.186}$$

Using the scaling of the boundary (Eq. (2.5)), the nodal displacement functions are expressed as

$$\{u(\xi)\} = \xi\{u_{b1}\} \tag{6.187}$$

It is first shown that a linear displacement field leads to constant strains in the scaled boundary finite element formulations. The displacement field $\{u\} = \{u(\xi, \eta)\}$ in the scaled boundary coordinates are obtained by interpolating the nodal displacement functions (Eq. (2.79)) as

$$\{u\} = \xi[N_u]\{u_{b1}\} \tag{6.188}$$

Using Eqs. (6.186) and (2.17), Eq. (6.188) is simplified as

$$\{u\} = \xi \left\{ \begin{array}{c} a_{11}x_b + a_{12}y_b \\ a_{21}x_b + a_{22}y_b \end{array} \right\} \tag{6.189}$$

To evaluate the strains using Eq. (2.81), the partial derivatives of the displacement field with respect to the scaled boundary coordinates are evaluated

$$\{u\}_{,\xi} = [N_u]\{u(\xi)\}_{,\xi} = \left\{ \begin{array}{c} a_{11}x_b + a_{12}y_b \\ a_{21}x_b + a_{22}y_b \end{array} \right\} \tag{6.190a}$$

$$\{u\}_{,\eta} = [N_u]_{,\eta} \{u(\xi)\} = \xi \left\{ \begin{array}{c} a_{11}x_{b,\eta} + a_{12}y_{b,\eta} \\ a_{21}x_{b,\eta} + a_{22}y_{b,\eta} \end{array} \right\} \tag{6.190b}$$

Substituting Eq. (6.190) and Eq. (2.67) for $[b_1]$ and $[b_2]$ into Eq. (2.81) leads to

$$\{\varepsilon\} = \frac{1}{|J_b|} \begin{bmatrix} y_{b,\eta} & 0 \\ 0 & -x_{b,\eta} \\ -x_{b,\eta} & y_{b,\eta} \end{bmatrix} \left\{ \begin{array}{c} a_{11}x_b + a_{12}y_b \\ a_{21}x_b + a_{22}y_b \end{array} \right\}$$

$$+ \frac{1}{|J_b|} \begin{bmatrix} -y_b & 0 \\ 0 & x_b \\ x_b & -y_b \end{bmatrix} \left\{ \begin{array}{c} a_{11}x_{b,\eta} + a_{12}y_{b,\eta} \\ a_{21}x_{b,\eta} + a_{22}y_{b,\eta} \end{array} \right\} \tag{6.191}$$

Equation (6.191) is simplified by grouping the terms according to the arbitrary constants a_{ij} and using the definition of $|J_b|$ in Eq. (2.31)

$$\{\varepsilon\} = \frac{1}{|J_b|} \left\{ \begin{array}{c} a_{11}|J_b| \\ a_{22}|J_b| \\ (a_{12} + a_{21})|J_b| \end{array} \right\} = \left\{ \begin{array}{c} a_{11} \\ a_{22} \\ a_{12} + a_{21} \end{array} \right\} \tag{6.192}$$

which are the constant strains corresponding to the linear term of the displacement field in Eq. (6.185). The constant stresses are equal to (Eq. (A.11) with vanishing initial strains)

$$\{\sigma\} = [D] \left\{ \begin{array}{c} a_{11} \\ a_{22} \\ a_{12} + a_{21} \end{array} \right\} \tag{6.193}$$

The nodal force functions (Eq. (6.159a)) along the radial lines are expressed for the linear term of displacement field (Eq. (6.187)) as

$$\{q(\xi)\} = \xi([E_0] + [E_1]^T)\{u_{b1}\} \tag{6.194}$$

Using Eqs. (6.187) and (6.194), Eq. (6.159b) will hold if the following relationship

$$([E_0] + [E_1]^T)\{u_{b1}\} - ([E_1] + [E_2])\{u_{b1}\} = 0 \tag{6.195}$$

is satisfied.

The term $([E_0] + [E_1]^T)\{u_{b1}\}$ is addressed. Using the definitions of $[E_0]$ (Eq. (2.110a)) and $[E_1]$ (Eq. (2.110b)) of one element and the stresses in Eq. (2.84) yields, for constant stresses $\{\sigma\}$,

$$([E_0] + [E_1]^T)\{u_{b1}\} = \int_{-1}^{+1} [B_1]^T |J_b| d\eta \{\sigma\} \qquad (6.196)$$

Similarly, using the definitions of $[E_1]$ (Eq. (2.110b)) and $[E_2]$ (Eq. (2.110c)) of one element and the stresses in Eq. (2.84), the second term $([E_1] + [E_2])\{u_{b1}\}$ of Eq. (6.195) is expressed as

$$([E_1] + [E_2])\{u_{b1}\} = \int_{-1}^{+1} [B_2]^T |J_b| d\eta \{\sigma\} \qquad (6.197)$$

To show that Eq. (6.195) is true, the right-hand side of Eq. (6.196) is expressed, using $[B_1]$ in Eq. (2.82a) and the identity in Eq. (2.68), as

$$\int_{-1}^{+1} [B_1]^T |J_b| d\eta = \int_{-1}^{+1} [N_u]^T |J_b| [b_1]^T d\eta = -\int_{-1}^{+1} [N_u]^T (|J_b| [b_2]^T)_{,\eta} \, d\eta \qquad (6.198)$$

Integrating by parts leads to

$$\int_{-1}^{+1} [B_1]^T |J_b| d\eta = -\left([N_u]^T |J_b| [b_2]^T\right)_{-1}^{+1} + \int_{-1}^{+1} [N_u]_{,\eta}^T [b_2]^T |J_b| d\eta$$

$$= -\left([N_u]^T |J_b| [b_2]^T\right)_{-1}^{+1} + \int_{-1}^{+1} [B_2]^T |J_b| d\eta \qquad (6.199)$$

The first term on the right-hand side is further expanded, using Eqs. (2.67b) and (2.80), as

$$\left([N_u]^T |J_b| [b_2]^T\right)_{-1}^{+1} = \begin{bmatrix} N_1 & 0 \\ 0 & N_1 \\ N_2 & 0 \\ \vdots & \vdots \\ 0 & N_p \\ N_{p+1} & 0 \\ 0 & N_{p+1} \end{bmatrix} \begin{bmatrix} -y_b & 0 & x_b \\ 0 & x_b & -y_b \end{bmatrix} \Bigg|_{-1}^{+1} \qquad (6.200)$$

Evaluating this expression at the limits and using the Kronecker delta property of the shape functions yields

$$\left([N_u]^T |J_b| [b_2]^T\right)_{-1}^{+1}$$

$$= \begin{bmatrix} 0 & 0 & 0 \\ 0 & 0 & 0 \\ 0 & 0 & 0 \\ \vdots & \vdots & \vdots \\ 0 & 0 & 0 \\ -y_{p+1} & 0 & x_{p+1} \\ 0 & x_{p+1} & -y_{p+1} \end{bmatrix} - \begin{bmatrix} y_1 & 0 & -x_1 \\ 0 & -x_1 & y_1 \\ 0 & 0 & 0 \\ \vdots & \vdots & \vdots \\ 0 & 0 & 0 \\ 0 & 0 & 0 \\ 0 & 0 & 0 \end{bmatrix} = \begin{bmatrix} y_1 & 0 & -x_1 \\ 0 & -x_1 & y_1 \\ 0 & 0 & 0 \\ \vdots & \vdots & \vdots \\ 0 & 0 & 0 \\ -y_{p+1} & 0 & x_{p+1} \\ 0 & x_{p+1} & -y_{p+1} \end{bmatrix} \qquad (6.201)$$

Note that the entries corresponding to all the intermediate nodes vanish. When the line elements on the boundary of the S-domain form a closed loop, one end node connects two line elements. It occurs twice in the assemblage, i.e. once as the first node where $\eta = -1$ and once as the last node where $\eta = +1$. The contributions from the two connecting elements cancel. Alternatively, the closed loop can be regarded as one element assemblage where every node is an intermediate node. Considering the assemblage,

$$\left([N_u]^T |J_b| [b_2]^T\right)_{-1}^{+1} = 0 \tag{6.202}$$

applies. Based on Eq. (6.196) with Eqs. (6.199) and (6.202), and Eq. (6.197), Eq. (6.195) holds. Therefore, the scaled boundary finite element equation is satisfied by an arbitrary linear displacement field when the line elements on the boundary of an S-element form a close loop.

The scaled boundary finite element equation in displacement (Eq. (6.160)) also admits any arbitrary linear displacement field. Substituting Eq. (6.187) into Eq. (6.160) with $s = 2$ results in

$$([E_0] - [E_1] + [E_1]^T - [E_2])\{u_{b1}\} = 0 \tag{6.203}$$

which is simply Eq. (6.195).

To examine the scaled boundary finite element equation expressed as first-order differential equations (Eq. (6.161) with Eqs. (6.156) and (6.158)), an eigenvector is constructed from the nodal displacements $\{u_{b1}\}$ (Eq. (6.186)) of the linear displacement field and the corresponding nodal forces (Eq. (6.186)) on the boundary ($\xi = 1$) as

$$\{\phi\} = \left\{ \begin{array}{c} \{u_{b1}\} \\ ([E_0] + [E_1]^T)\{u_{b1}\} \end{array} \right\} \tag{6.204}$$

Substituting Eq. (6.204) and the coefficient matrix $[Z_p]$ (Eq. (6.158) with $s = 2$) into the left-hand side of Eq. (6.182) and using Eq. (6.195) yields

$$[Z_p]\{\phi\} = \begin{bmatrix} -[E_0]^{-1}[E_1]^T & [E_0]^{-1} \\ [E_2] - [E_1][E_0]^{-1}[E_1]^T & [E_1][E_0]^{-1} \end{bmatrix} \left\{ \begin{array}{c} \{u_{b1}\} \\ ([E_0] + [E_1]^T)\{u_{b1}\} \end{array} \right\}$$

$$= \left\{ \begin{array}{c} \{u_{b1}\} \\ ([E_2] + [E_1])\{u_{b1}\} \end{array} \right\} = \left\{ \begin{array}{c} \{u_{b1}\} \\ ([E_0] + [E_1]^T)\{u_{b1}\} \end{array} \right\} = \{\phi\} \tag{6.205}$$

Comparing with the right-hand side of Eq. (6.182), $\lambda = 1$ is the eigenvalue for the linear displacement field.

In three dimensions, the linear displacement field is expressed as

$$\{u(x,y,z)\} = \left\{ \begin{array}{c} u_x(x,y,z) \\ u_y(x,y,z) \\ u_z(x,y,z) \end{array} \right\} = \left\{ \begin{array}{c} a_{11}x + a_{12}y + a_{13}z \\ a_{21}x + a_{22}y + a_{23}z \\ a_{31}x + a_{32}y + a_{33}z \end{array} \right\} \tag{6.206}$$

The nodal displacement functions are expressed in Eq. (6.187), where the vector $\{u_{b1}\}$ containing the nodal displacements on the boundary follows from Eq. (6.206) as

$$\{u_{b1}\} = \left\{ \begin{array}{c} u_x(x_1,y_1,z_1) \\ u_y(x_1,y_1,z_1) \\ u_z(x_1,y_1,z_1) \\ u_x(x_2,y_2,z_2) \\ \vdots \end{array} \right\} = \left\{ \begin{array}{c} a_{11}x_1 + a_{12}y_1 + a_{13}z_1 \\ a_{21}x_1 + a_{22}y_1 + a_{23}z_1 \\ a_{31}x_1 + a_{32}y_1 + a_{33}z_1 \\ a_{11}x_2 + a_{12}y_2 + a_{13}z_2 \\ \vdots \end{array} \right\} \tag{6.207}$$

The proof of linear completeness follows the same procedure as in the two-dimensional case. The scaled boundary finite element formulations lead to the correct constant strains and stresses. To prove that the linear displacement field satisfies the scaled boundary finite element equation (Eq. (160) with $s=3$), it is sufficient to show

$$2([E_0] + [E_1]^T)\{u_{b1}\} - ([E_1] + [E_2])\{u_{b1}\} = 0 \tag{6.208}$$

holds (which corresponds to Eq. (6.195) in the two-dimensional case).

Only the key equations are given below. The first term of Eq. (6.208) is expressed, using Eqs.(6.140a) and (6.140b), as

$$2([E_0] + [E_1]^T)\{u_{b1}\} = 2 \int_{S^\xi} [B_1]^T |J_b| \mathrm{d}\eta \mathrm{d}\zeta \{\sigma\} \tag{6.209}$$

$$([E_1] + [E_2])\{u_{b1}\} = \int_{S^\xi} [B_2]^T |J_b| \mathrm{d}\eta \mathrm{d}\zeta \{\sigma\} \tag{6.210}$$

where $\{\sigma\}$ represents the constant stress field. Using the identity in Eq. (6.80), the integration on the right-hand side is written as

$$2 \int_{S^\xi} [B_1]^T |J_b| \mathrm{d}\eta \mathrm{d}\zeta = 2 \int_{S^\xi} [N_u]^T |J_b| [b_1]^T \mathrm{d}\eta \mathrm{d}\zeta$$
$$= -\int_{S^\xi} [N_u]^T \left((|J_b| [b_2])_{,\eta} + (|J_b| [b_3])_{,\zeta} \right) \mathrm{d}\eta \mathrm{d}\zeta \tag{6.211}$$

Applying Gauss's theorem, it is reformulated as

$$2 \int_{S^\xi} [B_1]^T |J_b| \mathrm{d}\eta \mathrm{d}\zeta = -\int_{\Gamma^\xi} [N_u]^T \left(|J_b| [b_2] \mathrm{d}\zeta + |J_b| [b_3] \mathrm{d}\eta \right) + \int_{S^\xi} [B_2]^T |J_b| \mathrm{d}\eta \mathrm{d}\zeta \tag{6.212}$$

where $[B_2]$ in Eq. (6.141b) has been substituted. It is observed from Eqs. (6.81b) and (6.81c) that $|J_b| [b_2]$ and $|J_b| [b_3]$ on Γ^ξ are formed by the normal vectors on the side faces of the pyramid obtained by scaling the surface element. On the common faces of two adjacent pyramids, the normal vectors will have the same values but opposite directions. When the elements on the boundary of the S-domain form a closed surface,

$$\int_{\Gamma^\xi} [N_u]^T \left(|J_b| [b_2] \mathrm{d}\zeta + |J_b| [b_3] \mathrm{d}\eta \right) = 0 \tag{6.213}$$

applies after assemblage. Substituting Eq. (6.212) with Eq. (6.213) into Eq. (6.209) is expressed as

$$2([E_0] + [E_1]^T)\{u_{b1}\} = \int_{S^\xi} [B_2]^T |J_b| \mathrm{d}\eta \mathrm{d}\zeta \{\sigma\} \tag{6.214}$$

Subtracting Eq. (6.210) from Eq. (6.214), Eq. (6.208) is obtained. Therefore, an arbitrary linear displacement field is a solution of the scaled boundary finite element equation.

The eigenvalue (Eq. (6.182)) corresponding to the linear displacement field is $\lambda = 1.5$ in three dimensions.

6.12   Scaled Boundary Finite Element Equation in Stiffness

The physical property of an S-domain should be independent of the boundary conditions. In the conventional finite element method for dynamic analysis, the static stiffness and mass matrices are used to represent the physical property of an element. The finite element approach is based on the assumption that the effect on the inertial force on the internal displacements is negligible and the displacements can be interpolated with the shape functions from the nodal values. In a practical analysis, the finite element mesh should be divided in such a way that there are sufficient number of nodes in one wavelength at the maximum frequency of interest. Typically, around 12 nodes per wavelength are used for meshes of linear elements. The application of this approach of static and mass matrices to the scaled boundary finite element method is presented in Section 3.10.

In the scaled boundary finite element method, an equation for the dynamic stiffness matrix of an S-domain can be derived. This provides an alternative approach for dynamic analysis (Song, 2009).

The derivation of the scaled boundary finite element equation in dynamic stiffness is performed in the frequency domain. A harmonic motion $\{u(\xi)\} = \{U(\xi)\}e^{+i\omega t}$ is considered, where ω is the angular frequency of excitation, and $\{U(\omega)\}$ the amplitudes of nodal displacement functions. The acceleration follows as $\{\ddot{u}(\xi)\} = -\omega^2\{U(\xi)\}e^{+i\omega t}$. For simplicity in nomenclatures, $\{u(\xi)\}$ is also used in place of $\{U(\xi)\}$ for displacement amplitudes. Since the physical property of an S-element is independent of external loads, the contributions of initial stresses $\{q_0(\xi)\}$ and other external load $\{P(\xi)\}$ will not be considered in this section. Equation (6.147) is written in the frequency domain as

$$\{q(\xi)\} = \xi^{s-2}\left([E_0]\xi\{u(\xi)\}_{,\xi} + [E_0]^T\{u(\xi)\}\right) \tag{6.215a}$$

$$\xi^{3-s}\{q(\xi)\}_{,\xi} - [E_1]\xi\{u(\xi)\}_{,\xi} - [E_2]\{u(\xi)\} + \{P(\xi)\} + (\omega\xi)^2[M_0]\{u(\xi)\} = 0 \tag{6.215b}$$

Note that there is no change from Eq. (6.146) to Eq. (6.215), except that $\{u(\xi)\}$ and $\{q(\xi)\}$ stand for the amplitudes of nodal displacement and force functions in the latter. Similarly, Eq (6.147) is expressed in the frequency domain as

$$[E_0]\xi^2\{u(\xi)\}_{,\xi\xi} + ((s-1)[E_0] - [E_1] + [E_1]^T)\xi\{u(\xi)\}_{,\xi}$$
$$+ ((s-2)[E_1]^T - [E_2])\{u(\xi)\} + \{P(\xi)\} + (\omega\xi)^2[M_0]\{u(\xi)\} = 0 \tag{6.216}$$

A part of an S-domain bounded by a coordinate surface with a constant ξ is considered. The nodal displacements and forces are $\{u(\xi)\}$ and $\{q(\xi)\}$ (Eq. (6.146a)), respectively. The dynamic stiffness matrix is defined as

$$\{q(\xi)\} = [S(\omega, \xi)]\{u(\xi)\} \tag{6.217}$$

It is a function of the excitation frequency ω and the radial coordinate ξ. The dynamic stiffness matrix of the S-domain on the boundary ($\xi = 1$) is denoted as $[S(\omega)] = [S(\omega, \xi = 1)]$.

Equating the right-hand sides of Eq. (6.215a) and Eq. (6.217) yields

$$[S(\omega, \xi)]\{u(\xi)\} = [E_0]\xi^{s-1}\{u(\xi)\}_{,\xi} + [E_1]^T\xi^{s-2}\{u(\xi)\} \tag{6.218}$$

From Eqs. (6.216) and (6.218), an equation for $[S(\omega, \xi)]$ can be derived by eliminating the displacement amplitudes $\{u(\xi)\}$. Differentiating Eq. (6.218) with respect to ξ leads to

$$[S(\omega, \xi)]_{,\xi} \{u(\xi)\} + [S(\omega, \xi)]\{u(\xi)\}_{,\xi} - [E_0]\xi^{s-1}\{u(\xi)\}_{,\xi\xi}$$
$$- ((s-1)[E_0] + [E_1]^T)\xi^{s-2}\{u(\xi)\}_{,\xi} - (s-2)[E_1]^T\xi^{s-3}\{u(\xi)\} = 0 \qquad (6.219)$$

Summing Eq. (6.216) and Eq. (6.219) multiplied with ξ^{3-s} results in

$$\xi^{3-s}[S(\omega, \xi)]_{,\xi}\{u(\xi)\} + (\xi^{2-s}[S(\omega, \xi)] - [E_1])\xi\{u(\xi)\}_{,\xi}$$
$$- [E_2]\{u(\xi)\} + (\omega\xi)^2[M_0]\{u(\xi)\} = 0 \qquad (6.220)$$

Solving Eq. (6.218) for $\xi\{u(\xi)\}_{,\xi}$ leads to

$$\xi\{u(\xi)\}_{,\xi} = [E_0]^{-1}(\xi^{2-s}[S(\omega, \xi)] - [E_1]^T)\{u(\xi)\} \qquad (6.221)$$

Substituting Eq. (6.221) into Eq. (6.220) yields for arbitrary $\{u(\xi)\}$

$$(\xi^{2-s}[S(\omega, \xi)] - [E_1])[E_0]^{-1}(\xi^{2-s}[S(\omega, \xi)] - [E_1]^T) - [E_2]$$
$$+ \xi^{3-s}[S(\omega, \xi)]_{,\xi} + (\omega\xi)^2[M^0] = 0 \qquad (6.222)$$

Evaluating the derivative of $\xi^{2-s}[S(\omega, \xi)]$ with respect to ξ, i.e. $(\xi^{2-s}[S(\omega, \xi)])_{,\xi}$, by the product rule and rearranging the terms leads to

$$\xi^{2-s}[S(\omega, \xi)]_{,\xi} = (\xi^{2-s}[S(\omega, \xi)])_{,\xi} + (s-2)\xi^{1-s}[S(\omega, \xi)] \qquad (6.223)$$

Substituting Eq. (6.223) into Eq. (6.222), a first-order differential equation for $\xi^{2-s}[S(\omega, \xi)]$ is obtained

$$(\xi^{2-s}[S(\omega, \xi)] - [E_1])[E_0]^{-1}(\xi^{2-s}[S(\omega, \xi)] - [E_1]^T) - [E_2]$$
$$+ (s-2)\xi^{2-s}[S(\omega, \xi)] + \xi(\xi^{2-s}[S(\omega, \xi)])_{,\xi} + (\omega\xi)^2[M_0] = 0 \qquad (6.224)$$

It can be found by inspection that the independent variables ω and ξ always appear as a product. Introducing the variable

$$\bar{\omega} = \omega\xi \qquad (6.225)$$

and replacing the function with

$$[\bar{S}(\bar{\omega})] = \xi^{s-2}[S(\omega, \xi)] \qquad (6.226)$$

Eq. (6.224) is rewritten as

$$([\bar{S}(\bar{\omega})] - [E_1])[E_0]^{-1}([\bar{S}(\bar{\omega})] - [E_1]^T) - [E_2]$$
$$+ (s-2)[\bar{S}(\bar{\omega})] + \bar{\omega}([\bar{S}(\bar{\omega})])_{,\bar{\omega}} + \bar{\omega}^2[M_0] = 0 \qquad (6.227)$$

where only one independent variable $\bar{\omega}$ remains. The second-last term of Eq. (6.227), $\bar{\omega}([\bar{S}(\bar{\omega})])_{,\bar{\omega}}$, can be interpreted either for varying ξ with fixed ω or for varying ω with fixed ξ

$$\bar{\omega}[\bar{S}(\bar{\omega})]_{,\bar{\omega}} = \xi[\bar{S}(\bar{\omega})]_{,\xi} = \omega[\bar{S}(\bar{\omega})]_{,\omega} \qquad (6.228)$$

To evaluate the dynamic stiffness matrix of an S-element, a fixed value of the radial coordinate ξ is considered. Using Eq. (6.228), the term of derivation with respect to ξ in Eq. (6.224) is substituted by the derivative with respect to ω, leading to

$$(\xi^{2-s}[S(\omega, \xi)] - [E_1])[E_0]^{-1}(\xi^{2-s}[S(\omega, \xi)] - [E_1]^T) - [E_2]$$
$$+ (s-2)\xi^{2-s}[S(\omega, \xi)] + \omega(\xi^{2-s}[S(\omega, \xi)])_{,\omega} + (\omega\xi)^2[M_0] = 0 \qquad (6.229)$$

Note that, in this equation, ξ is a specified value and not a variable any more. On the boundary of the S-domain ($\xi = 1$), the equation of the dynamic stiffness matrix is written as

$$([S(\omega)] - [E_1])[E_0]^{-1}([S(\omega)] - [E_1]^T) - [E_2] + (s - 2)[S(\omega)]$$
$$+ \omega[S(\omega)]_{,\omega} + \omega^2[M_0] = 0 \qquad (6.230)$$

This represents the *scaled boundary finite-element equation in dynamic stiffness* formulated in the frequency domain. It is a system of nonlinear first-order ordinary differential equations with the excitation frequency ω as the independent variable.

Setting $\omega = 0$ in Eq. (6.230) yields the scaled boundary finite-element equation for static stiffness matrix $[K] = [S(\omega = 0)]$

$$([K] - [E_1])[E_0]^{-1}([K] - [E_1]^T) - [E_2] + (s - 2)[K] = 0 \qquad (6.231)$$

This can be reformulated as an algebraic Riccati equation (Golub and van Loan, 1996)

$$[K][E_0]^{-1}[K] - ([E_1][E^0]^{-1} - 0.5(s - 2)[I])[K]$$
$$- [K]([E_0]^{-1}[E_1]^T - 0.5(s - 2)[I]) - [E_2] + [E_1][E_0]^{-1}[E_1]^T = 0 \qquad (6.232)$$

Solution of this equation results in the static stiffness matrix of the S-element with degrees of freedom on the boundary only. It is worthwhile noting that the displacement solution is not directly required.

It can be proven that the stiffness matrix $[S(\omega)]$ or $[K]$ is symmetric. For example, the transpose of Eq. (6.230) for the dynamic stiffness matrix $[S(\omega)]$ is expressed as

$$([S(\omega)]^T - [E_1])[E_0]^{-1}([S(\omega)]^T - [E_1]^T) - [E_2]$$
$$+ [S(\omega)]^T + \omega[S(\omega)]_{,\omega}^T + \omega^2[M_0] = 0 \qquad (6.233)$$

since the coefficient matrices $[E_0]$, $[E_2]$ and $[M_0]$ are symmetric (Section 6.10). It is observed from Eqs. (6.230) and (6.233), $[S(\omega)]$ and $[S(\omega)]^T$ can be interchanged. Therefore, the dynamic stiffness matrix is symmetric

$$[S(\omega)] = [S(\omega)]^T \qquad (6.234)$$

Similarly, using Eq. (6.231), it can be shown that the static stiffness matrix $[K]$ is also symmetric

$$[K] = [K]^T \qquad (6.235)$$

7

Solution of the Scaled Boundary Finite Element Equation in Statics by Schur Decomposition

7.1   Introduction

This chapter presents a robust solution procedure for the scaled boundary finite element equation in statics based on the real Schur decomposition. The solution of the displacement functions are expressed as matrix power functions.

A solution procedure for the scaled boundary finite element equation is presented in Chapter 3. The eigenvalue decomposition is employed to decouple the system of ordinary equations. The solution of displacement field inside an S-element is obtained as a power series of the dimensionless radial coordinate. The eigenvalues determine the exponents of the power functions in the series. The corresponding eigenvectors describe the angular variations of displacements and forces, and can be regarded as modes of displacement and force. The static stiffness matrix of the S-domain is obtained directly from the eigenvectors.

Mathematically, the eigenvalue decomposition of a matrix is defective when repeated eigenvalues with parallel eigenvectors exist and the matrix is not diagonalizable. In such a case, the solution procedure based on eigenvalue decomposition will lead to the loss of solutions and the numerical algorithm will break down. This is demonstrated in Section 7.2 below using a simple example.

In the theory of elasticity, the solutions of some problems are known to contain not only power functions but also logarithmic functions, for which the solution procedure based on eigenvalue decomposition fails to reproduce. One example that admits logarithmic functions in its solution is a two-dimensional unbounded domain subject to loads that are not self-equilibrating in the translational directions. Such a case can easily be identified *a priori* and special treatments can be adopted to recover the logarithmic functions in the solution (Deeks and Wolf, 2002a). Another example in elasticity is the solution of the stress field around the vertex of a wedge (Sternberg and Koiter, 1958; Dempsey and Sinclair, 1979). A logarithmic function may occur under certain combinations of boundary condition, material composition and opening angle of the wedge. The dominant stress singularity can be power-logarithmic. In this case, the existence of logarithmic functions in the solutions is not easily known *a priori* and special treatments are more complex to design. Simply neglecting the logarithmic function leads to the loss of solution.

From the point of view of numerical analysis, difficulties also exist when the eigenvectors are nearly parallel. The eigenvector matrix will become ill-conditioned and the

The Scaled Boundary Finite Element Method: Introduction to Theory and Implementation,
First Edition. Chongmin Song.

accuracy of the eigenvalue decomposition deteriorates. This situation happens in the adjacent region of existence of logarithmic functions in the solution. It also tends to occur when the number of nodes of an S-element is large.

A robust solution procedure for the scaled boundary finite element equation in statics, proposed in Song (2004a), is presented in this chapter. In this procedure, the block-diagonal real Schur decomposition replaces the eigenvalue decomposition employed in Chapter 3, leading to a set of well-conditioned basis vectors. The stiffness matrix of an S-element is computed directly from the basis vectors. The system of ordinary differential equations for displacement are decoupled into a series of smaller systems corresponding to the diagonal blocks. The displacement solution of each of the smaller systems is expressed as a matrix exponential function. This solution procedure is robust. It is capable of representing not only the logarithmic functions but also the transition between power functions and logarithmic functions. The existence of logarithmic functions in the solution is not required to be known *a priori*. In addition, the real Schur decomposition has analytical and numerical advantages in comparison with the eigenvector decomposition (Golub and van Loan, 1996).

In this chapter, the solution procedure based on Schur decomposition is applied to the the system of homogeneous first-order ordinary differential equations in Section 6.9. This system of equations results from a static problem with vanishing body loads. For easy reference, the relevant equations are reproduced here. The system of first-order ordinary differential equations is written as

$$\xi\{X(\xi)\}_{,\xi} = [Z_p]\{X(\xi)\} \tag{7.1}$$

where the unknown functions $\{X(\xi)\}$ (Eq. (6.156) with vanishing initial stresses, i.e. $\{q_0\} = 0$) consist of the nodal displacement functions $\{u(\xi)\}$ and force functions $\{q(\xi)\}$

$$\{X(\xi)\} = \left\{ \begin{array}{c} \xi^{+0.5(s-2)}\{u(\xi)\} \\ \xi^{-0.5(s-2)}\{q(\xi)\} \end{array} \right\} \tag{7.2}$$

The number of degrees of freedom of the nodes on the boundary is denoted by n. The lengths of the vectors $\{u(\xi)\}$ and $\{q(\xi)\}$ are equal to n. The length of the vector of the unknown function $\{X(\xi)\}$ is thus $2n$. The constant matrix $[Z_p]$ (Eq. (6.158)) is equal to

$$[Z_p] = \begin{bmatrix} -[E_0]^{-1}[E_1]^T + 0.5(s-2)[I] & [E_0]^{-1} \\ [E_2] - [E_1][E_0]^{-1}[E_1]^T & [E_1][E_0]^{-1} - 0.5(s-2)[I] \end{bmatrix} \tag{7.3}$$

which are obtained from the coefficient matrices $[E_0]$, $[E_1]$ and $[E_2]$ of the S-domain given in Eq. (2.110) in two dimensions and Eq. (6.140) in three dimensions. The size of the constant matrix $[Z_p]$ is $2n \times 2n$.

The body loads and inertial forces are treated in the same way as in the solution procedure based on eigenvalue decomposition in Chapter 3. The shape functions of the S-element are constructed from the solution of the nodal displacement functions (Section 7.8.3). Using the shape functions, standard finite element techniques are applied to handle body forces (Section 7.9) and inertial forces (represented by the mass matrix in Section 7.10).

7.2 Basics of Matrix Exponential Function

The method of eigenvalue decomposition is a standard technique for the solution of systems of ordinary differential equations such as Eq. (7.1). As presented in Section 3.1, a similarity transformation using the eigenvectors fully decouples the system of equations into independent ones. The general solution of the system of equations is obtained as a sum of the solutions of the decoupled equations with the eigenvectors forming the basis vectors.

The eigenvalue decomposition does not provide the complete general solution when the coefficient matrix of the equations has repeated eigenvalues with parallel eigenvectors. When the eigenvectors are nearly parallel, this procedure leads to results of poor accuracy. The deficiency of the eigenvalue method is demonstrated using the following system of two equations

$$\xi u_{,\xi} = \lambda_1 u + av \tag{7.4a}$$

$$\xi v_{,\xi} = \lambda_2 v \tag{7.4b}$$

where $u = u(\xi)$ and $v = v(\xi)$ are the two functions, λ_1, λ_2 and $a \neq 0$ are real numbers.

Equation (7.4) can be solved analytically. The solution of the second equation (Eq. (7.4b)) is equal to

$$v = c_2 \xi^{\lambda_2} \tag{7.5}$$

where c_2 is an integration constant. Substituting it into Eq. (7.4a) yields a nonhomogeneous equation for u

$$\xi u_{,\xi} = \lambda_1 u + c_2 a \xi^{\lambda_2} \tag{7.6}$$

Applying the technique of variation of parameters, the solution of Eq. (7.6) is postulated as

$$u = c \xi^{\lambda_1} \tag{7.7}$$

where $c = c(\xi)$ is a function of ξ and to be determined. Differentiating Eq. (7.7) with respect to ξ results in

$$u_{,\xi} = c_{,\xi} \xi^{\lambda_1} + \lambda_1 c \xi^{\lambda_1 - 1} \tag{7.8}$$

Substituting Eqs. (7.7) and (7.8) into Eq. (7.6) yields, after simplification, an equation for the function c

$$c_{,\xi} = c_2 a \xi^{\lambda_2 - \lambda_1 - 1} \tag{7.9}$$

Its solution is expressed as

$$c = \begin{cases} c_1 + c_2 \dfrac{a}{\lambda_2 - \lambda_1} \xi^{\lambda_2 - \lambda_1} & \text{when} \quad \lambda_1 \neq \lambda_2 \\[2mm] c_1 + c_2 a \ln \xi & \text{when} \quad \lambda_1 = \lambda_2 \end{cases} \tag{7.10}$$

Substituting Eq. (7.10) into Eq. (7.7) leads to the solution for the function u

$$u = \begin{cases} c_1 \xi^{\lambda_1} + c_2 \dfrac{a}{\lambda_2 - \lambda_1} \xi^{\lambda_2} & \text{when} \quad \lambda_1 \neq \lambda_2 \\[2mm] c_1 \xi^{\lambda} + c_2 a \xi^{\lambda} \ln \xi & \text{when} \quad \lambda_1 = \lambda_2 = \lambda \end{cases} \tag{7.11}$$

Note that there is a logarithmic function in the solution when $\lambda_1 = \lambda_2$.

To solve Eq. (7.4) by the eigenvalue decomposition, Eq. (7.4) is rewritten in the matrix form of Eq. (7.1) with $\{X\} = [u \quad v]^T$ and the 2×2 coefficient matrix

$$\xi \left\{ \begin{matrix} u \\ v \end{matrix} \right\}_{,\xi} = \underbrace{\begin{bmatrix} \lambda_1 & a \\ 0 & \lambda_2 \end{bmatrix}}_{[Z_p]} \left\{ \begin{matrix} u \\ v \end{matrix} \right\} \tag{7.12}$$

The eigenvalues of $[Z_p]$ are equal to λ_1 and λ_2. The eigenvector matrix can be written as

$$[\Phi] = \begin{bmatrix} 1 & \dfrac{a}{\lambda_1 - \lambda_2} \\ 0 & 1 \end{bmatrix} \tag{7.13}$$

In the eigenvalue decomposition method, the general solution of Eq. (7.12) is expressed as (Eq. (3.15))

$$\{X(\xi)\} = c_1 \xi^{\lambda_1} \left\{ \begin{matrix} 1 \\ 0 \end{matrix} \right\} + c_2 \xi^{\lambda_2} \left\{ \begin{matrix} \dfrac{a}{\lambda_2 - \lambda_1} \\ 1 \end{matrix} \right\} \tag{7.14}$$

This solution is identical to the one in Eqs. (7.11) and (7.5) when $\lambda_1 \neq \lambda_2$. As λ_1 and λ_2 approach each other, the two eigenvectors in Eq. (7.13) tend to be parallel, and the eigenvector matrix $[\Phi]$ becomes ill-conditioned. When $\lambda_1 = \lambda_2 = \lambda$ occurs, the two solutions in Eq. (7.14) collapse to one, leading to the loss of the solution with a logarithmic function in Eq. (7.11). The eigenvector matrix $[\Phi]$ becomes singular and the numerical inversion of $[\Phi]$ will break down.

Analytically, the lost solution can be recovered when this situation is known *a priori*. Expressing λ_1 and λ_2 as λ and $\lambda + \Delta_\lambda$, respectively, the solution in Eq. (7.14) is written as

$$\{X(\xi)\} = c_1 \xi^{-\lambda} \left\{ \begin{matrix} 1 \\ 0 \end{matrix} \right\} + c_2 \xi^{\lambda + \Delta_\lambda} \left\{ \begin{matrix} \dfrac{a}{\Delta_\lambda} \\ 1 \end{matrix} \right\} \tag{7.15}$$

At the limit of $\Delta_\lambda \to 0$

$$\lim_{\Delta_\lambda \to 0} \left(\frac{\xi^{\lambda + \Delta_\lambda}}{\Delta_\lambda} \right) = \xi^\lambda \ln \xi \tag{7.16}$$

applies, and the solution in Eq. (7.15) is expressed as

$$\{X(\xi)\} = c_1 \xi^{\lambda_1} \left\{ \begin{matrix} 1 \\ 0 \end{matrix} \right\} + c_2 \xi^\lambda \left\{ \begin{matrix} a \ln \xi \\ 1 \end{matrix} \right\} \tag{7.17}$$

which is simply the solution in Eqs. (7.5) and (7.11) for the case of $\lambda_1 = \lambda_2 = \lambda$. The analytical procedure is cumbersome to be implemented numerically for a large system of equations. More importantly, the poor accuracy caused by nearly-parallel eigenvectors is not improved, and this procedure is not useful if the cases leading to the loss of solution cannot be identified *a priori*.

The theory of matrix functions provides a robust procedure to avoid the above-illustrated deficiency of the eigenvalue decomposition method. Although matrix functions have not been commonly applied in the field of computational mechanics, the theory and numerical algorithm have been well established in mathematics. Only the contents that are essential to the topics in this book are covered here. For details on matrix functions, the reader is referred to Gantmacher (1977) for the theoretical background, and to Golub and van Loan (Section 11.3, 1996) and Moler and van Loan (2003) for the numerical algorithms. Some useful properties of matrix power function are given in Appendix B.

According to the theory of matrices, the solution of the homogeneous equation (7.1) is expressed as

$$\{X(\xi)\} = \xi^{[Z_p]}\{c\} \tag{7.18}$$

where $\{c\}$ denotes the integration constants. The fundamental matrix $\xi^{[Z_p]}$ is a matrix power function and can be expressed as a matrix exponential function

$$\xi^{[Z_p]} = e^{[Z_p]\ln\xi} \tag{7.19}$$

The solution of Eq. (7.1) is equivalent to the evaluation of the matrix exponential. The simplest technique (although not commonly used in actual numerical algorithms) is to use the Taylor series

$$\xi^{[Z_p]} = e^{[Z_p]\ln\xi} = [I] + [Z_p]\ln\xi + \frac{([Z_p]\ln\xi)^2}{2!} + \cdots \tag{7.20}$$

The application of the matrix power function to the solution of the system of ordinary differential equations in Eq. (7.4) with $\lambda_1 = \lambda_2 = \lambda$ is demonstrated. As previously shown with Eq. (7.12), the system of equations is expressed as

$$\xi\left\{\begin{matrix} u \\ v \end{matrix}\right\}_{,\xi} = \underbrace{\begin{bmatrix} \lambda & a \\ 0 & \lambda \end{bmatrix}}_{[Z_p]}\left\{\begin{matrix} u \\ v \end{matrix}\right\} \tag{7.21}$$

which is in the same form as the scaled boundary finite element equation (7.1). The fundamental matrix of Eq. (7.21) is equal to

$$\xi^{[Z_p]} = \xi^{\begin{bmatrix} \lambda & a \\ 0 & \lambda \end{bmatrix}} = \xi^{\left(\lambda\begin{bmatrix} 1 & 0 \\ 0 & 1 \end{bmatrix}+\begin{bmatrix} 0 & a \\ 0 & 0 \end{bmatrix}\right)} = \xi^\lambda \xi^{\begin{bmatrix} 0 & a \\ 0 & 0 \end{bmatrix}} \tag{7.22}$$

Applying the Taylor series in Eq. (7.20) and considering

$$\begin{bmatrix} 0 & a \\ 0 & 0 \end{bmatrix}^2 = \begin{bmatrix} 0 & a \\ 0 & 0 \end{bmatrix}\begin{bmatrix} 0 & a \\ 0 & 0 \end{bmatrix} = 0 \tag{7.23}$$

leads to

$$\xi^{[Z_p]} = \xi^\lambda \left(\begin{bmatrix} 1 & 0 \\ 0 & 1 \end{bmatrix}+\begin{bmatrix} 0 & a \\ 0 & 0 \end{bmatrix}\ln\xi\right) = \xi^\lambda\begin{bmatrix} 1 & a\ln\xi \\ 0 & 1 \end{bmatrix} \tag{7.24}$$

Using Eq. (7.18), the solution is expressed as

$$\left\{\begin{matrix} u \\ v \end{matrix}\right\} = \xi^\lambda\begin{bmatrix} 1 & a\ln\xi \\ 0 & 1 \end{bmatrix}\left\{\begin{matrix} c_1 \\ c_2 \end{matrix}\right\} \tag{7.25}$$

The first row gives the solution of u in Eq. (7.11) for the case of $\lambda_1 = \lambda_2 = \lambda$. The logarithmic function occurs naturally in the solution.

Generally, logarithmic functions occur in the solution of a system of ordinary differential equations when two or more of the real eigenvalues of a diagonal block $[Z_p]$ are identical but the adjacent off-diagonal terms are nonzero. As another example of the application of the matrix exponential, a $m \times m$ Jordan matrix with multiple repeated eigenvalues λ is considered

$$[Z_p] = \begin{bmatrix} \lambda & 1 & 0 & \cdots & 0 \\ 0 & \lambda & 1 & \cdots & 0 \\ \vdots & \vdots & \ddots & \ddots & \vdots \\ 0 & 0 & 0 & \ddots & 1 \\ 0 & 0 & 0 & \cdots & \lambda \end{bmatrix} = \lambda[I] + \begin{bmatrix} 0 & 1 & 0 & \cdots & 0 \\ 0 & 0 & 1 & \cdots & 0 \\ \vdots & \vdots & \ddots & \ddots & \vdots \\ 0 & 0 & 0 & \ddots & 1 \\ 0 & 0 & 0 & \cdots & 0 \end{bmatrix} \tag{7.26}$$

where $[I]$ is a $m \times m$ identity matrix. The matrix cannot be further reduced by eigenvalue decomposition as all eigenvectors are parallel. Applying Eq. (7.20), the matrix power function $\xi^{[Z_p]}$ is obtained straightforwardly as

$$\xi^{[Z_p]} = \xi^\lambda \begin{bmatrix} 1 & \ln\xi & (\ln\xi)^2 & \cdots & \dfrac{(\ln\xi)^{m-1}}{(m-1)!} \\ 0 & 1 & \ln\xi & \cdots & \dfrac{(\ln\xi)^{m-2}}{(m-2)!} \\ \vdots & \vdots & \ddots & \ddots & \vdots \\ 0 & 0 & 0 & \ddots & \ln\xi \\ 0 & 0 & 0 & \cdots & 1 \end{bmatrix} \tag{7.27}$$

Logarithmic functions, which would be lost when using the eigenvalue decomposition, occur in the matrix power function.

As illustrated by the above two examples, the matrix power function, or equivalently the matrix exponential function, is a robust technique for the solution of the scaled boundary finite element equation in Eq. (7.1). Theoretically, the matrix exponential can be computed using the Taylor series in Eq. (7.20). Only basic arithmetic operations of matrices are involved. The algorithm is applicable to any matrix. No explicit detection of the occurrence of the logarithmic functions, which is susceptible to numerical errors, is required.

In numerical analysis, the convergence of the Taylor series is slow especially at a large value of the argument. To improve the computational efficiency, the scaling and squaring method based on Padé approximation of the exponential function has been developed (Moler and van Loan, 2003). Detailed algorithms are available in Golub and van Loan (Section 11.3, 1996), and Al-Mohy and Higham (2009). In MATLAB, the function expm(x) computes the matrix exponential exp([x]). The function expm1(x) computes exp([x]) − 1 to avoid the round-off error in directly computing exp([x]) at small argument [x].

The matrix power function $\xi^{[Z_p]}$ (Eq. (7.20)) provides a general solution (Eq. (7.18)) of the system of ordinary differential equations in Eq. (7.1). However, $\xi^{[Z_p]}$ tends to infinity at the scaling centre ($\xi = 0$) as the real parts of some of its eigenvalues are negative (see Section 3.1). To enforce the condition of the finiteness of displacements at the scaling centre, Eq. (7.1) is decoupled by a Schur decomposition of the constant coefficient matrix $[Z_p]$.

7.3 Schur Decomposition

In the eigenvalue method (Section 3.1), the system of $2n$ ordinary differential equations is fully decoupled. The general solution is expressed as a series of $2n$ terms (Eq.(3.15)). The solution of a bounded S-element is determined by selecting the half of the terms that satisfy the boundary condition at the scaling centre. This simple procedure may lead to the loss of numerical accuracy or even solutions, as illustrated in Section 7.2. In the present method based on Schur decomposition, the decoupling of the ordinary differential equations is limited to the extent that the boundary condition at the scaling centre can be enforced so that the loss of solutions can be avoided.

7.3.1 Introduction

The Schur decomposition (Golub and van Loan, 1996) is an intermediate step in computing eigenvalues and eigenvectors using the QR algorithm, although the latter is much better known and more widely applied in computational mechanics. The Schur decomposition does not suffer from the numerical hazards inherent in computing eigenvectors associated with repeated or nearly-repeated eigenvalues. The characteristics of Schur decomposition represent a significant advantage in numerical analysis. Functions necessary for performing Schur decompositions can be found in LAPACK (Anderson et al., 1999) and MATLAB (schur). Only a brief introduction is provided here. For a comprehensive treatment, the reader is referred to the book by Golub and van Loan (1996).

To decouple the system of $2n$ homogeneous ordinary differential equations in Eq. (7.1), the Schur decomposition of the $2n \times 2n$ coefficient matrix $[Z_p]$ in Eq. (7.3) is performed

$$[Z_p] = [V][S][V]^T \tag{7.28}$$

where $[S]$ denotes a real Schur form matrix. The real transformation matrix $[V]$ is a orthogonal matrix satisfying

$$[V]^T[V] = [I] \tag{7.29}$$

where $[I]$ is an identity matrix.

The real Schur form matrix $[S]$ is a quasi-upper triangular matrix. Its diagonal consists of 1×1 blocks and 2×2 blocks. A real matrix made of two 1×1 diagonal blocks is shown as an example

$$\begin{bmatrix} \lambda_1 & b \\ 0 & \lambda_2 \end{bmatrix}$$

Obviously, the diagonals are the eigenvalues. A real 2×2 diagonal block

$$\begin{bmatrix} a & b \\ c & a \end{bmatrix}$$

satisfying the product $bc < 0$ is considered. The two eigenvalues are a pair of complex conjugates and equal to $a + i\sqrt{bc}$ and $a - i\sqrt{bc}$. Note that the diagonal entries are the real parts of the eigenvalues of both the 1×1 blocks and 2×2 blocks. Therefore, the diagonal entries of a real Schur form matrix $[S]$ are the real parts of the eigenvalues of the matrix.

Inversely, Eq. (7.28) can be expressed as

$$[V]^T[Z_p][V] = [S] \tag{7.30}$$

or equivalently

$$[Z_p][V] = [V][S] \tag{7.31}$$

To enforce boundary conditions analytically in the solution of the scaled boundary finite element equation, the behaviour of the solution is evaluated based on the sign of the eigenvalues. As demonstrated in Section 6.10, the eigenvalues of the Hamiltonian matrix $[Z_p]$ occur in pairs of $(\lambda, -\lambda)$. In three dimensions, all eigenvalues are different from 0. They can be divided into two groups of the equal size. One group consists of the eigenvalues that have positive real parts, and the other group consists of the eigenvalues that have negative real parts. In two dimensions, two pairs of eigenvalues, corresponding to the two translational rigid body motions, are equal to 0 (Section 3.1). The rest of eigenvalues can be divided into two groups of equal size according to their signs as in the three-dimensional case.

To facilitate the numerical implementation, the Schur matrix $[S]$ is sorted in such a way that the real parts of the eigenvalues (i.e. the diagonal entries) of $[S]$ are in descending order. An algorithm for ordering the transformation matrix $[V]$ according to the sorted eigenvalues can be found in Golub and van Loan (1996, Section 7.6.2). MATLAB has a built-in function `ordschur` to reorder the eigenvalues of a Schur decomposition. The ordering does not affect the properties of the Schur decomposition given above.

7.3.2 Treatment of the Diagonal Block of Eigenvalues of 0

When the coefficient matrix $[Z_p]$, and thus the Schur matrix $[S]$ (Eq. (7.30)), has eigenvalues of 0, an additional treatment of the diagonal blocks of eigenvalues of 0 is necessary in order to select the solutions satisfying boundary conditions.

It is shown in Example 3.1 on page 80 and Section 6.11 that the eigenvalue of $[Z_p]$ in Eq. (7.3) is equal to 0 when the corresponding eigenvector describes a translational rigid body motion in two dimensions. As the eigenvalues of a Hamiltonian matrix occur in pairs of $(\lambda, -\lambda)$, $[Z_p]$ of a two-dimensional S-element has two pairs of eigenvalues of 0 corresponding to the two independent translational rigid-body motions. Therefore, a 4×4 diagonal block with diagonal entries of 0 exists in the sorted real Schur form matrix $[S]$ in Eq. (7.28). The two translational motions cannot be identified by the eigenvalues alone. The Schur decomposition of such a matrix needs an additional treatment before the condition of finiteness of displacement at the scaling centre can be enforced.

The Schur decomposition that is directly obtained from a standard computer code such as the function `schur` in MATLAB is denoted by

$$[V^{(0)}]^T[Z_p][V^{(0)}] = [S^{(0)}] \tag{7.32}$$

The superscript '(0)' is add to the notations of the matrices so that the notations in the final decomposition derived later will be consistent with these in Eq. (7.28). The real Schur form matrix $[S^{(0)}]$ is partitioned as

$$[S^{(0)}]_{2n \times 2n} = \begin{bmatrix} [S_p]_{(n-2) \times (n-2)} & * & * \\ 0 & [S_z^{(0)}]_{4 \times 4} & * \\ 0 & 0 & [S_n]_{(n-2) \times (n-2)} \end{bmatrix} \tag{7.33}$$

where the asterisk symbol "$*$" stands for a non-zero real matrix. The transformation matrix $[V^{(0)}]$ is orthogonal

$$[V^{(0)}][V^{(0)}]^T = [I] \tag{7.34}$$

It is partitioned conformably into three column blocks

$$[V^{(0)}]_{2n \times 2n} = [[V_p]_{2n \times (n-2)} \ [V_z^{(0)}]_{2n \times 4} \ [V_n]_{2n \times (n-2)}] \tag{7.35}$$

The diagonal entries of the real Schur form matrix $[S^{(0)}]$ of the size $2n \times 2n$ are sorted in descending order. The first $(n-2)$ diagonal entries are positive. The diagonal block containing them is denoted as $[S_p]$ (subscript 'p' for positive). The corresponding column block of the transformation matrix $[V^{(0)}]$ is denoted as $[V_p]$. The last diagonal block with negative diagonal entries and the corresponding column block of $[V^{(0)}]$ are denoted as $[S_n]$ and $[V_n]$, respectively, (subscript 'n' for negative). The middle block $[S_z^{(0)}]$ is a 4×4 matrix in real Schur form with all the four eigenvalues equal to 0. The column block $[V_z^{(0)}]$ contains four column vectors including the two translational modes that have to be included in the solution.

An orthogonal transformation is devised to reduce the 4×4 diagonal block $[S_z^{(0)}]$ in Eq. (7.33) into a simple form so that the two translational rigid-body motions in the transformation matrix $[V]$ can be identified. The algorithm consists of the two steps.

In the first step, the diagonal block $[S_z^{(0)}]$ of the four eigenvalues of 0 is addressed. The objective is to devise an orthogonal matrix $[V_z^{(1)}]$ so that $[S_z^{(0)}]$ is transformed into

$$[S_z] = [V_z^{(1)}]^T [S_z^{(0)}][V_z^{(1)}] = \begin{bmatrix} [0]_{2 \times 2} & [T]_{2 \times 2} \\ [0]_{2 \times 2} & [0]_{2 \times 2} \end{bmatrix} = \begin{bmatrix} 0 & 0 & T_{11} & T_{12} \\ 0 & 0 & T_{21} & T_{22} \\ 0 & 0 & 0 & 0 \\ 0 & 0 & 0 & 0 \end{bmatrix} \tag{7.36}$$

where the only nonzero entries appear in the upper-right block and denoted as $[T]$.

The transformation matrix $[V_z^{(1)}]$ is determined from the eigenvalue decomposition of the matrix $[S_z^{(0)}]^T [S_z^{(0)}]$. Each translational rigid body motion reduces the rank of the coefficient matrix $[Z_p]$. The eigenvalue problem of $[S_z^{(0)}]^T [S_z^{(0)}]$, whose rank is 2, is expressed as

$$[V_z^{(1)}]^T [S_z^{(0)}]^T [S_z^{(0)}][V_z^{(1)}] = \text{diag}(0, 0, *, *) \tag{7.37}$$

It can also be written as

$$([S_z^{(0)}][V_z^{(1)}])^T [S_z^{(0)}][V_z^{(1)}] = \text{diag}(0, 0, *, *) \tag{7.38}$$

The matrix $[S_z^{(0)}]^T [S_z^{(0)}]$ is symmetric and semi-positive definite. All eigenvalues are real and sorted in ascending order in Eq. (7.37). The first two eigenvalues are equal to zero. The symbol '$*$' stands for a non-zero real number.

The eigenvector matrix $[V_z^{(1)}]$ of the symmetric matrix is orthogonal

$$[V_z^{(1)}]^T [V_z^{(1)}] = [I] \tag{7.39}$$

and selected as the transformation matrix to be used in Eq. (7.36).

It is illustrated in the following that the matrix product $[V_z^{(1)}]^T [S_z^{(0)}][V_z^{(1)}]$ indeed leads to the matrix $[S_z]$ in Eq. (7.36). To satisfy Eq. (7.38), the first two columns of $[S_z^{(0)}][V_z^{(1)}]$ must be equal to zero. In addition, all the four eigenvalues of $[S_z^{(0)}]$, and thus $[V_z^{(1)}]^T [S_z^{(0)}][V_z^{(1)}]$, are equal to zero. The nonzero entries of $[V_z^{(1)}]^T [S_z^{(0)}][V_z^{(1)}]$ can only appear in the following positions

$$[V_z^{(1)}]^T [S_z^{(0)}][V_z^{(1)}] = \begin{bmatrix} 0 & 0 & T_{11} & T_{12} \\ 0 & 0 & T_{21} & T_{22} \\ 0 & 0 & 0 & a_1 \\ 0 & 0 & a_2 & 0 \end{bmatrix} \tag{7.40}$$

with the product

$$a_1 a_2 = 0 \tag{7.41}$$

It is postulated that no square terms of a logarithmic function ($\ln^2 \xi$) will arise, the Taylor series in Eq. (7.20) results in

$$[S_z^{(0)}]^2 = 0 \tag{7.42}$$

Considering Eq. (7.42), the square of the left-hand side of Eq. (7.40) yields

$$[V_z^{(1)}]^T [S_z^{(0)}]^2 [V_z^{(1)}] = 0 \tag{7.43}$$

The square of the right-hand side of Eq. (7.40) should also vanish, which holds when

$$\begin{bmatrix} T_{11} & T_{12} \\ T_{21} & T_{22} \end{bmatrix} \begin{bmatrix} 0 & a_1 \\ a_2 & 0 \end{bmatrix} = \begin{bmatrix} 0 & 0 \\ 0 & 0 \end{bmatrix} \tag{7.44}$$

If either of the two coefficients a_1 and a_2 is nonzero, the rank of $[S_z^{(0)}]$ is one. Since the rank of $[S_z^{(0)}]$ is two, both a_1 and a_2 should be equal to zero. Therefore, Eq. (7.36) holds.

In the second step, an orthogonal transformation matrix is constructed as

$$[V^{(1)}] = \text{diag}([I]_{(n-2)\times(n-2)}, [V_z^{(1)}]_{4\times4}, [I]_{(n-2)\times(n-2)}) \tag{7.45}$$

and applied to the Schur decomposition in Eq. (7.33). Pre- and post-multiplying Eq. (7.32) with $[V^{(1)}]^T$ and $[V^{(1)}]$, respectively, the Schur decomposition in the same form as Eq. (7.30) is obtained

$$[V]^T [Z][V] = [S]$$

$$= \begin{bmatrix} [S_n] & * & * \\ 0 & [V_z^{(1)}]^T [S_z^{(0)}][V_z^{(1)}] & * \\ 0 & 0 & [S_p] \end{bmatrix} = \begin{bmatrix} [S_n] & * & * \\ 0 & [S_z] & * \\ 0 & 0 & [S_p] \end{bmatrix} \tag{7.46}$$

where the transformation matrix accumulated from Eqs. (7.35) and (7.45) is expressed as

$$[V] = [V^{(0)}][V^{(1)}] = [[V_p] \ [V_z^{(0)}][V_z^{(1)}] \ [V_n]] \tag{7.47}$$

Only $[V_z^{(0)}]$, the four columns of $[V^{(0)}]$ corresponding to eigenvalues of 0, are affected.

Since both $[V^{(0)}]$ and $[V_z^{(1)}]$ are orthogonal (see Eqs. (7.34) and (7.39)), $[V]$ is also an orthogonal matrix and Eq. (7.29) holds. In other words, the additional transformation to treat the block of eigenvalues of 0 does not affect the orthogonality of the Schur decomposition.

Substituting Eq. (7.36) into Eq. (7.46), the Schur form matrix is partitioned as

$$[V]^T[Z][V] = [S] = \begin{bmatrix} [S_p] & * & * & * \\ 0 & [0]_{2\times2} & [T]_{2\times2} & * \\ 0 & 0 & [0]_{2\times2} & * \\ 0 & 0 & 0 & [S_n] \end{bmatrix} \tag{7.48}$$

The two translational rigid-body motions will be identified from the transformation matrix $[V]$ (Eq. (7.47)) based on this structure of the Schur matrix $[S]$ in Section 7.4.

7.4 Solution Procedure for a Bounded S-element by Schur Decomposition

The solution for bounded S-elements, defined by $0 \leq \xi \leq 1$, is derived first. It is the most common case in engineering applications and can be solved without further block diagonalization of the Schur form matrix $[S]$.

The solution procedure consists of the following steps:

1) Transforming the system of first-order ordinary differential equations (Eq. 7.1) of the scaled boundary finite element method.
2) Enforcing the boundary condition at the scaling centre.
3) Determining the analytical solution for displacement and nodal force functions.
4) Determining the static stiffness matrix.

7.4.1 Transformation of the Scaled Boundary Finite Element Equation

Equation (7.1) is transformed into a form that allows the condition of finiteness of displacement at the scaling centre $\xi = 0$ to be enforced. The unknown functions $\{X(\xi)\}$ are expressed as

$$\{X(\xi)\} = [V]\{W(\xi)\} \tag{7.49}$$

where the columns of the orthogonal transformation matrix $[V]$ in Eq. (7.28) are the basis and $\{W(\xi)\}$ are the generalized-coordinate functions. Substituting Eq. (7.49) into Eq. (7.1) yields

$$[V]\xi\{W(\xi)\}_{,\xi} = [Z_p][V]\{W(\xi)\} \tag{7.50}$$

Pre-multiplying Eq. (7.49) with $[V]^T$ and using Eqs. (7.29) and (7.28) results in a system of differential equations of the generalized coordinates $\{W(\xi)\}$

$$\xi\{W(\xi)\}_{,\xi} = [S]\{W(\xi)\} \tag{7.51}$$

To decouple the system of equations for the purpose of satisfying the boundary condition at the scaling centre, the real Schur form matrix $[S]$ of $2n \times 2n$ is partitioned into four submatrices of equal size $(n \times n)$

$$[S] = \begin{bmatrix} [S_b] & * \\ 0 & [S_u] \end{bmatrix} \tag{7.52}$$

where the asterisk symbol '*' stands for a non-zero real matrix. As will become evident later, $[S_b]$ and $[S_u]$ in Eq. (7.52) are related to the solutions for the bounded domain (subscript b for bounded) defined by $0 \le \xi \le 1$ and for the unbounded domain (subscript u for unbounded) defined by $\xi \ge 1$, respectively. Conformable to the partition of $[S]$ in Eq. (7.52), the orthogonal transformation matrix $[V]$ in Eq. (7.29) is partitioned into four submatrices of size $n \times n$

$$[V] = \begin{bmatrix} [V_b^{(u)}] & [V_u^{(u)}] \\ [V_b^{(q)}] & [V_u^{(q)}] \end{bmatrix} \tag{7.53}$$

The first row block of $[V]$ in Eq. (7.53) is related to the solution of displacements (superscript (u) denoting the set of basis vectors of displacement) and the second row block to the nodal force functions (superscript (q) denoting the set of basis vectors of forces).

The system of $2n$ equations in Eq. (7.51) is partitioned into two subsystems of the same size n. The vector of $2n$ generalized-coordinate functions is partitioned as

$$\{W(\xi)\} = \left\{ \begin{array}{c} \{W_b(\xi)\} \\ \{W_u(\xi)\} \end{array} \right\} \tag{7.54}$$

Using Eqs. (7.53) and (7.2), Eq. (7.49) is rewritten as two sets of equations

$$\{u(\xi)\} = \xi^{-0.5(s-2)}([V_b^{(u)}]\{W_b(\xi)\} + [V_u^{(u)}]\{W_u(\xi)\}) \tag{7.55a}$$

$$\{q(\xi)\} = \xi^{+0.5(s-2)}([V_b^{(q)}]\{W_b(\xi)\} + [V_u^{(q)}]\{W_u(\xi)\}) \tag{7.55b}$$

providing the solutions for the nodal displacement and force functions. The solution of Eq. (7.1) with a general coefficient matrix $[Z_p]$ becomes the solution of Eq. (7.51) with a real Schur matrix $[S]$.

Using Eqs. (7.54) and (7.52), Eq. (7.51) is expressed as

$$\xi \frac{d}{d\xi} \left\{ \begin{array}{c} \{W_b(\xi)\} \\ \{W_u(\xi)\} \end{array} \right\} = \begin{bmatrix} [S_b] & * \\ 0 & [S_u] \end{bmatrix} \left\{ \begin{array}{c} \{W_b(\xi)\} \\ \{W_u(\xi)\} \end{array} \right\} \tag{7.56}$$

7.4.2 Enforcing the Boundary Condition at the Scaling Centre

Enforcing the condition of finiteness of solution at the scaling centre ($\xi = 0$) will lead to

$$\{W_u(\xi)\} = 0 \tag{7.57}$$

The proof of Eq. (7.57) differs depending on whether the Schur matrix $[S]$ has eigenvalues of 0 or not. The two cases are considered separately in the following:

1) $[Z_p]$ having no eigenvalues of 0

   The case where $[Z_p]$, and thus $[S]$, have no eigenvalue of 0 occurs in three-dimensional problems. It is shown in Section 6.10 that the eigenvalues of $[Z_p]$ occur in pairs of $(\lambda, -\lambda)$. As the real parts of the eigenvalues (i.e. the diagonal entries) of $[S]$ have been sorted in descending order and no eigenvalues of 0 exists, the real parts of all the eigenvalues of $[S_b]$ are positive and those of $[S_u]$ negative.

It is noted from Eq. (7.56) that the second set of this equation is independent of the first set

$$\xi\{W_u(\xi)\}_{,\xi} = [S_u]\{W_u(\xi)\} \tag{7.58}$$

Its general solution is expressed as

$$\{W_u(\xi)\} = \xi^{[S_u]}\{c_u\} \tag{7.59}$$

with the integration constants $\{c_u\}$. Since the real parts of all the eigenvalues of $[S_u]$ are negative, the matrix power function $\xi^{[S_u]}$ tends to infinity at the scaling centre $\xi = 0$ and is inadmissible to the solution of displacements in Eq. (7.55a). Therefore,

$$\{c_u\} = 0 \tag{7.60}$$

applies and

$$\{W_u(\xi)\} = 0 \tag{7.61}$$

is obtained.

2) $[Z_p]$ having eigenvalues of 0

In this case, the partitioning of the Schur matrix in Eq. (7.52) is expressed as

$$[S] = \begin{bmatrix} [S_b] & * \\ 0 & [S_u] \end{bmatrix} = \begin{bmatrix} \begin{bmatrix} [S_p] & * \\ 0 & [0]_{2\times2} \end{bmatrix} & \begin{bmatrix} * & * \\ [T]_{2\times2} & * \end{bmatrix} \\ 0 & \begin{bmatrix} [0]_{2\times2} & * \\ 0 & [S_n] \end{bmatrix} \end{bmatrix} \tag{7.62}$$

Both of the submatrices $[S_b]$ and $[S_u]$ include two eigenvalues of 0.

The system of equations in Eq. (7.56) is further partitioned conformably to the Schur matrix in Eq. (7.62). The generalized-coordinate functions (Eq. (7.54)) are partitioned as

$$\{W(\xi)\} = \left\{ \begin{array}{c} \{W_b(\xi)\} \\ \{W_u(\xi)\} \end{array} \right\} = \left\{ \begin{array}{c} \left\{ \begin{array}{c} \{W_p(\xi)\} \\ \{W_{z1}(\xi)\} \end{array} \right\} \\ \left\{ \begin{array}{c} \{W_{z2}(\xi)\} \\ \{W_n(\xi)\} \end{array} \right\} \end{array} \right\} \tag{7.63}$$

which defines the functions $\{W_b(\xi)\}$ and $\{W_u(\xi)\}$.

Using Eqs. (7.63) and (7.62), Eq. (7.56) is rewritten as

$$\xi\frac{d}{d\xi}\left\{ \begin{array}{c} \{W_p(\xi)\} \\ \{W_{z1}(\xi)\} \\ \{W_{z2}(\xi)\} \\ \{W_n(\xi)\} \end{array} \right\} = \begin{bmatrix} [S_p] & * & * & * \\ 0 & [0]_{2\times2} & [T]_{2\times2} & * \\ 0 & 0 & [0]_{2\times2} & * \\ 0 & 0 & 0 & [S_n] \end{bmatrix} \left\{ \begin{array}{c} \{W_p(\xi)\} \\ \{W_{z1}(\xi)\} \\ \{W_{z2}(\xi)\} \\ \{W_n(\xi)\} \end{array} \right\} \tag{7.64}$$

This system of equations is divided into four sets and solved by back-substitution. The last set of equations in Eq. (7.56) is independent of the others and expressed as

$$\xi\{W_n(\xi)\}_{,\xi} = [S_n]\{W_n(\xi)\} \tag{7.65}$$

Its general solution is expressed as

$$\{W_n(\xi)\} = \xi^{[S_n]}\{c_n\} \tag{7.66}$$

with the integration constants $\{c_n\}$. Since the real parts of all the eigenvalues of $[S_n]$ are negative, the matrix power function $\xi^{[S_n]}$ tends to infinity at the scaling centre $\xi = 0$ and should vanish from the solution, i.e.

$$\{c_n\} = 0; \qquad \{W_n(\xi)\} = 0 \tag{7.67}$$

Considering Eq. (7.67), the third set of equations is written as

$$\xi\{W_{z2}(\xi)\}_{,\xi} = 0 \tag{7.68}$$

Its generation solution is expressed as

$$\{W_{z2}(\xi)\} = \{c_{z2}\} \tag{7.69}$$

with the integration constants $\{c_{z2}\}$. Using Eqs. (7.67) and (7.69), the second set of equations is formulated as

$$\xi\{W_{z1}(\xi)\}_{,\xi} = [T]\{c_{z2}\} \tag{7.70}$$

The general solution is expressed as

$$\{W_{z1}(\xi)\} = [T]\ln\xi\{c_{z2}\} + \{c_{z1}\} \tag{7.71}$$

with the integration constants $\{c_{z1}\}$. As the function $\ln\xi$ in the first term of the solution tends to negative infinity at $\xi = 0$, the integration constants $\{c_{z2}\}$ must vanish for the solution to remain finite. Thus, the solutions in Eqs. (7.69) and (7.71) are obtained as

$$\{W_{z2}(\xi)\} = \{0\} \tag{7.72a}$$
$$\{W_{z1}(\xi)\} = \{c_{z1}\} \tag{7.72b}$$

Combining Eqs. (7.67) and (7.72), $\{W_u(\xi)\}$ as partitioned in Eq. (7.63) vanishes

$$\{W_u(\xi)\} = \left\{ \begin{array}{c} \{W_{z2}(\xi)\} \\ \{W_n(\xi)\} \end{array} \right\} = 0 \tag{7.73}$$

It can be seen that Eq. (7.57) holds whether the Schur matrix $[S]$ has eigenvalues of 0 or not.

7.4.3   Determining the Solution for Displacement and Nodal Force Functions

Substituting Eq. (7.57) into Eq. (7.56), the first set of equations of Eq. (7.56) is written as

$$\xi\{W_b(\xi)\}_{,\xi} = [S_b]\{W_b(\xi)\} \tag{7.74}$$

The general solution of Eq. (7.74) is expressed as

$$\{W_b(\xi)\} = \xi^{[S_b]}\{c_b\} \tag{7.75}$$

with the integration constants $\{c_b\}$.

Substituting Eqs. (7.57) and (7.75) into Eq. (7.55) yields the solutions for the nodal displacement and force functions

$$\{u(\xi)\} = \xi^{-0.5(s-2)}[V_b^{(u)}]\xi^{[S_b]}\{c_b\} \tag{7.76a}$$
$$\{q(\xi)\} = \xi^{+0.5(s-2)}[V_b^{(q)}]\xi^{[S_b]}\{c_b\} \tag{7.76b}$$

7.4.4 Determining the Static Stiffness Matrix

Eliminating the common term $\xi^{[S_p]}\{c_b\}$ from Eqs. (7.76a) and (7.76b) results in

$$\{q(\xi)\} = \xi^{s-2}[V_b^{(q)}][V_b^{(u)}]^{-1}\{u(\xi)\} \tag{7.77}$$

It relates the nodal forces with the nodal displacements at a constant radial coordinate ξ. Therefore, the static stiffness matrix of a bounded domain enclosed by the scaled boundary at ξ is equal to

$$[K(\xi)] = \xi^{s-2}[V_b^{(q)}][V_b^{(u)}]^{-1} \tag{7.78}$$

The static stiffness matrix is independent of the size of the S-element in two dimensions ($s = 2$) and proportional to the size of the domain in three dimensions ($s = 3$). At the boundary ($\xi = 1$), the static stiffness matrix of the S-element is equal to

$$[K] = [V_b^{(q)}][V_b^{(u)}]^{-1} \tag{7.79}$$

which is in the same form as Eq. (3.25) obtained in the eigenvalue decomposition method. Only the first half of the columns, denoted with the subscript b (for bounded domain), of the transformation matrix $[V]$ in Eq. (7.53) is needed. As the transformation matrix $[V]$ is orthogonal, this solution procedure based on Schur decomposition is numerically more robust.

It is worthwhile noting from Eq. (7.79) that the matrix power function is used in the derivation to select the basis vectors that satisfy the boundary condition at the scaled centre analytically. However, it is not directly evaluated in determining the stiffness matrix.

The assembly of the global system of equations from the equations of the S-elements follows the standard finite element procedures. It is described in Section 3.5 for the two-dimensional case. The nodal displacements are obtained after solving the global system of equations.

7.5 Solution of Displacement and Stress Fields of an S-element

After the nodal displacements of the global system is obtained, the post-processing of displacement and stress fields can be performed for an individual S-element independent of the others.

7.5.1 Integration Constants

The nodal displacements $\{u_b\} = \{u(\xi = 1)\}$ of an S-element are extracted from the global solution based on the element connectivity (Section 3.7). The nodal displacements allow the integration constants in the solution of the S-element (Eq. (7.76a)) to be determined.

The solution of the displacement functions Eq. (7.76a) is formulated at $\xi = 1$, leading to

$$\{u_b\} = \{u(\xi = 1)\} = [V_b^{(u)}]\{c\} \tag{7.80}$$

where $1^{[S]} = [I]$ has been applied. The integration constants are obtained as

$$\{c\} = [V_b^{(u)}]^{-1}\{u_b\} \tag{7.81}$$

The solutions of the nodal displacement and force functions are determined by Eq. (7.76a) with Eq. (7.81).

7.5.2 Stress Modes and Stresses on the Boundary

The strains of an S-element are determined from the displacement field using Eq. (6.92). Along the radial coordinate ξ emanating from the scaling centre, the displacement field is described analytically by the solution of the displacement functions. The solution of a bounded S-element in Eq. (7.76a) is addressed. Following Eq. (B.17) in Appendix B.2, the derivative of the matrix power function in Eq. (7.76a) is expressed as

$$(\xi^{-0.5(s-2)}\xi^{[S_b]})_{,\xi} = (\xi^{[S_b]-0.5(s-2)[I]})_{,\xi} = ([S_b] - 0.5(s-2)[I])\xi^{[S_b]-0.5s[I]} \tag{7.82}$$

The derivative of the displacement functions with respect to ξ is expressed as

$$\{u(\xi)\}_{,\xi} = [V_b^{(u)}]([S_b] - 0.5(s-2)[I])\xi^{[S_b]-0.5s[I]}\{c\} \tag{7.83}$$

Along the circumferential directions η and ζ, the displacement field is represented numerically by interpolating the displacement functions (Eq. (6.86)). The strain field in three dimensions is given in Eq. (6.92), which is the same expression as Eq. (2.83) for the two-dimensional case. As discussed in Section 3.7 for the two-dimensional case, the strains and stresses are evaluated element-by-element typically at the Gauss points of a lower-order rule than that for full integration. For a specified point on the element, the strains along a radial line passing through the specified point are obtained by substituting Eqs. (7.76a) and (7.83) into Eq. (6.92) as

$$\{\varepsilon\} = \left([B_1][V_b^{(u)}]([S_b] - 0.5(s-2)[I]) + [B_2][V_b^{(u)}]\right)\xi^{[S_b]-0.5s[I]}\{c\} \tag{7.84}$$

This equation is simplified as

$$\{\varepsilon\} = [\Phi_b^{(\varepsilon)}]\xi^{[S_b]-0.5s[I]}\{c\} \tag{7.85}$$

by introducing the strain modes

$$[\Phi_b^{(\varepsilon)}] = [B_1][V_b^{(u)}]([S_b] - 0.5(s-2)[I]) + [B_2][V_b^{(u)}] \tag{7.86}$$

In the two-dimensional case ($s = 2$), Eq. (7.86) has the same form as Eq. (3.66) obtained in the eigenvalue method except that the real Schur form matrix $[S_b]$ replaces the diagonal matrix of eigenvalues $\langle\lambda_b\rangle$.

The stress field is expressed as (Eq. (A.11) without the initial stresses)

$$\{\sigma\} = [\Phi_b^{(\sigma)}]\xi^{[S_b]-0.5s[I]}\{c\} \tag{7.87}$$

where the stress modes $[\Phi_b^{(\sigma)}]$ are defined as

$$[\Phi_b^{(\sigma)}] = [D][\Phi_b^{(\varepsilon)}]$$
$$= [D]\left([B_1][V_b^{(u)}]([S_b] - 0.5(s-2)[I]) + [B_2][V_b^{(u)}]\right) \tag{7.88}$$

On the boundary ($\xi = 1$) of the S-element, the matrix function $1^{[S_b]-0.5s[I]} = [I]$ applies. The strains and stresses are expressed as

$$\{\varepsilon_b\} = \{\varepsilon(\xi = 1)\} = [\Phi_b^{(\varepsilon)}]\{c\} \tag{7.89a}$$

$$\{\sigma_b\} = \{\sigma(\xi = 1)\} = [\Phi_b^{(\sigma)}]\{c\} \tag{7.89b}$$

They are combinations of the strain and stress modes, respectively.

The displacements and stresses inside an S-element can be evaluated directly from the semi-analytical solution Eq. (7.76) and Eq. (7.87), respectively. The matrix power function can be evaluated by the 'scaling and square' algorithm (expm or expm1 in MATLAB) at specified values of the radial coordinate ξ. In this approach the upper triangular part of the Schur form matrix is fully populated. Unlike in the eigenvalue method, all the displacement and stress modes are coupled with each other. As the number of times of evaluating the matrix functions increases, the computational efficiency can be improved by further block diagonalization of the Schur matrix as discussed in Section 7.6.

It is worthwhile mentioning that it is sufficient to obtain the displacements and stresses on the boundary in most of the practical applications. In this case, the matrix function, say $\xi^{[S_b]}$, appears in the derivations only and is not evaluated numerically. It is thus advantageous to use the Schur decomposition without block diagonalization. Since the transformation matrix is well conditioned, this solution procedure does not suffer from the deficiency of the eigenvalue method (Section 3.1) and is numerically robust.

7.6 Block-diagonal Schur Decomposition

In the Schur decomposition (Eq. (7.31)), the upper-triangular part of the real Schur form matrix $[S]$ is fully populated. In contrast to the eigenvalue decomposition (Section 3.1) where a diagonal matrix of eigenvalues is sought, the Schur decomposition leads to a system of coupled solutions of the ordinary differential equations.

In some practical applications, it is advantageous or even necessary to further block-diagonalize the Schur decomposition and separate certain displacement and stress modes from the others. The most notable case is the applications in fracture mechanics presented in Chapter 10. By block-diagonalizing the Schur matrix, the singular stress and T-stress terms are easily isolated from the solution of the complete stress field. The stress intensity factors and T-stress are evaluated directly from their definitions by standard stress recovery techniques of the finite element method. The block diagonalization of the Schur decomposition is also useful to reduce the computational cost of evaluating the matrix power functions when calculating the displacements and stresses at several specified values of the dimensionless radial coordinates ξ inside an S-element.

A general algorithm for the block diagonalization of a Schur matrix is detailed in Golub and van Loan (1996) with the analysis of error. It is summarized in the following. A more robust procedure controlling the condition number of the transformation matrix $[\Psi]$ can be found in Bavely and Stewart (1979). In MATLAB, the function bdschur perform the block-diagonal Schur decomposition.

For easy reference, the Schur decomposition in Eq. (7.30) is reproduced here

$$[V]^T[Z_p][V] = [S] \tag{7.90}$$

The diagonal entries, i.e. the real parts of the eigenvalues, of $[S]$ are sorted in descending order. To perform the block diagonalization, the real Schur form matrix $[S]$ is partitioned according to the sizes of the diagonal blocks $[S_{ii}]$ ($i = 1, 2, 3, \ldots$) as

$$[S] = \begin{bmatrix} [S_{11}] & [S_{12}] & [S_{13}] & \cdots \\ [0] & [S_{22}] & [S_{23}] & \cdots \\ [0] & [0] & [S_{33}] & \cdots \\ \vdots & \vdots & \vdots & \ddots \end{bmatrix} \tag{7.91}$$

The eigenvalues of the diagonal blocks must be sufficiently disjointed. As the diagonals of the Schur form matrix $[S]$ (i.e. the real parts of the eigenvalues) are sorted in descending order, the partition can be performed by examining the difference between two adjacent diagonal entries. When the difference is larger than a specified value, the two diagonal entries are partitioned into different blocks.

The transformation matrix $[V]$ in Eq. (7.90) is partitioned conformably to the partition of $[S]$ as

$$[V] = [[V_1] \quad [V_2] \quad [V_3] \quad \ldots] \tag{7.92}$$

In the following procedure, the decoupling of two diagonal blocks at a time is considered. It can be applied repeatedly to reduce a Schur form matrix into multiple diagonal blocks.

The real Schur form matrix $[S_B]$ consists of the two diagonal blocks $[S_{aa}]$ and $[S_{bb}]$ is expressed as

$$[S_B] = \begin{bmatrix} [S_{aa}]_{n_a \times n_a} & [S_{ab}]_{n_a \times n_b} \\ 0 & [S_{bb}]_{n_b \times n_b} \end{bmatrix} \tag{7.93}$$

where $[S_B]$ can be the whole Schur matrix $[S]$ in Eq. (7.90) or a diagonal block of $[S]$. In the latter case, $[S_B]$ has to be decoupled from all other diagonal blocks, i.e. the off-diagonal blocks related to $[S_B]$ are equal to 0. The integers n_a and n_b in Eq. (7.93) indicate the sizes of the two diagonal blocks. $[S_{ab}]$ is the off-diagonal block relating to the diagonal blocks $[S_{aa}]$ and $[S_{bb}]$. The Schur decomposition of the matrix $[Z_p]$ corresponding to the diagonal block $[S_B]$ (Eq. (7.93)) decoupled from other diagonal blocks in $[S]$ is expressed as

$$[Z_p][V_B] = [V_B][S_B] \tag{7.94}$$

where $[V_B]$ include the $(n_a + n_b)$ columns of the transformation matrix $[V]$. It is partitioned conformably to $[S_B]$

$$[V_B] = [[V_a] \quad [V_b]] \tag{7.95}$$

A transformation matrix $[Y_B]$ is selected as

$$[Y_B] = \begin{bmatrix} [I]_{n_a \times n_a} & [Y_{ab}]_{n_a \times n_b} \\ [0] & [I]_{n_a \times n_b} \end{bmatrix} \tag{7.96}$$

where $[I]$ denotes an identity matrix. $[Y_{ab}]$ is to be determined to block-diagonalize $[S_B]$. The inverse of $[Y_B]$ is equal to

$$[Y_B]^{-1} = \begin{bmatrix} [I] & -[Y_{ab}] \\ [0] & [I] \end{bmatrix} \tag{7.97}$$

Applying the transformation matrix $[Y_B]$ (Eq. (7.96) with Eq. (7.97)) to $[S_B]$ in Eq. (7.93) leads to

$$[Y_B]^{-1}[S_B][Y_B] = \begin{bmatrix} [S_{aa}] & [S_{aa}][Y_{ab}] - [Y_{ab}][S_{bb}] + [S_{ab}] \\ [0] & [S_{bb}] \end{bmatrix} \tag{7.98}$$

where the off-diagonal block will vanish when $[Y_{ab}]$ satisfies

$$[S_{aa}][Y_{ab}] - [Y_{ab}][S_{bb}] = -[S_{ab}] \tag{7.99}$$

This is a Sylvester matrix equation. Its has a unique solution as the eigenvalues of $[S_{aa}]$ and $[S_{bb}]$ are disjointed (Bartels and Stewart, 1972). Since both $[S_{aa}]$ and $[S_{bb}]$ are in the real Schur form, only a back substitution is necessary. The subroutine DTRSYL in LAPACK (Anderson et al., 1999) performs this task. This equation can also be solved by the function lyap in MATLAB. Applying the transformation defined in $[Y]$ (Eq. (7.96)) to the Schur decomposition in Eq. (7.94) yields

$$[\Psi_B]^{-1}[Z_p][\Psi_B] = \begin{bmatrix} [S_{aa}] & 0 \\ 0 & [S_{bb}] \end{bmatrix} \tag{7.100}$$

where the accumulated transformation matrix $[\Psi]$ is expressed as

$$[\Psi_B] = [V_B][Y_B] \tag{7.101}$$

Substituting Eqs. (7.95) and (7.96) into Eq. (7.101) results in

$$[\Psi_B] = [\,[V_a] \quad [V_a][Y_{ab}] + [V_b]\,] \tag{7.102}$$

Note that only the columns of the original transformation matrix $[V]$ corresponding to the position of the zeroed off-diagonal block of $[S_B]$ are modified. The two diagonal blocks in $[S_B]$ are not changed.

The block diagonalization of the real Schur form matrix $[S]$ in Eq. (7.91) may be performed using the above procedure repeatedly starting from the leading or last diagonal block. In the first iteration, $[S_B] = [S]$ applies. When starting from the leading diagonal block, $[S_{aa}]$ is taken as $[S_{11}]$ and $[S_{ab}]$ as $[S_{12}, S_{13}, ...]$. $[S_{bb}]$ is the submatrix including all the other blocks under $[S_{ab}]$. The first iteration will eliminate $[S_{ab}]$, i.e, all the off-diagonal blocks on the first row block. Analogously, in the i-th iteration, $[S_{aa}] = [S_{ii}]$ and $[S_{ab}] = [S_{i,i+1}, S_{i,i+2}, ...]$ apply. $[S_{bb}]$ contains again all the other entries under $[S_{ab}]$. This process is repeated until all off-diagonal blocks are eliminated.

At the end of the block diagonalization process, the Schur decomposition of the matrix $[Z_p]$ in Eq. (7.90) is converted to a block-diagonal Schur decomposition expressed as

$$[\Psi]^{-1}[Z_p][\Psi] = [S] \tag{7.103}$$

with the real block-diagonal Schur matrix

$$[S] = \mathrm{diag}\left([S_1], [S_2], [S_3], ...\right) \tag{7.104}$$

For simplicity in notations, the subscripts ii ($i = 1,\ 2,\ 3,\ \ldots$) of the diagonal blocks are replaced by i. The accumulated transformation matrix is denoted as $[\Psi]$. It is accumulated from Eq. (7.101) and can be written as the product between the original orthogonal transformation matrix $[V]$ (Eq. 7.90) and the accumulated transformation $[Y]$ for block diagonalization

$$[\Psi] = [V][Y] \tag{7.105}$$

The transformation matrix $[\Psi]$ will not be orthogonal. It can be partitioned conformably to $[S]$ as

$$[\Psi] = [\,[\Psi_1]\quad [\Psi_2]\quad [\Psi_3]\quad \ldots\,] \tag{7.106}$$

Equation (7.103) is equivalent to

$$[Z_p][\Psi] = [\Psi][S] \tag{7.107}$$

The block-diagonal Schur decomposition corresponding to a diagonal block $[S_i]$ is expressed as

$$[Z_p][\Psi_i] = [\Psi_i][S_i] \tag{7.108}$$

Note that Eq. (7.107) has the same formal expression as Eq. (7.31). For simplicity in notation, the same symbol $[S]$ is used in Eq. (7.107) for the block-diagonal Schur matrix and in Eq. (7.31) for the full Schur matrix. Equation (7.31) can be regarded as a special case of Eq. (7.107) where $[S]$ is a full quasi-diagonal matrix and $[\Psi]$ is an orthogonal matrix.

The coefficient matrix $[Z_p]$ of the first-order ordinary differential equations in the scaled boundary finite element method (Eq. (7.1)) is a Hamiltonian matrix. As shown on page 271 in Section 6.10, its eigenvalues occur in pairs of ($\lambda, -\lambda$). The diagonal blocks can be chosen accordingly.

When all the eigenvalues of $[Z_p]$ are nonzero, the real Schur form matrix $[S]$ is block-diagonalized as

$$[S] = \mathrm{diag}([S_1], \ldots,\ [S_{N-1}],\ [S_N],\ [S_{N+1}],\ [S_{N+2}], \ldots,\ [S_{2N}]) \tag{7.109}$$

The eigenvalues of the diagonal blocks $[S_i]$ ($i = 1,\ 2,\ \ldots, 2N$) are disjointed. The $2N$ diagonal blocks of $[S]$ form N pairs of conjugates ($[S_i], [S_{\bar{i}}]$) where the index of the block conjugate to the ith block is denoted as $\bar{i}$ and equal to

$$\bar{i} = 2N + 1 - i \tag{7.110}$$

The eigenvalues of the two conjugate blocks satisfy

$$\lambda([S_i]) = -\lambda([S_{\bar{i}}]) \tag{7.111}$$

when ordered in a suitable sequence (for example, the eigenvalues of $[S_i]$ are in descending order and the eigenvalues of $[S_{\bar{i}}]$ in ascending order). Since the real parts of the eigenvalues of $[S]$ has been sorted in descending order before the block diagonalization, the real parts of the eigenvalues of the first N blocks are positive and those of the second N blocks negative, i.e.

$$\begin{aligned} \mathrm{Re}(\lambda([S_i])) &> 0 \qquad (i = 1,\ 2,\ \ldots,\ N) \\ \mathrm{Re}(\lambda([S_i])) &< 0 \qquad (i = N+1,\ N+2,\ \ldots,\ 2N) \end{aligned} \tag{7.112}$$

The transformation matrix $[\Psi]$ is partitioned into column blocks conformably to $[S]$ in Eq. (7.109) as

$$[\Psi] = [[\Psi_1] \quad \dots \quad [\Psi_{N-1}] \quad [\Psi_N] \quad [\Psi_{N+1}] \quad [\Psi_{N+2}] \quad \dots \quad [\Psi_{2N}]] \tag{7.113}$$

When $[Z_p]$ has eigenvalues of 0, the above block diagonalization is expressed as

$$[S^{(1)}] = \mathrm{diag}\left([S_1], \dots, [S_{N-1}], \begin{bmatrix} [S_N] & [T] \\ 0 & [S_{N+1}] \end{bmatrix}, [S_{N+2}], \dots, [S_{2N}]\right)$$

where $[S_N] = [S_{N+1}] = 0 \tag{7.114}$

and the real parts of the eigenvalues of the diagonal blocks satisfy

$$\begin{aligned} \mathrm{Re}(\lambda([S_i])) &> 0 & (i = 1, 2, \dots, N-1) \\ \mathrm{Re}(\lambda([S_i])) &< 0 & (i = N+2, N+3, \dots, 2N) \end{aligned} \tag{7.115}$$

The transformation matrix is partitioned as

$$[\Psi^{(1)}] = [[\Psi_1] \quad \dots \quad [\Psi_{N-1}] \quad [\Psi_N^{(1)}] \quad [\Psi_{N+1}] \quad [\Psi_{N+2}] \quad \dots \quad [\Psi_{2N}]] \tag{7.116}$$

The diagonal block of eigenvalues of 0 cannot be diagonalized.

This block-diagonal Schur matrix in Eq. (7.114) is further simplified. The block-diagonal Schur decomposition expressed in Eq. (7.108) is formulated for the diagonal block of eigenvalues of 0 as

$$[Z_p][[\Psi_N^{(1)}], [\Psi_{N+1}]] = [[\Psi_N^{(1)}], [\Psi_{N+1}]]\begin{bmatrix} 0 & [T] \\ 0 & 0 \end{bmatrix} \tag{7.117}$$

Applying a transformation $\mathrm{diag}([T], [I])$ to this equation results in

$$[Z_p][[\Psi_N^{(1)}], [\Psi_{N+1}]]\begin{bmatrix} [T] & 0 \\ 0 & [I] \end{bmatrix} = [[\Psi_N^{(1)}], [\Psi_{N+1}]]\begin{bmatrix} 0 & [T] \\ 0 & 0 \end{bmatrix}\begin{bmatrix} [T] & 0 \\ 0 & [I] \end{bmatrix} \tag{7.118}$$

This equation is rewritten as

$$[Z_p][[\Psi_N^{(1)}][T], [\Psi_{N+1}]] = [[\Psi_N^{(1)}][T], [\Psi_{N+1}]]\begin{bmatrix} 0 & [I] \\ 0 & 0 \end{bmatrix} \tag{7.119}$$

It is easy to verify that the right-hand sides of Eqs. (7.118) and (7.119) are equal to $[[0], [\Psi_N^{(1)}][T]]$. Introducing the transformation

$$[\Psi_N] = [\Psi_N^{(1)}][T] \tag{7.120}$$

Eq. (7.119) is rewritten as

$$[Z_p][[\Psi_N], [\Psi_{N+1}]] = [[\Psi_N], [\Psi_{N+1}]]\begin{bmatrix} [0] & [I] \\ [0] & [0] \end{bmatrix} \tag{7.121}$$

Therefore, the block-diagonal decomposition in Eq. (7.114) can be expressed as

$$[S] = \mathrm{diag}\left([S_1], \dots, [S_{N-1}], \begin{bmatrix} [S_N] & [I] \\ 0 & [S_{N+1}] \end{bmatrix}, [S_{N+2}], \dots, [S_{2N}]\right)$$

$$\text{where } [S_N] = [S_{N+1}] = 0 \tag{7.122}$$

with the transformation matrices given in Eq. (7.113).

For later use in determining the stiffness matrix of an S-element, the block-diagonal Schur matrix in Eqs. (7.109) and (7.122) are partitioned as in Eq. (7.52) with

$$[S_b] = \text{diag}([S_1], \ldots, [S_{N-1}], [S_N]) \tag{7.123}$$

$$[S_u] = \text{diag}([S_{N+1}], \ldots, [S_{2N-1}], [S_{2N}]) \tag{7.124}$$

All diagonal entries of $[S_b]$ and $[S_u]$ are non-negative and non-positive, respectively. The following partitions of the transformation matrix $[\Psi]$ are introduced

$$[\Psi] = \begin{bmatrix} [\Psi_b^{(u)}] & [\Psi_u^{(u)}] \\ [\Psi_b^{(q)}] & [\Psi_u^{(q)}] \end{bmatrix} \tag{7.125}$$

$$= \begin{bmatrix} \begin{bmatrix} [\Psi_1^{(u)}] & \cdots & [\Psi_{N-1}^{(u)}] & [\Psi_N^{(u)}] \end{bmatrix} & \begin{bmatrix} [\Psi_{N+1}^{(u)}] & [\Psi_{N+2}^{(u)}] & \cdots & [\Psi_{2N}^{(u)}] \end{bmatrix} \\ \begin{bmatrix} [\Psi_1^{(q)}] & \cdots & [\Psi_{N-1}^{(q)}] & [\Psi_N^{(q)}] \end{bmatrix} & \begin{bmatrix} [\Psi_{N+1}^{(q)}] & [\Psi_{N+2}^{(q)}] & \cdots & [\Psi_{2N}^{(q)}] \end{bmatrix} \end{bmatrix} \tag{7.126}$$

The partition in Eq. (7.125) is analogous to that in Eq. (7.53) and the subscripts and superscripts have the same significance. In Eq. (7.126), the partition of columns follows that in Eq. (7.113) and is conformable to the partition of the Schur matrix $[S]$ in Eqs. (7.109) and (7.122). Each column block in Eq. (7.113) is divided into two sub-blocks of the same size, i.e.

$$[\Psi_i] = \begin{bmatrix} [\Psi_i^{(u)}] \\ [\Psi_i^{(q)}] \end{bmatrix} \qquad (i = 1, 2, \ldots, 2N) \tag{7.127}$$

The correspondence between the partitions in Eqs. (7.125) and (7.126) is self-explanatory.

The orthogonal transformation matrix $[V]$ of the original Schur decomposition and the transformation matrix $[\Psi]$ of the block-diagonal Schur decomposition are related as in Eq. (7.105). The transform matrix $[Y]$ is accumulated from the upper-triangular matrix in Eq. (7.96) during the recursive block-diagonalization process. $[Y]$ is also upper-triangular and can be partitioned as

$$[Y] = \begin{bmatrix} [Y_b] & [Y_{bu}] \\ 0 & [Y_u] \end{bmatrix} \tag{7.128}$$

Substituting the partitions of $[\Psi]$ in Eq. (7.125), of $[V]$ in Eq. (7.53) and of $[Y]$ in Eq. (7.128) into Eq. (7.105) leads to

$$\begin{bmatrix} [\Psi_b^{(u)}] & [\Psi_u^{(u)}] \\ [\Psi_b^{(q)}] & [\Psi_u^{(q)}] \end{bmatrix} = \begin{bmatrix} [V_b^{(u)}] & [V_u^{(u)}] \\ [V_b^{(q)}] & [V_u^{(q)}] \end{bmatrix} \begin{bmatrix} [Y_b] & [Y_{bu}] \\ 0 & [Y_u] \end{bmatrix}$$

$$= \begin{bmatrix} [V_b^{(u)}][Y_b] & [V_b^{(u)}][Y_{bu}] + [V_u^{(u)}][Y_u] \\ [V_b^{(q)}][Y_b] & [V_b^{(q)}][Y_{bu}] + [V_u^{(q)}][Y_u] \end{bmatrix} \tag{7.129}$$

It is observed that the transformation of the first column block is independent of the transformation of the second one

$$\begin{bmatrix} [\Psi_b^{(u)}] \\ [\Psi_b^{(q)}] \end{bmatrix} = \begin{bmatrix} [V_b^{(u)}][Y_b] \\ [V_b^{(q)}][Y_b] \end{bmatrix} = \begin{bmatrix} [V_b^{(u)}] \\ [V_b^{(q)}] \end{bmatrix} [Y_b] \tag{7.130}$$

It is proved in Song (2004c) that the column blocks of the transformation matrix $[\Psi]$ is $[J_{2n}]$-orthogonal.

$$[\Psi_i]^T[J_{2n}][\Psi_j] = (-[\Psi_j]^T[J_{2n}][\Psi_i])^T = \begin{cases} [H_i] & \text{when} \quad j = \bar{\imath} \\ 0 & \text{when} \quad j \neq \bar{\imath} \end{cases} \tag{7.131}$$

i.e. the block $[\Psi_i]$ is $[J_{2n}]$-orthogonal to all the other blocks, including itself, other than its conjugate block. The conjugate column blocks $[\Psi_i]$ and $[\Psi_{\bar{\imath}}]$ can be normalized to satisfy, for example,

$$\max(|[\Psi_i]|) = \max(|[\Psi_{\bar{\imath}}]|) \tag{7.132a}$$

$$\max(|[H_i]|) = 1 \tag{7.132b}$$

where $\max(| \bullet |)$ stands for the maximum absolute value in all the entries in the matrix.

7.7 Solution Procedure by Block-diagonal Schur Decomposition

In this section, the general solution of the system of first-order ordinary differential equations (Eq. (7.1)) of the scaled boundary finite element method is obtained. The solutions for displacement functions and stiffness matrices are determined for bounded and unbounded S-elements by enforcing the boundary conditions at the scaling centre and at infinity, respectively.

7.7.1 General Solution of the Scaled Boundary Finite Element Equation

The solution procedure follows that in Section 7.4. Introducing the generalized-coordinate functions $\{W(\xi)\}$

$$\{X(\xi)\} = [\Psi]\{W(\xi)\} \tag{7.133}$$

Equation (7.1) is expressed as

$$[\Psi]\xi\{W(\xi)\}_{,\xi} = [Z_p][\Psi]\{W(\xi)\} \tag{7.134}$$

Pre-multiplying this equation with $[\Psi]^{-1}$ yields a system of equations for the generalized-coordinate functions $\{W(\xi)\}$

$$\xi\{W(\xi)\}_{,\xi} = [\Psi]^{-1}[Z_p][\Psi]\{W(\xi)\} \tag{7.135}$$

Using the block-diagonal Schur decompostion (Eq. 7.103), Eq. (7.135) is rewritten as

$$\xi\{W(\xi)\}_{,\xi} = [S]\{W(\xi)\} \tag{7.136}$$

The fundamental matrix of this ordinary differential equation can be expressed as a matrix power function $\xi^{[S]}$ (Appendix B.2). The general solution is given by

$$\{W(\xi)\} = \xi^{[S]}\{c\} \tag{7.137}$$

with the integration constant $\{c\}$.

Since the constant matrix $[S]$ is a block diagonal matrix, Eq. (7.136) can be decoupled accordingly. The vector of generalized-coordinate functions $\{W(\xi)\}$ is partitioned conformably to $[S]$ (Eq. (7.109)) and $[\Psi]$ (Eq. (7.113))

$$\{W\} = \{\{W_1\}; \ \ldots ; \ \{W_{N-1}\}; \ \{W_N\}; \ \{W_{N+1}\}; \ \{W_{N+2}\}; \ \ldots ; \ \{W_{2N}\}\} \quad (7.138)$$

where the semi-colon symbol ';' stands for the vertical concatenation of vectors, and the argument ξ is omitted for the sake of brevity. Equation (7.136) is formulated as a series of independent equations for $\{W_i(\xi)\}$ $(i = 1, \ 2, \ \ldots, \ 2N)$. Substituting Eqs. (7.138) and (7.113) into Eq. (7.134) yields

$$\{X(\xi)\} = [\Psi]\{W(\xi)\} = \sum_{i=1}^{2N} [\Psi_i]\{W_i(\xi)\} \quad (7.139)$$

The procedure to determine $\{W_i(\xi)\}$ and to enforce the boundary condition at the scaling centre varies depending on whether $[Z_p]$ has eigenvalues of 0 or not. The two cases are treated separately in the following.

7.7.1.1 $[Z_p]$ Having No Eigenvalues of Zero

The case of $[Z_p]$ having no eigenvalues of 0 is considered first. The Schur matrix $[S]$ in Eq. (7.109) consists of $2N$ decoupled diagonal blocks. Using Eqs. (7.109) and (7.138), Eq. (7.136) is written as a series of $2N$ independent systems of equations

$$\xi\{W_i(\xi)\}_{,\xi} = [S_i]\{W_i(\xi)\} \qquad (i = 1, \ 2, \ \ldots, \ 2N) \quad (7.140)$$

Using the matrix power function (Appendix B.2), its general solution is expressed as

$$\{W_i(\xi)\} = \xi^{[S_i]}\{c_i\} \qquad (i = 1, \ 2, \ \ldots, \ 2N) \quad (7.141)$$

with integration constants $\{c_i\}$. Substituting Eq. (7.141) into Eq. (7.139) leads to the general solution of Eq. (7.1)

$$\{X(\xi)\} = \sum_{i=1}^{2N} [\Psi_i]\xi^{[S_i]}\{c_i\} \quad (7.142)$$

Equation (7.142) can also be obtained directly from Eq. (7.137). The matrix power function of the block diagonal matrix $[S]$ (Eq. (7.109)) is expressed as

$$\xi^{[S]} = \mathrm{diag}(\xi^{[S_1]}, \ldots, \ \xi^{[S_{N-1}]}, \ \xi^{[S_N]}, \ \xi^{[S_{N+1}]}, \ \xi^{[S_{N+2}]}, \ldots, \ \xi^{[S_{2N}]}) \quad (7.143)$$

Substituting Eq. (7.143) into Eq. (7.137) and using the partition of the transformation matrix in Eq. (7.113) leads to Eq. (7.142), where the integration constants are also partitioned conformably.

7.7.1.2 $[Z_p]$ Having Eigenvalues of Zero

When $[Z_p]$ has eigenvalues of 0, the block-diagonal Schur matrix is given in Eq. (7.122). It consists of $2N - 2$ decoupled diagonal blocks of non-zero eigenvalues and a diagonal block of eigenvalues of 0 in the middle. Substituting Eqs. (7.138) and (7.122), Eq. (7.136) is decoupled into $2N - 1$ independent systems of equations

$$\xi\{W_i(\xi)\}_{,\xi} = [S_i]\{W_i(\xi)\} \qquad (i = 1, 2, \ldots, N - 1, N + 2, \ldots, 2N)$$

(7.144a)

$$\xi \frac{d}{d\xi} \left\{ \begin{array}{c} \{W_N(\xi)\} \\ \{W_{N+1}(\xi)\} \end{array} \right\} = \begin{bmatrix} 0 & [I] \\ 0 & 0 \end{bmatrix} \left\{ \begin{array}{c} \{W_N(\xi)\} \\ \{W_{N+1}(\xi)\} \end{array} \right\}$$

(7.144b)

Equation (7.144a) is solved in the same way as Eq. (7.140), leading to the general solution

$$\{W_i(\xi)\} = \xi^{[S_i]}\{c_i\} \qquad (i = 1, 2, \ldots, N - 1, N + 2, \ldots, 2N)$$

(7.145)

The fundamental solution of Eq. (7.144a) is obtained by applying Eq. (B.2) as (see also Eqs. (B.6) to (B.8))

$$\xi^{\begin{bmatrix} 0 & [I] \\ 0 & 0 \end{bmatrix}} = \begin{bmatrix} [I] & \ln\xi[I] \\ 0 & [I] \end{bmatrix}$$

(7.146)

The general solution of Eq. (7.144b) is expressed as

$$\left\{ \begin{array}{c} \{W_N(\xi)\} \\ \{W_{N+1}(\xi)\} \end{array} \right\} = \begin{bmatrix} [I] & [I]\ln\xi \\ 0 & [I] \end{bmatrix} \left\{ \begin{array}{c} \{c_N\} \\ \{c_{N+1}\} \end{array} \right\} = \left\{ \begin{array}{c} \{c_N\} + \ln\xi\{c_{N+1}\} \\ \{c_{N+1}\} \end{array} \right\}$$

(7.147)

Substituting Eqs. (7.145) and (7.147) in Eq. (7.139) leads to the general solution of Eq. (7.1)

$$\{X(\xi)\} = \sum_{i=1}^{N-1} [\Psi_i]\xi^{[S_i]}\{c_i\} + [\Psi_N](\{c_N\} + \ln\xi\{c_{N+1}\})$$

$$+ [\Psi_{N+1}]\{c_{N+1}\} + \sum_{i=N+1}^{2N} [\Psi_i]\xi^{[S_i]}\{c_i\}$$

(7.148)

The integration constants $\{c_i\}$ $(i = 1, 2, \ldots, 2N)$ are determined by enforcing the boundary conditions. Half of constants are found for a bounded S-element using the conditions at the scaling centre, and for an unbounded domain using the conditions at infinity.

Similar to the previous case, Eq. (7.148) can also be obtained directly from Eq. (7.137) by using the matrix power function of the block-diagonal Schur matrix (Eq. (7.122))

$$\xi^{[S]} = \text{diag}\left(\xi^{[S_1]}, \ldots, \xi^{[S_{N-1}]}, \begin{bmatrix} [I] & \ln\xi[I] \\ [0] & [I] \end{bmatrix}, \xi^{[S_{N+2}]}, \ldots, \xi^{[S_{2N}]} \right)$$

(7.149)

7.7.2 Solution for Bounded S-elements

A bounded S-element is defined in radial coordinate ξ by $0 \le \xi \le 1$. The general solution is given in Eq. (7.142) when $[Z_p]$ does not have any eigenvalues of 0 and in Eq. (7.148) when $[Z_p]$ has eigenvalues of 0. It is shown below that to satisfy the finiteness of the solution at the scaling centre $\xi = 0$, the second half of the integration constants has to vanish, i.e.

$$\{c_i\} = \{0\} \qquad (i = N + 1, N + 2, \ldots, 2N)$$

(7.150)

The case for $[Z_p]$ having no eigenvalues of 0 is addressed first. The real parts of eigenvalues of the last N diagonal blocks $[S_i]$ ($i = N + 1,\ N + 2,\ \dots,\ 2N$) are negative (Eq. (7.112)). At $\xi = 0$, the matrix power functions $\xi^{[S_i]}$ ($i = N + 1,\ N + 2,\ \dots,\ 2N$) in the general solution Eq. (7.142). tend to infinity. These terms are not admissible to the solution, therefore, Eq. (7.150) applies.

Similarly, when $[Z_p]$ has eigenvalues of 0, the real parts of eigenvalues of the last $N - 1$ diagonal blocks $[S_i]$ ($i = N + 2,\ N + 3,\ \dots,\ 2N$) are negative (Eq. (7.115)) and $\xi^{[S_i]}$ ($i = N + 2,\ N + 3,\ \dots,\ 2N$) in the general solution Eq. (7.148) tend to infinity at $\xi = 0$. The eigenvalues of 0 lead to a logarithm term ($\ln \xi$) in the general solution. This term is also inadmissible to the solution as $\ln \xi$ tends to negative infinity at $\xi = 0$. To satisfy the finiteness of the solution, Eq. (7.150) must hold.

Considering Eq. (7.150), the solution of a bounded S-element is expressed, for both Eq. (7.142) and Eq. (7.148), as

$$\{X(\xi)\} = \sum_{i=1}^{N} [\Psi_i] \xi^{[S_i]} \{c_i\} \tag{7.151}$$

Note that $\xi^{[0]} = [I]$ applies when eigenvalues of 0 exist.

Using Eq. (7.2) and the partition of the transformation matrix $[\Psi]$ in Eq. (7.127), the solutions for the displacement and force functions are obtained

$$\{u(\xi)\} = \xi^{-0.5(s-2)} \sum_{i=1}^{N} [\Psi_i^{(u)}] \xi^{[S_i]} \{c_i\} \tag{7.152a}$$

$$\{q(\xi)\} = \xi^{+0.5(s-2)} \sum_{i=1}^{N} [\Psi_i^{(q)}] \xi^{[S_i]} \{c_i\} \tag{7.152b}$$

Using the partition of the transformation matrix $[\Psi]$ defined in Eq. (7.125), this equation is expressed in matrix form as

$$\{u(\xi)\} = \xi^{-0.5(s-2)} [\Psi_b^{(u)}] \xi^{[S_b]} \{c_b\} \tag{7.153a}$$

$$\{q(\xi)\} = \xi^{+0.5(s-2)} [\Psi_b^{(q)}] \xi^{[S_b]} \{c_b\} \tag{7.153b}$$

where $[S_b]$ is the block-diagonal matrix defined in Eq. (7.123) and $\{c_b\}$ is the vector of integration constants

$$\{c_b\} = \{\{c_1\}; \dots; \{c_{N-1}\}; \{c_N\}\}) \tag{7.154}$$

Note that Eq. (7.153) resembles Eq. (7.76). They are related by a transformation (Section 7.6) to block-diagonalize the Schur matrix in Eq. (7.76).

The static stiffness matrix of a bounded domain is determined following the steps on page 295 in Section 7.4. Eliminating the integration constants $\{c_b\}$ from Eqs. (7.153a) and (7.153b) results in

$$\{q(\xi)\} = \xi^{s-2} [\Psi_b^{(q)}][\Psi_b^{(u)}]^{-1} \{u(\xi)\} \tag{7.155}$$

The static stiffness matrix of a bounded S-element enclosed by the scaled boundary at ξ is expressed as

$$[K(\xi)] = \xi^{s-2} [\Psi_b^{(q)}][\Psi_b^{(u)}]^{-1} \tag{7.156}$$

The static stiffness of the bounded S-element at its boundary ($\xi = 1$) is equal to

$$[K] = [\Psi_b^{(q)}][\Psi_b^{(u)}]^{-1} \tag{7.157}$$

It can easily be shown that the stiffness matrix in Eq. (7.157) is analytically equal to the stiffness matrix in Eq. (7.79). Substituting the transformation Eq. (7.130) into Eq. (7.157) yields

$$[K] = [V_b^{(q)}][Y_b]([V_b^{(u)}][Y_b])^{-1} = [V_b^{(q)}][V_b^{(u)}]^{-1} \tag{7.158}$$

which is simply Eq. (7.79).

Similarly, it is demonstrated in Remark 2 on page 318, Section 7.11 that the stiffness matrix obtained by (block-diagonal) Schur decomposition is also analytically equal to the stiffness matrix obtained from the eigenvalue decomposition method.

7.7.3 Solution for Unbounded S-elements

The solution for an unbounded S-element, specified by $1 \leq \xi < \infty$, can be obtained using the same block-diagonal Schur decomposition for the solution of the corresponding bounded S-element specified by $0 \leq \xi \leq 1$. The boundary condition at infinity, i.e. $\xi \to \infty$, is enforced analytically.

The general solution of the scaled boundary finite element equation formulated as a system of first-order ordinary differential equations is given in Eq. (7.142) for the case of $[Z_p]$ having no eigenvalues of 0 and in Eq. (7.148) for the case of $[Z_p]$ having eigenvalues of 0. Although the displacement solutions for the two cases differ, the static stiffness matrices of both cases are expressed as

$$[K] = -[\Psi_u^{(q)}][\Psi_u^{(u)}]^{-1} \tag{7.159}$$

where $[\Psi_u^{(u)}]$ and $[\Psi_u^{(q)}]$ are the submatrices of the transformation matrix $[\Psi]$ defined in Eq. (7.126).

7.7.3.1 $[Z_p]$ Having No Eigenvalues of Zero

In this case, the real parts of the eigenvalues of the first N diagonal blocks $[S_i]$ ($i = 1, 2, \ldots, N$) in Eq. (7.109) are positive (see Eq. (7.112)). The matrix power functions $\xi^{[S_i]}$ ($i = 1, 2, \ldots, N$) in the general solution Eq. (7.142) tend to infinity as ξ approaches infinity and are not admissible in the solution. Therefore,

$$\{c_i\} = \{0\} \qquad (i = 1, 2, \ldots, N) \tag{7.160}$$

must hold, yielding

$$\{X(\xi)\} = \sum_{i=N+1}^{2N} [\Psi_i]\xi^{[S_i]}\{c_i\} \tag{7.161}$$

Using Eq. (7.2) and the partition of the transformation matrix $[\Psi]$ in Eq. (7.127), the solutions for the displacement and force functions are expressed as

$$\{u(\xi)\} = \xi^{-0.5(s-2)} \sum_{i=N+1}^{2N} [\Psi_i^{(u)}]\xi^{[S_i]}\{c_i\} \tag{7.162a}$$

$$\{q(\xi)\} = \xi^{+0.5(s-2)} \sum_{i=N+1}^{2N} [\Psi_i^{(q)}]\xi^{[S_i]}\{c_i\} \tag{7.162b}$$

Using the partition of the transformation matrix $[\Psi]$ defined in Eq. (7.126), this equation is expressed in matrix form as

$$\{u(\xi)\} = \xi^{-0.5(s-2)}[\Psi_u^{(u)}]\xi^{[S_u]}\{c_u\} \tag{7.163a}$$

$$\{q(\xi)\} = \xi^{+0.5(s-2)}[\Psi_u^{(q)}]\xi^{[S_u]}\{c_u\} \tag{7.163b}$$

where $[S_u]$ is the block-diagonal matrix defined in Eq. (7.124) and $\{c_u\}$ is the vector of integration constants

$$\{c_u\} = \{\{c_{N+1}\}; \ldots; \{c_{2N-1}\}; \{c_{sN}\}) \tag{7.164}$$

Eliminating the integration constants $\{c_u\}$ from Eq. (7.163) results in

$$\{q(\xi)\} = \xi^{s-2}[\Psi_u^{(q)}][\Psi_u^{(u)}]^{-1}\{u(\xi)\} \tag{7.165}$$

The static stiffness matrix of an unbounded S-element exterior to the scaled boundary at ξ is equal to

$$[K(\xi)] = -\xi^{s-2}[\Psi_u^{(q)}][\Psi_u^{(u)}]^{-1} \tag{7.166}$$

At the boundary ($\xi = 1$), the static stiffness matrix of the unbounded S-domain exterior to is given in Eq. (7.159).

7.7.3.2 $[Z_p]$ Having Eigenvalues of Zero

When $[Z_p]$ has a block of eigenvalues of 0, a term containing a logarithmic function $\ln \xi$ appears in the general solution Eq. (7.142). The behaviour of this term is further investigated.

The basis vectors of this term correspond to the block of eigenvalues of 0, i.e. $[S_N]$ and $[S_{N+1}]$, in Eq. (7.122). The Schur decomposition of this diagonal block is expressed, using the partition of $[\Psi]$ in Eq. (7.126), as

$$[Z_p]\begin{bmatrix} [\Psi_N^{(u)}] & [\Psi_{N+1}^{(u)}] \\ [\Psi_N^{(q)}] & [\Psi_{N+1}^{(q)}] \end{bmatrix} = \begin{bmatrix} [\Psi_N^{(u)}] & [\Psi_{N+1}^{(u)}] \\ [\Psi_N^{(q)}] & [\Psi_{N+1}^{(q)}] \end{bmatrix}\begin{bmatrix} 0 & [I] \\ 0 & 0 \end{bmatrix} \tag{7.167}$$

Expanding the first column block of Eq. (7.167) using $[Z_p]$ given in Eq. (7.3) with $s = 2$ results in

$$[E_0]^{-1}[E_1]^T[\Psi_N^{(u)}] - [E_0]^{-1}[\Psi_N^{(q)}] = 0 \tag{7.168a}$$

$$(-[E_2] + [E_1][E_0]^{-1}[E_1]^T)[\Psi_N^{(u)}] - [E_1][E_0]^{-1}[\Psi_N^{(q)}] = 0 \tag{7.168b}$$

Eliminating $[\Psi_N^{(q)}]$ by substituting $[E_0]^{-1}[\Psi_N^{(q)}]$ in Eq. (7.168a) into Eq. (7.168b) yields

$$[E_2][\Psi_N^{(u)}] = [0] \tag{7.169}$$

Considering Eq. (6.169) in Section 6.10, $[\Psi_N^{(u)}]$ represents a translational rigid motion. $[\Psi_N^{(q)}]$ is obtained from Eqs. (7.168a) and (6.179) as

$$[\Psi_N^{(q)}] = [E_1]^T[\Psi_N^{(u)}] = [0] \tag{7.170}$$

It corresponds to the vanishing nodal forces of the rigid body motion, which is consistent with Eq. (6.180) in Section (6.11). Note that $[\Psi_N^{(u)}]$ is not uniquely determined as any independent linear transformation of $[\Psi_N^{(u)}]$ satisfies Eq. (7.168a).

Expanding the second column block of Eq. (7.167) results in

$$[E_0]^{-1}[E_1]^T[\Psi_{N+1}^{(u)}] - [E_0]^{-1}[\Psi_{N+1}^{(q)}] = [\Psi_N^{(u)}] \tag{7.171a}$$

$$(-[E_2] + [E_1][E_0]^{-1}[E^1]^T)[\Psi_{N+1}^{(u)}] - [E_1][E_0]^{-1}[\Psi_{N+1}^{(q)}] = [0] \tag{7.171b}$$

Eliminating $[\Psi_{N+1}^{(q)}]$ yields

$$[E_2][\Psi_{N+1}^{(u)}] = [E_1][\Psi_N^{(u)}] \tag{7.172}$$

As $[E_2][\Psi_N^{(u)}] = 0$ (Eq. (7.169)), Eq. (7.172) still holds when a matrix proportional to $[\Psi_N^{(u)}]$ is added to $[\Psi_N^{(u)}]$. Thus, $[\Psi_{N+1}^{(u)}]$ is not uniquely determined, which is consistent with the statement in Deeks and Wolf (2002a).

Using the partition of $[\Psi]$ in Eq. (7.126) and Eq. (7.170), the general solutions of the nodal displacement and forces for two-dimensional problems ($s = 2$) is obtained from Eq. (7.148) as

$$\{u(\xi)\} = \sum_{i=1}^{N-1}[\Psi_i^{(u)}]\xi^{[S_i]}\{c_i\} + [\Psi_N^{(u)}]\{c_N\} + (\ln\xi[\Psi_N^{(u)}] + [\Psi_{N+1}^{(u)}])\{c_{N+1}\}$$

$$+ \sum_{i=N+2}^{2N}[\Psi_i^{(u)}]\xi^{[S_i]}\{c_i\} \tag{7.173a}$$

$$\{q(\xi)\} = \sum_{i=1}^{N-1}[\Psi_i^{(q)}]\xi^{[S_i]}\{c_i\} + [\Psi_{N+1}^{(q)}]\{c_{N+1}\} + \sum_{i=N+2}^{2N}[\Psi_i^{(q)}]\xi^{[S_i]}\{c_i\} \tag{7.173b}$$

The real parts of the eigenvalues of the diagonal blocks $[S_i]$ ($i = 1, 2, \ldots, N\text{-}1$) are all positive (Eq. (7.112)). The matrix power functions $\xi^{[S_i]}$ ($i = 1, 2, \ldots, N-1$) at $\xi \to \infty$ and are not admissible in the solution. Therefore, the integration constants

$$\{c_i\} = \{0\} \qquad (i = 1, 2, \ldots, N-1) \tag{7.174}$$

apply. As $\xi \to \infty$, the predominant term in the displacement solution (Eq. (7.173a)) is one of the product of the translational motion $[\Psi_N^{(u)}]$ with the logarithmic function $\ln\xi$. This result is consistent with analytical solutions of two-dimensional problems, for example the fundamental solution of the boundary element method (Brebbia et al., 1984). The integration constants $\{c_N\}$ are undetermined. In other words, only a solution for displacements admitting arbitrary translational motions can be determined. On the other hand, the solution for the force functions is uniquely determined. For simplicity, $\{c_N\} = 0$ is set in Eq. (7.173), leading to

$$\{u(\xi)\} = (\ln\xi[\Psi_N^{(u)}] + [\Psi_{N+1}^{(u)}])\{c_{N+1}\} + \sum_{i=N+2}^{2N}[\Psi_i^{(u)}]\xi^{[S_i]}\{c_i\} \tag{7.175a}$$

$$\{q(\xi)\} = [\Psi_{N+1}^{(q)}]\{c_{N+1}\} + \sum_{i=N+2}^{2N}[\Psi_i^{(q)}]\xi^{[S_i]}\{c_i\} \tag{7.175b}$$

The remaining integration constants $\{c_i\}$ ($i = N + 1, N + 2, \ldots, 2N$) are to be determined using boundary conditions at $\xi = 1$. It is shown in Song (2004a) that when the external loads on the unbounded S-element are self-equilibrating, i.e. $[\Psi_N^{(u)}]^T\{q(\xi)\} = 0$, the integration constants $\{c_N\} = 0$ applies and the logarithmic functions vanish from the solution in Eq. (7.175).

Using the partition of the transformation matrix $[\Psi]$ defined in Eq. (7.126), Eq. (7.175) is written as

$$\{u(\xi)\} = \ln \xi [\Psi_N^{(u)}]\{c_{N+1}\} + [\Psi_u^{(u)}]\xi^{[S_u]}\{c_u\} \tag{7.176a}$$

$$\{q(\xi)\} = [\Psi_u^{(q)}]\xi^{[S_u]}\{c_u\} \tag{7.176b}$$

where $[S_u]$ is the block-diagonal matrix with non-positive diagonal entries

$$[S_u] = \text{diag}([S_{N+1}], \dots, [S_{2N-1}], [S_{2N}]) \tag{7.177}$$

$\{c_u\}$ is the vector of integration constants

$$\{c_u\} = \{\{c_{N+1}\}; \dots; \{c_{2N-1}\}; \{c_{2N}\}\} \tag{7.178}$$

Formulating Eq. (7.176) at $\xi = 1$ leads to

$$\{u_b\} = \{u(\xi = 1)\} = [\Psi_u^{(u)}]\{c_u\} \tag{7.179a}$$

$$-\{R\} = \{q(\xi = 1)\} = [\Psi_u^{(q)}]\{c_u\} \tag{7.179b}$$

Eliminating the integration constants $\{c_u\}$ from Eq. (7.179) yields

$$\{R\} = -[\Psi_u^{(q)}][\Psi_u^{(u)}]^{-1}\{u_b\} \tag{7.180}$$

which is simply the stiffness matrix of the unbounded S-element in Eq. (7.159).

7.8 Displacements and Stresses of an S-element by Block-diagonal Schur Decomposition

The assemblage of the stiffness matrices of the S-elements to form the global stiffness matrix and the solution for the nodal displacements of the global system are described in Section 3.5. The evaluation of displacement and stress field of an S-element using the block-diagonal Schur decomposition follows the procedure in Section 7.5.

7.8.1 Integration Constants and Displacement Fields

After extracting the nodal displacements $\{u_b\}$ of an S-element, the integration constants in the solution of the S-element can be determined.

For a bounded S-element, Eq. (7.153a) is written on the boundary ($\xi = 1$) as

$$\{u_b\} = \{u(\xi = 1)\} = [\Psi_b^{(u)}]\{c_b\} \tag{7.181}$$

Since $1^{[S_b]} = [I]$. The integration constants $\{c_b\}$ are obtained as

$$\{c_b\} = [\Psi_b^{(u)}]^{-1}\{u_b\} \tag{7.182}$$

For an unbounded S-element, the solutions of the displacement functions are given in Eq. (7.163a) when the coefficient matrix $[Z_p]$ does not have eigenvalues of 0 and Eq. (7.176a) when $[Z_p]$ has eigenvalues of 0. On the boundary $\xi = 1$, both equations are expressed as

$$\{u_b\} = \{u(\xi = 1)\} = [\Psi_u^{(u)}]\{c_u\} \tag{7.183}$$

The integration constants $\{c_u\}$ are equal to

$$\{c_u\} = [\Psi_u^{(u)}]^{-1}\{u_b\} \tag{7.184}$$

After the integration constants are obtained, the solution of the displacement field of an S-element is fully determined. The solution is given in Eq. (7.152) for a bounded S-element and in Eqs. (7.163a) and (7.175a) for an unbounded S-element without and with eigenvalues of 0, respectively. These equations are written in a unified expression as

$$\{u(\xi)\} = \delta_{\ln}\ln\xi[\Psi_N^{(u)}]\{c_{N+1}\} + \xi^{-0.5(s-2)}\sum_{i=i_B}^{N+i_B-1}[\Psi_i^{(u)}]\xi^{[S_i]}\{c_i\} \tag{7.185}$$

with

$$\delta_{\ln} = \begin{cases} 1 & \text{for unbounded S-element with} \quad \lambda([S_N]) = [0] \\ 0 & \text{otherwise} \end{cases} \tag{7.186a}$$

and

$$i_B = \begin{cases} 1 & \text{for bounded S-element} \\ N+1 & \text{for unbounded S-element} \end{cases} \tag{7.186b}$$

Equation (7.185) can be evaluated term-by-term. The block diagonalization of each of the two blocks in Eq. (7.52) into several smaller diagonal blocks in Eq. (7.109) or Eq. (7.122) will reduce the computational effort on evaluating the matrix power functions. When the eigenvalues of the diagonal blocks are well separated, the loss of numerical accuracy is insignificant.

The column vectors of $[\Psi_i^{(u)}]$ (Eq. (7.126)) can be regarded as displacement modes. The displacement modes of a particular characteristics (such as rigid-body motion and constant stress modes) can be separated from the others using their eigenvalues.

7.8.2 Stress Modes and Stress Fields

As in Section 7.5, the strain field of an S-element is evaluated from the semi-analytical solution of the displacement field using Eq. (6.92).

The solution of the nodal displacement functions of an S-element is semi-analytical and expressed as a series in Eq. (7.185). It can be interpreted as the superposition of displacement modes $[\Psi_i^{(u)}]$ with varying participation factors along the radial direction. The differentiation of Eq. (7.185) with respect to the radial coordinates ξ is performed analytically, leading to

$$\{u(\xi)\}_{,\xi} = \delta_{\ln}\xi^{-1}[\Psi_N^{(u)}]\{c_{N+1}\} + \sum_{i=i_B}^{N+i_B-1}[\Psi_i^{(u)}]([S_i] - 0.5(s-2)[I])\xi^{[S_i]-0.5s[I]}\{c_i\} \tag{7.187}$$

Along the circumferential directions, the displacement field is described by interpolating the displacement functions (Eq. 6.87). When deriving the expression for strains (Eq. 6.92), the differentiation of the displacement field is performed numerically in the circumferential directions. To achieve high accuracy, the strains are typically evaluated element-by-element at the Gauss points of a lower-order rule than that for full integration. Similar to Section 3.7, strain modes $[\Psi_i^{(\varepsilon)}]$ corresponding to the displacement modes $[\Psi_i^{(u)}]$ in Eq. (7.126) are defined. For simplicity in notation, the same symbol $[\Psi_i^{(u)}]$ will refer to the values at the nodes of an element in the following part of this section and is extracted from the displacement modes of the S-element according to the element connectivity.

The strain modes corresponding to the term with $\ln \xi$ in Eq. (7.186a) are considered. Substituting $\ln \xi [\Psi_N^{(u)}]$ into Eq. (6.92), the strains of this term are expressed as

$$\{\varepsilon_{\ln}\} = \left(\xi^{-1}[B_1][\Psi_N^{(u)}] + \ln \xi [B_2][\Psi_N^{(u)}] \right) \{c_{N+1}\} \tag{7.188}$$

Since $[\Psi_N^{(u)}]$ represents translational rigid-body motions, $[B_2][\Psi_N^{(u)}] = \{0\}$ holds (Eq. 6.168). The strain modes are defined as

$$[\Psi_{\ln}^{(\varepsilon)}] = [B_1][\Psi_N^{(u)}] \tag{7.189}$$

The strain modes of all the other terms (with the matrix power function $\xi^{[S_i]-0.5s[I]}$) in Eq. (7.186a) are obtained by using Eq. (6.92) with Eqs. (7.186a) and (7.187)

$$[\Psi_i^{(\varepsilon)}] = [B_1][\Psi_i^{(u)}]([S_i] - 0.5(s-2)[I]) + [B_2][\Psi_i^{(u)}] \qquad (i = 1, 2, \ldots, 2N) \tag{7.190}$$

Substituting Eqs. (7.185) and (7.187) into Eq. (6.92) and using Eqs. (7.189) and (7.190), the strain field of the S-domain is expressed as

$$\{\varepsilon\} = \delta_{\ln}\xi^{-1}[\Psi_{\ln}^{(\varepsilon)}]\{c_{N+1}\} + \sum_{i=i_B}^{N+i_B-1} [\Psi_i^{(\varepsilon)}]\xi^{[S_i]-0.5s[I]}\{c_i\} \tag{7.191}$$

with $\delta_{\ln}$ and i_B defined in Eqs. (7.186a) and (7.186b), respectively.

Following Eq. (3.71) in Section 3.7, the stress modes are defined as

$$[\Psi_{\ln}^{(\sigma)}] = [D][\Psi_{\ln}^{(\varepsilon)}] \tag{7.192a}$$
$$[\Psi_i^{(\sigma)}] = [D][\Psi_i^{(\varepsilon)}] \tag{7.192b}$$

with the elasticity matrix of the material $[D]$. The stress field is expressed as

$$\{\sigma\} = \delta_{\ln}\xi^{-1}[\Psi_{\ln}^{(\sigma)}]\{c_{N+1}\} + \sum_{i=i_B}^{N+i_B-1} [\Psi_i^{(\sigma)}]\xi^{[S_i]-0.5s[I]}\{c_i\} \tag{7.193}$$

It can be regarded as a superposition of the stress modes.

7.8.3 Shape Functions of Polytope Elements

It is shown in Section 3.7.3 and Example 3.10 that the shape functions of a polygon element satisfying the scaling requirement can be constructed using the solution of displacement field obtained by the eigenvalue method. In this section, the shape functions of polytope elements (polygon in two dimensions and polyhedron in three dimensions) are constructed using the (block-diagonal) Schur decomposition. Only polytope elements that satisfy the scaling requirement (or in other words, the polytope elements are bounded S-domains) are considered.

When the Schur decomposition (Eq. (7.51) with Eqs. (7.52) and (7.53)) is used in the solution procedure, the solution of the displacement functions of the polytope element is given in Eq. (7.76a). The integration constants are determined by Eq. (7.81) using the nodal displacements $\{u_b\}$. Substituting Eq. (7.81) into Eq. (7.76a) results in

$$\{u(\xi)\} = [V_b^{(u)}]\xi^{[S_b]-0.5(s-2)[I]}[V_b^{(u)}]^{-1}\{u_b\} \tag{7.194}$$

The variation of displacements along the circumferential directions is represented by interpolating the displacement functions $\{u(\xi)\}$ using the shape functions $[N_u]$ defined on the surface elements. Substituting Eq. (7.194) into Eq. (6.87) leads to

$$\{u\} = [N_u][V_b^{(u)}]\xi^{[S_b]-0.5(s-2)[I]}[V_b^{(u)}]^{-1}\{u_b\} \tag{7.195}$$

The S-element is regarded as being composed of pyramid sectors obtained by scaling individual surface elements. The interpolation in Eq. (7.195) is defined in each pyramid sector covered by scaling one surface element (see Section 3.7.3). Only the corresponding rows of $[V_b^{(u)}]$ are involved. The superscript e is omitted for simplicity in notation. The displacement field inside the S-element is expressed as the interpolation of the nodal displacements $\{u_b\}$

$$\{u\} = [N_V]\{u_b\} \tag{7.196}$$

where the shape functions of the S-element are equal to

$$[N_V] = [N_u][V_b^{(u)}]\xi^{[S_b]-0.5(s-2)[I]}[V_b^{(u)}]^{-1} \tag{7.197}$$

On the boundary $\xi = 1$, the shape functions $[N_V]$ of the S-element are equal to the shape functions of the elements on the boundary.

The above derivations can also be performed with the block-diagonal Schur decomposition in Eq. (7.108). The solution of the displacement functions of the S-element is given in Eq. (7.153a) in matrix form with the submatrices of the Schur decomposition are defined in Eqs. (7.123) and (7.125). Substituting the integration constants in Eq. (7.182) into Eq. (7.153a) leads to

$$\{u(\xi)\} = [\Psi_b^{(u)}]\xi^{[S_b]-0.5(s-2)[I]}[\Psi_b^{(u)}]^{-1}\{u_b\} \tag{7.198}$$

The shape functions of the S-element are expressed as

$$[N_V] = [N_u][\Psi_b^{(u)}]\xi^{[S_b]-0.5(s-2)[I]}[\Psi_b^{(u)}]^{-1} \tag{7.199}$$

Note that the matrix $[S_b]$ is block-diagonal (Eq. (7.123)).

7.9 Body Loads

A constant body load applied to an S-element is considered in the eigenvalue method in Section 7.10. The derivation can be straightforwardly extended to the present solution procedure based on Schur decomposition.

The equation of consistent nodal forces obtained using the virtual work principle for a body load, Eq. (7.208), is reproduced for easy reference

$$\{F_B\} = \int_V [N_V]^T \{b\} dV \tag{7.200}$$

The summation symbol $\sum_e$ standing for the finite element assembly is omitted for conciseness in notation. It is implied that the operations are performed element-by-element on the surface and assembled for the S-element. The shape functions is defined in Eq. (7.197) and its transpose is equal to

$$[N_V]^T = [V_b^{(u)}]^{-T} \xi^{[S_b]^T - 0.5(s-2)[I]} [V_b^{(u)}]^T [N_u]^T \tag{7.201}$$

The infinitesimal volumes dV in two and three dimensions are given in Eqs. (2.47) and (6.73), respectively. They are expressed as

$$dV = \xi^{s-1} |J_b| d\xi dA \qquad \text{with} \quad dA = \begin{cases} d\eta & \text{in 2D} \\ d\eta d\zeta & \text{in 3D} \end{cases} \tag{7.202}$$

where A denotes domain of integration over a surface element in terms of the local coordinates η and ζ.

Substituting Eqs. (7.201) and (7.202) into Eq. (7.200) and simplifying yields

$$\{F_B\} = [V_b^{(u)}]^{-T} \int_A \int_0^1 \xi^{[S_b]^T + 0.5s[I]} [V_b^{(u)}]^T [N_u]^T \{b\} |J_b| d\xi dA \tag{7.203}$$

Grouping the variables for the integration over the surface element results in

$$\{F_B\} = [V_b^{(u)}]^{-T} \int_0^1 \xi^{[S_b]^T + 0.5s[I]} [V_b^{(u)}]^T \int_A [N_u]^T \{b\} |J_b| dA d\xi \tag{7.204}$$

When the body force is constant over the S-domain, the surface integration is expressed as

$$\{F_{B0}\} = \int_A [N_u]^T |J_b| d\eta \{b\} \tag{7.205}$$

and

$$\{F_B\} = [V_b^{(u)}]^{-T} \int_0^1 \xi^{[S_b]^T + 0.5s[I]} d\xi [V_b^{(u)}]^T \{F_{B0}\} \tag{7.206}$$

applies. Performing the integration analytically along the radial coordinate ξ (see Eq. (B.22)) yields the nodal force vector

$$\{F_B\} = [V_b^{(u)}]^{-T} ([S_b]^T + (0.5s + 1)[I]) [V_b^{(u)}]^T \{F_{B0}\} \tag{7.207}$$

Equation (7.207) with $s = 2$ is equivalent to Eq. (3.84) except that the diagonal eigenvalue matrix and eigenvectors in Eq. (3.84) are replaced by the Schur matrix and the transformation matrix. It can easily be shown that the block-diagonal Schur decomposition can be directly used in Eq. (7.207).

7.10   Mass Matrix

In Section 3.10.1, the mass matrix of an S-element is obtained from its shape functions determined by the eigenvalue method. The same procedure can be adopted to use the shape functions determined by the Schur decomposition. Based on the virtual work principle, Eq. (3.94) is obtained and reproduced below

$$[M] = \int_V [N_V]^T \rho [N_V] \mathrm{d}V \tag{7.208}$$

with the summation symbol omitted.

Substituting Eqs. (7.197), (7.201) and (7.202) into Eq. (7.208) and grouping the variables depending on the circumferential directions (η and ζ) yield

$$[M] = [V_b^{(u)}]^{-T} \int_0^1 \xi^{[S_b]^T + [I]} [V_b^{(u)}]^T \underbrace{\int_A [N_u]^T \rho [N_u] |J_b| \mathrm{d}A [V_b^{(u)}] \xi^{[S_b]} \mathrm{d}\xi [V_b^{(u)}]^{-1}}_{[M_0]}$$

$$\tag{7.209}$$

The integration over A is equal to the element coefficient matrix $[M_0]$ in Eq. (3.102) for two dimensions and in Eq. (6.121) for three dimensions. Equation (7.209) is rewritten as

$$[M] = [V_b^{(u)}]^{-T} [\bar{m}] [V_b^{(u)}]^{-1} \tag{7.210}$$

where $[\bar{m}]$ contains the integration over the radial direction ξ

$$[\bar{m}] = \int_0^1 \xi^{[S_b]^T + [I]} [m_0] \xi^{[S_b]} \mathrm{d}\xi \tag{7.211}$$

with the constant matrix

$$[m_0] = [V_b^{(u)}]^T [M_0] [V_b^{(u)}] \tag{7.212}$$

Equation (7.211) is integrated by parts. Using the property of the matrix power function (Eq. (B.23))

$$\left(\xi^{[S_b] + [I]} \right)' = \xi^{[S_b]} ([S_b] + [I]) \tag{7.213}$$

Eq. (7.211) is rewritten as

$$[\bar{m}] = \int_0^1 \xi^{[S_b]^T + [I]} [m_0] \mathrm{d}\xi^{[S_b] + [I]} ([S_b] + [I])^{-1} \tag{7.214}$$

Integration by parts results in

$$[\bar{m}] = \xi^{[S_b]^T + [I]} [m_0] ([S_b] + [I])^{-1} \Big|_0^1 - \int_0^1 \mathrm{d}\xi^{[S_b]^T + [I]} [m_0] \xi^{[S_b] + [I]} ([S_b] + [I])^{-1}$$

$$\tag{7.215}$$

Since all diagonal entries of the real Schur matrix $[S_b]$ are non-negative, all diagonal entries of $[S_b]^T + [I]$ are positive and the matrix function $\xi^{[S_b]^T + [I]}$ vanishing at $\xi = 0$. Considering Eq. (7.213) with $[S_b]^T$ replacing $[S_b]$, Eq. (7.215) is expressed as

$$[\bar{m}] = [m_0]([S_b] + [I])^{-1} - ([S_b]^T + [I]) \int_0^1 \xi^{[S_b]^T} [m_0] \xi^{[S_b] + [I]} d\xi ([S_b] + [I])^{-1}$$

(7.216)

The remaining integration is simply $[\bar{m}]$ as defined in Eq. (7.211). Post-multiplying Eq. (7.216) with $([S_b] + [I])$ and simplifying leads to a Lyapunov equation

$$([S_b]^T + [I])[\bar{m}] + [\bar{m}]([S_b] + [I]) = [m_0]$$

(7.217)

As $[S_b]$ is a Schur matrix, Eq. (7.217) is solved by a simple back substitution (Bartels and Stewart, 1972). As the real parts of the eigenvalues of $[S_b]$ are non-negative, a solution of Eq. (7.217) always exists. With the solution of $[\bar{m}]$, the mass matrix is obtained from Eq. (7.210).

The mass matrix can also be obtained when the block-diagonal Schur decomposition is used in determining the shape functions of the S-element. All the equations hold after replacing the matrix $[V]$ with the transformation matrix $[\Psi]$ of the block-diagonal Schur decomposition and using the block-diagonal Schur matrix $[S_b]$.

In the initial stage of the development of the scaled boundary finite element method, the mass matrix (Section 10.3, Wolf and Song, 1996) is obtained from the scaled boundary finite element equation in dynamic stiffness matrix (Eq. (6.233)). It is shown below that this approach and the above approach using the shape functions lead to the same mass matrix.

For easy reference, Eq. (6.233) is reproduced below

$$([S(\omega)]^T - [E_1])[E_0]^{-1}([S(\omega)]^T - [E_1]^T) - [E_2] + [S(\omega)]^T$$
$$+ \omega[S(\omega)]^T,_\omega + \omega^2[M_0] = 0$$

(7.218)

To determine the mass matrix $[M]$ of a bounded S-element, the low-frequency behaviour of this equation is addressed. The dynamic stiffness matrix $[S(\omega)]$ is approximated by

$$[S(\omega)] = [K] - \omega^2[M]$$

(7.219)

where terms of order in ω equal to or higher than 4 are neglected and this expansion is valid for low frequencies. The static stiffness matrix $[K]$ is obtained from Eq. (7.79) and satisfies Eq. (6.231).

Substituting Eq. (7.219) into Eq. (7.218) leads to a constant term independent of ω, a term in ω^2 and higher-order terms in ω which are neglected. The constant term, which is equal to the left-hand side of Eq. (6.231), vanishes. The coefficient matrix of the remaining term in ω^2 is written as

$$\left(([K] - [E_1])[E_0]^{-1} + 0.5s[I]\right)[M]$$
$$+ [M]\left([E_0]^{-1}([K] - [E_1]^T) + 0.5s[I]\right) - [M_0] = 0$$

(7.220)

This is a Lyapunov equation with the mass matrix $[M]$ as the unknown. It can be simplified by using the intermediate results of the Schur decomposition, i.e. Eq. (7.31) with matrix $[Z_p]$ in Eq. (7.3), the partition of the Schur matrix $[S]$ in Eq. (7.52) and the partition of transformation matrix $[V]$ in Eq. (7.53). These equations are grouped together and expressed as

$$
\begin{bmatrix} -[E_0]^{-1}[E_1]^T + 0.5(s-2)[I] & [E_0]^{-1} \\ [E_2] - [E_1][E_0]^{-1}[E_1]^T & [E_1][E_0]^{-1} - 0.5(s-2)[I] \end{bmatrix} \begin{bmatrix} [V_b^{(u)}] & [V_u^{(u)}] \\ [V_b^{(q)}] & [V_u^{(q)}] \end{bmatrix}
$$

$$
= \begin{bmatrix} [V_b^{(u)}] & [V_u^{(u)}] \\ [V_b^{(q)}] & [V_u^{(q)}] \end{bmatrix} \begin{bmatrix} [S_b] & * \\ 0 & [S_u] \end{bmatrix} \tag{7.221}
$$

The first submatrix of the product in Eq. (7.221), postmultiplied by $[V_b^{(u)}]^{-1}$, leads to

$$
[E_0]^{-1}([K] - [E_1]^T) + 0.5(s-2)[I] = [V_b^{(u)}][S_b][V_b^{(u)}]^{-1} \tag{7.222}
$$

with $[K]$ given in Eq. (7.79). Substituting Eq. (7.222) in Eq. (7.220) results in

$$
[V_b^{(u)}]^{-T}[S_b]^T[V_b^{(u)}]^T[M] + [M][V_b^{(u)}][S_b][V_b^{(u)}]^{-1} + 2[M] = [M_0] \tag{7.223}
$$

Premultiplying Eq. (7.223) by $[V_b^{(u)}]^T$ and postmultiplying by $[V_b^{(u)}]$ yields

$$
([S_b]^T + [I])[V_b^{(u)}]^T[M][V_b^{(u)}] + [V_b^{(u)}]^T[M][V_b^{(u)}]([S_b] + [I]) = [V_b^{(u)}]^T[M_0][V_b^{(u)}] \tag{7.224}
$$

Using Eqs. (7.210) and (7.212), Eq. (7.224) can be written as Eq. (7.217). Therefore, the consistent mass matrix of the S-element obtained using the shape functions is a low frequency expansion of the dynamic stiffness matrix.

7.11 Remarks

The functions to perform the Schur decomposition are available in MATLAB and LAPACK. In MATLAB, the built-in function `schur` computes the Schur decomposition and the function `ordschur` rearranges the eigenvalues in a specified order. The columns of the transformation matrix form the basis vectors of the solution.

It is worthwhile noting the following remarks:

1) The stiffness matrix can be determined by directly solving the Riccati equation (6.232). When no eigenvalues of 0 exist, a procedure to determine the solution of the Riccati equation by Schur decomposition is presented in Laub (1979). This procedure was used in the early development of the scaled boundary finite element method (Song and Wolf, 1997). The solution for stiffness matrix is identical to Eq. (7.79). When eigenvalues of 0 exist, a small positive value is added to the diagonal entries of the coefficient matrix $[E_2]$ so that the eigenvalues of 0 become pairs of small negative and positive numbers. The treatment of eigenvalues of 0 in Section 7.3.2 can be regarded as an extension of the solution procedure in Laub (1979).

2) The solutions of the stiffness matrix obtained by the Schur decomposition and by the eigenvalue decomposition are analytically equal to each other when the eigenvectors are all independent. This can be illustrated by considering the solutions of the bounded S-element in Eqs. (7.79) and (3.25).

The Schur decomposition in Eq. (7.31) is written for the basis vectors of the solution of the bounded S-element (Eq. (7.76)) as

$$[Z_p] \begin{bmatrix} [V_b^{(u)}] \\ [V_b^{(q)}] \end{bmatrix} = \begin{bmatrix} [V_b^{(u)}] \\ [V_b^{(q)}] \end{bmatrix} [S_b] \tag{7.225}$$

Assuming that the eigenvectors of the Schur matrix $[S_b]$ are independent, the eigenvalue decomposition of $[S_b]$ is expressed as

$$[S_b] = [Y_b] \langle \lambda_b \rangle [Y_b]^{-1} \tag{7.226}$$

where $[Y_b]$ is the matrix of eigenvectors. Substituting Eq. (7.226) into Eq. (7.225) leads to

$$[Z_p] \begin{bmatrix} [\Phi_b^{(u)}] \\ [\Phi_b^{(q)}] \end{bmatrix} = \begin{bmatrix} [\Phi_b^{(u)}] \\ [\Phi_b^{(q)}] \end{bmatrix} \langle \lambda_b \rangle \tag{7.227}$$

The eigenvectors of $[Z_p]$ are obtained as

$$\begin{bmatrix} [\Phi_b^{(u)}] \\ [\Phi_b^{(q)}] \end{bmatrix} = \begin{bmatrix} [V_b^{(u)}] \\ [V_b^{(q)}] \end{bmatrix} [Y_b] \tag{7.228}$$

Inversely,

$$\begin{bmatrix} [V_b^{(u)}] \\ [V_b^{(q)}] \end{bmatrix} = \begin{bmatrix} [\Phi_b^{(u)}] \\ [\Phi_b^{(q)}] \end{bmatrix} [Y_b]^{-1} \tag{7.229}$$

applies. Substituting Eq. (7.229) into Eq. (7.79) results in

$$[K] = [V_b^{(q)}][V_b^{(u)}]^{-1} = \left([\Phi_b^{(q)}][Y_b]^{-1} \right) \left([\Phi_b^{(u)}][Y_b]^{-1} \right)^{-1}$$
$$= [\Phi_b^{(q)}][\Phi_b^{(u)}]^{-1} \tag{7.230}$$

which is the solution of the stiffness matrix by the eigenvalue method, Eq. (3.25).

Following the above derivation, it is easy to show that the basis vectors of any well-defined similarity transformation of the Schur matrix $[S_b]$ can be used in computing the stiffness matrix and the results will be analytically identical. For example, the case of a block-diagonal Schur transformation is shown in Section 7.7.2.

3) For the point of view of numerical analysis, the solution based on the Schur decomposition with orthogonal transformation matrix is the most robust, accurate and efficient. It is sufficient for computing the stiffness matrix of an S-element and the displacements and stresses at the boundary and scaling centre. When it is required to separate certain displacement and stress modes (such as in computing the stress intensity factors), the block-diagonal Schur decomposition is required. If the eigenvalues of diagonal blocks in the block-diagonal Schur decomposition (Eq. (7.104)) are well separated, the condition of the transformation matrix $[\Psi]$ in Eq. (7.106) is

much better than that of the eigenvector matrix $[\Phi]$ in Eq. (3.12). The stiffness matrix obtained by using the block-diagonalized Schur form is less susceptible to numerical errors.

7.12 Examples

Only three examples are presented in this section to illustrate that the solution procedure based on Schur decomposition and the matrix function is able to robustly reproduce logarithmic functions in the solution. All the examples in the following chapters are computed using this solution procedure.

7.12.1 Circular Cavity in Full-plane

The plane strain problem of a circular cavity in a full-plane is shown in Figure 7.1. The radius of the cavity is r_0. The material properties of the full-plane are shear modulus G and Poisson's ratio $v = 0.2$. When a vertical force P is applied to a full-plane without the cavity at a location corresponding to the centre of the cavity, the analytical solution is available, for example, known as the fundamental solution of the boundary element method (Brebbia et al., 1984). The external load on the cavity wall is chosen as the traction on a surface in the full-plane matching the cavity

$$p_x(r, \theta) = \frac{P}{2\pi(1 - v)r} \cos(\theta) \sin(\theta) \tag{7.231a}$$

$$p_y(r, \theta) = \frac{P}{4\pi(1 - v)r} \left((1 - 2v) + 2\sin^2(\theta)\right) \tag{7.231b}$$

where r and θ are the polar coordinates indicated in Figure 7.1. The exact solution for displacement components is

$$u_x(r, \theta) = \frac{P}{8\pi(1 - v)G} \cos(\theta) \sin(\theta) \tag{7.232a}$$

$$u_y(r, \theta) = -\frac{P}{8\pi(1 - v)G} \left((3 - 4v) \ln(r) - \sin^2(\theta)\right) \tag{7.232b}$$

Figure 7.1 Circular cavity in full-plane subject to vertical loading.

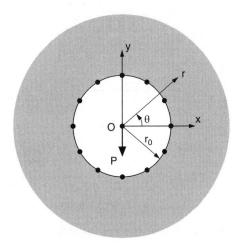

and for stress components is

$$\sigma_x(r,\theta) = -\frac{P}{4\pi(1-v)r}\left(-(1-2v) + 2\cos^2(\theta)\right)\sin(\theta) \tag{7.233a}$$

$$\sigma_y(r,\theta) = -\frac{P}{4\pi(1-v)r}\left((1-2v) + 2\sin^2(\theta)\right)\sin(\theta) \tag{7.233b}$$

$$\tau_{xy}(r,\theta) = -\frac{P}{4\pi(1-v)r}\left((1-2v) + 2\sin^2(\theta)\right)\cos(\theta) \tag{7.233c}$$

The fundamental solution in Eq. (7.232) has two terms in the polar coordinate r: one is $\ln(r)$ and the another is r^0, i.e. independent of r. Comparing Eq. (7.232) with the scaled boundary finite element solution in Eq. (7.185), it is found that the fundamental solution includes the zero-eigenvalue terms only. The translational motions are undetermined.

This problem is modelled as a single S-element. The scaling centre O is chosen at the centre of the circular cavity (Figure 7.1). Two uniformly spaced meshes of 4-node elements are used to discretize the cavity wall. The coarse mesh has 4 line elements as shown in Figure 7.1. The fine mesh consists of 8 line elements.

The exact solution for the displacement distribution on the cavity wall (Eq. (7.232)) is shown in Figure 7.2 by the continuous solid lines. The scaled boundary finite element solution for the nodal displacements on the wall is illustrated as dots for the coarse mesh and crosses for the fine mesh. The undetermined translational motions are constrained in all the solutions by shifting the displacements so that the averages of the displacements on the cavity wall in horizontal and vertical directions are equal to zero.

The comparison of the vertical displacement along the radial direction $\theta = 90°$ in Figure 7.3 illustrates that the matrix function solution accurately represents the logarithmic function in the exact solution (Eq. (7.232b)). It is observed in Figure 7.2 and 7.3 that the results of displacements agree well with the exact solution even for the coarse mesh with only 4 line elements.

The stresses on the cavity wall are shown in Figure 7.4. The scaled boundary finite element results are evaluated at order 3 Gauss integration points. Again, the coarse

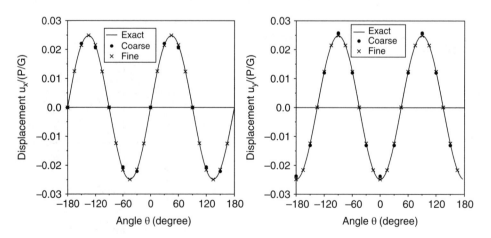

Figure 7.2 Displacement on wall of circular cavity subject to vertical loading.

Figure 7.3 Vertical displacement along the radial line at $\theta = 90°$ of circular cavity subject to vertical loading.

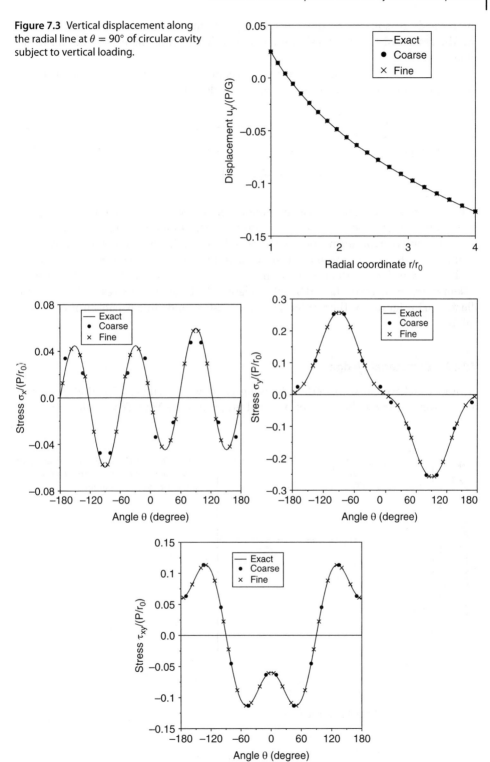

Figure 7.4 Stresses on wall of circular cavity subject to vertical loading.

mesh yields reasonably accurate results considering that σ_x is much smaller than σ_y. The results of the fine mesh are indistinguishable from the exact solution in the plots.

An error norm for the stresses at the Gauss points on a circle with a radius r (N_{gp} number of Gauss points for stress calculation) is defined as

$$\|e(r)\| = \frac{\displaystyle\sum_{i=1}^{N_{gp}} \left((\sigma_{xi} - \tilde{\sigma}_{xi})^2 + (\sigma_{yi} - \tilde{\sigma}_{yi})^2 + (\tau_{xyi} - \tilde{\tau}_{xyi})^2\right)}{\displaystyle\sum_{i=1}^{N_{gp}} \left(\sigma_{xi}^2 + \sigma_{yi}^2 + \tau_{xyi}^2\right)} \tag{7.234}$$

where $\tilde{\sigma}_{xi}$, $\tilde{\sigma}_{yi}$ and $\tilde{\tau}_{xyi}$ are the scaled boundary finite element solution for Gauss point stresses at the radial coordinate r. σ_{xi}, σ_{yi} and τ_{xyi} are the exact solutions at the same locations. The stress error norms for the coarse and fine meshes are plotted in Figure 7.5 as a function of the radial coordinate. It shows that the errors decay within a layer adjacent to the cavity wall and attend to constants. Along the radial line $\theta = 90°$, the stresses σ_x and σ_y ($\tau_{xy} = 0$) are evaluated by interpolating the Gauss point stresses. The result is illustrated in Figure 7.6. The results of the fine mesh are indistinguishable from the exact solution.

7.12.2 Bi-material Wedge

A bi-material wedge with perfect bonding is shown in Figure 7.7a. Both materials are isotropic. The combination of material properties are Young's moduli $E_1/E_2 = 10$

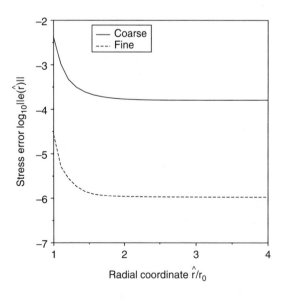

Figure 7.5 Variation of stress error norm in radial direction for circular cavity subject to vertical loading.

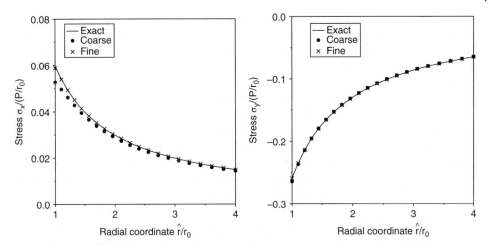

Figure 7.6 Stresses along radial line $\theta = 90°$ for circular cavity subject to vertical loading.

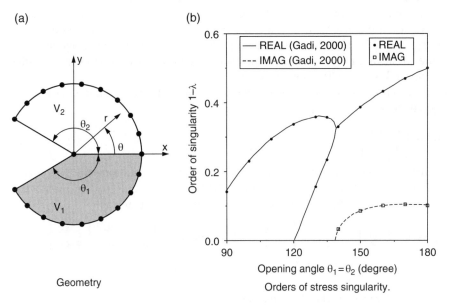

(a) Geometry

(b) Orders of stress singularity.

Figure 7.7 Bi-material wedge with traction-free wedge surfaces.

and Poisson's ratios $v_1 = v_2 = 0.2$. The wedge is under plane strain condition. The logarithmic stress singularity for this problem is studied analytically by Gadi et al. (2000). The asymptotic solution of the singular stresses is expressed as

$$\{\sigma(r,\theta)\} = \bar{K}_1 r^{-(1-\lambda)}\{h^1(\theta)\} + \bar{K}_2 r^{-(1+\lambda)}(-\ln(r)\{h^1(\theta)\} + \{h^2(\theta)\}) \qquad (7.235)$$

where $\bar{K}_1$ and $\bar{K}_2$ are two normalized stress intensity factors. The exponent $1 - \lambda$ is the order of stress singularity. The stresses tend to infinity at the vertex of the wedge (polar coordinate $r = 0$) when $1 - \lambda > 0$, i.e. $\lambda < 1$. The functions $\{h^1(\theta)\}$ and $\{h^2(\theta)\}$ are the stress modes describing the variation in the angular direction θ.

The bi-material wedge is modelled as one S-element. The scaling centre is chosen at the vertex. The boundary of the wedge is discretized with 6 cubic elements. The surfaces forming the wedge vertex and the material interface are obtained by scaling the nodes on the boundary. They are not discretized. The analytical solution of the order of stress singularity $1 - \lambda$ is presented in Gadi et al. (2000). The orders of stress singularity are calculated at several opening angles $\theta_1 = \theta_2$ using the scaled boundary finite element method. The real and imaginary parts are plotted in Figure 7.7b as dots and squares, respectively. The results agree well with the analytical solutions.

At the opening angle $\theta_1 = \theta_2 = 138.7718767°$, where the two real orders of singularity are transiting to a complex conjugate pair, a logarithmic term occurs in the solution with $1 - \lambda = 0.3214741102$. In the scaled boundary finite element solution, the block-diagonal Schur decomposition is performed. The diagonal block of the Schur matrix leading to stress singularities is isolated. It is denoted as $[s]$ in the following. The eigenvalues of $[s]$ satisfy $0 < \mathrm{Re}(\lambda) < 1$. The terms of $[s]$ are listed in Table 7.1 for three uniform meshes on the boundary. The diagonal terms s_{11} and s_{22}, which are equal to the eigenvalues, are converging to the exact value of $\lambda = 0.678526$. As the difference between s_{11} and s_{22} is three orders of magnitude smaller than the off-diagonal term s_{12} for the mesh of 24 elements, the eigenvalue method will lead to nearly parallel eigenvectors and results with poor accuracy.

Although it is not performed in the actual calculation in the matrix function solution, the diagonal block $[s]$ can be approximated by

$$[s] = \begin{bmatrix} \lambda & s_{12} \\ 0 & \lambda \end{bmatrix} \tag{7.236}$$

for the purpose of comparison with the analytical solution. Denoting the two corresponding columns in the matrix $[\Psi^{(u)}]$ in Eq. (7.185) as $\{\psi_1\}$ and $\{\psi_2\}$ and following Eq. (7.22), the displacements leading to singular stresses are written as

$$\{u(\xi)\} = [\{\psi_1\}\ \{\psi_2\}]\xi^{[s]} \begin{Bmatrix} c_1 \\ c_2 \end{Bmatrix} = [\{\psi_1\}\ \{\psi_2\}]\xi^\lambda \begin{bmatrix} 1 & s_{12}\ln\xi \\ 0 & 1 \end{bmatrix} \begin{Bmatrix} c_1 \\ c_2 \end{Bmatrix}$$

$$= c_1\xi^\lambda\{\psi_1\} + c_2\xi^\lambda (s_{12}\{\psi_1\}\ln\xi + \{\psi_2\}) \tag{7.237}$$

Table 7.1 Matrix $[s]$ with eigenvalues leading to stress singularities in bi-material wedge.

Elements	s_{11}	s_{12}	s_{21}	s_{22}
6	0.6801	0.2654	0	0.6769
12	0.6787	−0.2734	0	0.6783
24	0.6786	0.2764	0	0.6785

As the off-diagonal term s_{12} appears in the product with $\{\psi_1\}$ only, the difference in the sign of s_{12} observed in Table 7.1 for varying number of elements is insignificant by itself to the results. The singular stresses are obtained using Eq. (7.193). Transforming the scaled boundary coordinates to polar coordinates using Eq. (2.6), the stresses are expressed as

$$\{\sigma(r,\theta)\} = c_1 r^{-(1-\lambda)}\{\psi_{\sigma1}(\theta)\} + c_2 r^{-(1-\lambda)}(-s_{12}\ln(r)\{\psi_{\sigma1}(\theta)\} + \{\psi_{\sigma2}(\theta)\}) \quad (7.238)$$

The stress modes $\{\psi_{\sigma1}(\theta)\}$ and $\{\psi_{\sigma2}(\theta)\}$ represent the angular distribution of the singular stresses. They are related to the stress modes in the analytical solution (Eq. (7.235)) by

$$\{h^1(\theta)\} = a_1\{\psi_{\sigma1}(\theta)\} \quad (7.239a)$$

$$\{h^2(\theta)\} = \frac{a_1}{s_{12}}(\{\psi_{\sigma2}(\theta)\} + a_2\{h_1(\theta)\}) \quad (7.239b)$$

The two constants a_1 and a_2 are determined by normalizing the stress modes as in the analytical solution

$$h^1_{\theta\theta}(\theta = 0) = 1 \quad (7.240a)$$

$$h^2_{\theta\theta}(\theta = 0) = 0 \quad (7.240b)$$

$\{\psi_{\sigma1}(\theta)\}$ and $\{\psi_{\sigma2}(\theta)\}$ are calculated at order 3 Gauss integration points. Their values at $\theta = 0$ are obtained by interpolation. The normalized stress modes $\{h^1(\theta)\}$ and $\{h^2(\theta)\}$ calculated at the Gauss points for the mesh with 6 line elements are shown in Figure 7.8. The results agree well with the analytical solutions (Gadi et al., 2000).

7.12.3 Interface Crack in Anisotropic Bi-material Full-plane

A crack along the interface of two anisotropic half-planes is shown in Figure 7.9. A concentrated force P is applied perpendicularly to and on the upper crack face at a distance a from the crack tip. The analytical solution for this problem is available in Ma and Luo (1996). The material properties used in this example are: $E_{11} = 0.09852$, $E_{22} = 0.58140$, $G_{12} = 0.07813$, $v_{12} = 0.0857$ for material 1 and $E_{11} = 0.06452$, $E_{22} = 1.70358$, $G_{12} = 0.08696$, $v_{12} = 0.02129$ for material 2. Plane strain condition is considered. The full-plane is divided into a bounded S-element and an unbounded S-element by introducing a circular boundary of radius a with its centre at the crack tip as shown in Figure 7.9. This circular boundary is discretized with 12 cubic elements. Both the bounded S-element defined with $0 \leq \xi \leq 1$ and the unbounded S-element defined with $1 \leq \xi < \infty$ are modelled with the same mesh of line elements on boundary. A common scaling centre is chosen at the crack tip. The block-diagonalized Schur decomposition is performed only once and used for the solution of both S-elements (Eq. (7.153b) for the bounded S-element and Eq. (7.175a) for the unbounded S-element). The stress singularity at the crack tip in the bounded S-element and the logarithmic displacement distribution in the unbounded S-element

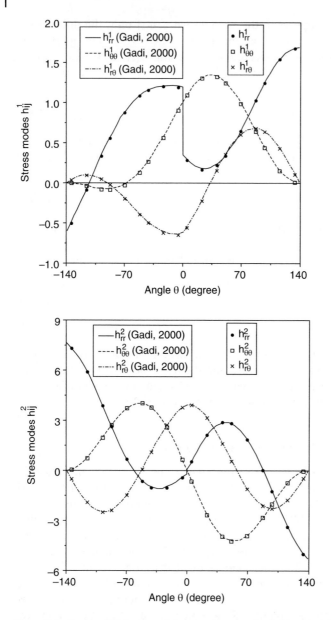

Figure 7.8 Angular distribution of singular stress modes in bi-material wedge at opening angle $\theta_1 = \theta_2 = 138.7718767°$.

are represented by analytical functions in the scaled boundary finite element solution. The exact value of the order of stress singularity $1 + \lambda = 0.5 + i0.012$ is obtained. The angular distribution of stresses at $r = 0.01a$ is very close to that of the analytical solution (Ma and Luo, 1996) as shown in Figure 7.10.

Figure 7.9 Interface crack in anisotropic bi-material full-plane.

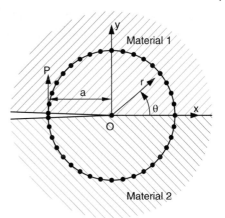

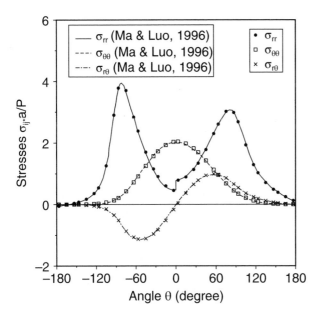

Figure 7.10 Angular distribution of stresses at $r = 0.01a$ for interface crack in anisotropic bi-material full-plane.

7.13 Summary

In this chapter, the Schur decomposition is introduced to replace the eigenvalue decomposition in the solution of the scaled boundary finite element equation. The solution is expressed as a matrix power function. This solution procedure is numerically robust.

1) In computing the stiffness matrix, only the transformation matrix of the Schur decomposition is used. The real Schur matrix is employed to select the basis vectors of the solution that satisfies the boundary conditions at the scaling centre (for

bounded domains) or at the infinity (for unbounded domains), but is not explicitly involved in the computation. The displacement solution appears in the derivation and is not explicitly evaluated.

2) In the solution of a two-dimensional elasticity problem, a block of four eigenvalues of 0 exists (Eq. (7.33)). This diagonal block and the corresponding columns of the transformation matrix span the solutions of the bounded and unbounded S-elements. The additional transformation in Section 7.3.2 is introduced to separate the basis vectors.

3) When the static stiffness matrix of an unbounded S-element is of interest, a block diagonalization of the Schur matrix (Section 7.6) is performed to separate the solution of the bounded S-element from that of the unbounded S-element. The conditions of finiteness of displacements inside the S-element can be enforced simply and analytically. If the solution is sought only for the bounded domain ($0 \leq \xi \leq 1$), the block diagonalization is unnecessary, although it can still be performed.

8

High-order Elements

In the scaled boundary finite element method, a problem domain is meshed into S-elements, in the same manner as in the finite element method. An S-element may have arbitrary number of faces and edges and can assume more complex shapes than standard finite elements do. Only the boundary of the S-element is discretized. The volume is described by scaling the elements on boundary. Combining the versatility in shape with the simplicity in generating a boundary mesh, the S-elements significantly reduce the efforts required for mesh generation in comparison with the standard finite elements.

To improve the accuracy of a scaled boundary finite element simulation, the mesh of the S-elements can be refined. Figure 8.1 illustrates the refinement schemes of an S-element mesh using the simple initial mesh in Figure 8.1a:

1) $h-$refinement (Figure 8.1b): Reducing the sizes (h) of the S-elements. This refinement scheme leads to an increase in the number of S-elements.
2) $p-$refinement (Figure 8.1c): Increasing the orders (p) of the elements on the boundaries of the S-elements. The number of nodes increases without changing the sizes and number of S-elements.
3) h_s-refinement (Figure 8.1d): Reducing the sizes (h) of the elements on the boundaries of the S-element by subdivision. The sizes and number of S-elements do not change. This can be regarded as an $h-$refinement performed at the S-element level.

The first two ($h-$ and $p-$) schemes are standard in the finite element method. The last scheme (h_s-refinement) is an extra possibility provided by the scaled boundary finite element method. The three refinement schemes can also be mixed in a mesh refinement. Two examples of mixed refinement are depicted in Figures 8.1e and 8.1f. This illustrates the flexibility of S-elements in mesh generation in addition to those shown in Chapter 4.

High-order elements for $p-$refinement are presented in this chapter. Their formulations are based on the Lagrange interpolation of displacements. The $p-$refinement scheme, when used with spectral elements, has been shown to be highly effective in improving the accuracy of a scaled boundary finite element analysis especially for fracture problems where the stress fields vary rapidly around the crack tips.

The Scaled Boundary Finite Element Method: Introduction to Theory and Implementation,
First Edition. Chongmin Song.
© 2018 John Wiley & Sons Ltd. Published 2018 by John Wiley & Sons Ltd.
Companion website: www.wiley.com/go/author/scaledboundaryfiniteelementmethod

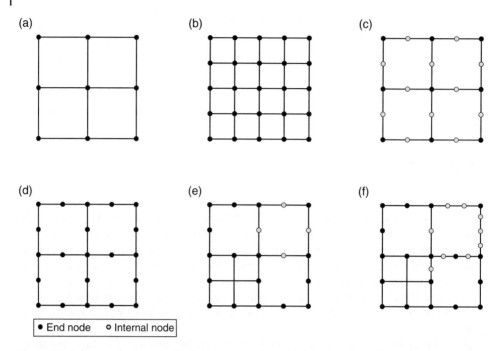

Figure 8.1 Refinement schemes of S-element mesh. The solid dots (•) indicate the end nodes and the circles (∘) the internal nodes of line elements (a) Initial mesh. (b) *h*-refinement by subdividing S-elements into smaller ones. (c) *p*-refinement by increasing the order of elements on boundary of S-elements. (d) h_s-refinement by subdividing the elements on the boundary of S-elements into smaller ones. (e) and (f) examples of mixed types of refinement.

8.1 Lagrange Interpolation

Lagrange interpolation provides a straightforward technique to construct explicit shape functions of high-order elements. The shape functions expressed as Lagrange polynomials are simple and efficient for differentiation and integration.

In Lagrange interpolation, as depicted in Figure 8.2, the polynomial of the least degree is constructed to interpolate a given set of distinct points. To interpolate $p + 1$ points $(x_1, u_1), (x_2, u_2), \ldots, (x_p, u_p), (x_{p+1}, u_{p+1})$, the least degree of the polynomial is equal to or lower than p. Since the interpolating polynomial of the least degree is unique, it is

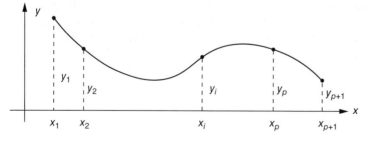

Figure 8.2 Lagrange interpolation.

sufficient to find one expression of the polynomial. The general formula of the Lagrange interpolating polynomial is expressed as

$$u(x) = \sum_{i=1}^{p=1} N_i(x)u_i = N_1(x)u_1 + N_2(x)u_2 + \ldots + N_p(x)u_p + N_{p+1}(x)u_{p+1} \tag{8.1}$$

where $N_i(x)$, ($i = 1, 2, \ldots, p, p+1$) are the Lagrange basis polynomials. It is the same type of interpolation as what is used in an isoparametric finite element. The Lagrange basis polynomials are equivalent to the shape functions.

For Eq. (8.1) to be interpolative,

$$u(x_j) = \sum_{i=1}^{p=1} N_i(x_j)u_i = u_j \tag{8.2}$$

has to hold. For example, Eq. (8.2) is expanded at the first data point ($j = 1$) as

$$N_1(x_1)u_1 + N_2(x_1)u_2 + \ldots + N_p(x_1)u_p + N_{p+1}(x_1)u_{p+1} = u_1$$

Since this equation is valid for arbitrary values of u_i ($i = 1, 2, \ldots, p, p+1$), $N_1(x_1) = 1$ and $N_i(x_1) = 0$ with $i = 2, \ldots, p, p+1$ have to be satisfied. Generalizing the argument to other data points, the Lagrange basis polynomials have the Kronecker delta property

$$N_i(x_j) = \delta_{ij} = \begin{cases} 1 & \text{if } i = j \\ 0 & \text{if } i \neq j \end{cases} \tag{8.3}$$

As $u(x)$ is a combination of the Lagrange basis polynomials (Eq. (8.1)), the Lagrange basis polynomials have to be of degree p or lower.

The Kronecker delta property provides a convenient way to construct the Lagrange basis polynomials. If a polynomial is equal to 0 at a point x_j, i.e. has a root at $x = x_j$, the polynomial must have the factor $(x - x_j)$. The Lagrange basis polynomials that satisfy Eq. (8.3) have p roots. Therefore, they are expressed as

$$N_i(x) = c_i \prod_{k=1,k\neq i}^{p+1} (x - x_k) = c_i(x - x_1)(x - x_2) \ldots (x - x_p)(x - x_{p+1}) \tag{8.4}$$

where c_i is a constant. Using the Kronecker delta property at $x = x_i$

$$N_i(x_i) = c_i \prod_{k=1,k\neq i}^{p+1} (x_i - x_k) = c_i(x_i - x_1) \ldots (x_i - x_{i-1})(x_i - x_{i+1}) \ldots (x_i - x_{p+1}) = 1 \tag{8.5}$$

the constant c_i is obtained as

$$c_i = \prod_{k=1,k\neq i}^{p+1} \frac{1}{(x_i - x_k)} = \frac{1}{(x_i - x_1) \ldots (x_i - x_{i-1})(x_i - x_{i+1}) \ldots (x_i - x_{p+1})} \tag{8.6}$$

Substituting Eq. (8.6) into Eq. (8.4) yields the Lagrange basis functions

$$N_i(x) = \prod_{k=1,k\neq i}^{p+1} \frac{(x - x_k)}{(x_i - x_k)} = \frac{(x - x_1)(x - x_2) \ldots (x - x_p)(x - x_{p+1})}{(x_i - x_1) \ldots (x_i - x_{i-1})(x_i - x_{i+1}) \ldots (x_i - x_{p+1})} \tag{8.7}$$

The functions are p-th degree polynomials and uniquely determined.

It can easily be observed that the Lagrange basis functions are directly applicable to construct the shape functions of high-order elements.

■ Example 8.1 Cubic Element with Uniformly Distributed Nodes

A cubic ($p = 3$) element is shown in Figure 8.3 in its natural coordinate η. The four nodes are uniformly spaced with the nodal coordinates: $\eta_1 = -1$, $\eta_2 = -1/3$, $\eta_3 = 1/3$ and $\eta_4 = 1$. Determine its shape functions directly from the Lagrange basis polynomials in Eq. (8.7) with x replaced by η.

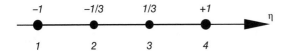

Figure 8.3 A cubic element with uniformly spaced nodes.

At node 1, the shape function $N_1(\eta)$ is obtained from Eq. (8.7) with $i = 1$ as

$$
\begin{aligned}
N_1(\eta) &= \frac{(\eta - \eta_2)(\eta - \eta_3)(\eta - \eta_4)}{(\eta_1 - \eta_2)(\eta_1 - \eta_3)(\eta_1 - \eta_4)} \\
&= \frac{(\eta - (-1/3))(\eta - 1/3)(\eta - 1)}{(-1 - (-1/3))(-1 - 1/3)(-1 - 1)} \\
&= -\frac{1}{16}(9\eta^2 - 1)(\eta - 1) \\
&= \frac{1}{16}(-9\eta^3 + 9\eta^2 + \eta - 1)
\end{aligned}
\tag{8.8a}
$$

Similarly, the shape functions at the other three nodes are expressed as

$$
\begin{aligned}
N_2(\eta) &= \frac{(\eta - \eta_1)(\eta - \eta_3)(\eta - \eta_4)}{(\eta_2 - \eta_1)(\eta_2 - \eta_3)(\eta_2 - \eta_4)} \\
&= \frac{(\eta - (-1))(\eta - 1/3)(\eta - 1)}{(-1/3 - (-1))(-1/3 - 1/3)(-1/3 - 1)} \\
&= \frac{9}{16}(3\eta - 1)(\eta^2 - 1) \\
&= \frac{9}{16}(3\eta^3 - \eta^2 - 3\eta + 1)
\end{aligned}
\tag{8.8b}
$$

$$
\begin{aligned}
N_3(\eta) &= \frac{(\eta - \eta_1)(\eta - \eta_2)(\eta - \eta_4)}{(\eta_3 - \eta_1)(\eta_3 - \eta_2)(\eta_3 - \eta_4)} \\
&= \frac{(\eta - (-1))(\eta - (-1/3))(\eta - 1)}{(1/3 - (-1))(1/3 - (-1/3))(1/3 - 1)} \\
&= -\frac{9}{16}(3\eta + 1)(\eta^2 - 1) \\
&= \frac{9}{16}(-3\eta^3 - \eta^2 + 3\eta + 1)
\end{aligned}
\tag{8.8c}
$$

$$
\begin{aligned}
N_4(\eta) &= \frac{(\eta - \eta_1)(\eta - \eta_2)(\eta - \eta_3)}{(\eta_4 - \eta_1)(\eta_4 - \eta_2)(\eta_4 - \eta_3)} \\
&= \frac{(\eta - (-1))(\eta - (-1/3))(\eta - 1/3)}{(1 - (-1))(1 - (-1/3))(1 - 1/3)}
\end{aligned}
$$

$$= \frac{1}{16}(9\eta^2 - 1)(\eta + 1)$$

$$= \frac{1}{16}(9\eta^3 + 9\eta^2 - \eta - 1) \tag{8.8d}$$

The shape functions are plotted in Figure 8.4. It is observed that the maximum values of $N_2(\eta)$ and $N_3(\eta)$ occur between the nodes and are larger than 1.

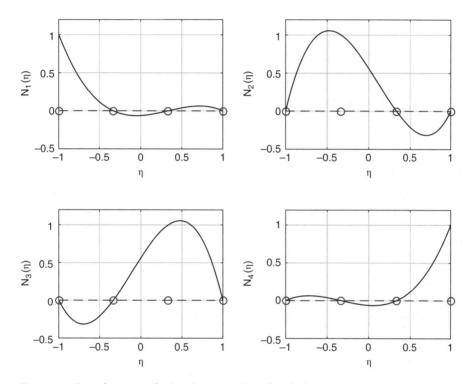

Figure 8.4 Shape functions of cubic elements with uniformly spaced nodes (indicated by circles).

8.2 One-dimensional Spectral Elements

As observed in Example 8.1, when the nodes of a high-order element are uniformed spaced, the maximum value of a Lagrange interpolating polynomial may occur between the nodes. As the number of nodes, i.e. the degree of the interpolating polynomial, increases, the Lagrange interpolating polynomial will exhibit greater oscillation between the nodes and may provide a poor approximation of the solution. In the spectral elements (Canuto et al., 1988), a set of non-uniformly spaced nodes are selected to achieve higher accuracy as the order of elements increases, much like the selection of Gauss points for numerical integrations. The spectral elements are introduced to the scaled boundary finite element method by Vu and Deeks (2006).

The Legendre polynomials (Kreyszig, 2011) are used in selecting the locations of the nodes in the natural coordinate $-1 \leq \eta \leq +1$. The p-th degree Legendre polynomial $P_p(\eta)$ is expressed as

$$P_p(\eta) = \frac{1}{2^p p!} \frac{d^p((\eta^2 - 1)^p)}{d\eta^p} \tag{8.9}$$

It is the solution to Legendre's differential equation

$$\frac{d}{d\eta}\left((1 - \eta^2)\frac{dP_p(\eta)}{d\eta}\right) + p(p + 1)P_p(\eta) = 0 \tag{8.10}$$

The Legendre polynomials can be generated recursively by

$$(p + 1)P_{p+1}(\eta) = (2p + 1)\eta P_p(\eta) - pP_{p-1}(\eta) \tag{8.11}$$

starting from the first two polynomials $P_0(\eta) = 1$ and $P_1(\eta) = \eta$. The first seven Legendre polynomials are listed in Table 8.1. The first-order derivatives are also shown for later use.

8.2.1 Shape Functions

A one-dimensional spectral element is defined in the natural coordinate η ($-1 \leq \eta \leq 1$). An order p spectral element has $p + 1$ nodes called the Gauss-Lobatto-Legendre points. The two end-nodes are located at the two extremities ($\eta = \pm 1$) of the element. The $p - 1$ internal nodes are located at the zeros of the derivative of the Legendre polynomial of degree p

$$\frac{dP_p(\eta)}{d\eta} = 0 \tag{8.12}$$

The locations of all the Gauss-Lobatto-Legendre points satisfy

$$(\eta^2 - 1)\frac{dP_p(\eta)}{d\eta} = 0 \tag{8.13}$$

Table 8.1 Legendre polynomials $P_p(\eta)$ from order 0 to order 6 and their first-order derivatives $dP_p(\eta)/d\eta$.

Order p	$P_p(\eta)$	$dP_p(\eta)/d\eta$
0	1	0
1	η	1
2	$\frac{1}{2}(3\eta^2 - 1)$	3η
3	$\frac{1}{2}(5\eta^3 - 3\eta)$	$\frac{1}{3}(5\eta^2 - 1)$
4	$\frac{1}{8}(35\eta^4 - 30\eta^2 + 3)$	$\frac{5}{2}\eta(7\eta^2 - 3)$
5	$\frac{1}{8}(63\eta^5 - 70\eta^3 + 15\eta)$	$\frac{15}{8}(21\eta^4 - 14\eta^2 + 1)$
6	$\frac{1}{16}(231\eta^6 - 315\eta^4 + 105\eta^2 - 5)$	$\frac{21}{8}\eta(33\eta^4 - 30\eta^2 + 5)$

Table 8.2 Natural nodal coordinates of one-dimensional spectral elements of orders 1–6 and weight factors for Gauss-Lobatto-Legendre quadrature.

Order p	Nodal coordinates (η_i)	Weight factors (w_i)
1	± 1	1
2	± 1, 0	$\dfrac{1}{3}$, $\dfrac{4}{3}$
3	± 1, $\pm\sqrt{\dfrac{1}{5}}$	$\dfrac{1}{6}$, $\dfrac{5}{6}$
4	± 1, $\pm\sqrt{\dfrac{3}{7}}$, 0	$\dfrac{1}{10}$, $\dfrac{49}{90}$, $\dfrac{32}{45}$
5	± 1, $\pm\sqrt{\dfrac{7+2\sqrt{7}}{21}}$, $\pm\sqrt{\dfrac{7-2\sqrt{7}}{21}}$	$\dfrac{1}{15}$, $\dfrac{14-\sqrt{7}}{30}$, $\dfrac{14+\sqrt{7}}{30}$
6	± 1	0.047619047619047619
	± 0.8302238962785669	0.276826047361565948
	± 0.4688487934707142	0.431745381209862623
	0	0.487619047619047619

The natural nodal coordinates of one-dimensional spectral elements of orders 1–6 are listed in Table 8.2. They are symmetric about 0. Up to order $p = 2$ (three-node elements), the nodes are uniformly spaced as those of the standard finite elements. For higher-order elements, the nodes are denser close to the two extremities than in the middle of the element.

The shape functions are Lagrange basis polynomials (Eq. 8.7) expressed in η as

$$N_i(\eta) = \prod_{k=1, k \neq i}^{p+1} \frac{\eta - \eta_k}{\eta_i - \eta_k} \tag{8.14}$$

■ Example 8.2 Fourth-order Spectral Element

Determine the shape functions of a 4th order ($p = 4$) spectral element.

From Eq. (8.12) and Table 8.1, the locations of the three internal nodes of the 4th order spectral element are the zeros of the polynomial

$$\frac{5}{2}\eta(7\eta^2 - 3) = 0$$

Its solutions are equal to

$$\eta = -\sqrt{\frac{3}{7}}, \quad 0, \quad +\sqrt{\frac{3}{7}}$$

Including the two extremities $\eta = -1$ and $+1$, the natural coordinates of the five nodes are

$$\eta_1 = -1, \quad \eta_2 = -\sqrt{\frac{3}{7}}, \quad \eta_3 = 0, \quad \eta_4 = +\sqrt{\frac{3}{7}}, \quad \eta_5 = +1$$

which are listed in Table 8.2. The spectral element is shown in Figure 8.5.

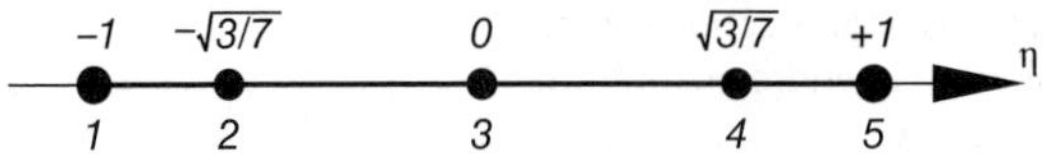

Figure 8.5 A 4th order spectral element.

Following Example 8.1, the shape functions are obtained using Eq. (8.14) and the nodal coordinates as

$$N_1(\eta) = \frac{(\eta - \eta_2)(\eta - \eta_3)(\eta - \eta_4)(\eta - \eta_5)}{(\eta_1 - \eta_2)(\eta_1 - \eta_3)(\eta_1 - \eta_4)(\eta_1 - \eta_5)}$$

$$= \frac{(\eta - (-\sqrt{3/7}))(\eta - 0)(\eta - \sqrt{3/7})(\eta - 1)}{(-1 - (-\sqrt{3/7}))(-1 - 0)(-1 - \sqrt{3/7})(-1 - 1)}$$

$$= \frac{1}{8}\eta(7\eta^2 - 3)(\eta - 1) \tag{8.15a}$$

$$N_2(\eta) = \frac{(\eta - \eta_1)(\eta - \eta_3)(\eta - \eta_4)(\eta - \eta_5)}{(\eta_2 - \eta_1)(\eta_2 - \eta_3)(\eta_2 - \eta_4)(\eta_2 - \eta_5)}$$

$$= \frac{(\eta - (-1))(\eta - 0)(\eta - \sqrt{3/7})(\eta - 1)}{(-\sqrt{3/7} - (-1))(-\sqrt{3/7} - 0)(-\sqrt{3/7} - \sqrt{3/7})(-\sqrt{3/7} - 1)}$$

$$= -\frac{7}{24}\eta(7\eta - \sqrt{21})(\eta^2 - 1) \tag{8.15b}$$

$$N_3(\eta) = \frac{(\eta - \eta_1)(\eta - \eta_2)(\eta - \eta_4)(\eta - \eta_5)}{(\eta_3 - \eta_1)(\eta_3 - \eta_2)(\eta_3 - \eta_4)(\eta_3 - \eta_5)}$$

$$= \frac{(\eta - (-1))(\eta - (-\sqrt{3/7}))(\eta - \sqrt{3/7})(\eta - 1)}{(0 - (-1))(0 - (-\sqrt{3/7}))(0 - \sqrt{3/7})(0 - 1)}$$

$$= \frac{1}{3}(7\eta^2 - 3)(\eta^2 - 1) \tag{8.15c}$$

$$N_4(\eta) = \frac{(\eta - \eta_1)(\eta - \eta_2)(\eta - \eta_3)(\eta - \eta_5)}{(\eta_4 - \eta_1)(\eta_4 - \eta_2)(\eta_4 - \eta_3)(\eta_1 - \eta_5)}$$

$$= \frac{(\eta - (-1))(\eta - (-\sqrt{3/7}))(\eta - 0)(\eta - 1)}{(\sqrt{3/7} - (-1))(\sqrt{3/7} - (-\sqrt{3/7}))(\sqrt{3/7} - 0)(\sqrt{3/7} - 1)}$$

$$= -\frac{7}{24}\eta(7\eta + \sqrt{21})(\eta^2 - 1) \tag{8.15d}$$

$$N_5(\eta) = \frac{(\eta - \eta_1)(\eta - \eta_2)(\eta - \eta_3)(\eta - \eta_4)}{(\eta_5 - \eta_1)(\eta_5 - \eta_2)(\eta_5 - \eta_3)(\eta_5 - \eta_4)}$$

$$= \frac{(\eta - (-1))(\eta - (-\sqrt{1/7}))(\eta - 0)(\eta - \sqrt{1/7})}{(1 - (-1))(1 - (-\sqrt{1/7}))(1 - 0)(1 - \sqrt{1/7})}$$

$$= \frac{1}{8}\eta(7\eta^2 - 3)(\eta + 1) \tag{8.15e}$$

The shape functions are depicted in Figure 8.6. Note the the maximum values of all the shape functions are equal to 1 and occur at the corresponding nodes. This holds for spectral elements of any orders.

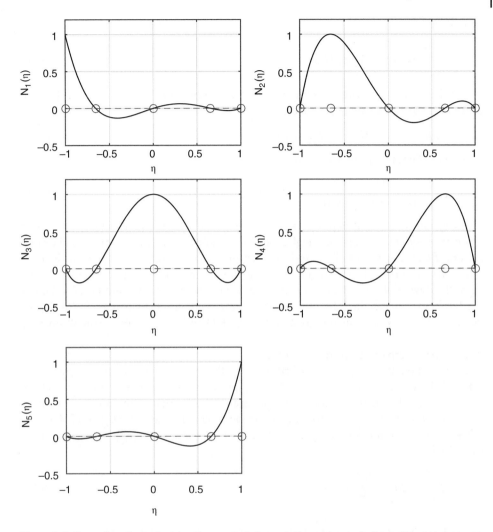

Figure 8.6 Shape functions of a 4th order spectral element. The nodes are indicated by circles.

8.2.2 Numerical Integration of Element Coefficient Matrices

The coefficient matrices of an S-element (Eqs. (2.110) and (3.102)) are given as integrations over the line elements on the boundary. For high-order elements, the integrations are performed numerically. Since the shape functions are polynomials (Eq. (8.14)), Gauss-Legendre quadrature can be used effectively. To explore the advantage of the spectral elements, Gauss-Lobatto-Legendre quadrature is employed.

8.2.2.1 Gauss-Legendre Quadrature

The Gauss-Legendre quadrature is widely used in the finite element method. The n-th order quadrature for the integration of a function $f(\eta)$ over the interval $-1 \leq \eta \leq +1$ is expressed as

$$\int_{-1}^{+1} f(\eta)d\eta = \sum_{k=1}^{n} w_k f(\eta_k) \tag{8.16}$$

where the locations of the sampling points (Gauss points) η_k are the roots of the n-th order Legendre polynomial (Table 8.1)

$$P_n(\eta_k) = 0 \qquad (k = 1, 2, \ldots, n) \tag{8.17}$$

The weight factors w_k are determined by

$$w_k = \frac{2(1 - \eta_k^2)}{(n+1)^2 (P_{n+1}(\eta_k))^2} \qquad (k = 1, 2, \ldots, n) \tag{8.18}$$

The locations of the Gauss points are symmetric about 0. The pair of Gauss points in symmetry, $\pm\eta_k$, have the same weight factor w_k. Table 8.3 lists the locations and weight factors of sampling points of Gauss-Legendre quadrature from the first to the fifth order. The n-th order Gauss-Legendre quadrature is capable of exactly integrating polynomials up to the degree of $(2n - 1)$. For a p-th order elements (with a constant determinant of the Jacobian matrix), the integrands of the element coefficient matrices are $2p$-degree polynomials. They are fully integrated by the $(p + 1)$-th Gauss-Legendre quadrature.

8.2.2.2 Gauss-Lobatto-Legendre Quadrature

The Gauss-Lobatto-Legendre quadrature uses the nodes of a spectral element, i.e. Gauss-Lobatto-Legendre points, as the sampling points for numerical integration. As is shown later in this section, this quadrature rule under-integrates the element coefficient matrices and leads to lumped diagonal element matrices $[E_0]$ and $[M_0]$.

The integration of a function $f(\eta)$ using the $(p + 1)$ nodes of a p-th order element is expressed as

$$\int_{-1}^{+1} f(\eta)d\eta = \sum_{k=1}^{p+1} w_k f(\eta_k) \tag{8.19}$$

Table 8.3 Sampling points and weight factors of Gauss-Legendre quadrature of orders 1 to 5.

Order n	Sampling point (η_k)	Weight factors (w_k)
1	0	2
2	$\pm\sqrt{\dfrac{1}{3}}$	1
3	$\pm\sqrt{\dfrac{3}{5}},\ 0$	$\dfrac{5}{9},\ \dfrac{8}{9}$
4	$\pm\sqrt{\dfrac{15 + 2\sqrt{30}}{35}},\ \pm\sqrt{\dfrac{15 - 2\sqrt{30}}{35}}$	$\dfrac{18 - \sqrt{30}}{36},\ \dfrac{18 - \sqrt{30}}{36}$
5	$\pm\sqrt{\dfrac{35 + 2\sqrt{70}}{63}},\ \pm\sqrt{\dfrac{35 - 2\sqrt{70}}{63}},\ 0$	$\dfrac{322 - 13\sqrt{70}}{900},\ \dfrac{322 + 13\sqrt{70}}{900},\ \dfrac{128}{225}$

where the weight factor at the node k (with natural coordinate η_k) is equal to

$$w_k = \frac{2}{p(p+1)(P_p(\eta_k))^2} \quad (k = 1, 2, \ldots, p+1) \tag{8.20}$$

The weight factors up to $p = 6$ are given in Table 8.2. The Gauss-Lobatto-Legendre quadrature is capable of exactly integrating polynomials up to the degree of $2(p+1) - 3 = 2p - 1$.

Using the Kronecker delta property of the shape functions (Eq. (8.3)), the numerical integration of the element coefficients can be simplified. Especially, it yields diagonal coefficient matrix $[M_0]$ and block-diagonal (or lumped) coefficient matrix $[E_0]$.

The coefficient matrix $[M_0]$ given in Eq. (3.102) is expressed for one element as

$$[M_0] = \int_{-1}^{+1} [N_u]^T \rho [N_u] |J_b| d\eta \tag{8.21}$$

The matrix of shape functions $[N_u]$ in Eq. (2.79) with Eq. (2.80) is rewritten as

$$[N_u] = [\, N_1[I] \; N_2[I] \; \ldots \; N_{p+1}[I] \,] \tag{8.22}$$

where $[I]$ is a 2×2 identity matrix. Using Eq. (8.22), $[M_0]$ in Eq. (8.21) can be divided into 2×2 submatrices each relating to two nodes. The submatrix related to nodes i and j is expressed as

$$[M_0]_{ij} = \int_{-1}^{+1} N_i(\eta) \rho N_j(\eta) |J_b(\eta)| d\eta [I] \quad (i, j = 1, 2, \ldots, p+1) \tag{8.23}$$

where the argument η is included for clarity. Applying the Gauss-Lobatto-Legendre quadrature (Eq. (8.19)) yields

$$[M_0]_{ij} = \sum_{k=1}^{p+1} w_k N_i(\eta_k) \rho N_j(\eta_k) |J_b(\eta_k)| [I] \quad (i, j = 1, 2, \ldots, p+1) \tag{8.24}$$

The sampling points η_k are the nodes of the element. Using the Kronecker delta property of the shape functions at the nodes (Eq. (8.3)), Eq. (8.24) is rewritten as

$$[M_0]_{ij} = \sum_{k=1}^{p+1} w_k \delta_{ik} \rho \delta_{jk} |J_b(\eta_k)| [I] \tag{8.25}$$

The Kronecker delta function δ_{jk} is considered. For the summation over the sampling points (i.e. nodes) $k = 1, 2, \ldots, p+1$, only the term $k = j$ is nonzero. Therefore, Eq. (8.25) is expressed as

$$[M_0]_{ij} = \delta_{ij} w_j \rho |J_b(\eta_j)| [I] \tag{8.26}$$

Considering the function δ_{ij}, the element coefficient matrix $[M_0]$ is diagonal with the diagonal entries computed from

$$[M_0]_{ii} = w_i \rho |J_b(\eta_i)| [I] \quad (i = 1, 2, \ldots, p+1) \tag{8.27}$$

The coefficient matrix $[E_0]$ in Eq. (2.110a) is integrated in the same way. The equation is written for one element as

$$[E_0] = \int_{-1}^{+1} [B_1]^T [D][B_1] |J_b| d\eta \tag{8.28}$$

Using $[B_1] = [b_1][N_u]$ (Eq. (2.82b)), Eq. (8.28) is expressed as

$$[E_0] = \int_{-1}^{+1} [N_u]^T [D_0][N_u]|J_b|\mathrm{d}\eta \tag{8.29}$$

where the abbreviation of a 2×2 matrix $[D_0] = [D_0(\eta)]$ is introduced

$$[D_0] = [b_1]^T [D][b_1] \tag{8.30}$$

Using Eq. (8.22), the 2×2 submatrix of $[E_0]$ (Eq. (8.29)) relating nodes i and j is expressed as

$$[E_0]_{ij} = \int_{-1}^{+1} N_i(\eta)[D_0(\eta)]N_j(\eta)|J(\eta)|\mathrm{d}\eta \qquad (i,j = 1,\, 2,\, \dots,\, p+1) \tag{8.31}$$

Integrating using the Gauss-Lobatto-Legendre quadrature (Eq. (8.19)) results in

$$[E_0]_{ij} = \sum_{k=1}^{p+1} w_k N_i(\eta_k)[D_0(\eta_k)]N_j(\eta_k)|J(\eta_k)| \tag{8.32}$$

Considering the Kronecker delta property of the shape functions (Eq. (8.3)) yields

$$[E_0]_{ij} = \sum_{k=1}^{p+1} w_k \delta_{ik}[D_0(\eta_k)]\delta_{jk}|J(\eta_k)| \tag{8.33}$$

Similar to Eq. (8.25), all the terms other than the one $k = j$ vanish in the summation leading to

$$[E_0]_{ij} = \delta_{ij} w_j [D_0(\eta_j)]|J(\eta_j)| \tag{8.34}$$

Therefore, $[E_0]$ is a block-diagonal matrix with the 2×2 diagonal blocks given by

$$[E_0]_{ii} = w_i |J(\eta_i)|[D_0(\eta_i)] \qquad (i = 1,\, 2,\, \dots,\, p+1) \tag{8.35}$$

The coefficient matrix $[E_1]$ in Eq. (2.110b) is considered. The element coefficient matrix is expressed as

$$[E_1] = \int_{-1}^{+1} [B_2]^T [D][B_1]|J_b|\mathrm{d}\eta \tag{8.36}$$

Using $[B_1] = [b_1][N_u]$ (Eq. (2.82a)), Eq. (8.28) is expressed as

$$[E_1] = \int_{-1}^{+1} [B_2]^T [D][b_1][N_u]|J_b|\mathrm{d}\eta \tag{8.37}$$

Using Eq. (8.22), the 2×2 submatrix of $[E_1]$ (Eq. (8.37)) relating nodes i and j is expressed as

$$[E_1]_{ij} = \int_{-1}^{+1} [B_2(\eta)]_i^T [D][b_1(\eta)]N_j(\eta)|J(\eta)|\mathrm{d}\eta \qquad (i,j = 1,\, 2,\, \dots,\, p+1) \tag{8.38}$$

where $[B_2(\eta)]_i$ is the submatrix of $[B_2]$ in Eq. (2.82b) related to node i

$$[B_2(\eta)]_i = [b_2(\eta)]N_i(\eta) \tag{8.39}$$

Performing the integration in Eq. (8.38) using the Gauss-Lobatto-Legendre quadrature (Eq. (8.19)) yields

$$[E_1]_{ij} = \sum_{k=1}^{n} w_k [B_2(\eta_k)]_i^T [D][b_1(\eta_k)]N_j(\eta_k)|J(\eta_k)| \tag{8.40}$$

Considering the Kronecker delta property of the shape functions $N_j(\eta_k) = \delta_{jk}$ (Eq. (8.3)), Eq. (8.40) is rewritten as

$$[E_1]_{ij} = \sum_{k=1}^{n} w_k [B_2(\eta_k)]_i^T [D][b_1(\eta_k)]\delta_{jk}|J(\eta_k)| \tag{8.41}$$

As only the term $k = j$ in the summation is non-zero,

$$[E_1]_{ij} = w_j|J(\eta_j)|[B_2(\eta_j)]_i^T [D][b_1(\eta_j)] \qquad (i, j = 1, 2, \ldots, p+1) \tag{8.42}$$

applies. Equation (8.42) can be expressed for the two columns related to node j as

$$[E_1]_{:j} = w_j|J(\eta_j)|[B_2(\eta_j)]^T [D][b_1(\eta_j)] \qquad (j = 1, 2, \ldots, p+1) \tag{8.43}$$

where the colon ':' stands for the whole range of $i = 1, 2, \ldots, p+1$. Equation (8.43) can be evaluated column-block by column-block and reduces the computational operations of standard Gauss quadrature.

The coefficient matrix $[E_2]$ in Eq. (2.110c) is integrated numerically by

$$[E_2] = \sum_{i=1}^{p+1} w_i[B_2(\eta_i)]^T [D][B_2(\eta_i)]|J_b(\eta_i)| \tag{8.44}$$

8.3 Two-dimensional Quadrilateral Spectral Elements

8.3.1 Shape Functions

The parent element of a two-dimensional quadrilateral spectral element is illustrated in Figure 8.7 using a cubic element ($p = 3$). It is a square of 2 units by 2 units defined by $-1 \le \eta \le +1$ and $-1 \le \zeta \le +1$. A p-th order element has a total of $(p+1)^2$ nodes forming a grid of $(p+1)$ columns and $(p+1)$ rows. The column number c and row number r are indicated in Figure 8.7 together with the natural coordinates. A node can be specified by its column and row numbers. The nodes are also numbered sequentially from left to right and from bottom to top starting with the node at the lower-left corner

Figure 8.7 Cubic spectral element ($p = 3$) in two dimensions.

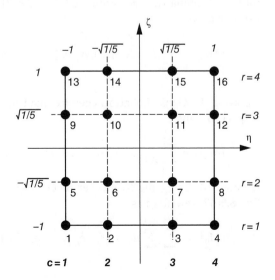

($\eta = \zeta = -1$). Any local nodal number i is thus related to its column number c_i and row number r_i by

$$i = (r_i - 1) \times (p+1) + c_i; \quad c_i, r_i = 1, 2, \ldots, p+1; \quad i = 1, 2, \ldots, (p+1)^2$$

(8.45)

The shape functions of the two-dimensional spectral element are constructed as the product of one-dimensional shape functions along η and ζ directions (Section 8.2.1)

$$N_i(\eta, \zeta) = N_{c_i}(\eta)N_{r_i}(\zeta); \quad c_i, r_i = 1, 2, \ldots, p+1; \quad i = 1, 2, \ldots, (p+1)^2$$

(8.46)

To show that the shape functions possess the Kronecker delta property, a node j at the column c_j and row r_j is considered (j follows from c_j and r_j according to Eq. (8.45)). Using Eq. (8.46), the Kronecker delta property of the one-dimensional shape functions and Eq. (8.45),

$$N_i(\eta_{c_j}, \zeta_{r_j}) = N_{c_i}(\eta_{c_j})N_{r_i}(\zeta_{r_j}) = \delta_{c_i c_j}\delta_{r_i r_j} = \delta_{ij}$$

$$(c_i, r_i, c_j, r_j = 1, 2, \ldots, p+1; \quad i, j = 1, 2, \ldots, (p+1)^2) \quad (8.47)$$

is obtained.

8.3.2 Integration of Element Coefficient Matrices by Gauss-Lobatto-Legendre Quadrature

The computation of the coefficient matrices by the Gauss-Lobatto-Legendre quadrature follows the procedure in Section 8.2.2.

The coefficient matrix $[M_0]$ in Eq. (6.140c) is addressed. For one quadrilateral spectral element on the surface of an S-element, it is expressed as

$$[M_0] = \int_{-1}^{+1}\int_{-1}^{+1} [N_u]^T \rho[N_u]|J_b|\mathrm{d}\eta\mathrm{d}\zeta$$

(8.48)

where the shape function matrix $[N_u]$ is defined in Eq. (6.88). Partition the matrix $[M_0]$ into 3×3 submatrices according to the nodes, the submatrix related to nodes i and j is expressed as

$$[M_0]_{ij} = \int_{-1}^{+1}\int_{-1}^{+1} N_i(\eta, \zeta)\rho N_j(\eta, \zeta)|J_b(\eta, \zeta)|\mathrm{d}\eta\mathrm{d}\zeta[I] \quad (i, j = 1, 2, \ldots, (p+1)^2)$$

(8.49)

Applying the Gauss-Lobatto-Legendre quadrature (Eq. (8.19)) to the integrations with respect to η and ζ

$$[M_0]_{ij} = \sum_{c_k=1}^{p+1}\sum_{r_k=1}^{p+1} w_{c_k}w_{r_k}N_i(\eta_{c_k}, \zeta_{r_k})\rho N_j(\eta_{c_k}, \zeta_{r_k})|J_b(\eta_{c_k}, \zeta_{r_k})|[I]$$

(8.50)

is obtained. Using the Kronecker delta property of the shape functions (Eq. (8.47)), Eq. (8.50) is rewritten as

$$[M_0]_{ij} = \sum_{c_k=1}^{p+1}\sum_{r_k=1}^{p+1} w_{c_k}w_{r_k}\delta_{c_i c_k}\delta_{r_i r_k}\rho\delta_{c_j c_k}\delta_{r_j r_k}|J_b(\eta_{c_k}, \zeta_{r_k})|[I]$$

(8.51)

Considering the function $\delta_{c_j c_k} \delta_{r_j r_k}$, only the term at the sampling point specified by the column $c_k = c_j$ and row $r_k = r_j$ is non-zero. Equating (8.51) is simplified as

$$[M_0]_{ij} = w_{c_j} w_{r_j} \delta_{c_i c_j} \delta_{r_i r_j} \rho J_b(\eta_{c_j}, \zeta_{r_j})|[I] \tag{8.52}$$

Similarly, the function $\delta_{c_i c_j} \delta_{r_i r_j}$ will vanish except when the column $c_j = c_i$ and row $r_j = r_i$, i.e. the node $j = i$ (Eq. (8.45)). Therefore, the coefficient matrix $[M_0]$ is diagonal with the diagonal entries given by

$$[M_0]_{ii} = w_{c_i} w_{r_i} \rho J_b(\eta_{c_i}, \zeta_{r_i})|[I] \qquad (i, j = 1, \ 2, \dots, \ (p+1)^2) \tag{8.53}$$

The above derivation can also be performed by regarding the two-dimensional grid of sampling points (which are the nodes in the Gauss-Lobatto-Legendre quadrature) as a series of $(p+1)^2$ sampling points denoted by

$$\mathbf{P}_i = (\eta_{c_i}, \zeta_{r_i}) \qquad (c_i, r_i = 1, \ 2, \dots, \ p+1; \quad i = 1, \ 2, \dots, \ (p+1)^2) \tag{8.54}$$

The weight factors at the sampling points are equal to

$$W_i = w_{c_i} w_{r_i} \qquad (c_i, r_i = 1, \ 2, \dots, \ p+1; \quad i = 1, \ 2, \dots, \ (p+1)^2) \tag{8.55}$$

The Gauss-Lobatto-Legendre quadrature for double integration is written as

$$\int_{-1}^{+1}\int_{-1}^{+1} f(\eta, \zeta) d\eta d\zeta = \sum_{i=1}^{(p+1)^2} W_i f(\mathbf{P}_i) \tag{8.56}$$

Employing Eqs. (8.54) and (8.55), Eq. (8.53) is expressed as

$$[M_0]_{ii} = W_i \rho |J_b(\mathbf{P}_i)|[I] \tag{8.57}$$

which is in the same format as Eq. (8.26) for the one-dimensional element.

The numerical integration of the coefficient matrix in $[E_0]$ (Eq. (6.140a)) is considered. The element coefficient matrix is expressed as

$$[E_0] = \int_{-1}^{+1}\int_{-1}^{+1} [B_1]^T [D][B_1]||J_b| d\eta d\zeta \tag{8.58}$$

Using Eq. (6.141a) and introducing the abbreviation denoting a 3×3 matrix

$$[D_0(\eta, \zeta)] = [b_1]^T [D][b_1] \tag{8.59}$$

The 3×3 submatrix of $[E_0]$ related to the i-th and j-th nodes are expressed as

$$[E_0]_{ij} = \int_{-1}^{+1}\int_{-1}^{+1} N_i(\eta, \zeta)[D_0(\eta, \zeta)]N_j(\eta, \zeta)|J_b(\eta, \zeta)| d\eta d\zeta \qquad (i, j = 1, \ 2, \dots, \ (p+1)^2) \tag{8.60}$$

Applying the Gauss-Lobatto-Legendre quadrature in Eq. (8.56) leads to

$$[E_0]_{ij} = \sum_{k=1}^{(p+1)^2} W_k N_i(\mathbf{P}_k)[D_0(\mathbf{P}_k)]N_j(\mathbf{P}_k)|J_b(\mathbf{P}_k)| \tag{8.61}$$

Using the Kronecker delta property of the shape functions (Eq. (8.47)), this equation is expressed as

$$[E_0]_{ij} = \sum_{k=1}^{(p+1)^2} W_k \delta_{ik}[D_0(\mathbf{P}_k)]\delta_{jk}|J_b(\mathbf{P}_k)| \tag{8.62}$$

which is simplified by considering the function δ_{jk} as

$$[E_0]_{ij} = \delta_{ij} W_j [D_0(\mathbf{P}_j)] |J_b(\mathbf{P}_j)| \tag{8.63}$$

Thus, the element coefficient matrix $[E_0]$ is block-diagonal. The 3×3 diagonal blocks, one block for each node, are equal to

$$[E_0]_{ii} = W_i [D_0(\mathbf{P}_i)] |J_b(\mathbf{P}_i)| \qquad (i = 1, \, 2, \ldots, \, (p+1)^2) \tag{8.64}$$

The coefficient matrix $[E_1]$ in Eq. (6.140b) is expressed for a quadrilateral spectral element as

$$[E_1] = \int_{-1}^{+1} \int_{-1}^{+1} [B_2]^T [D][B_1] |J_b| \, d\eta d\zeta \tag{8.65}$$

Substituting Eq. (6.141a) into Eq. (8.65) yields

$$[E_1] = \int_{-1}^{+1} \int_{-1}^{+1} [B_2]^T [D][b_1][N_u] |J_b| \, d\eta d\zeta \tag{8.66}$$

The three columns of $[E_1]$ related to node j are expressed as

$$[E_1]_{:j} = \int_{-1}^{+1} \int_{-1}^{+1} [B_2(\eta, \zeta)]^T [D][b_1(\eta, \zeta)] N_j(\eta, \zeta) |J_b(\eta, \zeta)| \, d\eta d\zeta$$

$$(j = 1, \, 2, \ldots, \, (p+1)^2) \tag{8.67}$$

where the colon ':' stands for the whole range of $i = 1, 2, \ldots, (p+1)^2$. Applying the Gauss-Lobatto-Legendre quadrature in Eq. (8.19) leads to

$$[E_1]_{:j} = \sum_{k=1}^{(p+1)^2} W_k [B_2(\mathbf{P}_k)]^T [D][b_1(\mathbf{P}_k)] N_j(\mathbf{P}_k) |J_b(\mathbf{P}_k)| \tag{8.68}$$

Considering the Kronecker delta property of the shape functions (Eq. (8.3)), only the term $k = j$ in the summation is not equal to zero. Equation (8.68) is reduced to

$$[E_1]_{:j} = W_j [B_2(\mathbf{P}_j)]^T [D][b_1(\mathbf{P}_j)] |J_b(\mathbf{P}_j)| \qquad (j = 1, \, 2, \ldots, \, (p+1)^2) \tag{8.69}$$

At each sample point (node), three columns of $[E_1]$ is completed. No summation over the sampling points are required.

The coefficient matrix $[E_2]$ in Eq. (6.140c) is integrated numerically by

$$[E_2] = \sum_{i=1}^{(p+1)^2} W_i [B_2(\mathbf{P}_i)]^T [D][B_2(\mathbf{P}_i)] |J_b(\mathbf{P}_i)| \tag{8.70}$$

8.4 Examples

Examples are presented in this section to evaluate the accuracy and convergence behaviour of high-order elements. High-order patch tests are also performed. Generally speaking, high-order elements converge at the theoretically estimated optimal rates and lead to significantly higher accuracy than linear elements.

8.4.1 A Cantilever Beam Subject to End Loading

A cantilever beam of height H and length L is shown in Figure 8.8. The elastic property of the material is given by the Young's modulus E and Poisson's ratio v. It is subject to a vertical force P at the free end. The plane stress state is assumed.

An approximate analytical solution of this problem is available in the theory of elasticity. The stresses are expressed as

$$\sigma_x = -\frac{P(L-x)y}{I}$$

$$\sigma_y = 0$$

$$\tau_{xy} = \frac{P}{2I}\left(\frac{H^2}{4} - y^2\right)$$

where the moment of inertia is equal to

$$I = \frac{H^3}{12}$$

The normal stress σ_x varies linearly along the depth of the beam. The resultant of the shear stress τ_{xy}, which follows a parabolic distribution as depicted in Figure 8.8, is equal to the vertical force P. The displacements are expressed as

$$u_x = -\frac{Py}{6EI}\left((6L - 3x)x + (2 + v)\left(y^2 - \frac{H^2}{4}\right)\right)$$

$$u_y = \frac{P}{6EI}\left(3v\left(L\left(y^2 - \frac{H^2}{4}\right) - xy^2\right) + (4 + 5v)\frac{H^2x}{4} + (3L - x)x^2\right)$$

Note that the displacements at the left end $(x = 0)$ are not equal to 0 but are negligible for slender beams $(H/L \ll 1)$.

To use the above analytical solution for error estimation, the surface tractions t_x, t_y, corresponding to analytical solution are applied as the boundary conditions in the following analyses:

Left side $(x = 0)$:
$$t_x = \frac{PL}{I}y; \quad t_y = -\frac{P}{2I}\left(\frac{H^2}{4} - y^2\right)$$

Right side $(x = L)$:
$$t_x = 0; \quad t_y = \frac{P}{2I}\left(\frac{H^2}{4} - y^2\right)$$

Top and bottom sides $(y = \pm H/2)$: $\quad t_x = t_y = 0$

The rigid-body motions are constrained by specifying $u_x(x = 0, y = \pm H/2) = 0$ and $u_y(x = 0, y = -H/2) = 0$ obtained from the above analytical solution. A consistent set

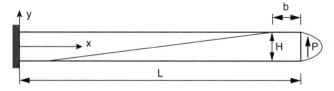

Figure 8.8 Cantilever beam subject to a vertical force at the free end.

of units is used in the analysis with the parameters: $H = 1$, $L = 10$, $P = 1$, $E = 100$ and $v = 0.3$.

Two types of meshes are used. In the first type, the beam is divided into two rectangular regions as illustrated in Figure 8.8 with $b = L/2$. In the second one, the two regions are skewed to trapezoids with $b = 0.1L$ as shown in Figure 8.8. Each of the regions is divided into a mesh of $n \times n$ S-elements as shown in Figure 8.9 at $n = 2$. Each edge of an S-element is modelled by a quadratic ($p = 2$) element. The element coefficient matrices are integrated numerically by standard Gauss quadrature.

A convergence study is performed using both types of meshes. The meshes are refined successively following the sequence $n = 1, 2, 4, 8, 16, 32$. The characteristic element size is chosen as ($h = H/n$). It is halved at each refinement. The relative error norm of displacement (Eq. (4.12)) versus the element size is plotted in Figure 8.10. It is observed that rectangular meshes lead to more accurate results than the distorted trapezoidal meshes. The convergence rates of both types of meshes are close to the theoretically predicted optimal rate of 3 for quadratic elements.

(a)   Rectangular S-elements.

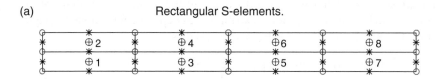

(b)   Trapezoidal S-elements.

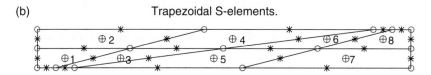

Figure 8.9 Mesh of a cantilever beam with quadratic line elements ($p = 2$) on edges of S-elements.

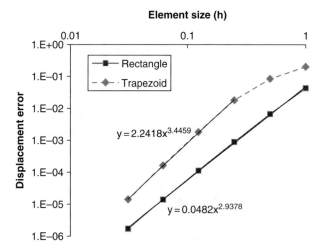

Figure 8.10 Convergence of displacement of a cantilever beam subject to vertical force at free end.

Figure 8.11 Mesh of a cantilever beam with cubic line elements ($p = 3$) on edges of S-elements.

A high-order patch test of the cantilever beam is performed. A mesh of two highly-distorted S-elements is shown in Figure 8.11. The first S-element has four edges. The second one is a concave polygon with six sides. Since the analytical solution is a third-degree polynomial, all edges are modelled with one cubic element. The relative error norm of displacement is obtained as 3×10^{-10}, which is considered as passing the patch test.

8.4.2 A Circular Hole in an Infinite Plate

The problem of a circular hole in an infinite plate in Section 4.4.2 on page 193 is addressed. The radius of the hole is $a = 0.4$ m. A square body of 2 m $\times 2$ m around the circular hole (Figure 4.19) is considered. Using symmetry, only a quarter of the square body is modelled by one S-element as shown in Figure 8.12. One line element is employed on each of the five edges of the S-element. Fifth-order elements are shown in the figure. The scaling centre is selected at $(0.6$ m$, 0.6$ m$)$. The vertical displacement at the bottom edge and the horizontal displacement at the left edge are fixed. The arc is traction-free. The exact solution of the displacement are prescribed at the right and top edges.

The convergence behaviour of the spectral elements is investigated. p-refinement of the discretization, i.e. the element order is increased while the number of elements remains the same, is performed starting from the order $p = 3$. The element coefficient matrices are integrated by the Gauss-Lobatto-Legendre quadrature. The relative error norm of displacement (Eq. (4.12)) is evaluated at each order of elements. The results are shown by the solid line marked 'GLL integration' in Figure 8.13a. In this semi-log plot, the error is in logarithmic scale and the element order is in linear scale. The

Figure 8.12 Mesh of a quarter of a square body with a circular hole using one S-domain with one spectral elements on each edge.

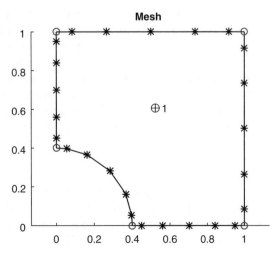

(a)

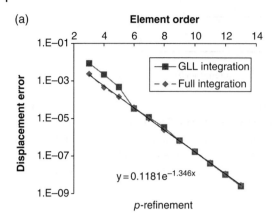

(b)

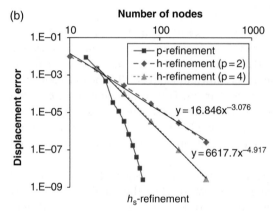

Figure 8.13 Convergence of displacements of a square body with a circular hole with increasing order of spectral elements.

straight line indicates that exponential convergence of displacement is achieved with *p*-refinement.

The effect of the schemes of numerical integration of the element coefficient matrices $[E_0]$, $[E_1]$ and $[E_2]$ (Eq. (2.110)) on the accuracy of the result is evaluated. For straight elements, the coefficient matrices are fully integrated by the standard Gauss quadrature. The coefficient matrix $[E_0]$ will be fully populated. The relative error norm of displacement is shown in Figure 8.13a by the dashed line marked 'Full integration'. When the order of element is lower than 6, full integration leads to more accurate results. For higher-order elements, the effect of the integration schemes is negligible.

The convergence behaviour of h_s-refinement (the elements are subdivided into smaller one while the order of element is maintained) is examined. Orders of elements $p = 2$ and $p = 4$ are chosen. The initial mesh consists of one line element per edge of the S-elements as shown in Figure 8.12. In each step of the h_s-refinement, one element is divided into two elements of equal length. Full integration is performed for the 2nd-order elements, and Gauss-Lobatto-Legendre integration for the 4th-order elements. The relative error norms versus the total number of nodes in the mesh are shown in the log-log plot in Figure 8.13b by the dashed line and the dotted line, respectively. The convergence rates are close to the theoretically predicted optimal values (3 for the 2nd-order element and

5 for the 4th-order element). It is observed that higher-order elements are significantly more accurate per node than lower-order elements.

8.4.3 An L-shaped Panel

The static analysis of the L-shaped panel shown in Figure 4.23 on page 199 (Section 4.4.3) is performed with spectral elements. The panel is modelled as one single S-element as depicted in Figure 8.14. The scaling centre is selected at the re-entrant corner. The boundary of the S-element is divided into six spectral elements of equal length. The mesh of 4th-order elements is shown in Figure 8.14. The dimensions, boundary conditions and material properties are the same as those in Section 4.4.3.

The convergence of p-refinement is evaluated. The vertical displacement v_A at the upper-right corner (point A in Figure 4.23) obtained at several selected orders of elements are presented in Table 8.4. Both the results using Gauss-Lobatto-Legendre quadrature (denoted as 'GLL integration') and the results using the standard Gauss quadrature (denoted as 'Full integration') are shown in Table 8.4. The displacement v_A converges to 1.05603839 at element order $p = 10$. This value is used to evaluate the relative error e_v of the results obtained with elements of lower orders. The errors are shown in Table 8.4 for selected orders of elements. The convergence results are plotted in semi-logarithmic scale in Figure 8.15a. It is observed from Figure 8.15a that the error decreases exponentially. The stress singularity at the re-entrant corner does not affect the convergence of the spectral elements since it is represented analytically. The stresses vary smoothly in the circumference direction and are modelled effectively by spectral elements.

Different from the problem of a circular hole in an infinite plate in Section 8.4.2, all elements are straight lines in this problem. Both numerical integration schemes lead to results of similar accuracy. It is only at $p = 1$ that the result obtained with the Gauss-Lobatto-Legendre quadrature is appreciably lower than that obtained with standard Gauss quadrature.

The convergence behaviour of h_s-refinement is investigated for elements of orders $p = 2$ and $p = 4$. Starting from the mesh of six elements in Figure 8.14, the number of elements is doubled in each refinement, while the orders of the elements remain the same.

Figure 8.14 Mesh of an L-shaped panel by one S-element with spectral elements on the edges.

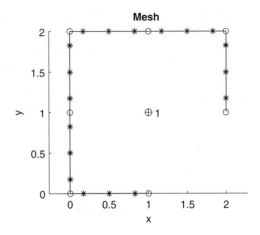

Table 8.4 Convergence of vertical displacement v_A at point A of an L-shaped panel with increasing order of elements: p-refinement.

Element order	Number of nodes	GLL integration		Full integration	
		v_A $(10^{-4}$ m$)$	Error e_v	v_A $(10^{-4}$ m$)$	Error e_v
1	7	0.80141003	2.4×10^{-1}	0.96961443	8.2×10^{-2}
2	13	1.05451565	1.4×10^{-3}	1.05720734	1.1×10^{-3}
3	19	1.05523152	7.6×10^{-4}	1.05555362	4.6×10^{-4}
4	25	1.05600709	3.0×10^{-5}	1.05599515	4.1×10^{-5}
6	37	1.05603581	2.4×10^{-6}	1.05603617	2.1×10^{-6}
8	49	1.05603849	9.1×10^{-8}	1.05603849	9.1×10^{-8}
10	61	1.05603839		1.05603839	
12	73	1.05603839		1.05603839	

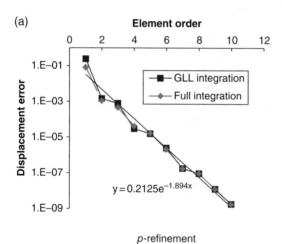

(a)

$y = 0.2125e^{-1.894x}$

p-refinement

Figure 8.15 Convergence of vertical displacement at point A of a L-shaped panel.

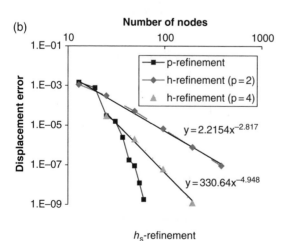

(b)

$y = 2.2154x^{-2.817}$

$y = 330.64x^{-4.948}$

h_s-refinement

Table 8.5 Convergence of vertical displacement v_A at point A of an L-shaped panel with increasing order of elements: h_s-refinement with 2nd- and 4th-order elements.

Number of nodes	Element order $p = 2$		Number of nodes	Element order $p = 4$	
	v_A (10^{-4} m)	Error e_v		v_A (10^{-4} m)	Error e_v
13	1.05720734	1.1×10^{-3}	25	1.05600709	3.0×10^{-5}
25	1.05635961	3.0×10^{-4}	49	1.05604049	2.0×10^{-6}
49	1.05609204	5.1×10^{-5}	97	1.05603846	6.3×10^{-8}
97	1.05604538	6.6×10^{-6}	193	1.05603840	1.2×10^{-9}
193	1.05603923	7.9×10^{-7}			
385	1.05603849	9.2×10^{-8}			

The results of displacement v_A and the relative error e_v are given in Table 8.5. The errors versus the number of nodes are plotted in Figure 8.15b in logarithmic scale. The convergence rates are close to the optimum values. The 4th-order element is much more accurate than the 2nd-order element with the same number of nodes. It can also be seen that p-refinement leads to significantly faster convergence than h_s-refinement.

8.4.4 A 3D Cantilever Beam Subject to End-shear Loading

A 3D cantilever beam of length L and rectangular cross-section of width $2a$ and height $2b$ is shown in Figure 8.16a with the coordinate system x, y and z. The z-axis lies on the neutral axis of the beam. The end at $z = L$ is fixed. A transverse shear force P is applied at the end $z = 0$ in the negative y-direction. The moment of inertia about the x-axis is equal to $I = 4ab^3/3$. The elastic property of the material is characterized by the Young's modulus E and Poisson's ratio ν.

(a) (b)

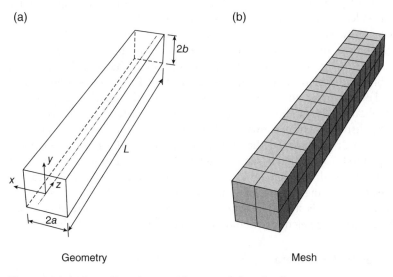

Geometry Mesh

Figure 8.16 A 3D cantilever beam subject to end-shear loading.

The exact solution of this problem can be found in Bishop (2014). The stress solution is expressed as

$$\sigma_x = \sigma_y = \tau_{xy} = 0$$

$$\sigma_z = \frac{P}{I}yz$$

$$\tau_{zx} = \frac{P}{I}\frac{v}{1+v}\frac{2a^2}{\pi^2}\sum_{n=0}^{\infty}\frac{(-1)^n}{n^2}\sin\left(\frac{n\pi x}{a}\right)\frac{\sinh\left(\frac{n\pi y}{a}\right)}{\cosh\left(\frac{n\pi b}{a}\right)}$$

$$\tau_{zy} = \frac{P}{I}\frac{b^2-y^2}{2} + \frac{P}{I}\frac{v}{1+v}\left(\frac{3x^2-a^2}{6} - \frac{2a^2}{\pi^2}\sum_{n=0}^{\infty}\frac{(-1)^n}{n^2}\cos\left(\frac{n\pi x}{a}\right)\frac{\cosh\left(\frac{n\pi y}{a}\right)}{\cosh\left(\frac{n\pi b}{a}\right)}\right)$$

The displacement solution is written as

$$u_x = -\frac{Pv}{EI}xyz$$

$$u_y = -\frac{P}{EI}\left(\frac{z^3}{6} - \frac{v(x^2-y^2)z}{2}\right)$$

$$u_z = \frac{P}{EI}\left(\frac{y(vx^2+z^2)}{2} + \frac{vy^3}{6} + (1+v)\left(b^2y - \frac{y^3}{3}\right) - \frac{va^2y}{3}\right.$$
$$\left. -\frac{4va^3}{\pi^3}\sum_{n=0}^{\infty}\frac{(-1)^n}{n^3}\cos\left(\frac{n\pi x}{a}\right)\frac{\sinh\left(\frac{n\pi y}{a}\right)}{\cosh\left(\frac{n\pi b}{a}\right)}\right)$$

In the following analysis, a consistent set of unit is used. The dimensions are $L = 80$, $a = b = 5$. The Young's modulus is equal to $E = 100$. The Poisson's ratio is chosen as $v = 0$. The shear force $P = 1$. The surface tractions consistent with the exact stress solution are prescribed at the ends $z = 0$ and $z = L$ as boundary conditions. Displacement constraints are provided to prevent the rigid-body motions only.

A convergence study is performed using linear ($p = 1$) and quadratic ($p = 2$) elements. The S-element mesh is shown in Figure 8.16b. The beam is divided into S-elements of size $a \times a \times a$. The mesh is refined uniformly by halving the elements size, i.e. subdividing one S-element into eight smaller ones. Each face is modelled as one 4-node ($p = 1$) element or one 9-node ($p = 2$) element. The relative error norms of displacement obtained from the convergence study are listed in Table 8.6. The convergence of the displacement is shown in Figure 8.17 as a function of the number of nodes in the mesh. It is observed that convergence rate of the linear element is approximately equal to 0.738, which is slightly higher than the optimum rate 2/3. The convergence rate of the quadratic element is about 1.4, higher than the optimum rate 1.

When each face of an S-element is modelled as one cubic ($p = 3$) element, the solution reaches machine accuracy. It is shown by repeating the analysis on other loading cases that the scaled boundary finite element method passes the 3rd-order patch test.

8.4.5 A Pressurized Hollow Sphere

A hollow sphere with an inner radius $a = 10$ mm and an out radius $b = 50$ mm is shown in Figure 8.18a. The material properties of the hollow sphere are Young's modulus

Table 8.6 Relative error norm of displacement of 3D cantilever beam subject to end-shear loading with mesh refinement.

Element	Linear element		Quadratic element	
size	Nodes	Error e_v	Nodes	Error e_v
5	153	9.3×10^{-2}	761	1.0×10^{-4}
2.5	825	2.5×10^{-2}	4753	1.3×10^{-5}
1.25	5265	6.6×10^{-3}	33185	1.7×10^{-6}
0.625	37281	1.7×10^{-3}	247105	1.6×10^{-7}

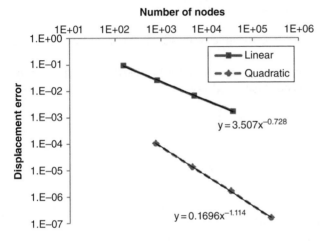

Figure 8.17 Convergence of displacement of 3D cantilever beam subject to end-shear loading.

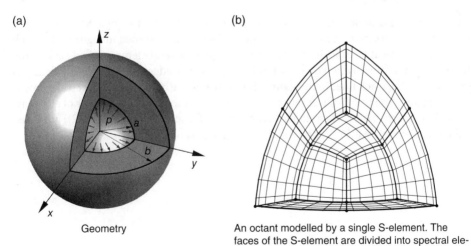

(a) Geometry

(b) An octant modelled by a single S-element. The faces of the S-element are divided into spectral elements. The thick lines show the edges of the spectral element.

Figure 8.18 A pressurized hollow sphere under uniform inner pressure.

Element order	Number of nodes	Displacement error norm e_v
1	14	2.9×10^{-1}
2	54	2.7×10^{-2}
3	124	3.1×10^{-3}
4	224	3.7×10^{-4}
5	354	6.7×10^{-5}
6	514	1.2×10^{-5}
7	704	2.4×10^{-6}
8	924	3.5×10^{-7}

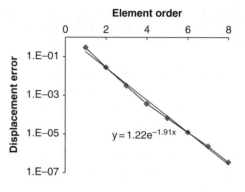

Figure 8.19 Convergence of displacement of a pressurized hollow sphere.

$E = 200\,\text{GPa}$ and Poisson's ratio $v = 0.3$. The inner wall is subject to a uniform pressure $p = 10\,\text{GPa}$.

The exact solution of this problem can be found in Bower (2009). The displacement solution is expressed in spherical coordinates r, θ, ϕ as

$$u_r = \frac{p}{2E} \frac{a^3}{b^3 - a^3} \left(2(1 - 2v)r + (1 + v)\frac{b^3}{r^2} \right)$$

$$u_\theta = u_\phi = 0$$

and the stress solution is

$$\sigma_r = p \frac{a^3}{b^3 - a^3} \frac{r^3 - b^3}{r^3}$$

$$\sigma_\theta = \sigma_\phi = \frac{p}{2} \frac{a^3}{b^3 - a^3} \frac{2r^3 + b^3}{r^3}$$

$$\tau_{\theta\phi} = \tau_{\phi r} = \tau_{r\theta} = 0$$

Due to symmetry, an octant of the hollow sphere is modelled as a single S-element. The boundary discretization of the boundary of the S-element is shown in Figure 8.18. Each of the two spherical faces is discretized by three quadrilateral elements, and each of the three annular sectors by two quadrilateral elements. The edges of the spectral elements are shown by the thick lines in the figure. The convergence behaviour of spectral element is investigated. The order of elements varies from $p = 1$ to $p = 8$. (The order of elements is $p = 6$ in Figure 8.18b). The relative error norms of displacement (Eq. (4.12)) are listed and plotted in Figure 8.19. Exponential convergence is observed.

9

Quadtree/Octree Algorithm of Mesh Generation for Scaled Boundary Finite Element Analysis

9.1 Introduction

In this chapter, a hierarchical-tree (quadtree in two dimensions and octree in three dimensions) algorithm for the generation of scaled boundary finite element meshes directly from geometric models is presented. The quadtree/octree algorithm and the scaled boundary finite element method are highly complementary. The quadtree/octree algorithm is not only capable of handling common Computer-Aided Design (CAD) models but also extremely simple and efficient to mesh digital images and STL (STere-oLithography) models. A cell of a quadtree/octree mesh satisfies the scaling requirement and is modelled directly by an S-element. The issue of hanging nodes, which hinders the application of quadtree/octree algorithm in the finite element method, is easily avoided owing to the highly flexible shapes of S-elements. The use of S-elements also allows curved boundaries to be modelled by trimming the quadtree/octree cells. The meshing process is tolerant of defects smaller than the size of the surrounding S-elements. The computational effort can also be reduced by considering that quadtree/octree cells have a small number of unique patterns only. The mesh generation algorithms are developed for the generation of S-element mesh. Numerical examples involving digital images, STL models and CAD models are presented for validation and for demonstration of the effectiveness and versatility of the algorithms.

9.1.1 Mesh Generation

Nowadays, engineering designs are commonly performed on CAD platforms, where curves and surfaces are typically represented by non-uniform rational B-spline (NURBS). The automatic mesh generation from CAD files is essential to engineering analysis. In the finite element method, a problem domain is discretized into a mesh of elements of simple geometries. The types of conventional finite elements are triangles and quadrilaterals in 2D, and tetrahedrons, hexahedrons, wedges and pyramids in 3D. A mesh of conventional finite elements has to conform to the boundary of the problem domain. In practical engineering applications, element size gradation is indispensable in the modelling of problems with complex geometry (Frey and George, 2008).

During the last few decades, the research on automatic mesh generation has led to the development of various meshing techniques. A vast amount of literature is

The Scaled Boundary Finite Element Method: Introduction to Theory and Implementation,
First Edition. Chongmin Song.
© 2018 John Wiley & Sons Ltd. Published 2018 by John Wiley & Sons Ltd.
Companion website: www.wiley.com/go/author/scaledboundaryfiniteelementmethod

available (Watson, 1981; Löhner and Parikh, 1988; Blacker and Meyers, 1993; Owen, 1998, to name a few). Triangular and tetrahedral elements are more flexible than quadrilateral and hexahedral elements in terms of representing geometries. Two main approaches for triangular and tetrahedral mesh generation are based on the Delaunay triangulation (Watson, 1981) and the advancing front technique (Löhner and Parikh, 1988). Quadrilateral and hexahedral elements are preferable from the point of view of solution accuracy and robustness, but the generation process of an all-hexahedral mesh is complicated. A number of methods have been proposed to generate hexahedral mesh, including plastering (Blacker and Meyers, 1993), whisker weaving (Tautges et al., 1996), sweeping (Staten et al., 1999), and octree-based approaches (Yerry and Shephard, 1984), to name a few. Wedge and pyramid elements can be used together with hexahedral and tetrahedral elements in hybrid meshes (Owen, 1998). In spite of the tremendous progress achieved in automatic mesh generation, considerable human interventions are still needed in the mesh generation of CAD models of engineering designs (Hughes et al., 2005). With increasingly affordable computer power, the human effort required in the mesh generation becomes more and more critical in terms of both cost and time.

Concurrently, the progress in digital technologies and their applications in engineering have posed new challenges to established approaches of numerical simulation. Owing to the advances in digital imaging technologies (Lengsfeld et al., 1998; Lian et al., 2013) and 3D printing (Bassoli et al., 2007; Rengier et al., 2010), digital images and STL (STereoLithography) format are being increasingly utilised in Computer-Aided Engineering (CAE) to describe the geometric models. Most traditional techniques for mesh generation of CAD models are not directly applicable to handle digital images and STL models.

In a digital image of an object, the colour/grayscale intensity of the pixels provides the information on geometric model and material distribution (see Figure 1.16 and Figure 1.17 on page 24). Several approaches have been developed for meshing digital images for finite element analysis. Most of them can be classified into two types. In one type of approaches, a digital image is converted to a CAD model. This requires the detection of the boundary of each region of interest in the image. The above-mentioned mesh generation techniques for handling CAD models can be applied. In the other type, the mesh is generated directly from the image.

One boundary detection technique is the marching cubes algorithm (analogous to marching squares in 2D) proposed by Lorensen and Cline (1987). This algorithm produces the iso-surfaces of the sub-regions. It is modified in Wang et al. (2005) to treat ambiguity and mesh incompatibility. A detailed survey on the development of the marching cubes algorithm can be found in Newman and Yi (2006).

The simplest way to directly use digital images is the voxel-based approach (Keyak et al., 1990), where each voxel is simply modelled as a hexahedral finite element. This method typically leads to a large number of elements (the model of the 3D image of 256 voxels per side in Figure 1.17 on page 24 would have 16.7 million elements) and excessive computational burden. Curved boundaries are also approximated by jagged surfaces in the mesh. While the jagged surfaces lead to localized stress concentrations in small elements (Lengsfeld et al., 1998; Ulrich et al., 1998; Lian et al., 2013), the average stresses remain close to the stresses obtained using a mesh generated from a CAD model with smooth boundaries. In the volumetric marching cubes method (VoMaC) (Müller and

Rüegsegger, 1995), the iso-surfaces detected from marching cubes algorithm are used to trim the elements along the boundary to reduce the jagged edges. The VoMaC is subsequently improved to consider up to eight sub-regions meeting at a cube vertex in the extended volumetric marching cubes method (EVoMaC) approach (Young et al., 2008). An octree mesh structure is also created by collecting a cluster of $2 \times 2 \times 2$ identical cubic elements into a single larger cube.

STL models are described by unstructured triangular facets. A typical STL model has ill-shaped, overlapped, self-intersecting facets and may not be watertight. It is thus not suitable for analysis. Repairing an STL model is not a trivial task (Attene et al., 2013). A number of methods to generate analysis-suitable surface meshes from STL models have been proposed (Béchet et al., 2002; Wang et al., 2007). To generate a quality mesh from an STL model is challenging. At the present state of development, well-known commercial finite element software (e.g. ANSYS and Abaqus) cannot robustly perform analyses using STL models.

The S-element, as a type of polytope element constructed using the scaled boundary finite element method, provides a high degree of flexibility in mesh generation. Its shape needs to satisfy only the scaling requirement. In contrast to the limited shapes of finite elements, the shape of an S-element can be a polygon/polyhedron with an arbitrary number of edges/faces and vertices. Furthermore, only the boundary of the S-element is discretized, which reduces the mesh generation effort. Although it is possible to generate a polyhedral mesh from a finite element mesh by following the approach for generating a polygon mesh from a triangular mesh in Section 4.3.3, such an approach will inherit the limitations of the chosen finite element generation technique. It is thus desirable to develop a mesh generation technique that utilises the flexibility of S-elements and does not need a finite element mesh as input.

In this chapter, a mesh generation technique based on a quadtree/octree algorithm is presented. This technique generates a mesh of S-elements directly from a geometric model without requiring a finite element mesh. It handles, in a unified approach, CAD models as well as digital images and STL models.

9.1.2 The Quadtree/Octree Algorithm

The quadtree/octree algorithm for mesh generation is a hierarchical-tree based technique (Yerry and Shephard, 1984; Greaves and Borthwick, 1999). The square/cubic bounding box of the problem domain is the root cell. Starting from the root cell, a cell is recursively divided by bisecting the edges until specified stopping criteria are met. In a subdivision, newly generated cells are called the children of the parent cell.

The meshing process of a circular domain is illustrated in Figure 9.1. The square bounding box of the circle is the root cell. At each level, the cells intersected by the circle are subdivided into (2^2) four by bisecting the sides to improve the representation of the circle to a specified accuracy. It is a common practice in mesh generation to limit the maximum difference in the division levels between two adjacent cells to one. This is referred to the 2:1 rule and the resulting mesh is called a balanced (GVS et al., 2001) or restricted quadtree mesh (Tabarraei and Sukumar, 2005).

In three dimensions, one cubic cells is split into eight (2^3) cells by bisecting the sides. A three-level octree is shown in Figure 9.2. The cell information is efficiently stored in a tree-type data structure, in which the root cell is at the highest level.

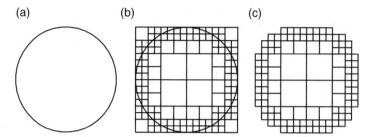

Figure 9.1 Generation of quadtree mesh on a circular domain. (a) Problem domain. (b) Quadtree mesh covering the problem domain. The cells are refined to fit the bonudary (c) Quadtree mesh after removing external cells.

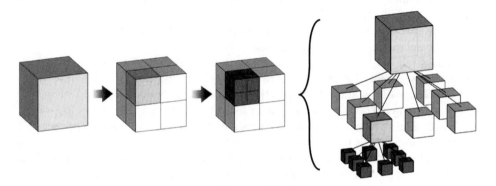

Figure 9.2 An octree with three levels.

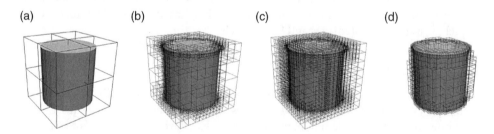

Figure 9.3 Generation of an octree mesh of a cylinder: (a) Problem domain and initial division of bounding box. (b) Cells are refined to fit the boundary. (c) Balanced octree mesh (2:1 rule). (d) Octree mesh after removing external cells.

The generation of an octree mesh of a cylinder is illustrated in Figure 9.3. Initially, the bounding box is divided into 8 cells (Figure 9.3a). The cells are subdivided to increase the accuracy in representing the boundary (Figure 9.3b). Further subdivisions are performed to enforce the 2:1 rule. The balanced octree is shown in Figure 9.3c. After removing the cells outside of the problem domains (external cells), the octree mesh in Figure 9.3d is obtained.

Figure 9.4 Quadtree mesh of the top-right quadrant of a circular domain. Demonstration of subdivision (dashed lines) is given in two quadtree cells with hanging nodes on their sides.

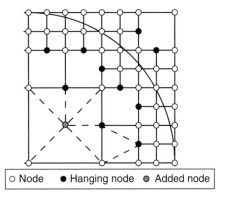

○ Node ● Hanging node ◉ Added node

The process of generating hierarchical trees is highly efficient. However, as illustrated in Figure 9.4 showing the upper-right quadrant of the quadtree in Figure 9.1, its direct use for mesh generation in the finite element method is hindered by two major issues:

1) *Hanging nodes.* As shown by solid dots in the quadtree in Figure 9.4, midside nodes exist at the common edges between two adjacent cells with different division levels. When conventional quadrilateral finite elements are used, a midside node is connected to the two smaller elements (lower level) but not to the larger element (higher level). This leads to displacement incompatibility along the edges and the midsize nodes are commonly referred to as hanging nodes (Greaves and Borthwick, 1999).

2) *Fitting of curved boundaries.* The edges at a vertex of quadtree/octree cells are perpendicular to each other. As shown in Figure 9.4, the curved boundary is represented by using small cells, leading to poor quality of fitting. Additionally, the boundary of the mesh is jagged, which may result in unrealistically high stresses.

A number of approaches to use a hierarchical tree for finite element mesh generation have been developed. In Yerry and Shephard (1983, 1984); Bern et al. (1994), the cells are subdivided into triangular/tetrahedral finite elements as illustrated by the dashed lines in Figure 9.4. Additional nodes may be added to improve the mesh quality and/or reduce the number of element types. Similarly, a quadtree can be subdivided into predominately quadrilateral elements (Ebeida et al., 2010). The issue of hanging nodes can be resolved by applying multi-point constraints or using elements with special shape functions to ensure the displacement compatibility (Gupta, 1978; Mcdill et al., 1987). Curved boundaries are modelled by trimming the boundary cells into polygons/polyhedra before further subdividing into finite elements. In Ebeida et al. (2010), a buffer zone was introduced between the boundary and the internal cells. A compatible mesh is then constructed to fill up this zone. An additional optimization step is usually performed to improve the quality of the final mesh.

The scaled boundary finite element method and the quadtree/octree algorithm are highly complementary techniques in numerical simulation. Figure 9.5 illustrates an S-element mesh generated by the quadtree algorithm. A cell is directly modelled as an S-element. Since the boundary of an S-element can be divided into an arbitrary number of line elements, a common edge between cells at different levels is modelled by multiple line elements representing the edges of the smaller cells. The displacement

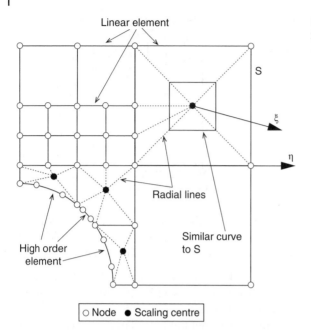

Figure 9.5 A quadtree mesh of S-elements.

compatibility is satisfied automatically and no hanging nodes exist. A cell trimmed by the boundary is also directly modelled as an S-element. Thus, the scaled boundary finite element method alleviates the difficulties encountered in using the quadtree/octree algorithm for finite element analysis, while the quadtree/octree algorithm provides an efficient approach for the generation of S-element meshes. Furthermore, the internal cells of a quadtree/octree mesh have a limited number of patterns of boundary discretization. Of all the S-elements of the same patterns, only one needs to be solved explicitly and the solutions of the others can be obtained by scaling (see Section 9.4).

9.2 Data Structure of S-element Meshes

S-elements are polygons/polyhedrons of arbitrary number of faces with boundary discretization only. The discretization may also form an open loop or surface (see Figure 2.6 on page 36). The increased flexibility in shape requires a data structure different from that of standard finite elements. In the accompanying program Platypus (see Section 3.5 on page 87), the data structure of a two-dimensional polygon element mesh has three types of objects, namely, node, edge and S-element.

The data structure of a three-dimensional polyhedral mesh has four types of objects: node, edge, face and S-element as illustrated in Figure 9.6.

A location of a node is specified by its coordinates (Figure 9.6a). Same as in the finite element method, every node in the system has a unique nodal number. An edge of a polyhedron is a line element interpolating the nodal values using shape functions. The direction of the edge (line element) is specified by the sequence of the nodes in its connectivity. An edge defined by a 2-node line element is shown in Figure 9.6b. A face of the polyhedron is a polygon formed by connecting the edges (Figure 9.6c). The directions of all the edges and the outward normal of the face (i.e. the normal pointing to the outside

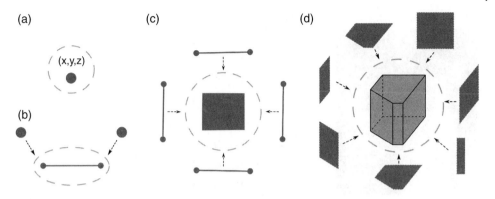

Figure 9.6 Types of objects in a three-dimensional polyhedral mesh. (a) Node. (b) Edge. (c) Face. (d) S-element.

of the polyhedron) must follow the right-hand rule. A polyhedral S-element is defined by a set of polygonal faces as shown in Figure 9.6d. The sequence of the faces in the data structure is clockwise.

9.3 Quadtree/Octree Mesh Generation of Digital Images

A digital image describes the geometry and material distribution of a problem by the colour/grey intensity of the pixels (in two dimensions) and voxels (in three dimensions). It is assumed that the image has been segmented and stored in a digital file format. A hierarchical quadtree/octree algorithm is well known for image encoding (Meagher, 1982). In the quadtree mesh generation presented below, the MATLAB function qtde-comp is used to decompose the images. The quadtree cells are further balanced and a quadtree mesh is generated.

9.3.1 Illustration of Quadtree Decomposition of Two-dimensional Images by an Example

The quadtree decomposition of an image aims to split the image into square cells of uniform (within a specified tolerance) colours. To present the quadtree decomposition process, the digital image shown in Figure 9.7a is employed as an example. The size of the image is 8×7 pixels. Each pixel in the image is assumed to represent a square domain of size $h \times h$ where h is the image sampling interval. This image is represented by the colour intensity value in each pixel. The colour intensity is specified in a range from 0 (darkest colour) to 15 (lightest colour). The colours of all the pixels are stored in an image colour matrix $[I_1]$ depicted in Figure 9.7b.

The quadtree algorithm is most efficient when the size of the image matrix is an integer power of 2, i.e. $2^n \times 2^n$ with the positive integer $n \geq 0$. To use the MATLAB function qtdecomp, the image matrix $[I_1]$ is padded with a background colour to construct a square matrix called $[I_2]$. The size of $[I_2]$ must be an integer power of 2. Padding the matrix $[I_1]$ in Figure 9.7b leads to the 8×8 image matrix $[I_2]$ shown in Figure 9.8. The background colour, which is 100 in this example, can be chosen arbitrarily as long as it is

(a)   Digital image.

(b)   Image matrix $[I_1]$

3	3	3	3	3	3	3
3	3	3	3	3	3	5
3	3	3	3	3	5	5
3	3	3	3	5	5	5
3	12	12	12	11	12	5
10	12	12	11	11	3	3
10	10	12	12	3	2	2
10	10	11	11	3	3	3

Figure 9.7 A digital image of 8 × 7 pixels for use as an example to demonstrate quadtree decomposition. The colour intensity is given in a range between 0 (darkest colour) and 15 (lightest colour).

3	3	3	3	3	3	3	100
3	3	3	3	3	3	5	100
3	3	3	3	3	5	5	100
3	3	3	3	5	5	5	100
5	12	12	12	11	12	5	100
10	12	12	11	11	3	3	100
10	10	12	12	3	2	2	100
10	10	11	11	3	3	3	100

Figure 9.8 The square image matrix $[I_2]$ of dimension of an integer power of 2 obtained by padding matrix $[I_1]$ in Figure 9.7b with a background colour (100) that is outside the colour range of the image.

sufficiently different from all the colours appearing in the original image. The quadtree decomposition is performed on the square matrix $[I_2]$.

The Image Processing Toolbox of MATLAB provides a function `qtdecomp.m` to perform the quadtree decomposition of a digital image. The command "`doc qtdecomp`" displays detailed documentation of this function. The source code is also available. This function is called up with the following input arguments:

- The square image matrix to be decomposed. In the present example, it is the padded matrix $[I_2]$ in Figure 9.8.
- A parameter `threshold` specifies the relative colour threshold with a value between 0 and 1 to control the homogeneity of colour intensity in a quadtree cell. `threshold` is converted inside the function `qtdecomp.m` to the actual threshold value of colour intensity according to the variable type of the image matrix. If the difference in the colour intensity within a quadtree cell is greater than the actual threshold value, the cell is split into four.
- An array [`mindim maxdim`] specifying the minimum and maximum dimensions of the edges of quadtree cells in pixels. Both `mindim` and `maxdim` must be an integer power of 2. A quadtree cell of a dimension of `mindim` will not be split even if it does not meet the homogeneity requirement of colour intensity. On the other hand, a quadtree cell of a dimension larger than `maxdim` will be split even if it meets the colour threshold condition.

The recursive process of quadtree decomposition is initialized by splitting the square image into cells of dimension `maxdim`. In the function `qtdecomp.m`, the homogeneity of a cell is measured by the difference between the maximum and minimum colour intensities of the pixels in the cell. When the difference is greater than the actual threshold value (converted from the input `threshold`), the cell is split further into four equal-sized cells by bisecting the edges. This process is performed recursively until all the cells satisfy the criterion of homogeneity or reaches the minimum dimension specified by `mindim`. In this algorithm, other criteria to control the cell size (such as material property, estimated error in post-processing, etc.) can be built in.

The process of quadtree decomposition of the image matrix in Figure 9.8 is illustrated in Figure 9.9. The parameters are chosen as follows. The actual threshold of colour intensity is equal to 2. The minimum dimension of a cell is `mindim=1` pixel and maximum dimension `maxdim=4` pixels. The initial decomposition into cells of the maximum dimension is shown in Figure 9.9a. It consists of cells (submatrices) of the same size (4×4 pixels). The difference between the maximum and minimum colour intensities of each cell (i.e. the maximum and minimum values of entries of each submatrix) is

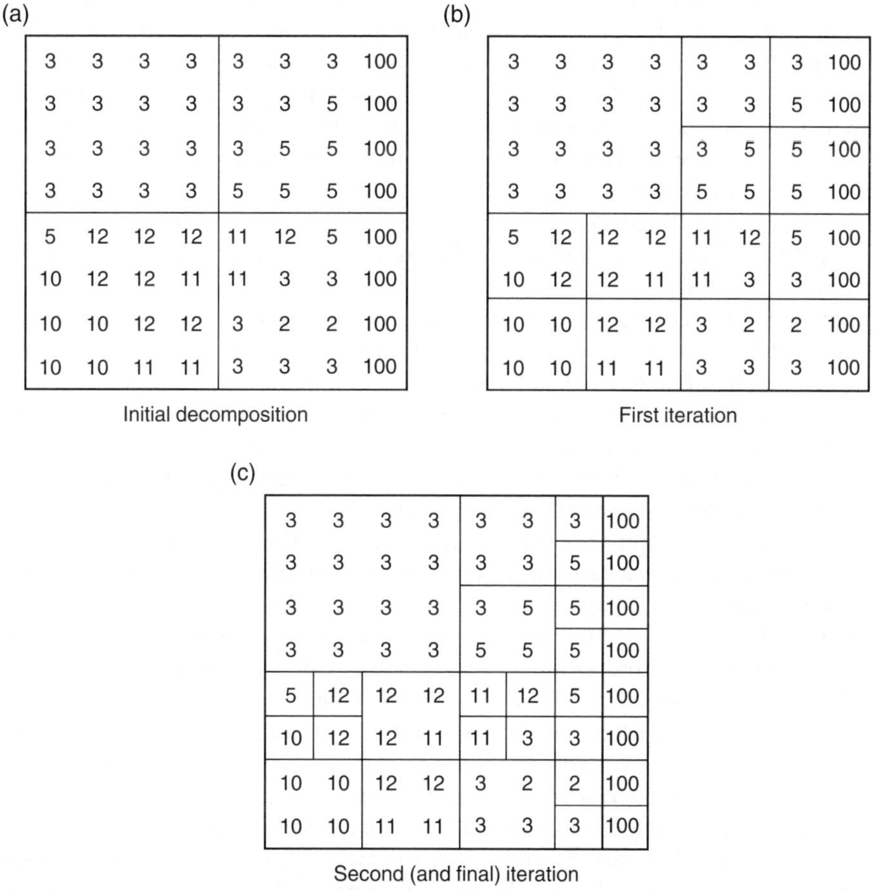

(a)

3	3	3	3	3	3	3	100
3	3	3	3	3	3	5	100
3	3	3	3	3	5	5	100
3	3	3	3	5	5	5	100
5	12	12	12	11	12	5	100
10	12	12	11	11	3	3	100
10	10	12	12	3	2	2	100
10	10	11	11	3	3	3	100

Initial decomposition

(b)

3	3	3	3	3	3	3	100
3	3	3	3	3	3	5	100
3	3	3	3	3	5	5	100
3	3	3	3	5	5	5	100
5	12	12	12	11	12	5	100
10	12	12	11	11	3	3	100
10	10	12	12	3	2	2	100
10	10	11	11	3	3	3	100

First iteration

(c)

3	3	3	3	3	3	3	100
3	3	3	3	3	3	5	100
3	3	3	3	3	5	5	100
3	3	3	3	5	5	5	100
5	12	12	12	11	12	5	100
10	12	12	11	11	3	3	100
10	10	12	12	3	2	2	100
10	10	11	11	3	3	3	100

Second (and final) iteration

Figure 9.9 Quadtree decomposition of the image matrix in Figure 9.8. The parameters are: actual threshold of colour intensity is equal to 2, `mindim=1` pixel and `maxdim=4` pixels.

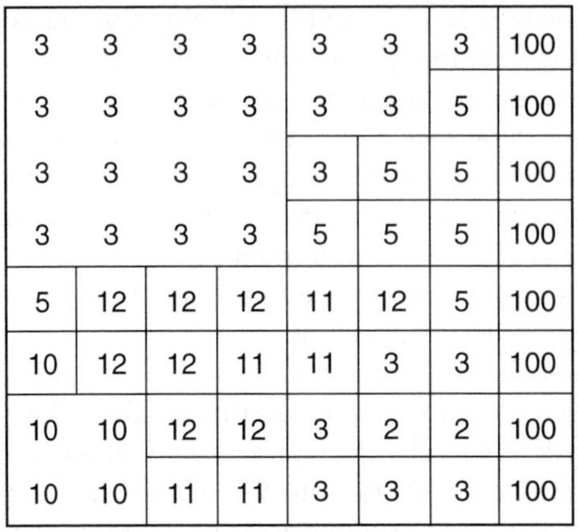

3	3	3	3	3	3	3	100
3	3	3	3	3	3	5	100
3	3	3	3	3	5	5	100
3	3	3	3	5	5	5	100
5	12	12	12	11	12	5	100
10	12	12	11	11	3	3	100
10	10	12	12	3	2	2	100
10	10	11	11	3	3	3	100

Figure 9.10 Quadtree decomposition of the image matrix in Figure 9.8 into cells of uniform colour intensity (`threshold=0`). The maximum dimension of the edge is `maxdim=4` pixels and the minimum dimension is `mindim=1` pixel.

determined and the cells to be split are identified. In Figure 9.9a, all the cells except for the one at the upper-left corner have greater difference in colour intensity than the threshold. They are split into four cells of 2×2 pixels as shown in Figure 9.9b depicting the decomposition after the first iteration. Repeating the same operations, the decomposition after the second iteration is shown in Figure 9.9c, where the homogeneities of the colour intensities of all the cells meet the required threshold (difference between maximum and minimum values is not greater than 2). The iteration terminates and the resulting decomposition leads to loss of resolution specified by the threshold value.

The value of the threshold can be set as `threshold=0` to maintain the full resolution of the original image. The decomposition of the image matrix in Figure 9.8 with the parameter `threshold=0` is illustrated in Figure 9.10. It is noted that all cells have a uniform colour intensity.

A quadtree/octree decomposition for mesh generation is usually balanced by limiting the length ratio between two adjacent cells to 2 (2 : 1 rule). The balanced decomposition reduces the number of unique cell patterns in the quadtree/octree decomposition. This simplifies mesh generation and can also be advantageous for numerical analysis (see Section 9.4 on page 370). The balancing can be enforced by recursively splitting cells that are connected to more than two cells at any of their edges. Figure 9.11 is obtained by balancing the quadtree decomposition in Figure 9.9c.

The final quadtree decomposition is obtained after deleting the cells in the padded area, i.e. those with the background colour. For the present example, the quadtree decomposition of image in Figure 9.7a is obtained by deleting the cells with the background colour intensity (100) in Figure 9.11. The final quadtree decomposition is illustrated in Figure 9.12. The geometry of a quadtree cell is characterized by its size (controlled by the level of splitting) and the pattern of connection to adjacent cells.

In a balanced quadtree decomposition, the number of unique patterns of quadtree cells is limited and typically much smaller than the number of cells in the quadtree decomposition. A cell in a balanced quadtree decomposition is depicted in Figure 9.13. The location and size of the cell are defined by its four corners shown as solid dots (●). When a side of the cell is directly connected to two smaller cells, a midpoint exists on the side. The locations of the possible midpoints are indicated by the circles (○) in Figure 9.13. When a side of the cell is directly connected to one cell only or is on the

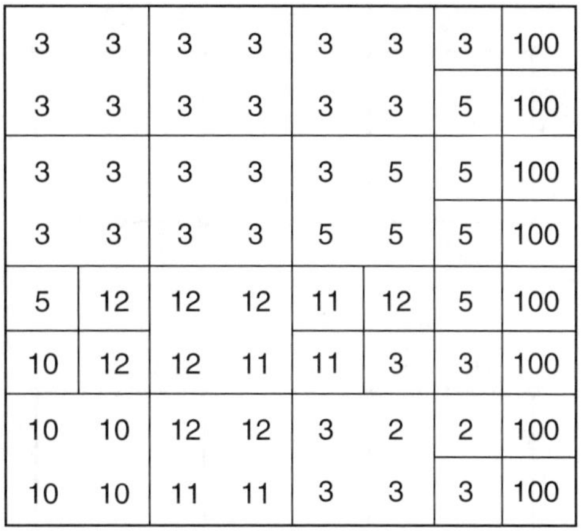

Figure 9.11 The balanced quadtree decomposition of the image matrix in Figure 9.8 obtained by balancing the decomposition in Figure 9.9c.

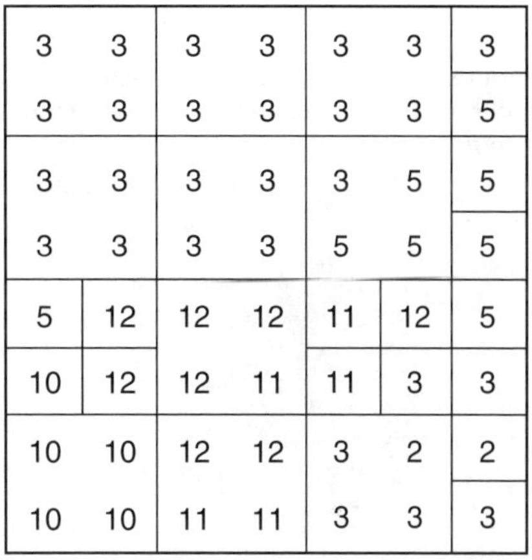

3	3	3	3	3	3	3	100
3	3	3	3	3	3	5	100
3	3	3	3	3	5	5	100
3	3	3	3	5	5	5	100
5	12	12	12	11	12	5	100
10	12	12	11	11	3	3	100
10	10	12	12	3	2	2	100
10	10	11	11	3	3	3	100

Figure 9.12 Balanced quadtree mesh of the image in Figure 9.7a obtained with the actual threshold of colour intensity being equal to 2, `mindim=1` pixel and `maxdim=4` pixels.

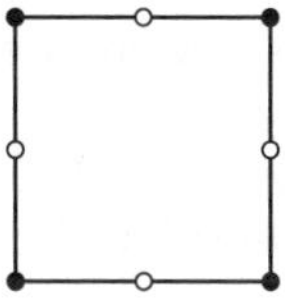

3	3	3	3	3	3	3
3	3	3	3	3	3	5
3	3	3	3	3	5	5
3	3	3	3	5	5	5
5	12	12	12	11	12	5
10	12	12	11	11	3	3
10	10	12	12	3	2	2
10	10	11	11	3	3	3

Figure 9.13 A balanced quadtree cell. The solid dots (•) indicate the four corners and the circles (○) the locations of possible midpoints connected to adjacent cells.

boundary (i.e. is not directly connected to any cells), the midpoint does not appear. The patterns of the cells are determined by the existence or not of the midpoints on the sides and the number of unique patterns is thus equal to $2^4 = 16$. This number can be reduced by splitting cells that have more than a specified number of midpoints.

When rotations of the quadtree cells are considered, the unique patterns are reduced further to the six shown in Figure 9.14. All the quadtree cells can be generated by rotating and scaling these six cells.

The quadtree decomposition is directly treated as a mesh of S-elements in the scaled boundary finite element method. Each quadtree cell is modelled as an S-element. The discretization of the boundary of the six unique patterns using 2-node line elements is illustrated in Figure 9.14. Note that two line elements are used on a side connected

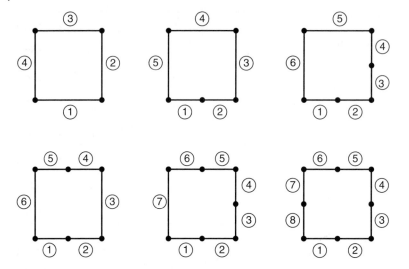

Figure 9.14 Unique patterns of S-elements in a balanced quadtree mesh. The numbers in circles indicate the element numbers on the boundary of the S-elements.

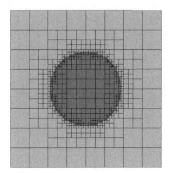

Figure 9.15 A balanced quadtree decomposition of a circular inclusion in a square domain.

with two adjacent S-elements, and the displacement compatibility across the boundary is always satisfied.

Another example of a balanced quadtree mesh is shown in Figure 9.15. The image is a square domain with a circular inclusion. The decomposition results in fine cells around the circular boundary and coarse cells away from it. The dimension of quadtree cells decreases exponentially as refinement progresses. After r times of refinement of a cell of length h, the dimension of the smallest cell becomes $h/2^r$. The transition of element size is thus very rapid. For example, the size of the smaller cells after 10 refinements ($r = 10$) will be $h/2^{10} = h/1024$, which is one-thousandth of the original size h.

9.3.2 Octree Decomposition

A 3D image of an object consists of a stack of 2D images, each of which is a slice of the 3D object. Each voxel in a 3D image is assumed to represent a cubic domain of size $h \times h \times h$, where h is the image sampling interval.

The algorithm of an octree decomposition of a 3D image is analogous to that of the quadtree decomposition. The image is fitted into a cubic bounding box of $2^n \times 2^n \times 2^n$

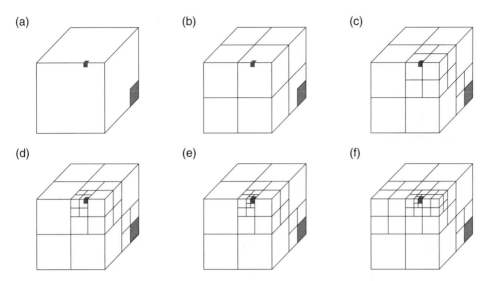

Figure 9.16 Example of octree decomposition of a 3D digital image into cells of uniform colour intensity (`threshold=0`). The maximum dimension of the edge is `maxdim=8` voxels and the minimum dimension is `mindim=1` voxel. (a) A cube of $16 \times 16 \times 16$ voxels with a grey spot of 1 voxel and a grey spot of $4 \times 4 \times 4$ voxels. (b)–(e) The first, second, third and final decomposition. (f) Balanced octree decomposition.

voxels where n is an integer. The image in the bounding box is recursively divided by splitting every cubic cell that does not meet the criterion of homogeneity of colour intensity into 8 equal-sized cubic cells. The process terminates when all cells either meet the criterion of homogeneity or reach the smallest size specified by the user. To reduce the number of unique cell patterns, the octree decomposition is also balanced by enforcing the 2 : 1 rule on the sizes of adjacent cells.

The process of octree decomposition is illustrated in Figure 9.16. The 3D image of $16 \times 16 \times 16$ voxels in Figure 9.16a contains a grey spot of 1 voxel and a grey spot of $4 \times 4 \times 4$ voxels. The image is decomposed into cells of uniform colour intensity (`threshold=0` and `mindim=1`). The imagine is initially decomposed into cells of dimension `maxdim=8` voxels (Figure 9.16b). The colour intensities of the two cells containing the grey spots are not uniform. They are split into 8 cells of dimension of 4 voxels (Figure 9.16c). To model the spot of 1 voxel, the cell containing it is recursively split twice (Figure 9.16d and Figure 9.16e). In the octree decomposition in Figure 9.16e, the colour intensity is uniform in every cell. After splitting the cells further to enforce the 2 : 1 rule, the balanced octree decomposition in Figure 9.16f is obtained.

The octree decomposition of one octant of a spherical inclusion in a cube is shown in Figure 9.17 as an example. The sizes of the cells are the smallest around the boundary of the sphere and increase gradually towards regions of homogeneous colour intensities.

The unique patterns of a balanced octree are identified by examining the connectivity with adjacent cells. A cell is shown in Figure 9.18, where the 8 corners are indicated by solid dots (•). Other than the 8 cells connected to one corner only, 18 adjacent cells are connected to this cell through the 12 edges and 6 faces. The patterns of the cells are determined by the presence (or not) of midpoints on the 12 edges and centre points of the 6 faces. When an edge is connected to 2 cells of smaller size, a midpoint

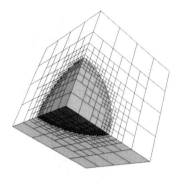

Figure 9.17 Balanced octree decomposition of one octant of a spherical inclusion in a cube.

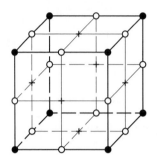

Figure 9.18 A balanced octree cell. The solid dots (•) indicate the corners. The circles (o) and the crosses (+) show, respectively, the locations of possible centre points of square faces and midpoints of edges connected to adjacent cells.

occurs on the edge. The 12 possible locations of the midpoints are shown as circles (o) in Figure 9.18. In total, there are $2^{12} = 4096$ unique patterns of configures of midpoints. When a face is connected to 4 cells of smaller size, a centre point, as shown by a cross (+), will be introduced. In this case, the 4 midpoints on the same face will appear together with the centre point. The number of unique patterns is not affected by the centre points of the faces.

The number of unique patterns can be reduced considering the rotations of the cells and by splitting the cells that have more midpoints and/or centre points than specified values.

An octree decomposition can be directly used as a mesh of S-elements. Each octree cell is modelled as an S-element in the scaled boundary finite element method. The scaling centre of the cubic domain is chosen at the geometric centre, from which the whole boundary of the cell is clearly visible. A node is introduced at every location where one or more corners of octree cells exists.

Only a boundary discretization of an S-element is required in the scaled boundary finite element method. The discretization must ensure the displacement compatibility with adjacent S-elements. In the following, the faces of octree cells are subdivided into standard triangular and quadrilateral elements. The use of polygon elements without subdivision of the faces is also possible. Interested readers are referred to Natarajan et al. (2017) and Zou et al. (2017).

It is sufficient to consider one face of an octree cell in the boundary discretization. The 7 unique patterns of node configuration are depicted in Figure 9.19 with the nodes indicated by solid dots (•). The first six patterns (Figures 9.19a–f) have corner and edge nodes only and are the same as the patterns of quadtree cells in Figure 9.14. The last one (Figure 9.19g) has a centre node. The chosen divisions of the square faces are illustrated by the thin solid lines. The elements consist of isosceles triangles, 1:2 rectangles and

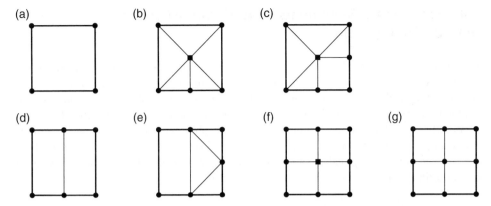

Figure 9.19 Unique patterns of configuration of nodes, indicated by solid dots (•), on a face of an octree cell. The faces are discretized by (isosceles) triangular and rectangular (square) elements as depicted by the additional thin solid lines. A centre node, shown as a square (■), is introduced in the configurations shown in Figures b, c and f.

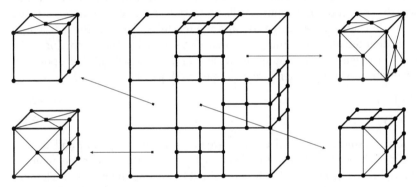

Figure 9.20 The surface discretization of S-elements in a balanced octree mesh with hanging nodes.

squares. Note that a centre node, indicated by a square (■), is introduced in the configurations shown in Figures 9.19b, 9.19c and 9.19f. Since one face can be shared by only two adjacent S-elements and the edges are not modified, this operation does not affect the meshes of any other adjacent S-elements. Note that other possibilities of surface discretization exists.

An example of surface discretization of octree cells using the seven patterns to construct S-elements is shown in Figure 9.20. The middle part of the figure illustrates the octree decomposition consisting of cells of two different sizes. Each face of the smaller cells is modelled as a square element. The S-elements obtained by discretizing four larger cells are depicted in the left and right parts of the figure. It can be seen that the discretization is performed face-by-face according to the patterns in Figure 9.19.

Note, in comparison with the use of octree mesh in the finite element method, that

- The use of S-elements directly ensures the displacement compatibility between adjacent octree cells. No hanging nodes exist in the mesh.
- The boundary discretization of octree cells is much simpler than the volume discretization requied by the finite element method.

9.4 Solutions of S-elements with the Same Pattern of Node Configuration

In a quadtree/octree mesh, the properties of an S-element depend on its size and node configuration. The number of unique patterns of node configurations of S-elements is limited and does not increase with the size of the problem. In a large-scale problem, many S-elements of the same patterns but different sizes will exist. Geometrically, those S-elements are similar and can be obtained by scaling one S-element according to their size ratios. The solutions of the S-elements of the same pattern can also be obtained by scaling, although not always linear. This will lead to saving in computational time.

9.4.1 Two-dimensional S-elements

The scaling of S-elements of the same quadtree cell pattern is illustrated in Figure 9.21. The four corner nodes are shown as solid dots (•). The locations of possible midside nodes are indicated by circles (○). For a given pattern, the midside nodes may or may not appear and the corresponding boundary discretizations are given in Figure 9.14. The S-element on the left side on the figure is treated as a master S-element. The half-length of the square is denoted as $\bar{b}$ (overline ($\bar{}$) for master S-element). The vectors of nodal coordinates of a line element are denoted as $\{\bar{x}\}$ and $\{\bar{y}\}$. The coordinates of a point on the boundary of the master S-element are expressed as (Eq. (2.17))

$$\bar{x}_b = [N]\{\bar{x}\} \tag{9.1a}$$
$$\bar{y}_b = [N]\{\bar{y}\} \tag{9.1b}$$

The formulation and solution of the S-element are described in Chapters 2 and 3.

An S-element of the same pattern of node configuration as the master S-element is shown on the right side of Figure 9.21. It is obtained by scaling the master S-element with a ratio denoted as R. The half-length of the S-element is equal to $R\bar{b}$. The nodal coordinates are expressed as

$$\{x\} = R\{\bar{x}\} \tag{9.2a}$$
$$\{y\} = R\{\bar{y}\} \tag{9.2b}$$

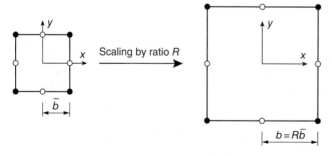

Figure 9.21 Scaling of S-elements generated from the same quadtree cell pattern. The solid dots (•) indicate the corner nodes. The circles (○) indicate the possible midside nodes at the midpoints of the edges.

Considering Eq. (9.1), the coordinates of a point on the boundary are related to those of the master cell by

$$x_b = [N]\{x\} = R[N]\{\bar{x}\} = R\bar{x}_b \tag{9.3a}$$

$$y_b = [N]\{y\} = R[N]\{\bar{y}\} = R\bar{y}_b \tag{9.3b}$$

Using Eq. (9.3), the solution of the S-element can be obtained from the solution of the master S-element. Only the key equations necessary for computer implementation are given below.

The Jacobian matrix (Eq. (2.28)) of the scaled boundary transformation is expressed on the boundary as

$$[J_b] = R[\bar{J}_b] \tag{9.4}$$

and its determinant (Eq. (2.31)) is equal to

$$|J_b| = R^2|\bar{J}_b| \tag{9.5}$$

The differential operator for 2D elasticity is expressed in Eq. (2.66) and, using Eqs. (9.3) and (9.5), the matrices in Eq. (2.67) are written as

$$[b_1] = R^{-1}[\bar{b}_1] \tag{9.6a}$$

$$[b_2] = R^{-1}[\bar{b}_2] \tag{9.6b}$$

The strain-displacement matrices in Eq. (2.82) are obtained as

$$[B_1] = R^{-1}[\bar{B}_1] \tag{9.7a}$$

$$[B_2] = R^{-1}[\bar{B}_2] \tag{9.7b}$$

Using Eqs. (9.5) and (9.7), the coefficient matrices in Eq. (2.110) are equal to those of the master S-element

$$[E_0] = [\bar{E}_0] \tag{9.8a}$$

$$[E_1] = [\bar{E}_1] \tag{9.8b}$$

$$[E_2] = [\bar{E}_2] \tag{9.8c}$$

Thus, the scaled boundary finite element equation (Eq. (2.112)) and the equivalent system of first-order differential equations (Eq. (3.4) with the coefficient matrix $[Z_p]$ in Eq. (3.5)) are the same as those of the master S-element

$$[Z_p] = [\bar{Z}_p] \tag{9.9}$$

In the S-element solution by eigenvalue decomposition (Section 3.1), the eigenvalue problem in Eq. (3.12) is the same as that of the master S-element. The eigenvalues (Eq. (3.14)) and eigenvector (Eq. (3.13)) of the master S-element can be directly used

$$\lambda_i = \bar{\lambda}_i \tag{9.10a}$$

$$\{\phi_i\} = \{\bar{\phi}_i\} \tag{9.10b}$$

In the S-element solution by block-diagonal Schur decomposition (Section 7.6), the decomposition in Eq. (7.108) is identical to that of the master S-element leading to

$$[S_i] = [\bar{S}_i] \tag{9.11a}$$

$$[\Psi_i] = [\bar{\Psi}_i] \tag{9.11b}$$

The stiffness matrix (Eq. (3.25) for eigenvalue decomposition or Eq. (7.157) for block-diagonal Schur decomposition) is also equal to that of the master S-element

$$[K] = [\bar{K}] \tag{9.12}$$

The strain modes in the eigenvalue decomposition (Eq. (3.62)) or in the block-diagonal Schur decomposition (Eq. (7.190)) are expressed as

$$\{\phi_i^{(\varepsilon)}\} = R^{-1}\{\bar{\phi}_i^{(\varepsilon)}\}; \quad \text{or} \quad [\Psi_i^{(\varepsilon)}] = R^{-1}[\bar{\Psi}_i^{(\varepsilon)}]$$

The coefficient matrix $[M_0]$ related to the mass matrix in Eq. (3.102) is expressed, using Eq. (9.5), as

$$[M_0] = R^2[\bar{M}_0]$$

and mass matrix in Eq. (3.105) as

$$[M] = R^2[\bar{M}]$$

Therefore, it is sufficient to compute the solution of the six master S-elements. The solutions of all the S-elements of the same pattern of node configuration can be obtained by scaling the solution of the master S-elements according to the relationships above. This will reduce the computational effort.

9.4.2 Three-dimensional S-elements

Three-dimensional S-elements of the same octree cell pattern are handled following the above procedure for two-dimensional S-elements. A master S-element is chosen for the cell pattern. The half-size of this cube is denoted as $\bar{b}$. Its faces are discretized as shown in Figure 9.19. The nodal coordinates of a surface element are denoted as $\{\bar{x}\}$, $\{\bar{y}\}$, $\{\bar{z}\}$. The solution of the master element is obtained by either the eigenvalue decomposition (Chapter 3) or the block-diagonal Schur decomposition (Chapter 7). In the notations in this section, the overline ($^-$) denotes the master S-element.

An S-element of the same pattern but with half-length $R\bar{b}$ is considered, where R is the scaling ratio. The nodal coordinates are expressed as

$$\{x\} = R\{\bar{x}\} \tag{9.13a}$$
$$\{y\} = R\{\bar{y}\} \tag{9.13b}$$
$$\{z\} = R\{\bar{z}\} \tag{9.13c}$$

and the coordinates of a point on boundary are written as

$$x_b = R\bar{x}_b \tag{9.14a}$$
$$y_b = R\bar{y}_b \tag{9.14b}$$
$$z_b = R\bar{z}_b \tag{9.14c}$$

The Jacobian matrix (Eq. (6.51)) on the boundary is expressed as

$$[J_b] = R[\bar{J}_b] \tag{9.15}$$

The determinant (Eq. (6.53a)) is obtained as

$$|J_b| = R^3|\bar{J}_b| \tag{9.16}$$

The matrices (Eq. (6.79)) defining the differential operator of 3D elasticity (Eq. (6.78)) are written as

$$[b_1] = R^{-1}[\bar{b}_1] \tag{9.17a}$$

$$[b_2] = R^{-1}[\bar{b}_2] \tag{9.17b}$$

$$[b_3] = R^{-1}[\bar{b}_3] \tag{9.17c}$$

The strain-displacement matrices (Eq. (6.91)) are given by

$$[B_1] = R^{-1}[\bar{B}_1] \tag{9.18a}$$

$$[B_2] = R^{-1}[\bar{B}_2] \tag{9.18b}$$

Using Eqs. (9.16) and (9.18), the coefficient matrices (Eq. (6.140)) are obtained as

$$[E_0] = R[\bar{E}_0] \tag{9.19a}$$

$$[E_1] = R[\bar{E}_1] \tag{9.19b}$$

$$[E_2] = R[\bar{E}_2] \tag{9.19c}$$

$$[M_0] = R[\bar{M}_0] \tag{9.19d}$$

To solve the scaled boundary finite element equation, a system of first-order ordinary differential equations (Eq. 7.3) are formulated. The constant matrix $[Z_p]$ (Eq. 7.3) is related to that of the master S-element $[\bar{Z}_p]$ by

$$[Z_p] = \begin{bmatrix} [I] & \\ & R[I] \end{bmatrix} [\bar{Z}_p] \begin{bmatrix} [I] & \\ & R^{-1}[I] \end{bmatrix} \tag{9.20}$$

where $[I]$ is an identity matrix of conformable size. The real Schur decomposition of $[Z_p]$ (Eq. (7.31)) is expressed as

$$\begin{bmatrix} [I] & \\ & R[I] \end{bmatrix} [\bar{Z}_p] \begin{bmatrix} [I] & \\ & R^{-1}[I] \end{bmatrix} [V] = [V][S] \tag{9.21}$$

with the transformation matrix $[V]$ and Schur matrix $[S]$. It is rewritten as

$$[\bar{Z}_p] \begin{bmatrix} [I] & \\ & R^{-1}[I] \end{bmatrix} [V] = \begin{bmatrix} [I] & \\ & R^{-1}[I] \end{bmatrix} [V][S] \tag{9.22}$$

Comparing with the Schur decomposition of the master S-element

$$[\bar{Z}_p][\bar{V}] = [\bar{V}][\bar{S}] \tag{9.23}$$

the following relationships with the transformation matrix $[\bar{V}]$ and Schur matrix $[\bar{S}]$ of the master cell are identified

$$[\bar{V}] = \begin{bmatrix} [I] & \\ & R^{-1}[I] \end{bmatrix} [V] \tag{9.24a}$$

$$[\bar{S}] = [S] \tag{9.24b}$$

Note that the Schur matrix is independent of the size of the S-element. Equation (9.24a) is rewritten as

$$[V] = \begin{bmatrix} [I] & \\ & R[I] \end{bmatrix} [\bar{V}] \tag{9.25}$$

Using the partitioning of the transformation matrix $[V]$ in Eq. (7.53), Eq. (9.25) is expressed as

$$\begin{bmatrix} [V_b^{(u)}] & [V_u^{(u)}] \\ [V_b^{(q)}] & [V_u^{(q)}] \end{bmatrix} = \begin{bmatrix} [\bar{V}_b^{(u)}] & [\bar{V}_u^{(u)}] \\ R[\bar{V}_b^{(q)}] & R[\bar{V}_u^{(q)}] \end{bmatrix} \tag{9.26}$$

Using Eq. (9.26), the stiffness matrix of the S-element (Eq. (7.79)) is obtained as

$$[K] = R[\bar{V}_b^{(q)}][\bar{V}_b^{(u)}]^{-1} = R[\bar{K}] \tag{9.27}$$

It is proportional to the size of the S-element.

Following the above procedure, it can also be shown that the scaling of the transformation matrix in Eq. (9.25) is also applicable to the block-diagonal Schur decomposition (Eq. (7.107)) and the eigenvalue decomposition (Eq. (3.12)). The block-diagonal Schur matrix and the eigenvalues are also independent of the size of the S-element.

The strain modes (Eq. (7.190)) corresponding to the partitioned transformation matrix in Eq. (7.126)) are obtained as

$$[\Psi_i^{(\varepsilon)}] = R^{-1}[\bar{\Psi}_i^{(\varepsilon)}] \tag{9.28}$$

where $[\bar{\Psi}_i^{(\varepsilon)}]$ are the strain modes of the master S-element.

9.5 Examples of Image-based Analysis

Linear elastic analysis of digital images of concrete specimens in two and three dimensions are presented to illustrate the use of quadtree/octree mesh in the scaled boundary finite element method. The meshes are generated directly from the digital images. The mesh generation process is fully automatic.

9.5.1 A 2D Concrete Specimen

A 2D image of a concrete specimen with a size of 381×383 pixels is shown in Figure 9.22a. This image is obtained from Figure 2b in Ren et al. (2013). In the image, the black, grey and white colours represent the aggregates, mortar and voids. Each pixel represents a square of 0.1 mm $\times 0.1$ mm. The elastic material properties are $E = 70$ GPa and $v = 0.2$ for the aggregates, and $E = 25$ GPa and $v = 0.2$ for the mortar. Plane strain condition is assumed.

In the quadtree decomposition of the image, the maximum and minimum sizes of the cells are 8 pixels and 1 pixel, respectively. The balanced quadtree decomposition is shown in Figure 9.22b. It consists of $40,738$ S-elements, leading to a total of $100,374$ degrees of freedom. The smallest S-elements appear at material interfaces. In homogeneous areas, larger S-elements are used, which reduces the number of S-elements necessary to achieve the specified resolution of the image. The ratio of number of S-elements in the quadtree mesh to the total number of pixels in the digital image is around 27.9%.

Using the unique patterns of node configurations, the S-element solutions are computed for the 16 master S-elements for each material, i.e. 32 in total, only. This represents less than 0.08% of the total number of S-elements in the mesh. The solutions of all the S-elements are obtained by scaling the solutions of the corresponding master S-elements.

(a)

(b)

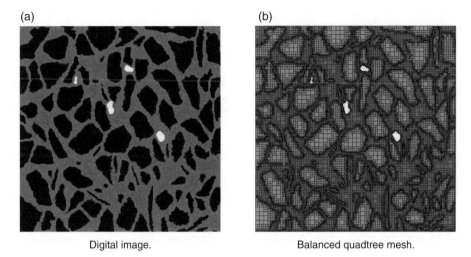

Digital image.   Balanced quadtree mesh.

Figure 9.22 Modelling of 2D concrete specimen.

The uniaxial extension of the specimen in the vertical direction is modelled. Displacement boundary conditions corresponding to a uniaxial tensile strain of 1×10^{-5} are enforced at the top and bottom of the concrete specimen. The whole process is fully automatic, eliminating the human effort in mesh generation, and computationally very efficient. The computer time is measured on a Microsoft Surface Pro 3 laptop computer with an i7-4650U CPU. The computer code is written in MATLAB. It takes less than 2 s to create the numerical model (generating mesh, enforcing boundary conditions, etc.), 1.5 s to perform the analysis to obtain nodal displacements and 3 s for stress calculation and graphical output.

The vertical displacement is plotted in Figure 9.23a for the aggregates and in Figure 9.23b for the mortar. The maximum principal stress is shown in Figure 9.24a for the aggregates and in Figure 9.24b for the mortar. Stress concentrations can be observed around the voids.

(a)   (b)

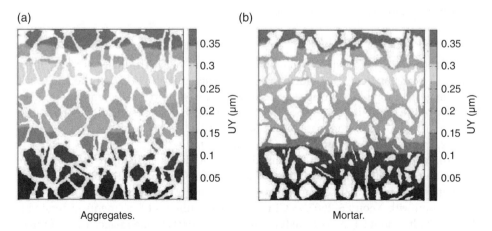

Aggregates.   Mortar.

Figure 9.23 Vertical displacement (μm) of 2D concrete specimen under uniaxial extension.

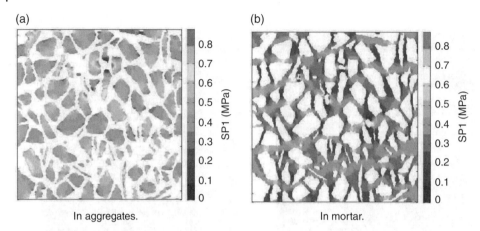

In aggregates.

In mortar.

Figure 9.24 Maximum principal stress (MPa) for the 2D concrete specimen under uniaxial extension.

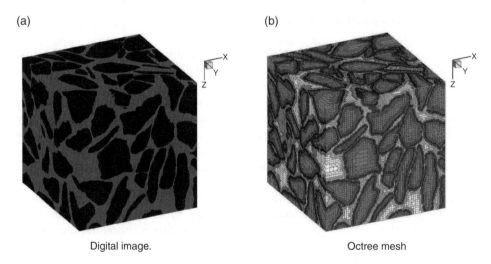

Digital image.

Octree mesh

Figure 9.25 Model of 3D concrete specimen.

9.5.2 A 3D Concrete Specimen

A 3D image of a concrete specimen is shown in Figure 9.25a. The image is a part of the X-ray CT scans of a concrete block analysed in Huang et al. (2015). In the figure, the black material represents the aggregates and the grey material represents the mortar. The size of the image is $256 \times 256 \times 256$ voxels. Each voxel represents 0.08 mm $\times$ 0.08 mm $\times$ 0.08 mm. The elastic material properties of the aggregates are $E = 70$ GPa and $v = 0.2$, and of the mortar are $E = 25$ GPa and $v = 0.2$.

In the octree decomposition, the maximum cell size is limited to 8^3 voxels. The minimum cell size is 1 voxel to use the full resolution of the image. The balanced octree decomposition takes 350 seconds using an in-house MATLAB code running on a workstation with Xeon E5-2667 v3 CPU. The total number of S-elements generated is $3,002,329$. It is about 17.9% of the total number of voxels in the image. The boundary discretization of the S-elements takes around 84 seconds. The octree mesh is shown in

Figure 9.26 Vertical displacement (in μm) 3D concrete specimen under uniaxial extension.

(a)

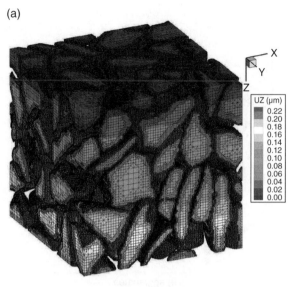

In aggregates.

(b)

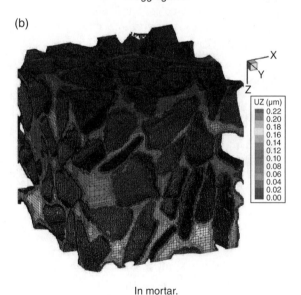

In mortar.

Figure 9.25b. It has a total of $17, 360, 049$ degrees of freedom. The two materials have a combined total of $2, 920$ master S-elements, which represent only about 0.1% of the total number of S-elements.

A uniaxial tensile strain of 1×10^{-5} is enforced by prescribing vertical displacements at the top and bottom faces of the specimen. The vertical displacement distributions in aggregates and in mortar are shown in Figures 9.26a and 9.26b, respectively. The displacement distribution inside the specimen is visualized using 3 yz−slices. Figures 9.27a and Figure 9.27b show the results in the aggregates and mortar, respectively.

(a)

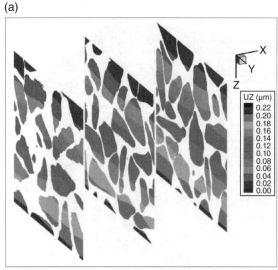

In aggregates.

Figure 9.27 Vertical displacement (μm) on 3 yz−slices of 3D concrete specimen under uniaxial extension.

(b)

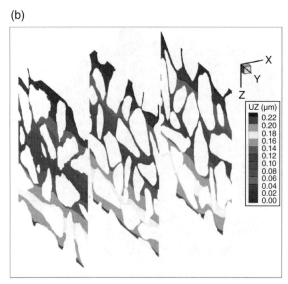

In mortar.

The maximum principal stress in aggregates and mortar are plotted in Figures 9.28a and 9.28b, respectively. The vertical stress distribution inside the specimen is depicted in Figures 9.29a and 9.29b, respectively.

9.6   Quadtree/Octree Mesh Generation for CAD Models

As briefly described in Section 9.1.2, the quadtree/octree algorithm has been applied to mesh CAD models. This technique is highly complementary with the scaled boundary

Figure 9.28 Maximum principal stress (MPa) of 3D concrete specimen.

(a)

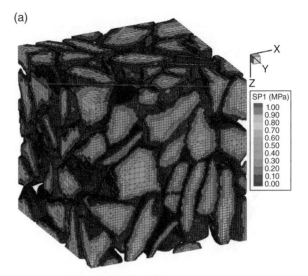

In aggregates.

(b)

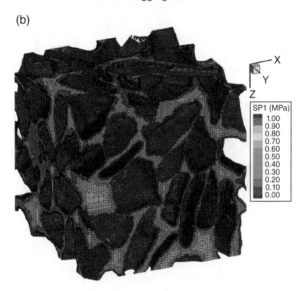

In mortar.

finite element method owing to the topological flexibility of S-elements. The issue of hanging nodes does not occur. It is also unnecessary to further split the quadtree/octree cells into finite elements that have limited topological shapes. The major steps in generating the S-element mesh of a CAD model consist of:

1) Create a quadtree/octree grid fitting the CAD model.
2) Trim the quadtree/octree cells by the surface of the CAD model.
3) Create the S-element mesh by discretizing the quadtree/octree cells (including trimmed ones).

(a)

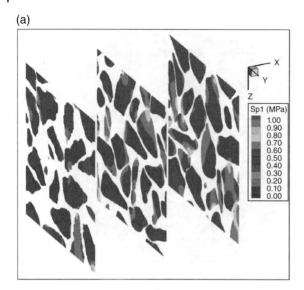

In aggregates

Figure 9.29 Maximum principal stress (in MPa) on 3 *yz*–slices of 3D concrete specimen under uniaxial extension.

(b)

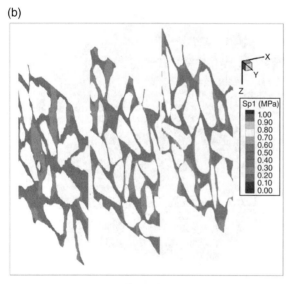

In mortar.

In this section, the key aspects of the process are described. The representation of sharp features (such as corners and edges) will not be discussed.

9.6.1 Quadtree/Octree Grid

The meshing process starts with identifying a square/cubic bounding box of the CAD model. Similar to mesh generation from digital images, a maximum cell size and a minimum cell sizes are specified by the user. The boundary box is divided into cells of uniform size not larger than the maximum cell size.

Figure 9.30 Quadtree grid of a square domain. The density of cells are controlled by the mesh seeds given on a inclined straight line and two circles.

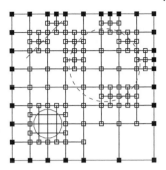

The density of cells representing the CAD model is controlled via mesh seed points, which is a technique commonly employed in finite element mesh generation. The inputs required from the user include:

- Maximum number of seed points allowed to appear in one cell,
- Seed points on the boundary of the CAD model and/or in regions of the model. The density of seed points can be determined based on the geometry (such as maximum chord height to chord length ratio, curvature, etc.) and the behaviour of solution (such as error indicators).

The cells containing the seed points can easily be determined at each level of quadtree/octree decomposition. The coordinates of a seed point are denoted as x_s, y_s and z_s, where the origin of the coordinate system is at the lower-left corner of the bounding box. At a given level of decomposition with cells size h, the indices of the seed point are given by

$$C_i = \mathrm{ceil}(x_s/h); \qquad C_j = \mathrm{ceil}(y_s/h); \qquad C_k = \mathrm{ceil}(z_s/h) \tag{9.29}$$

where the function 'ceil' rounds the argument to the nearest integer above its current value. The indices C_i, C_j and C_k of the cell start from 1.

After the seeds points are generated, the mesh process is very similar to that in the image-based analysis. The number of seeds points in the cells are counted. If the number in a cell is greater than the specified maximum number, the cell is split by bisecting its edges. This process is applied recursively until all cells contain no more seed points than the specified value. The resulting quadtree/octree decomposition is then balanced in the same way as in the decomposition of digital images (see Section 9.3). Figure 9.30 illustrates the quadtree grid of a square region where the seed points are given on a straight lines and two circles. A three-dimensional example is illustrated in Figure 9.3 in Section 9.1.2.

9.6.2 Trimming of Boundary Cells

The quadtree/octree grid consists of horizontal and vertical lines (see Figures 9.3 and 9.30). To improve the representation of the boundary without using excessive numbers of cells, cells intersecting with the boundary are trimmed. The trimming process of an octree follows the sequence of edges, faces and cells (Figure 9.31).

It is possible that some grid points are located very close to the boundary in comparison with the lengths of their edges. After trimming, short edges and poorly shaped

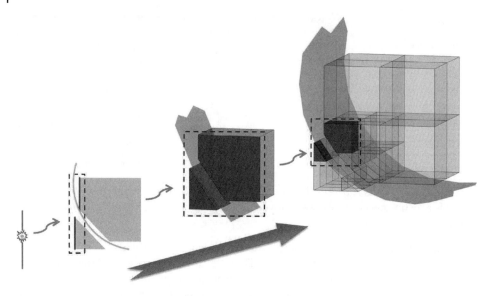

Figure 9.31 Sequence of operation in trimming an octree.

Figure 9.32 Splitting an edge. The signed distance function of a point is positive, zero or negative when the point is located outside the domain, on the boundary and inside the domain.

S-elements may occur. This is avoided by shifting the grid points within a specified distance either to the boundary or away from the boundary. For examples, when the distance of a grid point to the boundary is less than 1/10 of the shortest edge connected to it, it will be shifted to the closed point on the boundary. When the distance is within 1/10 to 1/5 of the shortest edge, it is shifted away from the boundary to a distance of 1/5 of the shortest edge.

The edges intersecting the boundary can be identified. One technique is to use the signed distance functions of the grid points (Section 4.2). At the resolution of the hierarchical-tree decomposition, an edge is considered as intersecting the boundary when the product of the signed distance functions at its two endpoints are negative as shown in Figure 9.32. The intercepting point, where the signed distance function is equal to 0, is determined and the edge is split into two lines.

When trimming a face, it is assumed that the boundary will cut no more than two edges of a square face. This assumption is true when the face is sufficiently small and the segment of the boundary within the face is approximated by a straight line. This condition can be met by controlling the placement of seed points. The trimming process is illustrated in Figure 9.33. A face cut by the boundary on two edges is shown in Figure 9.33a. The lines inside the domain are identified as those having one or two endpoints with negative values of the signed distance function (Figure 9.33b). The boundary is approximated by a line connecting the two intersection points (Figure 9.33). Following the connectivity of the face, the trimmed face is formed by connecting these lines. The trimmed face becomes a polygon.

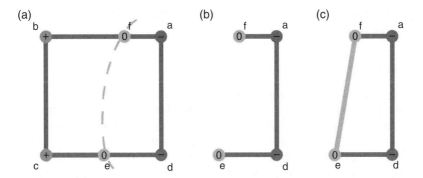

Figure 9.33 Trimming a face. (a) a face with two edges cut by boundary. (b) Lines inside the domain. (c) Trimmed face.

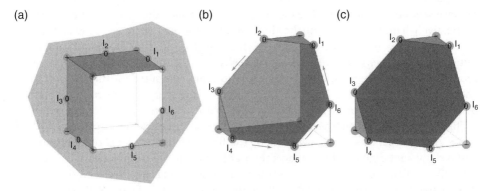

Figure 9.34 Trimming a cell. (a) Cell with faces trimmed by boundary. (b) Construction of boundary face. (c) Trimmed cell formed by trimmed faces and boundary face.

The trimming of a cell is illustrated in Figure 9.34. The cell contains faces trimmed by the boundary (Figure 9.34a). A face is constructed by connecting the boundary edges of the trimmed faces (Figure 9.34b). The boundary face closes the trimmed faces to form a trimmed cell, which is a polyhedron.

The polyhedrons resulting from trimming the cubic octree cells, as well as all other cells, are modelled as S-elements. The boundary of a polyhedron is discretized by triangular and quadrilateral surface elements. An example of boundary discretization is shown in Figure 6.2. When a face has more than four vertices, the face is divided into triangular elements. Optionally, a node can be inserted at the centroid of the face if this operation improves the quality of the triangular elements.

The boundary discretization of the octree cells that are not trimmed by the boundary are described in Section 9.3.2.

9.7 Examples Using Quadtree/Octree Meshes of CAD Models

Examples of two- and three-dimensional CAD models are analysed to demonstrate the use of the quadtree/octree algorithm for generating S-element meshes. After the parameters of element sizes (used to create mesh seeds) are specified, the mesh generation process is fully automatic.

9.7.1 Square Body with Multiple Holes

A unit square body ($L = 1$ m) with nine randomly distributed holes of different sizes, shown in Figure 9.35a, is considered. The material properties are Young's modulus $E = 100$ GPa and Poisson's ratio $v = 0.25$. The displacements at the bottom edge of the square are fully constrained and a uniform tension $P = 1$ MPa is applied at the top edge of the square. This example highlights flexibility of S-elements and quadtree algorithm in handling mesh transition between features with very different dimensions. The ability of high-order elements in accurately modelling problems with curved boundaries is also demonstrated.

The quadtree mesh shown in Figure 9.35b is generated by placing the same number of mesh seeds on every circle. The mesh transition between the holes of difference sizes is effectively handled without further human interventions. There are 1169 cells in the mesh. Among them, 352 are polygonal cells resulting from shifting grid points and trimming the square cells. The rest are square cells with 12 unique patterns. Each cell is modelled as one S-element. The element solutions are computed for the 352 polygonal and 12 master S-elements. The element solutions of the other S-elements are obtained by scaling these of the master S-elements of the same pattern of node configuration.

This problem is also analysed using the commercial finite element software ANSYS v14.5. A convergence study is performed. The finite element mesh is refined until the first 6 significant digits of the displacements at Point A in Figure 9.35a have converged.

In the scaled boundary finite element analysis using quadtree mesh, h- and p-refinements are performed starting from the mesh shown in Figure 9.35 with 3-node

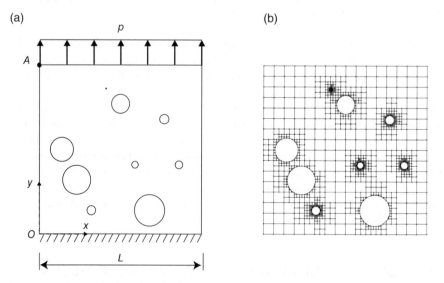

Figure 9.35 Square body with multiple holes under uniaxial tension. (a) Geometry and boundary conditions. (b) Quadtree mesh.

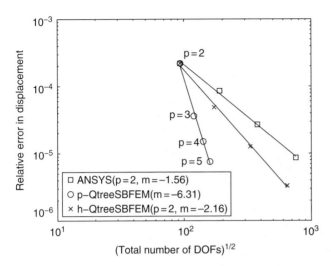

Figure 9.36 Convergence of displacements at Point A of a square body with multiple holes under uniaxial tension with *h*- and *p*-refinements. *p* is the element order and *m* is the slope of the trendline representing the rate of convergence.

line elements ($p = 2$) on the boundary of S-elements. The converged finite element result is used as the reference solution. The relative error of the displacements at Point A is plotted in Figure 9.36. The results converge for both types of refinements. It is observed that high-order elements leads to higher accuracy at the same number of degrees of freedom (DOFs). *p*-refinement converges faster than *h*-refinement.

The contour plots of the stress component σ_y obtained from the scaled boundary finite element analysis and the finite element analysis by ANSYS are shown in Figure 9.37. Good agreement is observed.

9.7.2 An Evolving Void in a Square Body

A square body with a void evolving from a 'star' to a circle is shown in Figure 9.38. The size of the square plate is $L = 1$m. The material properties are Young's modulus $E = 100$ GPa and Poisson's ratio $v = 0.25$. The boundary conditions are indicated in Figure 9.38 with $p = 1$MPa.

The stress distributions at different stages of the evolving void are studied. Figure 9.39 shows the quadtree meshes of S-elements around the void at three selected stages. The boundaries of S-elements are modelled with 4th-order line elements. The density of the S-elements around the void is controlled by the number of seed points placed according to the curvature of the boundary. Stage 1 (Figure 9.39a) has higher curvature than Stage 3 (Figure 9.39c) and thus denser S-elements on the boundary. The change of the meshes from one stage to the other is mostly localized around the void boundary. The part of mesh away from the void is not affected by the change in the void, which is advantageous in handling moving boundary problems. The von Mises stress distributions are shown in Figure 9.40. The stress concentrators at the tips of the star shape in Stage 1 disappear as the boundary of the void becomes smooth in Stage 3.

(a)

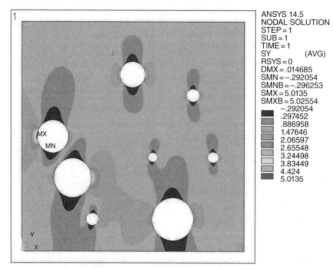

Finite element analysis using ANSYS

(b)

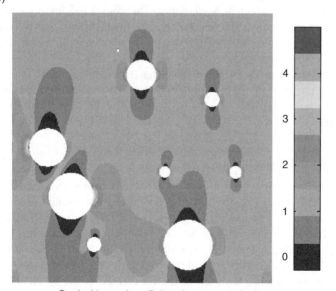

Scaled boundary finite element analysis

Figure 9.37 Contour plots of stress σ_y in of a square body with multiple holes under uniaxial tension.

9.7.3 Adaptive Analysis of an L-shaped Panel

An L-shaped panel is shown in Figure 9.41 with the boundary conditions. The uniform traction is equal to $p = 1$ MPa. The material properties are Young's modulus $E = 1$ GPa and Poisson's ratio $\nu = 0.3$. Plane stress states are considered. A stress singularity occurs at the re-entrant corner.

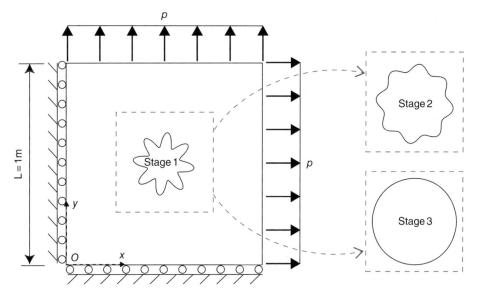

Figure 9.38 An evolving void in a squared body.

An adaptive analysis is performed using both linear and quadratic elements on the boundary of S-elements. The mesh refinement is based on *a posteriori* error estimator. The sequence of mesh refinements using quadratic elements on the boundaries of S-elements is shown in Figure 9.42. The initial mesh is coarse and uniform (Figure 9.42a). As the adaptive analysis progresses, the mesh refinement is localized around the re-entrant corner. The S-elements and quadtree mesh are straightforward and simple in handling local mesh refinement.

The convergence behaviour of the adaptive analysis is examined. A reference solution is obtained using a commercial finite element software. The errors in strain energy of the adaptive analysis steps are plotted in Figure 9.43. A series of analyses by uniform refinement are also performed and the errors are shown for comparison. It is observed that adaptive refinements lead to faster convergence than uniform refinements. Quadratic elements are more accurate than linear elements at the same number of degrees of freedom.

9.7.4 A Mechanical Part

A mechanical part in Wang (2006) is shown in Figure 9.44a. The dimensions of the bounding box is $50 \times 50 \times 100 \text{ mm}^3$. In the following analysis, the material properties are Young's modulus $E = 200 \text{ GPa}$ and Poisson's ratio $v = 0.3$. The bottom of the mechanical part is fixed and the top is subject to a uniform traction $p = 0.2 \text{ GPa}$.

The octree mesh shown in Figure 9.44b is generated with the minimum cell size of $S_{min} = 0.78 \text{ mm}$ and maximum cell size $4S_{min}$. The contour of vertical displacement is also shown in Figure 9.44b. To verify the result, a finite element analysis is performed using Abaqus. The results of vertical displacement along the dotted line in Figure 9.44a are compared in Figure 9.45. The largest percentage difference between the two sets of results is 1.4%.

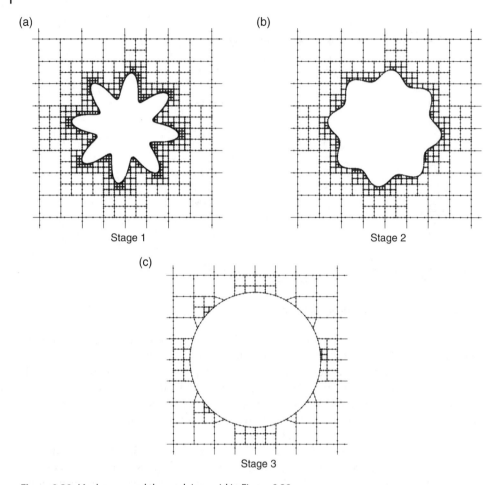

Figure 9.39 Meshes around the evolving void in Figure 9.38.

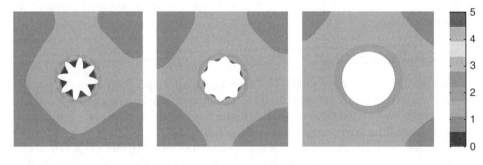

Figure 9.40 Von Mises stress in a square body with an evolving void.

Figure 9.41 L-shaped panel under a tensile load.

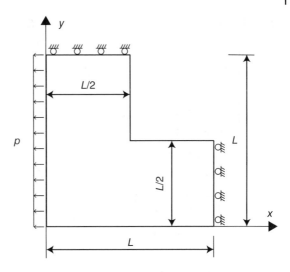

9.7.5 STL Models

Two STL models are analysed in this section. The STL format is native to stereolithography CAD software used by 3D printers. An STL file describes the surface of a solid object by the unit normal and vertices of unstructured triangles. The STL format of a triangular facet is shown in Figure 9.46. The STL syntax does not enforce any water tightness nor contain information on connectivity of triangles, which makes the format itself extremely simple.

The octree mesh generation algorithm described in Section 9.6.2 requires the intersection points between lines and surfaces. These operations can also be performed on STL models. Since this algorithm relies on cutting lines by the boundary, it tolerates the existence of certain flaws (such as self-intersecting parts and even small holes), typically less than the size of the smallest octree cell.

The STL model of the sphinx downloaded from the website http://www.thingiverse .com is shown in Figure 9.47a. The size of the model is $420 \times 1000 \times 600 \, \text{mm}^3$. It can be observed from the close-up view that some triangles are extremely elongated and unsuitable to analysis, The minimum and maximum sizes of octree cells are $S_{min} = 1.95 \, \text{mm}$ and $16 \times 1.95 \, \text{mm} = 31.2 \, \text{mm}$, respectively. The mesh is depicted in Figure 9.47b. The close-up view of the mesh in the same region as in Figure 9.47a is also shown.

The deformation of the sculpture under its self-weight is evaluated. The material properties are chosen as Young's modulus $E = 50 \, \text{MPa}$, Poisson's ratio $v = 0.3$ and mass density $2,350 \, \text{kg/m}^3$. The base of the sculpture is fixed. The vertical displacement is shown in Figure 9.48b. The largest value happens on the nose of the sphinx. The displacement on a section of the sculpture is revealed in Figure 9.48a. The transition of the element size in the octree mesh can be observed.

The STL model of the status of Lucy downloaded from https://www.rocq.inria.fr/ gamma is shown in Figure 9.49. This model is a coarser version of the well-known Lucy of the Stanford 3D scanning repository (http://graphics.stanford.edu/data/3Dscanrep). A few flaws exist in the model and one of them is shown in the close-up view in

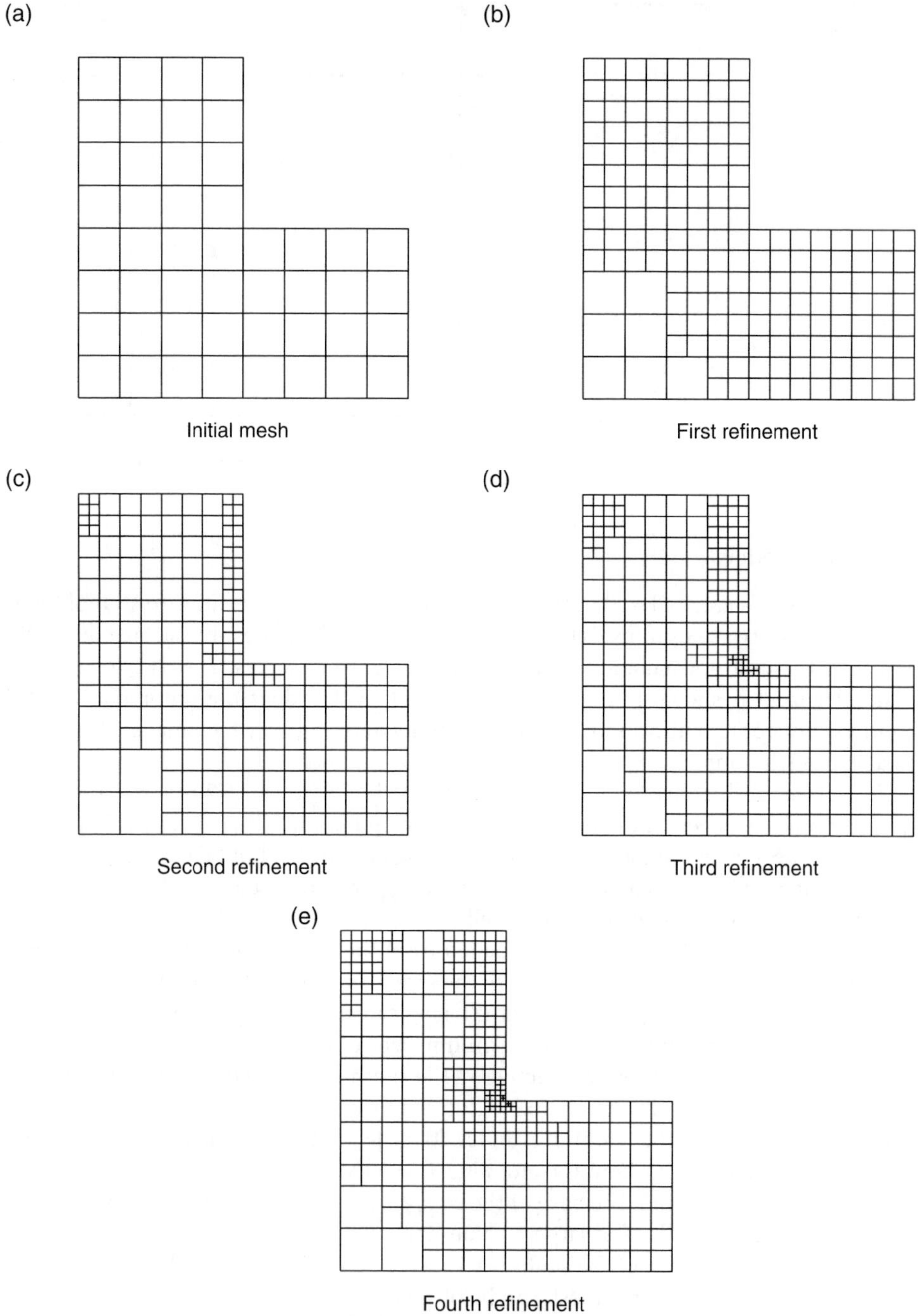

(a)

(b)

Initial mesh

First refinement

(c)

(d)

Second refinement

Third refinement

(e)

Fourth refinement

Figure 9.42 Adaptive mesh refinement of L-shaped panel using quadratic elements.

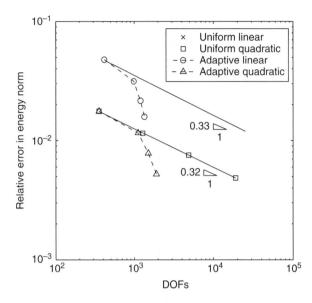

Figure 9.43 Relative error in the energy norm for the L-shaped domain for both uniform and adaptive refinement.

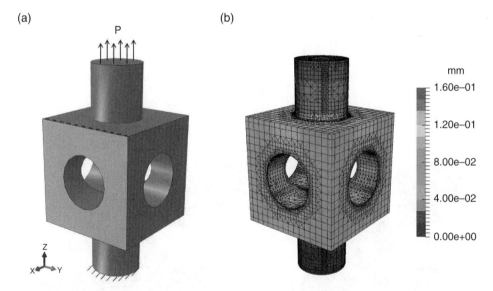

Figure 9.44 A mechanical part subject to tension. (a) Geometry and boundary conditions. (b) Mesh and contour plot of vertical displacement.

Figure 9.49. The Lucy has a size of $580 \times 330 \times 1000\,\text{mm}^3$. The minimum size of octree cell is chosen as 0.98 mm to capture details in the geometry. The maximum cell size is $16 \times 0.98\,\text{mm} = 15.68\,\text{mm}$. A stress analysis under its self-weight is performed. The material parameters are the same as those used for the sphinx. The von Mises stress and the maximum principal stress are displayed in Figure 9.50. The locations of maximum values are shown in the close-up view.

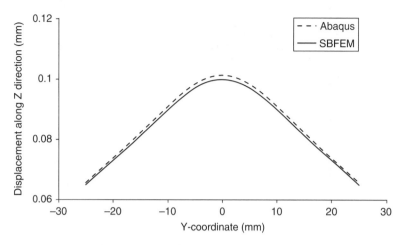

Figure 9.45 Vertical displacement of a mechanical part along the dotted line shown in Figure 9.44a.

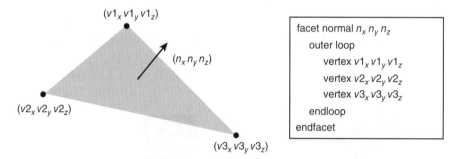

Figure 9.46 A triangular facet of STL and its ASCII representation.

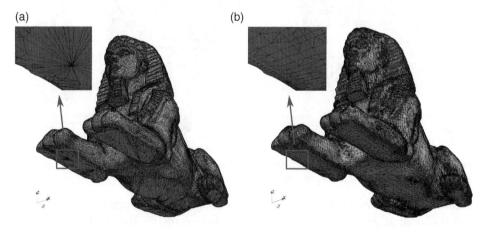

Figure 9.47 Mesh generation of the sphinx. (a) STL model. The close-up view shows some elongated triangles. (b) Mesh of S-elements generated by octree algorithm. The close-up view of mesh is at the same location as the close-up view in the STL model.

(a)

(b)

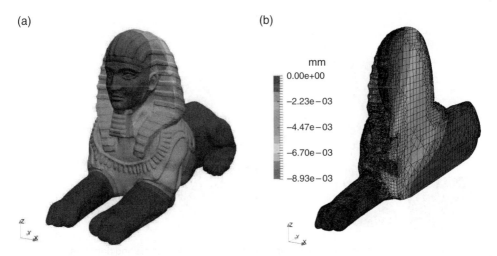

Figure 9.48 Vertical displacement of a sphinx under self-weight. (a) Contour on the whole model. (b) Contour on half of the model showing interior of mesh.

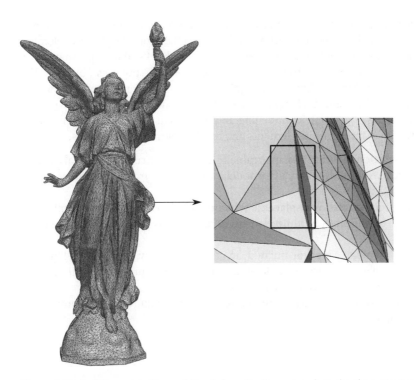

Figure 9.49 An STL model of Lucy. A flaw is found in the rectangle in the close-up view.

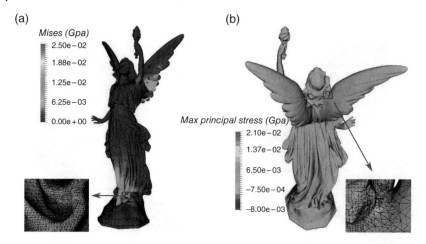

Figure 9.50 Stress analysis of Lucy under self-weight. Left is von Mises stress and right is the maximum principal stress. The locations of maximum values are shown in the close-up views.

9.8 Remarks

The hierarchical quadtree/octree is applied to the mesh generation of S-elements. This type of hierarchical-tree algorithm and the scaled boundary finite element method are highly complementary. The S-elements are highly flexible in geometry and require the discretization of only the boundary, which reduces the mesh burden. The quadtree/octree algorithm is simple and highly efficient. This framework combining quadtree/octree meshes and S-elements provides a promising technique for fully automatic integrating of geometric modelling and numerical anlaysis. Its key features, in comparison with standard finite elements and mesh generation algorithms, include:

1) The mesh generation algorithm is applicable to the native formats of digital images, NURBS-based CAD models and STL models. Conversion from one format to the other is not essential.
2) The issue of hanging nodes or displacement incompatibility hindering the application of hierarchical-tree meshes in finite element analysis is simply eliminated using the S-elements.
3) High-order S-elements are formulated in the same way as linear elements. The high-order elements are advantageous in modelling problems with curved boundaries and exhibit superior convergence rates.
4) Remeshing in adaptive analysis and modelling of moving boundary problems is significantly simplified.

10

Linear Elastic Fracture Mechanics

10.1 Introduction

Cracks of various sizes often exist in engineering structures of components. The regions around crack tips are often the weakest points where failure initiates. Some cases of structural failure due to fracture can be found in Anderson (2005).

Based on the theory of linear elasticity , the stresses around a crack tip are proportional to $1/\sqrt{r}$, where r is the distance to the crack tip (Williams, 1957). At the crack tip $r = 0$, the stresses tend to infinity and a (square-root) singularity exists. In this case, stress-based failure criteria are no longer applicable by themselves. The theory of fracture mechanics was developed to study cracks and crack propagation. Many textbooks on fracture mechanics (Anderson, 2005; Kanninen, 1985, to name a few) have been published. In the theory of linear elastic fracture mechanics, small-scale yielding is assumed, i.e. the deviations from linearity occur only over a region that is small compared to geometrical dimensions of the problem (Rice, 1968a). Linear elastic fracture mechanics has been widely used in fracture analysis when the materials are brittle or when the load is low enough to fulfil the small-scale yielding assumption (for example, in the study of fatigue failure and corrosions).

The solution of the stress field around a crack tip is vital to fracture analysis. In the linear elastic fracture mechanics, the stress intensity factors uniquely characterize the stress singularity and are commonly used in fracture criteria governing crack extension. Extensive literature exists on the computation of stress intensity factors. Handbooks of stress intensity factors of common cracked specimens, for example, Tada et al. (1985), are also available. In addition to the stress intensity factors, the non-singular T-stress acting parallel to the crack plane is also of physical significance. It is shown by Cotterell (1966) that the T-stress affects crack path stability.

Stress singularities also exist at interfacial cracks, V-notches and free edges formed by dissimilar materials (Rice, 1988; Dempsey and Sinclair, 1979; Lindemann and Becker, 2002). In these situations, the singular stress fields are generally more complex and not proportional to $1/\sqrt{r}$. Other parameters in addition to the stress intensity factors are required to describe the singular stress field.

The presence of singularities poses challenges to the finite element method since the standard interpolation functions are polynomials and do not resemble the singular stress field. In order to obtain accurate results for stresses, the mesh needs to be refined locally

The Scaled Boundary Finite Element Method: Introduction to Theory and Implementation,
First Edition. Chongmin Song.
© 2018 John Wiley & Sons Ltd. Published 2018 by John Wiley & Sons Ltd.
Companion website: www.wiley.com/go/author/scaledboundaryfiniteelementmethod

in the vicinity of the singular point. Furthermore, additional post-processing techniques need to be applied to extract the stress intensity factors for engineering purposes.

To improve the efficiency in computing the singular stress fields and extracting stress intensity factors, *a priori* knowledge of the asymptotic solution of the singular field is often used in the finite element and boundary element methods to design special techniques. For example, quarter-point crack elements (e.g. Barsoum, 1974; Henshell and Shaw, 1975 and Tan and Gao, 1990) introduce a square-root singularity as occurring in the asymptotic solution of a crack in a homogeneous body. Elements enriched with the asymptotic stress series (e.g. Tong et al., 1973) have been developed. To accurately extract the stress intensity factors and the T-stress, path-independent integrals (Rice, 1968b; Sladek and Sladek, 1997) are often performed.

The asymptotic solutions of singular stress fields may be very complex in cases such as notches formed by multiple anisotropic materials. The effectiveness of techniques requiring asymptotic solutions such as element enrichment is compromised. A large number of recent publications and novel techniques for singularity problems are available. A literature review on this topic is beyond the scope of this book.

In the scaled boundary finite element method, a semi-analytical solution of the stress field around a singularity point is obtained as demonstrated in Examples 3.6 and 3.9. This semi-analytical nature is highly advantageous in analysing problems of linear fracture mechanics. The singular stress fields at crack tips and multi-material corners are addressed in Song and Wolf (2002). No local mesh refinement or asymptotic enrichment is required. The singularity is represented analytically without *a priori* knowledge of the asymptotic solution. The scaled boundary finite element method has been further developed and applied to evaluate the free-edge stresses around holes in laminates (Lindemann and Becker, 2002), to compute the orders of singularity and stress intensity factors for multi-material plates under static loading and dynamic loading (Song, 2004b), and to predict the directions of cracks emerging from notches at bimaterial junctions (Müller et al., 2005). It is extended to include the power-logarithmic singularities, and to calculate the T-stress, higher-order terms and angular distributions of stresses in Song (2005) and Chidgzey and Deeks (2005). Three-dimensional fracture analyses have also been performed (Goswami and Becker, 2012; Saputra et al., 2015; Hell and Becker, 2015).

The application of the scaled boundary finite element method is not limited to the evaluation of stress intensity factors. Based on the semi-analytical solution of the singular stress field obtained by the scaled boundary finite element method, a definition of generalized stress intensity factors at a multi-material corner is proposed in Song et al. (2010a). It is consistent with the classical definitions of stress intensity factors at crack tips and valid for all types of singularity. The generalized stress intensity factors can be determined directly from the scaled boundary finite element solution by following a standard stress recovery procedure.

The scaled boundary finite element method has evolved to become an attractive alternative for fracture analysis and possesses the following salient features:

1) No analytical asymptotic solutions are required in the definition and evaluation of generalized stress intensity factors. It is thus applicable to cracks, V-notches and multi-material corners composed of any number of isotropic and anisotropic materials.

2) The mesh burden is significantly reduced by modelling the region around the singularity point as an S-element. Only the boundary of the S-element, excluding the straight crack faces and material interfaces, is discretized. No local mesh refinement around the singularity point is required.

3) All types of stress singularity (real power singularity $r^{-\lambda}$, complex power singularity $r^{-(\lambda_R \pm i\lambda_I)}$ and power-logarithmic singularity $r^{-\lambda} \ln r$) are modelled analytically in a unified expression.

4) A unified definition of generalized stress intensity factors for all types of stress singularities is proposed. This definition is consistent with standard definitions of stress intensity factors for cracks in homogeneous materials and on interfaces of dissimilar isotropic materials.

5) The generalized stress singular intensity factors and T−stress are determined by standard stress recovery techniques in the finite element method without addressing any singular functions numerically.

6) The use of S-elements of arbitrary number of edges greatly simplifies the modelling of crack propagation.

10.2 Basics of Fracture Analysis: Asymptotic Solutions, Stress Intensity Factors, and the T-stress

Some key concepts and equations in linear elastic fracture mechanics that are directly used in the following sections are provided here.

10.2.1 Crack in Homogeneous Isotropic Material

The linear elastic stress analysis of cracks in two-dimensional homogeneous isotropic materials is covered comprehensively in many textbooks on fracture mechanics (Rice (1968a); Anderson (2005); Kanninen (1985), to name a few). A crack is depicted in Figure 10.1. The isotropic material is characterized by the shear modulus G and Poisson's ratio v. For convenience, the origin of the Cartesian coordinates x, y is chosen at the crack tip. Without loss of generality, the case of a crack opening to the left and lying on the negative x-axis is considered. The asymptotic solution near the crack tip is given in the polar coordinates (r, θ) as a series of power functions with real exponents (Xiao et al., 2004). The displacement field is expressed as

$$u_x(r, \theta) = \sum_{i=1}^{\infty} \frac{1}{2G} r^{\frac{i}{2}} \left\{ a_i \left[\left(\kappa + \frac{i}{2} + (-1)^i \right) \cos \frac{i}{2}\theta - \frac{i}{2} \cos \left(\frac{i}{2} - 2 \right)\theta \right] \right.$$
$$\left. + b_i \left[\left(\kappa + \frac{i}{2} - (-1)^i \right) \sin \frac{i}{2}\theta - \frac{i}{2} \sin \left(\frac{i}{2} - 2 \right)\theta \right] \right\} \tag{10.1a}$$

$$u_y(r, \theta) = \sum_{i=1}^{\infty} \frac{1}{2G} r^{\frac{i}{2}} \left\{ a_i^1 \left[\left(\kappa - \frac{i}{2} - (-1)^i \right) \sin \frac{i}{2}\theta + \frac{i}{2} \sin \left(\frac{i}{2} - 2 \right)\theta \right] \right.$$
$$\left. - b_i \left[\left(\kappa - \frac{i}{2} + (-1)^i \right) \cos \frac{i}{2}\theta + \frac{i}{2} \cos \left(\frac{i}{2} - 2 \right)\theta \right] \right\} \tag{10.1b}$$

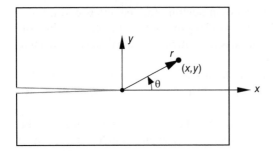

Figure 10.1 A crack in a homogeneous isotropic material.

where the coefficients a_i and b_i are real numbers. The angle θ is measured counterclockwise from the positive x-direction and κ is the Kolosov's constant (Eq. (A.52)). Note that an arbitrary rigid boundary motion can be added to the displacement field in Eq. (10.1) and the result remains a valid asymptotic solution.

The stress field is expressed as

$$
\sigma_x(r,\theta) = \sum_{i=1}^{\infty} \frac{i}{2} r^{\frac{i}{2}-1} \left\{ a_i \left[\left(2 + \frac{i}{2} + (-1)^i\right) \cos\left(\frac{i}{2} - 1\right)\theta - \left(\frac{i}{2} - 1\right)\cos\left(\frac{i}{2} - 3\right)\theta \right] \right.
$$
$$
\left. + b_i \left[\left(2 + \frac{i}{2} - (-1)^i\right) \sin\left(\frac{i}{2} - 1\right)\theta - \left(\frac{i}{2} - 1\right)\sin\left(\frac{i}{2} - 3\right)\theta \right] \right\}
$$
(10.2a)

$$
\sigma_y(r,\theta) = \sum_{i=1}^{\infty} \frac{i}{2} r^{\frac{i}{2}-1} \left\{ a_i \left[\left(2 - \frac{i}{2} - (-1)^i\right) \cos\left(\frac{i}{2} - 1\right)\theta + \left(\frac{i}{2} - 1\right)\cos\left(\frac{i}{2} - 3\right)\theta \right] \right.
$$
$$
\left. + b_i \left[\left(2 - \frac{i}{2} + (-1)^i\right) \sin\left(\frac{i}{2} - 1\right)\theta + \left(\frac{i}{2} - 1\right)\sin\left(\frac{i}{2} - 3\right)\theta \right] \right\}
$$
(10.2b)

$$
\tau_{xy}(r,\theta) = \sum_{i=1}^{\infty} \frac{i}{2} r^{\frac{i}{2}-1} \left\{ a_i \left[\left(\frac{i}{2} - 1\right) \sin\left(\frac{i}{2} - 3\right)\theta - \left(\frac{i}{2} + (-1)^i\right) \sin\left(\frac{i}{2} - 1\right)\theta \right] \right.
$$
$$
\left. - b_i \left[\left(\frac{i}{2} - 1\right) \cos\left(\frac{i}{2} - 3\right)\theta - \left(\frac{i}{2} - (-1)^i\right) \cos\left(\frac{i}{2} - 1\right)\theta \right] \right\}
$$
(10.2c)

The stress field is independent of the material properties. Note that the sign of b_i in Eqs. (10.1) and (10.2) is opposite to that in Xiao et al. (2004) so that the sign in Eq. (10.4) below is positive.

The asymptotic solution satisfies the governing differential equations of elasticity and the traction-free boundary conditions at the crack faces. The coefficients in the asymptotic solution are not determined. Generally speaking, the conditions on the boundary of a finite body are not satisfied. The asymptotic solution is valid in only a small region around the crack tip.

It is observed from Eqs. (10.1) and (10.2) that each term of the series consists of a power function of r describing the distribution along the radial direction and a function (inside the curly brackets) of θ describing the angular variation. The first five terms

($i = 1, 2, \ldots, 5$) of the power functions in the stress field (Eq. (10.2)) are $r^{-1/2}$, $r^0 = 1$, $r^{1/2}$, r and $r^{3/2}$. As $r \to 0$, the leading term $r^{-1/2} = 1/\sqrt{r}$ becomes dominant. At the crack tip $r = 0$, the stresses are singular. There are two independent singular stress modes associated with the coefficients a_1 and b_1. In the leading term of the displacement field, the power function is $\sqrt{r}$.

The stress intensity factors are defined by analytically removing the singularity in the solution

$$K_I = \lim_{r \to 0} \sqrt{2\pi r} \sigma_y(r, \theta = 0) \tag{10.3a}$$

$$K_{II} = \lim_{r \to 0} \sqrt{2\pi r} \tau_{xy}(r, \theta = 0) \tag{10.3b}$$

so that the intensity of the stress field is measured by finite values. The dimension of the stress intensity factor is Stress $\times \sqrt{\text{Length}}$. Each of the stress intensity factors K_I and K_{II} corresponds to one singular stress mode as indicated by the subscripts I and II.

Substituting Eq. (10.2b) into Eq. (10.3a) and Eq. (10.2c) into Eq. (10.3b), respectively, the stress intensity factors are expressed as

$$K_I = \sqrt{2\pi} a_1 \tag{10.4a}$$

$$K_{II} = \sqrt{2\pi} b_1 \tag{10.4b}$$

All the terms other than the singular stress term ($i = 1$) vanish as the limit $r \to 0$ is taken. It is thus convenient to consider only the singular stress terms in the definition of the stress intensity factors. Denoting the stresses of the singular terms as $\sigma_{xx}^{(s)}(r, \theta)$, $\sigma_y^{(s)}(r, \theta)$ and $\tau_{xy}^{(s)}(r, \theta)$ (superscript (s) for singular), the singular stresses in Eq. (10.2) are expressed as

$$
\begin{Bmatrix} \sigma_x^{(s)}(r,\theta) \\ \sigma_y^{(s)}(r,\theta) \\ \tau_{xy}^{(s)}(r,\theta) \end{Bmatrix} = \frac{K_I}{\sqrt{2\pi r}} \frac{1}{4} \begin{Bmatrix} 3\cos\frac{1}{2}\theta + \cos\frac{5}{2}\theta \\ 5\cos\frac{1}{2}\theta - \cos\frac{5}{2}\theta \\ \sin\frac{5}{2}\theta - \sin\frac{1}{2}\theta \end{Bmatrix} + \frac{K_{II}}{\sqrt{2\pi r}} \frac{1}{4} \begin{Bmatrix} 7\sin\frac{1}{2}\theta + \sin\frac{5}{2}\theta \\ \sin\frac{1}{2}\theta - \sin\frac{5}{2}\theta \\ \cos\frac{5}{2}\theta + 3\cos\frac{1}{2}\theta \end{Bmatrix}
\tag{10.5}
$$

where Eq. (10.4) has been substituted. In the front of the crack tip ($\theta = 0$), the two stress components $\sigma_y^{(s)}(r, \theta)$ and $\tau_{xy}^{(s)}(r, \theta)$ are expressed as

$$
\begin{Bmatrix} \sigma_y^{(s)}(r,0) \\ \tau_{xy}^{(s)}(r,0) \end{Bmatrix} = \frac{1}{\sqrt{2\pi r}} \begin{Bmatrix} K_I \\ K_{II} \end{Bmatrix}
\tag{10.6}
$$

The definition of the stress intensity factors is expressed as

$$
\begin{Bmatrix} K_I \\ K_{II} \end{Bmatrix} = \sqrt{2\pi r} \begin{Bmatrix} \sigma_y^{(s)}(r,0) \\ \tau_{xy}^{(s)}(r,0) \end{Bmatrix}
\tag{10.7}
$$

The displacement field corresponding to the singular stress field is expressed in terms of the stress intensity factors. Using Eq. (10.4), the term $i = 1$ in Eq. (10.1) is written as

$$
\begin{Bmatrix} u_x^{(s)}(r,\theta) \\ u_y^{(s)}(r,\theta) \end{Bmatrix} = \frac{K_I}{2G} \sqrt{\frac{r}{2\pi}} \begin{Bmatrix} \left(\kappa - \frac{1}{2}\right) \cos\frac{1}{2}\theta - \frac{1}{2} \cos\frac{3}{2}\theta \\ \left(\kappa + \frac{1}{2}\right) \sin\frac{1}{2}\theta - \frac{1}{2} \sin\frac{3}{2}\theta \end{Bmatrix}
$$

$$
+ \frac{K_{II}}{2G} \sqrt{\frac{r}{2\pi}} \begin{Bmatrix} \left(\kappa + \frac{3}{2}\right) \sin\frac{1}{2}\theta + \frac{1}{2} \sin\frac{3}{2}\theta \\ -\left(\kappa - \frac{3}{2}\right) \cos\frac{1}{2}\theta - \frac{1}{2} \cos\frac{3}{2}\theta \end{Bmatrix} \tag{10.8}
$$

The crack opening displacements, i.e. the relative displacements between the upper crack face ($\theta = +\pi$) and the lower crack face ($\theta = -\pi$), are expressed as

$$
\begin{Bmatrix} \delta_x^{(s)}(r) \\ \delta_y^{(s)}(r) \end{Bmatrix} = \begin{Bmatrix} u_x^{(s)}(r,+\pi) \\ u_y^{(s)}(r,+\pi) \end{Bmatrix} - \begin{Bmatrix} u_x^{(s)}(r,-\pi) \\ u_y^{(s)}(r,-\pi) \end{Bmatrix}
$$

$$
= \frac{\kappa+1}{G} \sqrt{\frac{r}{2\pi}} \begin{Bmatrix} K_{II} \\ K_I \end{Bmatrix} \tag{10.9}
$$

Inversely, it is written as

$$
\begin{Bmatrix} K_I \\ K_{II} \end{Bmatrix} = \frac{G}{\kappa+1} \sqrt{\frac{2\pi}{r}} \begin{Bmatrix} \delta_y^{(s)}(r) \\ \delta_x^{(s)}(r) \end{Bmatrix} \tag{10.10}
$$

which provides a way to evaluate the stress intensity factors.

The second term ($i = 2$) in Eq. (10.2) is related to the T-stress. It is independent of the radial coordinate r and expressed as

$$
\begin{Bmatrix} \sigma_x^{(T)}(\theta) \\ \sigma_y^{(T)}(\theta) \\ \tau_{xy}^{(T)}(\theta) \end{Bmatrix} = a_2 \begin{Bmatrix} 4 \\ 0 \\ 0 \end{Bmatrix} \tag{10.11}
$$

b_2 vanishes in the expression for stresses. The only nonzero stress $\sigma_x^{(T)}$ in Eq. (10.11) is a constant and independent of r and θ. The coefficient a_2 is related to the T-stress defined as

$$
T = \sigma_x^{(T)}(\theta = 0) = 4a_2 \tag{10.12}
$$

The displacement field of the T-stress term ($i = 2$) is expressed as

$$
\begin{Bmatrix} u_x^{(T)}(r,\theta) \\ u_y^{(T)}(r,\theta) \end{Bmatrix} = \frac{a_2}{2G} r \begin{Bmatrix} (\kappa+1)\cos\theta \\ (\kappa-3)\sin\theta \end{Bmatrix} + \frac{b_2(\kappa+1)}{2G} r \begin{Bmatrix} \sin\theta \\ -\cos\theta \end{Bmatrix} \tag{10.13}
$$

The coefficient b_2 is associated with the rigid-body rotation about the crack tip. Substituting the coefficient a_2 obtained from Eq. (10.12) into Eq. (10.11), the displacement component $u_x^{(T)}(r,\theta)$ is expressed along the direction $\theta = 0$ as

$$
u_x^{(T)}(r,\theta = 0) = \frac{(\kappa+1)}{2G} \frac{T}{4} r = \frac{\kappa+1}{8G} Tr \tag{10.14}
$$

Figure 10.2 An interfacial crack in a bimaterial material.

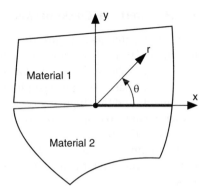

The T-stress can be determined by

$$T = \frac{8G}{\kappa + 1} \frac{1}{r} u_x^{(T)}(r, \theta = 0) \tag{10.15}$$

10.2.2 Interfacial Cracks between Two Isotropic Materials

A crack on the interface between two dissimilar materials is shown in Figure 10.2 with the Cartesian coordinates x, y (or the polar coordinates r, θ). The origin of the coordinates is located at the crack tip. The x-coordinate (or the angle $\theta = 0$) is chosen on the material interface. The crack lies on the negative x-axis. The material constants are the shear modulus G_1 and Poisson's ratio v_1 for material 1 and G_2 and v_2 for material 2.

Near the crack tip, two singular stress modes varying in proportion to r^λ exist. The orders of stress singularity of the two singular stress modes are a pair of complex conjugates $-0.5 \pm i\epsilon$ (Rice, 1988), where the oscillatory index ϵ depends on the material properties and is equal to

$$\epsilon = \frac{1}{2\pi} \ln \frac{G_2/G_1 \kappa_1 + 1}{\kappa_2 + G_2//G_1} \tag{10.16}$$

where κ_i ($i = 1, 2$) is the Kolosov's constant (Eq. (A.52)) of material i.

The stress intensity factors are defined by

$$\sigma_y^{(s)}(r, 0) + i\tau_{xy}^{(s)}(r, 0) = \frac{1}{\sqrt{2\pi r}} \left(\frac{r}{L}\right)^{i\epsilon} (K_I + iK_{II}) \tag{10.17}$$

where L is a characteristic length. Equation (10.17) can be expressed in matrix form as

$$\left\{ \begin{array}{c} \sigma_y^{(s)}(r, 0) \\ \tau_{xy}^{(s)}(r, 0) \end{array} \right\} = \frac{1}{\sqrt{2\pi r}} \begin{bmatrix} c(r) & -s(r) \\ s(r) & c(r) \end{bmatrix} \left\{ \begin{array}{c} K_I \\ K_{II} \end{array} \right\} \tag{10.18}$$

with

$$c(r) = \cos(\epsilon \ln(r/L)); \qquad s(r) = \sin(\epsilon \ln(r/L)) \tag{10.19}$$

When the oscillatory index ϵ is equal to zero, the definition in Eq. (10.18) is identical to that in Eq. (10.6).

It should be noted that not only the stress intensity factors K_I and K_{II} but also the oscillatory index ϵ are required to describe the singular stress field near the tip of a crack on the interface between two isotropic materials.

10.2.3 Interfacial Cracks between Two Anisotropic Materials

When the materials in Figure 10.2 are anisotropic, the definition of stress intensity factors in Eq. (10.17) will not completely remove the stress singularity at the crack tip and is no longer valid. A definition of generalized stress intensity factors is proposed by Cho et al. (1992) based on the asymptotic solution of Suo (1990) for the crack opening displacements and the singular stresses in front of the crack tip. The errors in the formulas presented in Cho et al. (1992) are corrected in Song et al. (2010b). The key formulas are summarized here.

The singular stresses $\{\sigma^{(s)}(r, \theta = 0)\}$ on the bimaterial interface and at a distance r in front of the crack tip are expressed by using the complex stress intensity factor $K = K_1 + iK_2$ as

$$\begin{Bmatrix} \tau_{xy}^{(s)}(\hat{r}, 0) \\ \sigma_{y}^{(s)}(\hat{r}, 0) \end{Bmatrix} = \frac{1}{\sqrt{2\pi r}} \operatorname{Re}\left(K\{W\} \left(\frac{r}{L}\right)^{i\epsilon} \right) \tag{10.20}$$

where $\operatorname{Re}(\bullet)$ denotes the real part of a variable. Other than the oscillatory index ϵ, a complex vector $\{W\}$ appears in the solution. It also depends on the material constants of the two materials and are given below in Eq. (10.34). L is a characteristic length. Note that the generalized stress intensity factors are denoted as K_1 and K_2 to differentiate them from the classical stress intensity factors K_I and K_{II} since this definition is not always consistent with the classical definition in Eq. (10.6).

The crack opening displacements $\{\delta^{(s)}(r)\}$ corresponding to the singular stress field are expressed as

$$\begin{Bmatrix} \delta_{x}^{(s)}(r) \\ \delta_{y}^{(s)}(r) \end{Bmatrix} = \sqrt{\frac{r}{2\pi}} \operatorname{Re}\left(\frac{\cosh(\pi\epsilon)}{1 + 2i\epsilon} K\{V\} \left(\frac{r}{L}\right)^{i\epsilon} \right) \tag{10.21}$$

where the crack opening displacements $\{\delta^{(s)}(r)\}$ at a distance r behind the crack tip are obtained from the displacement field $\{u^{(s)}(r, \theta)\}$ corresponding to the singular stress field

$$\{\delta^{(s)}(r)\} = \{u^{(s)}(r, \theta = +\pi)\} - \{u^{(s)}(r, \theta = -\pi)\} \tag{10.22}$$

The complex vector $\{V\}$ is given below in Eq. (10.35).

The oscillatory index ϵ and the complex vectors $\{W\}$ and $\{V\}$ are obtained by the following procedure. Only the in-plane motions are considered. The stress-strain relationship of an anisotropic material is expressed as

$$\{\sigma\} = [D]\{\epsilon\}; \qquad \{\epsilon\} = [a]\{\sigma\} \tag{10.23}$$

where the stresses $\{\sigma\}$, strains $\{\epsilon\}$, symmetric material stiffness matrix $[D]$ and symmetric material flexibility matrix $[a]$ are expressed as

$$\{\sigma\} = [\sigma_x, \sigma_y, \tau_{xy}]^T; \qquad \{\epsilon\} = [\epsilon_x, \epsilon_y, \gamma_{xy}]^T$$

$$[D] = \begin{bmatrix} D_{11} & D_{12} & D_{13} \\ D_{12} & D_{22} & D_{23} \\ D_{13} & D_{23} & D_{33} \end{bmatrix}; \qquad [a] = \begin{bmatrix} a_{11} & a_{12} & a_{13} \\ a_{12} & a_{22} & a_{23} \\ a_{13} & a_{23} & a_{33} \end{bmatrix} \tag{10.24}$$

The characteristic equation for in-plane deformation is written as

$$a_{11}\mu^4 - 2a_{13}\mu^3 + (2a_{12} + a_{33})\mu^2 - 2a_{23}\mu + a_{22} = 0 \tag{10.25}$$

It has two pairs of complex conjugate roots, denoted as

$$\mu_1 = \mu_{1R} + i\mu_{1I}; \quad \mu_2 = \mu_{2R} + i\mu_{2I}; \quad \mu_3 = \mu_{1R} - i\mu_{1I}; \quad \mu_4 = \mu_{2R} - i\mu_{2I} \quad (10.26)$$

where μ_1 and μ_2 are the two roots having positive imaginary parts (i.e. $\mu_{1I} > 0$ and $\mu_{2I} > 0$ apply). μ_3 and μ_4 are the complex conjugates of μ_1 and μ_2, respectively. A positive-definite Hermitian matrix $[H]$ involving bimaterial elastic constants is defined as (Eq. (23), Cho et al., 1992)

$$[H] = \begin{bmatrix} H_{11} & H_{12} - ig \\ H_{12} + ig & H_{22} \end{bmatrix} \quad (10.27)$$

where the entries are given by

$$H_{11} = \left[a_{11}(\mu_{1I} + \mu_{2I})\right]_1 + \left[a_{11}(\mu_{1I} + \mu_{2I})\right]_2$$

$$H_{22} = \left[a_{22}\left(\frac{\mu_{1I}}{\mu_{1R}^2 + \mu_{1I}^2} + \frac{\mu_{2I}}{\mu_{2R}^2 + \mu_{2I}^2}\right)\right]_1 + \left[a_{22}\left(\frac{\mu_{1I}}{\mu_{1R}^2 + \mu_{1I}^2} + \frac{\mu_{2I}}{\mu_{2R}^2 + \mu_{2I}^2}\right)\right]_2$$

$$H_{12} = \left[a_{11}(\mu_{1R}\mu_{2I} + \mu_{2R}\mu_{1I})\right]_1 + \left[a_{11}(\mu_{1R}\mu_{2I} + \mu_{2R}\mu_{1I})\right]_2$$

$$g = \left[a_{11}(\mu_{1R}\mu_{2R} + \mu_{1I}\mu_{2I}) - a_{12}\right]_1 - \left[a_{11}(\mu_{1R}\mu_{2R} + \mu_{1I}\mu_{2I}) - a_{12}\right]_2 \quad (10.28)$$

The subscripts 1 and 2 outside of the square brackets denote the material numbers shown in Figure 10.2.

The oscillatory index ϵ and the vector $\{V\}$ are determined by the eigenvalue problem (Eq. (16), Cho et al., 1992)

$$[H]^{-1}\{V\} = e^{2\pi\epsilon}[\bar{H}]^{-1}\{V\} \quad (10.29)$$

where $[\bar{H}]$ is the complex conjugate of $[H]$. The vector $\{W\}$ is expressed as (Eq. (15), Cho et al., 1992)

$$([H]^{-1} + [\bar{H}]^{-1})\{V\} = \{W\} \quad (10.30)$$

$\{W\}$ and $\{V\}$ are normalised such that the first entry of $\{W\}$ is equal to $-i$. Equations (10.29) and (10.30) are equivalent to the following eigenvalue problem

$$[\bar{H}]\{W\} = e^{2\pi\epsilon}[H]\{W\} \quad (10.31)$$

Using $[H]$ given in Eq. (10.27), the oscillatory index ϵ is obtained as

$$\epsilon = \frac{1}{2\pi}\ln\frac{1-\beta}{1+\beta} \quad (10.32)$$

with

$$\beta = \frac{g}{h}$$

$$h = \sqrt{H_{11}H_{22} - H_{12}^2} \quad (10.33)$$

The corresponding normalized eigenvector $\{W\}$ is expressed as

$$\{W\} = \left\{\begin{array}{c} -i \\ \dfrac{h + iH_{12}}{H_{22}} \end{array}\right\} = \left\{\begin{array}{c} -i \\ W_1 - iW_2 \end{array}\right\} \quad (10.34)$$

where W_1 and W_2 are defined for later use. The vector $\{V\}$ is obtained from Eq. (10.30) as

$$\{V\} = \frac{(1-\beta^2)h}{2H_{22}} \left\{ \begin{matrix} H_{12} - ih \\ H_{22} \end{matrix} \right\} = \left\{ \begin{matrix} G_1 - iP_1 \\ G_2 \end{matrix} \right\} \tag{10.35}$$

where G_1, G_2 and P_1 are defined for later use.

Substituting Eq. (10.34) into Eq. (10.20) and changing the sequence of $\tau_{xy}^{(s)}(r)$ and $\sigma_y^{(s)}(r)$ in the stress vector result in

$$\left\{ \begin{matrix} \sigma_y^{(s)}(r) \\ \tau_{xy}^{(s)}(r) \end{matrix} \right\} = \frac{1}{\sqrt{2\pi r}} \begin{bmatrix} W_1 & W_2 \\ 0 & 1 \end{bmatrix} \begin{bmatrix} c(r) & -s(r) \\ s(r) & c(r) \end{bmatrix} \left\{ \begin{matrix} K_1 \\ K_2 \end{matrix} \right\} \tag{10.36}$$

with $c(r)$ and $s(r)$ given in Eq. (10.19). The stress intensity factors are obtained from Eq. (10.36) as

$$\left\{ \begin{matrix} K_1 \\ K_2 \end{matrix} \right\} = \frac{\sqrt{2\pi r}}{W_1} \begin{bmatrix} c(r) & s(r) \\ -s(r) & c(r) \end{bmatrix} \begin{bmatrix} 1 & -W_2 \\ 0 & W_1 \end{bmatrix} \left\{ \begin{matrix} \sigma_y^{(s)}(r) \\ \tau_{xy}^{(s)}(r) \end{matrix} \right\} \tag{10.37}$$

When the materials are isotropic, both $W_1 = 1$ and $W_2 = 0$ applies. When the materials are orthotropic, only $W_2 = 0$ holds. For a crack in homogeneous orthotropic material the oscillatory index ϵ vanishes and $c(r) = 1$ and $s(r) = 0$ applies. It is shown by Cho et al. (1992) that the definition in Eq. (10.37) is expressed as

$$K_1 = \frac{\sqrt{2\pi r}}{W_1} \sigma_y^{(s)}(\hat{r}) = \frac{K_I}{W_1} \tag{10.38}$$

which is different from the classical definition in Eq. (10.6).

When both materials are isotropic ($W_1 = 1$ and $W_2 = 0$), the definition in Eq. (10.36) is consistent with the definition in Eq. (10.6) for a crack in a homogeneous material and with the definition in Eq. (10.18) for an interface crack between two isotropic materials.

It should also be noted from Eq. (10.36) that, for an interfacial crack between two anisotropic materials ($\epsilon \neq 0$ and $W_2 \neq 0$), the two constants W_1 and W_2 dependent on the material proprieties are involved in the definition of the stress intensity factors so that the stress intensity factors are independent of the radial coordinate r. This is in contrast to the definition in Eq. (10.18) (or Eq. (10.17)) for an interficial crack between two isotropic materials where only the orders of singularity $0.5 \pm i\epsilon$ are required.

The stress intensity factors can also be determined from the crack opening displacements. Substituting Eq. (10.35) into Eq. (10.21) and using Eq. (10.19) leads to

$$\left\{ \begin{matrix} \delta_x^{(s)}(r) \\ \delta_y^{(s)}(r) \end{matrix} \right\} = \frac{4\cosh \pi \epsilon}{1 + 4\epsilon^2} \sqrt{\frac{r}{2\pi}} \begin{bmatrix} d_{11} & d_{12} \\ d_{21} & d_{22} \end{bmatrix} \left\{ \begin{matrix} K_1 \\ K_2 \end{matrix} \right\} \tag{10.39}$$

where the matrix (the argument r is omitted for conciseness in notation)

$$[d] = \begin{bmatrix} d_{11} & d_{12} \\ d_{21} & d_{22} \end{bmatrix} = \begin{bmatrix} G_1 & P_1 \\ G_2 & 0 \end{bmatrix} \begin{bmatrix} c(r) + 2\epsilon s(r) & -(s(r) - 2\epsilon c(r)) \\ s(r) - 2\epsilon c(r) & c(r) + 2\epsilon s(r) \end{bmatrix} \tag{10.40}$$

is calculated at a specified distance r behind the crack tip by using Eqs. (10.35) and (10.19). The stress intensity factors follow from Eq. (10.39) as

$$\left\{ \begin{matrix} K_1 \\ K_2 \end{matrix} \right\} = \frac{1 + 4\epsilon^2}{4\cosh \pi \epsilon} \sqrt{\frac{2\pi}{r}} \frac{1}{d_{11}d_{22} - d_{12}d_{21}} \begin{bmatrix} d_{22} & -d_{12} \\ -d_{21} & d_{11} \end{bmatrix} \left\{ \begin{matrix} \delta_x^{(s)}(r) \\ \delta_y^{(s)}(r) \end{matrix} \right\} \tag{10.41}$$

The magnitude of the complex stress intensity factor is expressed as (Eq. (38), Cho et al., 1992)

$$K_i = \sqrt{K_1^2 + K_2^2}$$

$$= \frac{1 + 4\epsilon^2}{4\cosh \pi\epsilon} \sqrt{\frac{2\pi}{r}} \frac{\sqrt{\left(d_{22}\delta_x^{(s)}(r) - d_{12}\delta_y^{(s)}(r)\right)^2 + \left(d_{11}\delta_y^{(s)}(r) - d_{21}\delta_x^{(s)}(r)\right)^2}}{d_{11}d_{22} - d_{12}d_{21}}$$

$$(10.42)$$

K_i does not oscillate as $r \to 0$. The ratio between K_2 and K_1 is equal to

$$\frac{K_2}{K_1} = \frac{d_{11}\delta_y^{(s)}(r) - d_{21}\delta_x^{(s)}(r)}{d_{22}\delta_x^{(s)}(r) - d_{12}\delta_y^{(s)}(r)} \tag{10.43}$$

Equations (10.41) and (10.43) form the basis of the displacement extrapolation method in Cho et al. (1992).

10.2.4 Multi-material Wedges

Stress singularities also occur at a V-notch, a crack terminating at a material interface and a free edge formed by dissimilar materials. From the point of view of in-plane stress analysis, they can be regarded as special cases of a multi-material wedge as illustrated in Figure 10.3.

The asymptotic solutions for multi-material wedges depend on the geometrical configurations and properties of the constituent materials. The orders of stress singularity can be real and/or complex. There is a considerable amount of literature on the orders of singularity. However, the numbers of papers reporting on the solutions of the singular stress fields are limited. The definition and evaluation of the stress intensity factors are formulated in a manner similar to the case of interfacial cracks.

When two real orders of singularity (ω_1, ω_2) exist, it is customary to express the singular stress field as (Munz and Yang, 1993; Labossiere and Dunn, 1999)

$$\left\{ \begin{array}{c} \sigma_\theta^{(s)}(r, 0) \\ \tau_{r\theta}^{(s)}(r, 0) \end{array} \right\} = \frac{K_1}{\sqrt{2\pi L}} \left(\frac{r}{L}\right)^{-\omega_1} \left\{ \begin{array}{c} 1 \\ f_{r\theta1}(0) \end{array} \right\} + \frac{K_2}{\sqrt{2\pi L}} \left(\frac{r}{L}\right)^{-\omega_2} \left\{ \begin{array}{c} f_{\theta\theta2}(0) \\ 1 \end{array} \right\}$$

$$(10.44)$$

Figure 10.3 A multi-material wedge.

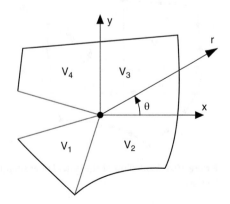

where K_1 and K_2 are the generalized stress intensity factors. The normalization of the two eigenfunctions $f_1(\theta)$ and $f_2(\theta)$ is chosen in such a way that the components $f_{\theta\theta1}(0) = 1$ and $f_{r\theta2}(0) = 1$ hold.

When the orders of singularity are a pair of complex conjugates $\omega \pm i\epsilon$, the singular stress field is expressed in terms of the generalized stress intensity factors K_1 and K_2 as (Yang and Munz, 1995)

$$\sigma_{ij}^{(s)}(r,0) = \sum_{k=1}^{2} \frac{K_k}{\sqrt{2\pi L}} \left(\frac{r}{L}\right)^{-\omega} \left(\cos\left(\epsilon \ln \frac{r}{L}\right) f_{ijk}^c(\theta) + \sin\left(\epsilon \ln \frac{r}{L}\right) f_{ijk}^s(\theta)\right) \quad (10.45)$$

with $i, j = r, \theta$ and the eigenfunctions $f_{ijk}^c(\theta)$ and $f_{ijk}^s(\theta)$.

The above two equations differ from the original definitions in the references by a factor $\sqrt{2\pi L}$. This trivial difference is introduced so that it is convenient to compare with the definitions based on the scaled boundary finite element solution. Other normalizations of the generalized stress intensity factors can be used without affecting the conclusions.

When the orders of singularity change between two real numbers and a pair of complex conjugates, a power-logarithmic singularity exists (Dempsey and Sinclair, 1979; Song, 2005). The definition of stress intensity factors for power-logarithmic stress singularity is rarely addressed (Pageau et al., 1996).

10.3 Modelling of Singular Stress Fields by the Scaled Boundary Finite Element Method

The modelling of stress fields around singularity points is illustrated in Figure 10.4 using an interface crack and a multi-material wedge as examples. Only the S-domain containing a singularity point is considered. When generating the scaled boundary finite element mesh, the S-domain has to be small enough so that the crack faces and material interfaces can be approximated by straight lines. The scaling centre is selected at the singularity point: the crack tip in Figure 10.4a and the apex of the wedge

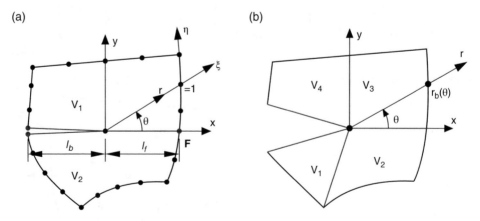

Figure 10.4 Modelling of stress field near singularity points. (a) Interfacial crack. (b) Multi-material wedge.

in Figure 10.4b. The representation of the geometry of the S-domain is discussed in Item 3a on page 35. The straight lines passing through the scaling centre of an S-domain (the so-called 'side-face') are not discretized. The boundary that is directly visible from the scaling centre (the so-called 'defining curve') is divided into elements as shown in Figure 10.4a. The elements do not form a closed loop. The resulting S-element can be assembled with other S-elements or finite elements to model cracks of complex geometry.

At the crack mouth, two independent nodes exist (Figure 10.4a). The crack faces are produced by scaling these two nodes towards the scaling centre. The traction-free boundary condition on the crack faces is satisfied automatically. At a material interface, the two elements are connected by a common node. The material interface is formed by scaling the common node. When computing the element coefficient matrices, each element assumes the material in the corresponding side of the interface. The displacement compatibility and the force equilibrium are enforced during the assemblage process.

All the other modelling and solution procedures are the same as those for a normal (uncracked) S-domain. The scaled boundary finite element equation resulting from the assemblage of all the elements on boundary can be solved by either the eigenvalue method in Chapter 3 or the Schur decomposition method in Chapter 7.

The scaled boundary finite element method leads to a semi-analytical solution of the stress field near a singular point similar to the asymptotic solution in Eq. (10.2). This feature provides a simple and accurate way to extract the stress intensity factors and T-stress. The use of eigenvalue method is relatively straightforward and is explained firstly in Section 10.4. A numerical examples of a crack in a homogeneous isotropic material is presented. The use of Schur decomposition method is explained in Section 10.5. This technique provides a unified solution to all types of singular stress fields. Based on this solution, generalized stress intensity factors are defined (Section 10.5). This definition includes the classical definitions in Section 10.2 as special cases.

10.4 Stress Intensity Factors and the T-stress of a Cracked Homogeneous Body

In the eigenvalue method, the solution of the displacement field of an S-element is given by (Eq. (3.18a))

$$\{u(\xi)\} = \sum_{i=1}^{n} c_i \xi^{\lambda_i} \{\phi_i^{(u)}\} \tag{10.46}$$

and the solution of the stress field by (Eq. (3.68))

$$\{\sigma(\xi, \eta)\} = \sum_{i=1}^{n-2} c_i \xi^{\lambda_i - 1} \{\phi_i^{(\sigma)}\} \tag{10.47}$$

where λ_i and $\{\phi_i^{(u)}\} = \{\phi_i^{(u)}(\eta)\}$ $(i = 1, 2, \dots n)$ are obtained by solving the eigenvalue problem in Eq. (3.10) and selected to satisfy the finiteness of displacement at the scaling centre (see Section 3.1). $\{\phi_i^{(\sigma)}\}$ $(i = 1, 2, \dots n)$ are the stress modes obtained from the corresponding displacement modes (Eq. (3.69)). The modes are sorted according to the descending order of the real parts of the eigenvalues. For a two-dimensional problem, the last two eigenvalues $(\lambda_{n-1}, \lambda_n)$ are equal to 0 and the displacement modes represent the rigid-body translational motions. The corresponding stress modes vanish and do not contribute to the stresses. They are not included in the summation in Eq. (10.47).

The semi-analytical solution of stresses in Eq. (10.47) resembles the asymptotic solution near the crack tip in Eq. (10.2). Both are expressed as a series of power functions. As $\xi = r/r_b$ (Eq. (2.6a)), the power functions $\xi^{\lambda_i-1} = (r/r_b)^{\lambda_i-1}$ are proportional to $(r)^{\lambda_i-1}$. Like the power functions $r^{\frac{i}{2}-1}$ in Eq. (10.2), they describe the variation of stresses along the radial direction. In the circumferential direction, the solution of stresses is numerical and can be handled using standard stress recovery techniques of the finite element method. The stress modes $\{\phi_i^{(\sigma)}\}$ are evaluated at Gauss points that all have unique values of angular coordinate $\theta = \theta(\eta)$ (Eq. 2.21). Grouping the angular coordinates of the Gauss points in a vector $\{\theta(\eta)\}$, $(\{\theta(\eta)\}, \{\phi_i^{(\sigma)}(\eta)\})$ describes the angular variation of the stresses and can be interpreted as a discrete representation of the trigonometric functions within the square brackets in Eq. (10.2).

As will be shown in Example 10.1 on page 411, the eigenvalues of the scaled boundary finite element solution converge as the mesh on the boundary of an S-element is refined. In the ascending order of the converged eigenvalues, the leading eigenvalues are equal to 0.5, 0.5, 1, 1, 1.5, 1.5, ... except for the two eigenvalues of 0. This is the same sequence as the exponents $i/2 - 1$ $(i = 1, 2, 3, ...)$ of the power function of r in Eq. (10.2).

The scaled boundary finite element solution can be used similarly to the asymptotic solution in a fracture analysis. The two singular stress terms, denoted as I and II, are identified by their eigenvalues $0 < \lambda_i < 1$, $(i = I, II)$. The singular stress field is expressed as

$$\{\sigma^{(s)}(\xi, \eta)\} = \sum_{i=I,II} c_i \xi^{\lambda_i-1} \{\phi_i^{(\sigma)}\} \tag{10.48}$$

Replacing the eigenvalues by $\lambda_i = 0.5$, $(i = I, II)$ yields

$$\{\sigma^{(s)}(\xi, \eta)\} = \xi^{-0.5} \sum_{i=I,II} c_i \{\phi_i^{(\sigma)}\} \tag{10.49}$$

The stresses on the boundary $(\xi = 1)$ are equal to

$$\{\sigma^{(s)}(\xi = 1, \eta)\} = \sum_{i=I,II} c_i \{\phi_i^{(\sigma)}\} \tag{10.50}$$

Equation (10.49) is rewritten as

$$\{\sigma^{(s)}(\xi, \eta)\} = \xi^{-0.5} \{\sigma^{(s)}(\xi = 1, \eta)\} \tag{10.51}$$

To determine the stress intensity factors, the stress distribution at the direction $\theta = 0$ is addressed. Same as the stress modes $\{\phi_i^{(\sigma)}\}$, the boundary stresses $\{\sigma^{(s)}(\xi = 1, \eta)\}$ are evaluated at the Gauss points of the elements. Together with the vector $\{\theta(\eta)\}$ of the angular coordinates of the Gauss points, the boundary stresses are interpolated to determine their values at $\theta = 0$, denoted as $\{\sigma^{(s)}(\xi = 1, \theta = 0)\}$. Equation (10.51) is written for the stress component σ_y and τ_{xy} as

$$\left\{ \begin{matrix} \sigma_y^{(s)}(\xi, \theta = 0) \\ \tau_{xy}^{(s)}(\xi, \theta = 0) \end{matrix} \right\} = \xi^{-0.5} \left\{ \begin{matrix} \sigma_y^{(s)}(\xi = 1, \theta = 0) \\ \tau_{xy}^{(s)}(\xi = 1, \theta = 0) \end{matrix} \right\} \tag{10.52}$$

To match with the definition of the stress intensity factors, the scaled boundary coordinate ξ in Eq. (10.52) is changed to the polar coordinate r by applying Eq. (2.6a) at $\theta = 0$

$$\xi = \frac{r}{l_f} \tag{10.53}$$

where, l_f, as shown in Figure 10.4a, denotes the distance from the crack tip to the boundary along the radial line $\theta = 0$. Substituting Eq. (10.53) into Eq. (10.52), the singular stresses are expressed as

$$\left\{ \begin{array}{c} \sigma_y^{(s)}(r, \theta = 0) \\ \tau_{xy}^{(s)}(r, \theta = 0) \end{array} \right\} = \sqrt{\frac{l_f}{r}} \left\{ \begin{array}{c} \sigma_y^{(s)}(\xi = 1, \theta = 0) \\ \tau_{xy}^{(s)}(\xi = 1, \theta = 0) \end{array} \right\} \tag{10.54}$$

The stress intensity factors are obtained, after substituting Eq. (10.54) into the definition in Eq. (10.7), as

$$\left\{ \begin{array}{c} K_I \\ K_{II} \end{array} \right\} = \sqrt{2\pi l_f} \left\{ \begin{array}{c} \sigma_y^{(s)}(\xi = 1, \theta = 0) \\ \tau_{xy}^{(s)}(\xi = 1, \theta = 0) \end{array} \right\} \tag{10.55}$$

Note that the stress singularity is handled analytically in the scaled boundary finite element method. It is sufficient to evaluate the singular modes, which can be regarded as evaluating the singular stresses on the boundary. This operation involves only standard stress recovery techniques.

If the asymptotic solution of the displacement field is available, the stress intensity factors can be determined from the crack opening displacements using Eq. (10.10). In this case, even the stress recovery is not needed. The displacement field of the singular stress terms is obtained by considering only the terms $\lambda_i = 0.5$, $(i = I, II)$ in Eq. (10.46)

$$\{u^{(s)}(\xi, \eta)\} = \xi^{0.5} \sum_{i=I,II} c_i \{\phi_i^{(u)}\} \tag{10.56}$$

At the boundary $\xi = 1$, the displacement field of the singular stress term is expressed as

$$\{u^{(s)}(\xi = 1, \eta)\} = \sum_{i=I,II} c_i \{\phi_i^{(u)}\} \tag{10.57}$$

Note that only a combination of displacement modes is needed in computing $\{u^{(s)}(\xi = 1, \eta)\}$. Equation (10.56) is rewritten as

$$\{u^{(s)}(\xi, \eta)\} = \xi^{0.5} \{u^{(s)}(\xi = 1, \eta)\} \tag{10.58}$$

The variation of the displacements of the singular stress terms on the upper crack face $(\theta = +\pi)$ and the lower crack face $(\theta = -\pi)$ is addressed. As shown in Figure 10.4a, the distance from the scaling centre to the crack mouth is denoted as l_b. The displacements of the two nodes at the crack mouth $(\xi = 1, \text{i.e. } r = l_b)$ are extracted from $\{u^{(s)}(\xi = 1, \eta)\}$ and the crack opening displacements at the boundary, $\delta_x^{(s)}(l_b)$ and $\delta_y^{(s)}(l_b)$, are determined using Eq. (10.9). On the crack faces, Eq. (2.6a) is expressed as

$$\xi = \frac{r}{l_b} \tag{10.59}$$

Considering Eqs. (10.58) and (10.59), the variation of the crack opening displacements is expressed as

$$\begin{Bmatrix} \delta_y^{(s)}(r) \\ \delta_x^{(s)}(r) \end{Bmatrix} = \sqrt{\frac{r}{l_b}} \begin{Bmatrix} \delta_y^{(s)}(l_b) \\ \delta_x^{(s)}(l_b) \end{Bmatrix} \tag{10.60}$$

Substituting Eq. (10.60) into Eq. (10.10) leads to the stress intensity factors

$$\begin{Bmatrix} K_I \\ K_{II} \end{Bmatrix} = \frac{G}{\kappa + 1} \sqrt{\frac{2\pi}{l_b}} \begin{Bmatrix} \delta_y^{(s)}(l_b) \\ \delta_x^{(s)}(l_b) \end{Bmatrix} \tag{10.61}$$

The T-stress can also be determined directly from the scaled boundary finite element solution. In the solution for the stress field (Eq. (10.47)), the two terms with their eigenvalues $\lambda_i = 1$ are linear combinations of the rotational rigid-body term that does not generate stresses and the T-stress term. Denoting the two terms by $i = T_1$, T_2, the T-stress field is expressed as

$$\{\sigma^{(T)}(\eta)\} = \sum_{i=T_1, T_2} c_i \{\phi_i^{(\sigma)}\} \tag{10.62}$$

Similar to the singular stress field, the T-stress field $\{\sigma^{(T)}(\eta)\}$ is evaluated at the Gauss points. The T-stress, i.e. the x-component of the T-stress field at $\theta = 0$ (Eq. (10.63)), is determined by interpolation the T-stress term

$$T = \sigma_x^{(T)}(\theta = 0) \tag{10.63}$$

The T-stress can be obtained from the displacement solution using Eq. (10.15) derived from the asymptotic solution. The T-stress terms are extracted from the solution in Eq. (10.46) by including only the terms $\lambda_i = 1$ ($i = T_1$, T_2). The displacement functions are expressed as

$$\{u^{(T)}(\xi)\} = \xi \sum_{i=T_1, T_2} c_i \{\phi_i^{(u)}\} \tag{10.64}$$

At the boundary $\xi = 1$, it leads to the nodal displacements

$$\{u^{(T)}(\xi = 1)\} = \sum_{i=T_1, T_2} c_i \{\phi_i^{(u)}\} \tag{10.65}$$

Equation (10.64) is rewritten as

$$\{u^{(T)}(\xi)\} = \xi \{u^{(T)}(\xi = 1)\} \tag{10.66}$$

The displacements at angle $\theta = 0$ in front of the crack tip, i.e. point F in Figure 10.4a, are obtained by interpolating the nodal displacements $\{u^{(T)}(\xi = 1)\}$. The x-displacement component of the displacement at point F is denoted as $u_{Fx}^{(T)}$. Using Eqs. (10.66) and (10.53), x-displacement along the line $\theta = 0$, is expressed as

$$u_x^{(T)}(r, \theta = 0) = \frac{r}{l_f} u_{Fx}^{(T)} \tag{10.67}$$

The T-stress is obtained using Eq. (10.46) as

$$T = \frac{8G}{\kappa + 1} \frac{1}{l_f} u_{Fx}^{(T)} \tag{10.68}$$

The same procedure can be used to determine the coefficients of higher-order terms in the asymptotic expansion by matching the expansion with the corresponding terms in the scaled boundary finite element solution in Eq. (10.47) (Song, 2005).

■ **Example 10.1   An Edge-cracked Homogeneous Square Body**

An edge-crack square body in plane strain condition is shown in Figure 10.5. The width of the square is $2b$. The length of the crack is $a = b$. The material is isotropic with Young's modulus E and Poisson's ratio $v = 0.25$. The analytical solution for this square plate is constructed from the asymptotic expansion of stresses in Eq. (10.2) with a pre-selected set of coefficients a_i and b_i. The analysis is performed by applying the surface traction obtained from this series as the boundary condition on the edges of the square. Displacement boundary conditions are only prescribed to constrain the three rigid body motions.

The square boundary is discretized with cubic (4-node) elements given in Example 8.1. To perform a convergence study, five meshes with 8, 16, 24, 32 and 48 elements are used. The nodes are evenly spaced along the edges. The mesh with 8 elements is shown in Figure 10.5.

The first 12 non-zero eigenvalues obtained by the scaled boundary finite element method are compared with their exact values in Table 10.1. The scaled boundary finite element results converge to the exact values as the mesh is refined. The 24-element mesh leads to results exact up to the first 7 digits for all the 12 eigenvalues. The eigenvalues of the lower modes, which are generally more important to the stresses near the crack tip, converge earlier than the higher modes. The integer eigenvalues up to 3 are accurate up to the machine accuracy when 48 uniformly-spaced cubic elements are used.

The stress modes obtained in the solution are investigated. They are evaluated at the sampling points of the third-order Gauss quadrature (the natural coordinates are $\eta = -\sqrt{3/5}, 0$ and $+\sqrt{3/5}$). To compare with the asymptotic solution in Eq. (10.2), the scaled boundary finite element solution in Eq. (10.47) is transformed into the polar coordinates r, θ. Substituting ξ defined in Eq. (2.6a) into Eq. (10.47), one term of the series is expressed as

$$\{\sigma_i(r, \eta)\} = c_i r^{\lambda - 1} \left(r_b^{-\lambda_i + 1} \{\phi_i^{(\sigma)}\} \right) \tag{10.69}$$

Figure 10.5 An edge-cracked square body: Geometry and boundary discretization with cubic elements.

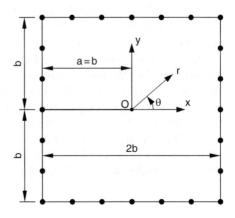

Table 10.1 Edge-cracked square body: Eigenvalues (λ_i).

Mode i	Number of elements			
	8	16	24	Exact $i/2$
1	0.499983	0.500000	0.500000	0.5
	0.499986	0.500000	0.500000	0.5
2	1.000000	1.000000	1.000000	1
	1.000000	1.000000	1.000000	1
3	1.500046	1.499999	1.500000	1.5
	1.500005	1.500000	1.500000	1.5
4	2.000000	2.000000	2.000000	2
	2.000000	2.000000	2.000000	2
5	2.499713	2.499999	2.500000	2.5
	2.500135	2.500001	2.500000	2.5
6	3.000000	3.000000	3.000000	3
	3.000000	3.000000	3.000000	3

where $r_b = r_b(\eta)$ is the radial coordinate of a (Gauss) point on the boundary. The terms $\left(r_b^{-\lambda_i+1}\{\phi_i^{(\sigma)}\}\right)$ in Eq (10.69) and $\theta(\eta)$ in Eq. (2.6c) are the parametric equations (with the natural coordinate η as the parametric variable) of a stress mode describing the angular stress distribution at a constant radial coordinate r. The stress modes associated with the stress intensity factors K_I, K_{II} and the T-stress are computed at the Gauss points. The results of the 8-element mesh are shown in Figure 10.6 as discrete points. The stress modes (the functions inside the square brackets) of the exact solution in Eq. (10.2) are plotted as continuous lines. Each stress mode is normalized with the value of its component σ_x at $\theta = 0$. Good agreement between the two sets of results is observed. The stress modes calculated with other four denser meshes are indistinguishable from the exact results and not shown.

The surface traction applied to the boundary is obtained from the asymptotic solution in Eq. (10.2). The coefficients a_1, b_1 and a_2 are selected according to Eqs. (10.4) and (10.12) in such a way that $K_I/(p\sqrt{a}) = 1$, $K_{II}/(p\sqrt{a}) = 1$ and $T/(K_I/\sqrt{a}) = 1$ apply (p is a constant with the unit of stress). The coefficients a_i, ($i = 3, 4, 5$) and b_i, ($i = 2, 3, 4, 5$) are arbitrarily chosen as $1 \times pa^{1-\frac{i}{2}}$ to increase the complexity of the stress field.

The singular stresses are obtained from Eq. (10.50) using the stress modes given at the Gauss points (see the plots of stress modes in Figure 10.6). The angular coordinates are stored in the vectors $\{\theta(\eta)\}$. The stresses at Gauss points of all elements on the boundary are stored in $\{\sigma^{(s)}(\xi = 1, \eta)\}$ for the singular stress terms and in $\{\sigma^{(T)}(\eta)\}$ for the T-stress terms, respectively. They are sorted in the ascending order of angular coordinate and interpolated by the MATLAB function interp1 to obtain the stresses at $\theta = 0$. The stress intensity factors and T-stress are evaluated using Eqs. (10.55) and (10.62), respectively. The results are normalized and shown in Table 10.2a. The error of the results

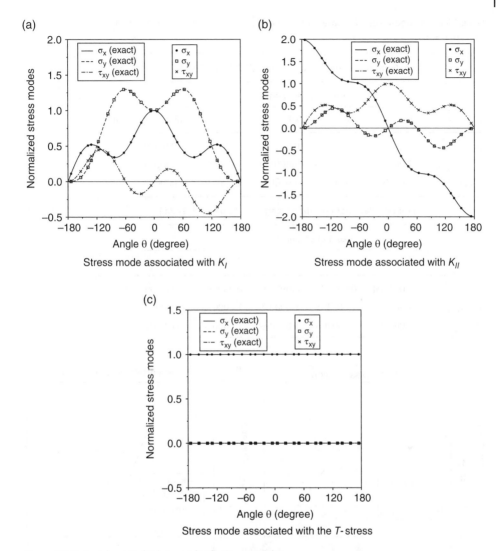

Figure 10.6 An edge-cracked square body: Stress modes.

obtained with the coarsest mesh (8 elements) is already less than 1%. When 48 cubic elements are used, the error decreases to 10^{-5}. The largest errors occur in K_{II}.

The stress intensity factors and T-stress are also obtained from the displacement field corresponding to the singular stress terms using Eqs. (10.61) and (10.68), respectively. The results are given in Table 10.2b. The errors are generally smaller than those obtained from the stress solution in Table 10.2a.

The relative errors of the results obtained from the stress solution and the displacement solution are plotted versus the number of degrees of freedom in Figure 10.7a and Figure 10.7b, respectively. The slopes of the curves in Figure 10.7a are also approximately equal to -4, showing a super-convergent rate.

Table 10.2 An edge-cracked square body: Normalized stress intensity factors K_I, K_{II} and the *T*-stress.

a) From stress solution using Eqs. (10.55) and (10.63).

	Number of elements					
	8	16	24	32	48	Exact
$K_I/(p\sqrt{a})$	1.004415	1.000011	0.999994	0.999997	0.999999	1
$K_{II}/(p\sqrt{a})$	1.007734	1.000835	1.000181	1.000060	1.000009	1
$T/(K_I/\sqrt{a})$	1.005591	1.000208	1.000034	1.000010	1.000002	1

b) From displacement solution using Eqs. (10.61) and (10.68).

	Number of Elements					
	8	16	24	32	48	Exact
$K_I/(p\sqrt{a})$	1.003004	0.999983	0.999991	0.999997	0.999999	1
$K_{II}/(p\sqrt{a})$	1.005073	1.000408	1.000079	1.000025	1.000005	1
$T/(K_I/\sqrt{a})$	1.005591	1.000208	1.000034	1.000010	1.000002	1

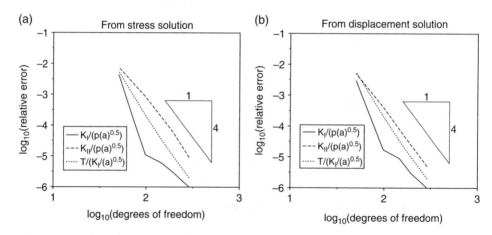

Figure 10.7 An edge-cracked square body: Convergence of stress intensity factors and the *T*-stress with increasing numbers of cubic elements.

It is worth mentioning that the computation of the complete stress field (Eq. 10.47 including all the terms) is not performed when determining the stress intensity factors and the *T*-stress. The extraction of the *T*-stress is independent of the extraction of the stress intensity factors and, thus, essentially free of the numerical pollution caused by the singular stresses.

Spectral elements (Section 8.2) are applied to model this example (Figure 10.5). The Gauss-Lobatto-Legendre quadrature is used for the integration of the element coefficient matrices. The boundary is divided into eight elements of the same length *b*. A convergence study is performed by increasing the order of element from 3 to 9.

Table 10.3 An edge-cracked square body modelled by spectral elements: Normalized stress intensity factors K_I, K_{II} and the T-stress.

a) From stress solution and Eqs. (10.55) and (10.63).

	Order of elements					
	3	4	5	7	9	Exact
$K_I/(p\sqrt{a})$	1.001168	0.999675	0.999916	1.000005	1.000000	1
$K_{II}/(p\sqrt{a})$	1.004753	1.002575	1.000319	0.999979	1.000001	1
$T/(K_I/\sqrt{a})$	1.000613	0.999899	1.000006	1.000000	1.000000	1

b) From displacement solution and Eqs. (10.61) and (10.68)

	Order of elements					
	3	4	5	7	9	Exact
$K_I/(p\sqrt{a})$	0.998180	0.999868	0.999996	1.000000	1.000000	1
$K_{II}/(p\sqrt{a})$	0.999679	0.999927	0.999996	1.000000	1.000000	1
$T/(K_I/\sqrt{a})$	1.000613	0.999899	1.000006	1.000000	1.000000	1

The stress intensity factors and the T-stress are evaluated from the stress solution using Eqs. (10.55) and from the displacement solution using Eqs. (10.61) and (10.68). The results are shown in Tables 10.3a and 10.3b, respectively.

The errors of K_{II} are larger than the errors of K_I and T. The results obtained from the displacement solution show higher accuracy than the results obtained from the stress solution. For both sets of results, rapid convergence is observed. The results obtained from the displacement solution have converged up to the first six digits at the element order $p = 7$. The errors of the results obtained from the stress solution and from the displacement solution are plotted in Figure 10.8a and Figure 10.8b, respectively.

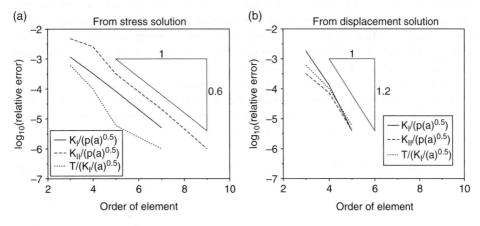

Figure 10.8 An edge-cracked square body modelled by spectral elements: Convergence of stress intensity factors and the T-stress.

Exponential convergence is observed. This example shows the superior performance of spectral elements in the scaled boundary finite element method.

Finally, it is worthwhile to mention the computational efficiency. The analyses are performed in MATLAB interactive mode on a laptop computer with an Intel i7-4650U CPU and 8GB of RAM. The analyses using 32 cubic elements take less than 0.4s of wall-clock time and the error is less than 0.01%. The same accuracy can be obtained with eight 7th-order elements taking less than 0.2s of wall-clock time.

10.5 Definition and Evaluation of Generalized Stress Intensity Factors

When the behaviour of a singular stress field is known analytically, for example, as an asymptotic solution, stress intensity factors are defined to describe the stress field (Section 10.2). The singular functions of the radial coordinate r are separated analytically. The stress intensity factors remain finite as r tends to zero.

It is shown in Section 10.4 that the semi-analytical solution of the scaled boundary finite element method can be used to extract stress intensity factors and the T-stress. As the stress variation in the radial direction, including the singularity, is represented analytically as a power series, the operation is as simple as the standard stress recovery in the finite element method. If the asymptotic solution of displacement is available, the stress intensity factors and the T-stress can be directly extracted from the displacement solution. Accurate results can be obtained efficiently with a small number of elements. This approach is extended to the case of interfacial cracks between two isotropic materials (Song and Wolf, 2002).

The scaled boundary finite element method is directly applicable to model a multi-material wedge as shown in Figure 10.4b, Section 10.3. A semi-analytical solution of the singular stress field is obtained. In the radial direction emanating from the apex of the wedge, the solution is analytical and represents the stress singularity. It is employed similarly to an asymptotic solution in Song et al. (2010a) to define generalized stress intensity factors by separating the singular functions analytically. At the same time, a simple procedure to evaluate the generalized stress intensity factors is derived. For the case of a crack in a homogeneous body and the case of an interfacial crack between two isotropic materials, this definition of generalized stress intensity factors reduces to the classical definition of the stress intensity factors.

The matrix function solution using the block-diagonal Schur decomposition in Section 7.6 is employed for the solution of the singular stress field to avoid the loss of solution of the eigenvalue method. The solution of the displacement field is given in Eq. (7.152). In two dimensions ($s = 2$), it is written as

$$\{u(\xi)\} = \sum_{i=1}^{N} [\Psi_i^{(u)}] \xi^{[S_i]} \{c_i\} \tag{10.70}$$

where $[S_i]$ given in Eq. (7.114) and $[\Psi_i^{(u)}]$ given in Eq. (7.113) are the diagonal blocks of the Schur matrix and the corresponding column blocks of the block-diagonal Schur decomposition in Eq. (7.108). $\{c_i\}$ are integration constants.

The solution of the stress field is given in Eq. (7.193). It is expressed for a two-dimensional bounded S-element ($s = 2$, $\delta_{ln} = 0$, $i_B = 1$ and $[S_N] = [0]$) as

$$\{\sigma(\xi, \eta)\} = \sum_{i=1}^{N-1} [\Psi_i^{(\sigma)}(\eta)]\xi^{[S_i]-[I]}\{c_i\} \tag{10.71}$$

where $[\Psi_i^{(\sigma)}(\eta)]$ are the stress modes. Using Eq. (7.192b) with the strain modes $[\Psi_i^{(\epsilon)}]$ in Eq. (7.190) and $s = 2$, the stress modes are obtained from the displacement modes $[\Psi_i^{(u)}]$ in Eq. (10.70) as

$$[\Psi_i^{(\sigma)}(\eta)] = [D]\left([B_1][\Psi_i^{(u)}][S_i] + [B_2][\Psi_i^{(u)}]\right) \tag{10.72}$$

The stress modes and stresses are evaluated element-by-element at Gauss points. For the convenience in defining the generalized stress intensity factors, the stress modes are transformed to polar coordinates r, θ (Figure 10.4) by following the standard procedure for stress transformation

$$\begin{Bmatrix} \sigma_r \\ \sigma_\theta \\ \tau_{r\theta} \end{Bmatrix} = \begin{bmatrix} \cos^2\theta & \sin^2\theta & 2\sin\theta\cos\theta \\ \sin^2\theta & \cos^2\theta & -2\sin\theta\cos\theta \\ -\sin\theta\cos\theta & \sin\theta\cos\theta & \cos^2\theta - \sin^2\theta \end{bmatrix} \begin{Bmatrix} \sigma_x \\ \sigma_y \\ \tau_{xy} \end{Bmatrix} \tag{10.73}$$

The singular stress terms are isolated from the solution. The eigenvalues of the diagonal blocks should be well separated in the block-diagonal Schur decomposition. When the real parts of the eigenvalues, i.e. the diagonal entries, of a diagonal block $[S_i]$ are between 0 and 1, the matrix power function in Eq. (10.71) leads to a stress singularity at $\xi = 0$. In the following, all such diagonal blocks are grouped into one block denoted as $[S^{(s)}]$ (superscript (s) for singular) and $0 < \lambda([S^{(s)}]) < 1$ applies. The corresponding stress modes in polar coordinates are denoted as $[\Psi^{(s)}]$. The integration constants are denoted as $\{c^{(s)}\}$. The singular stress field is expressed as

$$\{\sigma^{(s)}(\xi, \eta)\} = [\Psi^{(s)}(\eta)]\xi^{[S^{(s)}]-[I]}\{c^{(s)}\} \tag{10.74}$$

The stress components evaluated from the stress modes are also in polar coordinates.

The scaled boundary coordinates ξ and η are transformed to the polar coordinates r and θ. Using Eq. (2.6) and introducing a characteristic length L, ξ is expressed at a angle θ as

$$\xi = \frac{r}{r_b(\theta)} = \frac{L}{r_b(\theta)} \times \frac{r}{L} \tag{10.75}$$

where $r_b(\theta)$ is the distance from the scaling centre to the boundary along the radial line at angle θ (Figure 10.4b). Using the property in Eq. (B.19), the matrix power function of ξ is rewritten in the polar coordinates as

$$\xi^{[S^{(s)}]-[I]} = \left(\frac{L}{r_b(\theta)}\right)^{[S^{(s)}]-[I]} \left(\frac{r}{L}\right)^{[S^{(s)}]-[I]} \tag{10.76}$$

To avoid proliferation of notations, the same functions $\sigma^{(s)}$ is used to express singular stresses in polar coordinates. The singular stress field in Eq. (10.74) is expressed as

$$\{\sigma^{(s)}(r,\theta)\} = [\Psi_L^{(s)}(\theta)]\left(\frac{r}{L}\right)^{[S^{(s)}]-[I]}\{c^{(s)}\}$$

(10.77)

where the stress modes at the characteristic length L are expressed as

$$[\Psi_L^{(s)}(\theta)] = [\Psi^{(s)}(\eta(\theta))]\left(\frac{L}{r_b(\theta)}\right)^{[S^{(s)}]-[I]}$$

(10.78)

where a linear transformation is applied to the stress modes $[\Psi^{(s)}(\eta)]$ (for a given element, η and θ is related by a single-valued function in Eq. (2.6c)).

The case of two singular stress modes, i.e. $[S^{(s)}]$ is a 2×2 matrix, is considered. $\{c^{(s)}\}$ in Eq. (10.77) consists of two integration constants. Following the classical definition, two stress intensity factors are defined using the two stress components $\sigma_\theta^{(s)}(r,\theta)$ and $\tau_{r\theta}^{(s)}(r,\theta)$. Equation (10.77) still holds with the understanding that $\{\sigma^{(s)}(r,\theta)\} = [\sigma_\theta^{(s)}(r,\theta), \ \tau_{r\theta}^{(s)}(r,\theta)]^T$ applies and $[\Psi_L^{(s)}(\theta)]$ becomes a 2×2 matrix with only the entries corresponding to $\sigma_\theta^{(s)}(r,\theta)$ and $\tau_{r\theta}^{(s)}(r,\theta)$ retained. Equation (10.77) is further simplified by introducing a matrix of orders of singularity

$$[\tilde{S}^{(s)}(\theta)] = [\Psi_L^{(s)}(\theta)](-[S^{(s)}] + [I])[\Psi_L^{(s)}(\theta)]^{-1}$$

(10.79)

$[\tilde{S}^{(s)}(\theta)]$ and $(-[S^{(s)}] + [I])$ are similar matrices. Using the property of matrix power functions in Eq. (B.21),

$$\left(\frac{r}{L}\right)^{-[\tilde{S}^{(s)}(\theta)]} = [\Psi_L^{(s)}(\theta)]\left(\frac{r}{L}\right)^{[S^{(s)}]-[I]}[\Psi_L^{(s)}(\theta)]^{-1}$$

(10.80)

applies. Post-multiplying Eq. (10.80) with $[\Psi_L^{(s)}(\theta)]$ yields

$$\left(\frac{r}{L}\right)^{-[\tilde{S}^{(s)}(\theta)]}[\Psi_L^{(s)}(\theta)] = [\Psi_L^{(s)}(\theta)]\left(\frac{r}{L}\right)^{[S^{(s)}]-[I]}$$

(10.81)

Substituting Eq. (10.81) into Eq. (10.77) leads to

$$\begin{Bmatrix} \sigma_\theta^{(s)}(r,\theta) \\ \tau_{r\theta}^{(s)}(r,\theta) \end{Bmatrix} = \left(\frac{r}{L}\right)^{-[\tilde{S}^{(s)}(\theta)]}[\Psi_L^{(s)}(\theta)]\{c^{(s)}\}$$

(10.82)

The generalized stress intensity factors $\{K(\theta)\} = [K_I(\theta), \ K_{II}(\theta)]^T$ at angle θ are defined in

$$\begin{Bmatrix} \sigma_\theta^{(s)}(r,\theta) \\ \tau_{r\theta}^{(s)}(r,\theta) \end{Bmatrix} = \frac{1}{\sqrt{2\pi L}}\left(\frac{r}{L}\right)^{-[\tilde{S}^{(s)}(\theta)]}\begin{Bmatrix} K_I(\theta) \\ K_{II}(\theta) \end{Bmatrix}$$

(10.83)

Comparing Eq. (10.83) to Eq. (10.82), the generalized stress intensity factors can be evaluated directly from the singular stress modes by

$$\begin{Bmatrix} K_I(\theta) \\ K_{II}(\theta) \end{Bmatrix} = \sqrt{2\pi L}[\Psi_L^{(s)}(\theta)]\{c^{(s)}\}$$

(10.84)

The factor $\sqrt{2\pi L}$ is chosen so that the definition is identical to the classical definition in the case of a crack in a homogeneous material (this choice is trivial and any other constants are possible). The dimension of the generalized stress intensity factors is (Stress $\times \sqrt{\text{Length}}$) and independent of the orders of singularity. The generalized stress intensity factors can easily be evaluated by using Eq. (10.84) at any given angle θ, which is useful in determining the direction of crack propagation. The evaluation procedure is as simple as the stress recovery procedures in standard finite elements.

In front of a crack tip at angle $\theta = 0$ (Figure 10.4a), the generalized stress intensity factors in Eq. (10.83) are denoted as $\{K\} = [K_I, K_{II}]^T = \{K(0)\}$, and the matrix of order of singularity as $[\tilde{S}^{(s)}] = [\tilde{S}^{(s)}(0)]$. The stress components $\sigma_y^{(s)}(r, \theta = 0) = \sigma_\theta^{(s)}(r, \theta = 0)$ and $\tau_{xy}^{(s)}(r, \theta = 0) = \tau_{r\theta}^{(s)}(r, \theta = 0)$ apply (Eq. (10.73)). The generalized stress intensity factors K_I and K_{II} are defined by

$$\begin{Bmatrix} \sigma_y^{(s)}(r, 0) \\ \tau_{xy}^{(s)}(r, 0) \end{Bmatrix} = \frac{1}{\sqrt{2\pi L}} \left(\frac{r}{L}\right)^{-[\tilde{S}^{(s)}]} \begin{Bmatrix} K_I \\ K_{II} \end{Bmatrix} \tag{10.85}$$

The equation for evaluating K_I and K_{II} (Eq. (10.84)) is expressed as

$$\begin{Bmatrix} K_I \\ K_{II} \end{Bmatrix} = \sqrt{2\pi L}[\Psi_L^{(s)}(\theta = 0)]\{c^{(s)}\} \tag{10.86}$$

The relationships between the proposed definition of generalized stress intensity factors and those definitions summarized in Section 10.2 are examined in the following.

1) *Crack in a homogeneous body.* The matrix of orders of singularity should be equal to

$$[\tilde{S}^{(s)}] = 0.5[I] \tag{10.87}$$

Substituting Eq. (10.87) into the definition of the generalized stress intensity factors in Eq. (10.85) leads

$$\begin{Bmatrix} \sigma_y^{(s)}(r, 0) \\ \tau_{xy}^{(s)}(r, 0) \end{Bmatrix} = \frac{1}{\sqrt{2\pi L}} \left(\frac{r}{L}\right)^{-0.5} \begin{Bmatrix} K_I \\ K_{II} \end{Bmatrix} = \frac{1}{\sqrt{2\pi r}} \begin{Bmatrix} K_I \\ K_{II} \end{Bmatrix} \tag{10.88}$$

which is the same as the classical definition in Eq. (10.6).

2) *Interficial crack between two isotropic materials.* The matrix of orders of singularity obtained from the scaled boundary finite element solution should be in the form of

$$[\tilde{S}^{(s)}] = \begin{bmatrix} 0.5 & +\epsilon \\ -\epsilon & 0.5 \end{bmatrix} \tag{10.89}$$

where ϵ is the oscillatory index. Using Eqs. (B.20), the power matrix function $(r/L)^{-[\tilde{S}^{(s)}]}$ is obtained as

$$\left(\frac{r}{L}\right)^{-[\tilde{S}^{(s)}]} = \left(\frac{r}{L}\right)^{-0.5[I]-\begin{bmatrix} 0 & +\epsilon \\ -\epsilon & 0 \end{bmatrix}} = \left(\frac{r}{L}\right)^{-0.5}\left(\frac{r}{L}\right)^{-\begin{bmatrix} 0 & +\epsilon \\ -\epsilon & 0 \end{bmatrix}} \tag{10.90}$$

Using Eq. (B.10), Eq. (10.90) is written as

$$
\left(\frac{r}{L}\right)^{-[\tilde{S}^{(s)}]} = \left(\frac{r}{L}\right)^{-0.5} \begin{bmatrix} c(r) & -s(r) \\ s(r) & c(r) \end{bmatrix}
\tag{10.91}
$$

with $c(r)$ and $s(r)$ given in Eq. (10.19). Substituting Eq. (10.91) into the present definition in Eq. (10.85) results in

$$
\begin{aligned}
\left\{ \begin{matrix} \sigma_y^{(s)}(r,0) \\ \tau_{xy}^{(s)}(r,0) \end{matrix} \right\} &= \frac{1}{\sqrt{2\pi L}} \left(\frac{r}{L}\right)^{-0.5} \begin{bmatrix} c(r) & -s(r) \\ s(r) & c(r) \end{bmatrix} \left\{ \begin{matrix} K_I \\ K_{II} \end{matrix} \right\} \\
&= \frac{1}{\sqrt{2\pi r}} \begin{bmatrix} c(r) & -s(r) \\ s(r) & c(r) \end{bmatrix} \left\{ \begin{matrix} K_I \\ K_{II} \end{matrix} \right\}
\end{aligned}
\tag{10.92}
$$

which is the same as the definition in Eq. (10.18).

3) *Interficial crack between two anisotropic materials.* The definition of generalized stress intensity factors in Eq. (10.36) can be transformed to the present definition in Eq. (10.85). It is identified that the generalized stress intensity factors are related by

$$
\left\{ \begin{matrix} K_1 \\ K_2 \end{matrix} \right\} = \begin{bmatrix} W_1 & W_2 \\ 0 & 1 \end{bmatrix}^{-1} \left\{ \begin{matrix} K_I \\ K_{II} \end{matrix} \right\} = \frac{1}{W_1} \begin{bmatrix} 1 & -W_2 \\ 0 & W_1 \end{bmatrix} \left\{ \begin{matrix} K_I \\ K_{II} \end{matrix} \right\}
\tag{10.93}
$$

Using Eq. (10.93), Eq. (10.36) is expressed as

$$
\left\{ \begin{matrix} \sigma_y^{(s)}(r,0) \\ \tau_{xy}^{(s)}(r,0) \end{matrix} \right\} = \frac{1}{\sqrt{2\pi r}} \begin{bmatrix} W_1 & W_2 \\ 0 & 1 \end{bmatrix} \begin{bmatrix} c(r) & -s(r) \\ s(r) & c(r) \end{bmatrix} \begin{bmatrix} W_1 & W_2 \\ 0 & 1 \end{bmatrix}^{-1} \left\{ \begin{matrix} K_I \\ K_{II} \end{matrix} \right\}
\tag{10.94}
$$

Considering Eq. (10.91) with Eq. (10.89), it is rewritten as

$$
\left\{ \begin{matrix} \sigma_y^{(s)}(r,0) \\ \tau_{xy}^{(s)}(r,0) \end{matrix} \right\} = \frac{1}{\sqrt{2\pi L}} \begin{bmatrix} W_1 & W_2 \\ 0 & 1 \end{bmatrix} \left(\frac{r}{L}\right)^{-\begin{bmatrix} 0.5 & +\epsilon \\ -\epsilon & 0.5 \end{bmatrix}} \frac{1}{W_1} \begin{bmatrix} 1 & -W_2 \\ 0 & W_1 \end{bmatrix} \left\{ \begin{matrix} K_I \\ K_{II} \end{matrix} \right\}
\tag{10.95}
$$

Using the property of the matrix power function in Eq. (B.21), Eq. (10.95) is rewritten as Eq. (10.85), and the matrix of orders of singularity follows as

$$
\begin{aligned}
[\tilde{S}^{(s)}] &= \begin{bmatrix} W_1 & W_2 \\ 0 & 1 \end{bmatrix} \begin{bmatrix} 0.5 & +\epsilon \\ -\epsilon & 0.5 \end{bmatrix} \begin{bmatrix} W_1 & W_2 \\ 0 & 1 \end{bmatrix}^{-1} \\
&= 0.5[I] + \frac{\epsilon}{W_1} \begin{bmatrix} -W_2 & W_1^2 + W_2^2 \\ -1 & W_2 \end{bmatrix}
\end{aligned}
\tag{10.96}
$$

It is shown in Example 10.3 that the scaled boundary finite element solution indeed leads to this matrix of order of singularity.

4) *V-notch at a bimaterial interface.* The relationship between the present definition in Eq. (10.85) and the definition in Eqs. (10.44) and (10.45) is established by applying

the eigenvalue method in Section B.3 to evaluate the matrix power function. The following three types of eigenvalues may occur:

a) When two real orders of singularity exist, the eigenvalue decomposition of the matrix of orders of singularity is expressed as

$$[\tilde{S}^{(s)}] = [\Phi]\text{diag}(\omega_1, \omega_2)[\Phi]^{-1}$$
$$= \begin{bmatrix} 1 & \Phi_{12} \\ \Phi_{21} & 1 \end{bmatrix} \begin{bmatrix} \omega_1 & 0 \\ 0 & \omega_2 \end{bmatrix} \frac{1}{|\Phi|} \begin{bmatrix} 1 & -\Phi_{12} \\ -\Phi_{21} & 1 \end{bmatrix} \tag{10.97}$$

where the eigenvectors $[\Phi]$ are normalized in accordance with the normalization of the two eigenfunctions in Eq. (10.44). $|\Phi| = 1 - \Phi_{12}\Phi_{21}$ is the determinant of $[\Phi]$. Substituting Eq. (10.97) into Eq. (10.85) and applying the property of matrix power functions in Eqs. (B.24) and (B.25) yields

$$\left\{ \begin{matrix} \sigma_y^{(s)}(r,0) \\ \tau_{xy}^{(s)}(r,0) \end{matrix} \right\} = \frac{K_1}{\sqrt{2\pi L}} \left(\frac{r}{L}\right)^{-\omega_1} \left\{ \begin{matrix} 1 \\ \Phi_{21} \end{matrix} \right\} + \frac{K_2}{\sqrt{2\pi L}} \left(\frac{r}{L}\right)^{-\omega_2} \left\{ \begin{matrix} \Phi_{12} \\ 1 \end{matrix} \right\} \tag{10.98}$$

which is in the same form as Eq. (10.44). The generalized stress intensity factors K_1 and K_2 are related to K_I and K_{II} in Eq. (10.83) by

$$\left\{ \begin{matrix} K_1 \\ K_2 \end{matrix} \right\} = [\Phi]^{-1} \left\{ \begin{matrix} K_I \\ K_{II} \end{matrix} \right\} = \frac{1}{|\Phi|} \left\{ \begin{matrix} K_I - \Phi_{12}K_{II} \\ -\Phi_{21}K_I + K_{II} \end{matrix} \right\} \tag{10.99}$$

b) When the eigenvalues are a pair of complex conjugates $\omega \pm i\epsilon$, the eigenvalue decomposition of the matrix of orders of singularity is expressed as

$$[\tilde{S}^{(s)}] = [\Phi] \begin{bmatrix} \omega - i\epsilon & 0 \\ 0 & \omega + i\epsilon \end{bmatrix} [\Phi]^{-1} \tag{10.100}$$

where $[\Phi]$ contains a pair of complex conjugate eigenvectors. Substituting Eq. (10.100) into Eq. (10.85) and using $x^{\pm i\epsilon} = \cos(\epsilon \ln x) \pm i \sin(\epsilon \ln x)$ yield

$$\left\{ \begin{matrix} \sigma_y^{(s)}(r,0) \\ \tau_{xy}^{(s)}(r,0) \end{matrix} \right\} = \frac{1}{\sqrt{2\pi L}} \left(\frac{r}{L}\right)^{-\omega} [\Phi] \begin{bmatrix} c(r) & -s(r) \\ s(r) & c(r) \end{bmatrix} [\Phi]^{-1} \left\{ \begin{matrix} K_I \\ K_{II} \end{matrix} \right\} \tag{10.101}$$

This expression can be matched to Eq. (10.45).

c) When the two eigenvalues are equal ($\omega_1 = \omega_2 = \omega$) and correspond to parallel eigenvectors, which occurs at the transition point between complex and real eigenvalues, the eigenvalue method for calculating the matrix power functions breaks down (see Section B.3). The matrix of orders of singularity is transformed to a real Schur form (Golub and van Loan, 1996)

$$[\tilde{S}^{(s)}] = [\Phi] \begin{bmatrix} \omega & a \\ 0 & \omega \end{bmatrix} [\Phi]^T \tag{10.102}$$

with the orthogonal matrix $[\Phi]$ ($[\Phi]^{-1} = [\Phi]^T$). Using Eq. (B.20) and Eq. (B.8),

$$\left(\frac{r}{L}\right)^{-\begin{bmatrix} \omega & a \\ 0 & \omega \end{bmatrix}} = \left(\frac{r}{L}\right)^{-\omega} \begin{bmatrix} 1 & -a\ln(r/L) \\ 0 & 1 \end{bmatrix} \tag{10.103}$$

is obtained. Substituting Eq. (10.102) into Eq. (10.85) and using Eqs. (10.103) and (B.21) yield

$$\left\{ \begin{array}{c} \sigma_y^{(s)}(r,0) \\ \tau_{xy}^{(s)}(r,0) \end{array} \right\} = \frac{1}{\sqrt{2\pi L}} \left(\frac{r}{L}\right)^{-\omega} [\Phi] \left[\begin{array}{cc} 1 & -a\ln(r/L) \\ 0 & 1 \end{array} \right] [\Phi]^T \left\{ \begin{array}{c} K_I \\ K_{II} \end{array} \right\}$$

(10.104)

In this case, a power-logarithmic singularity exists.

The above cases demonstrate that the present definition includes all the definitions in Section 10.2 as special cases. Particularly, it is consistent with the definitions of classical stress intensity factors for crack problems (see Cases 1 and 2).

It should be emphasized that the eigenvalue method for evaluating the matrix function is used above solely for the purpose of illustrating the relationship between the present definition and the existing ones explicitly. It is well known that eigenvalue method is numerically unstable when multiple eigenvalues with nearly parallel eigenvectors exist (see Appendix B.3 or Moler and van Loan (2003)). In the solution for the singular stress field, this is revealed by the loss of the logarithmic function, which is shown to be the cause of the well-known 'paradox' or 'anomaly' (Sternberg and Koiter, 1958; Dempsey and Sinclair, 1979).

■ Example 10.2 An Angled Crack in a Rectangular Orthotropic Body

A rectangular orthotropic body with an angled crack is shown in Figure 10.9a. The dimensions are chosen as $b/h = 1$ and $a/h = 0.5$. The crack angle α varies in the analysis. A uniform tension p is applied at the upper and lower edges. The elastic properties of the material are $E_{11} = E_{22} = E_{33} = 15.4 \times 10^6$ psi, $G_{12} = G_{23} = G_{13} = 15.7 \times 10^6$ psi and $v_{12} = v_{23} = v_{13} = 0.4009$. Plane strain conditions exist. This example belongs to Case 1 on page 419. It has been studied in Banks-Sills et al. (2005).

The rectangular body is divided into two S-elements. Each of them contains one crack tip. The scaling centres of the S-elements are placed at the crack tips. As shown in Figure 10.9b for the angle of crack $\alpha = \pi/12$, the boundary of each S-element is divided into six spectral elements with Gauss-Lobatto-Legendre shape functions. The elements form an open loop with two independent nodes at the crack. The crack faces are obtained by scaling the nodes on the boundary. They are not discretized with elements and, thus, not shown in Figure 10.9.

A convergence study is performed for the crack angle $\alpha = \pi/12$ by increasing the element order. The four meshes with 6th, 8th, 10th and 14th order elements are shown in Figures 10.9b to 10.9e. The matrices of orders of singularity $[\tilde{S}^{(s)}]$ are shown in Table 10.4. The first 7 significant digits have converged to the exact solution $0.5[I]$ (Eq. 10.87).

The stress intensity factors in front of the crack tip (i.e. $\theta = 0$ measured from the extension of crack faces as shown in Figure 10.9a) are evaluated. The results normalized with $p\sqrt{\pi a}$ are listed in Table 10.5. The difference between the results obtained from the 10th order element mesh and the 14th order element mesh is very small. The 6th order element mesh with a total of 60 nodes leads to results with less than 1% error.

The present results of stress intensity factors are compared with the reference solutions obtained with M-integral and displacement extrapolation by Banks-Sills et al. (2005) for several angles of crack in Table 10.6. Good agreement is observed.

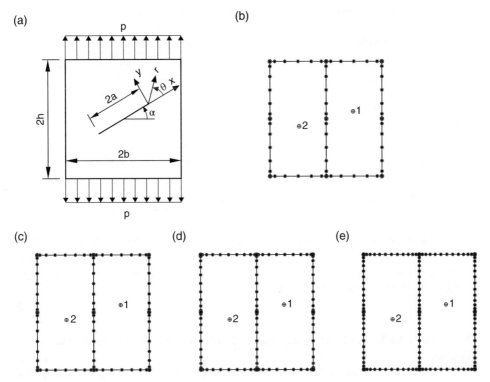

Figure 10.9 An angled crack a in rectangular orthotropic body. (a) Geometry. (b) Mesh of 6th order elements. Total number of nodes is 60. (c) Mesh of 8th order elements. Total number of nodes is 80. (d) Mesh of 10th order elements. Total number of nodes is 100. (e) Mesh of 14th order elements. Total number of nodes is 140.

Table 10.4 An angled crack in a rectangular orthotropic body with $\alpha = \pi/12$: Convergence of matrix of orders of singularity with increasing order of elements.

Element order	6	8	10	14
$\tilde{S}_{11}^{(s)}$	+0.5000380	+0.5000048	+0.5000004	+0.5000000
$\tilde{S}_{12}^{(s)}$	−0.0000580	−0.0000055	−0.0000005	−0.0000000
$\tilde{S}_{21}^{(s)}$	−0.0000060	+0.0000006	+0.0000000	+0.0000000
$\tilde{S}_{22}^{(s)}$	+0.5000087	+0.5000006	+0.5000000	+0.5000000

Table 10.5 An angled crack in a rectangular orthotropic body with $\alpha = \pi/12$: Convergence of stress intensity factors with increasing order of elements.

Element order	6	8	10	14
$K_I/(p\sqrt{\pi a})$	1.2658	1.2692	1.2695	1.2696
$K_{II}/(p\sqrt{\pi a})$	0.2925	0.2903	0.2911	0.2913

Table 10.6 Stress intensity factors of an angled crack in a rectangular orthotropic body.

	$K_I/(p\sqrt{\pi a})$			$K_{II}/(p\sqrt{\pi a})$		
	Present result	Banks-Sills et al. (2005)		Present result	Banks-Sills et al. (2005)	
α		M-integ.	Disp.		M-integ.	Disp.
0	1.3583	1.3577	1.3507	0	0	0
$\pi/12$	1.2696	1.2692	1.2644	0.2913	0.2912	0.2884
$\pi/6$	1.0270	1.0268	1.0239	0.5093	0.5092	0.5073
$\pi/4$	0.6944	0.6952	0.6926	0.5946	0.5807	0.5935
$\pi/3$	0.3580	0.3579	0.3575	0.5248	0.5248	0.5270
$5\pi/12$	0.0997	0.1095	0.0994	0.3118	0.3108	0.3105

Note that the mesh generation in the scaled boundary finite-element method is more flexible and simpler than in the finite element method. In the present problem, only the locations of the scaling centres have to be moved with the crack tip as the crack angle changes. The boundary is not remeshed.

■

■ **Example 10.3   An Interfacial Central Crack Between Two Anisotropic Materials**

A rectangular body made of two orthotropic materials is shown in Figure 10.10a. The height and the width of the rectangle are $2H$ and $W = H$, respectively. A central crack of length $2a = 0.4W$ is located on the material interface. A uniform tension p is applied. This example has been studied in Cho et al. (1992).

The rectangular body is divided into 4 S-elements as shown Figure 10.10b to improve the visibility of the boundary of S-elements from the scaling centres (see Item 4 on page 36). The scaling centres of the S-elements are indicated by the markers '⊕'. The crack tips are at the scaling centre of S-elements 1 and 2. The boundary visible from the scaling centres is discretized with 10-th order spectral elements.

When both materials are orthotropic, W_2 vanishes. The results in Cho et al. (1992) and the present results agree well. Only one case is presented here. The elastic properties are $E_{11} = 100$ GPa, $E_{22} = 50$ GPa, $G_{12} = 50.35426$ GPa and $v_{12} = 0.3$ for material 1 and $E_{11} = 100$ GPa, $E_{22} = 10$ GPa, $G_{12} = 27.034099$ GPa and $v_{12} = 0.3$ for material 2. Plane stress conditions are assumed. The characteristic length is chosen as $L = 2a$. The formulas in Section 10.2.3 lead to $\epsilon = -0.07035273$ (Eq. 10.32), $W_1 = 0.6373502$ and $W_2 = 0$ (Eq. 10.34). The matrix of orders of singularity obtained from the present analysis, denoted as $[\tilde{S}^{(s)}]_{\text{present}}$, and from the analytical solution, denoted as $[\tilde{S}^{(s)}]_{\text{exact}}$, in Eq. (10.96) are equal to

$$[\tilde{S}^{(s)}]_{\text{present}} = \begin{bmatrix} 0.500001 & -0.044840 \\ 0.110382 & 0.499999 \end{bmatrix}$$

$$[\tilde{S}^{(s)}]_{\text{exat}} = \begin{bmatrix} 0.500000 & -0.044839 \\ 0.110383 & 0.500000 \end{bmatrix}$$

The two results agree with each other up to the first five significant digits. The generalized stress intensity factors are obtained as $K_I = 1.12p\sqrt{\pi a}$, $K_{II} = -0.257p\sqrt{\pi a}$.

(a)   (b)

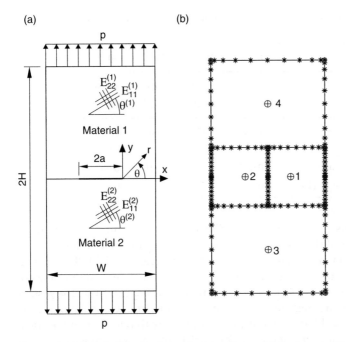

Figure 10.10 Interfacial central crack between two anisotropic materials. (a) Geometry. (b) Mesh.

Using Eq. (10.93), they are converted to the stress intensity factors defined in Cho et al. (1992) as $\sqrt{K_1^2 + K_2^2} = 1.776p\sqrt{\pi a}$ and $K_2/K_1 = -0.146$, which agree well with $\sqrt{K_1^2 + K_2^2} = 1.772p\sqrt{\pi a}$ and $K_2/K_1 = -0.143$ reported in Cho et al. (1992: Table 3).

An isotropic-anisotropic bimaterial is considered. Material 1 is orthotropic with $E_{11} = 100$ GPa, $E_{22} = 30$ GPa, $G_{12} = 23.5212$ GPa and $v_{12} = 0.3$. Material 2 is isotropic with $E = 100$ GPa and $v = 0.3$. The principal material axis 1 of material 1 is rotated from the horizontal x−axis by an angle $\theta^{(1)}$ (Figure 10.10a) to simulate anisotropic material behaviour with reference to the crack. The angles $\theta^{(1)} = 0$, 30°, 60° and 90° are considered. The right crack tip is addressed. The present results of matrix of orders of singularity and the analytical solution in Eq. (10.96) are compared in Table 10.7.

Table 10.7 An interfacial central crack between two anisotropic materials: Matrix of order of singularity.

$\theta^{(1)}$		$\tilde{S}_{11}^{(s)}$	$\tilde{S}_{12}^{(s)}$	$\tilde{S}_{21}^{(s)}$	$\tilde{S}_{22}^{(s)}$	ϵ
0	Present	0.5000000	0.0381745	−0.0562901	0.5000000	0.0463556
	Exact	0.5000000	0.0381745	−0.0562898	0.5000000	0.0463555
30°	Present	0.4921558	0.0427034	−0.0517610	0.5078442	0.0463556
	Exact	0.4921558	0.0427033	−0.0517610	0.5078442	0.0463555
60°	Present	0.4921564	0.0517592	−0.0427018	0.5078434	0.0463539
	Exact	0.4921558	0.0517610	−0.0427033	0.5078442	0.0463555
90°	Present	0.4999984	0.0562826	−0.0381699	0.5000004	0.0463498
	Exact	0.5000000	0.0562898	−0.0381745	0.5000000	0.0463555

Table 10.8 An interfacial central crack between two anisotropic materials: Generalized stress intensity factors.

$\theta^{(1)}$		0	30°	60°	90°
$K_I/(p\sqrt{\pi a})$		1.109	1.141	1.141	1.115
$K_{II}/(p\sqrt{\pi a})$		0.129	0.139	0.113	0.085
$\sqrt{K_1^2 + K_2^2}/(p\sqrt{\pi a})$	Present	1.352	1.259	1.039	0.922
	Cho et al. (1992)	1.35			0.9
K_2/K_1	Present	0.095	0.111	0.110	0.092
	Cho et al. (1992)	0.09			0.09

The error of present results is less than 0.0001. When 14th order elements are used, all the seven significant digits of the present solutions converge to the exact values.

The generalized stress intensity factors K_I and K_{II} are listed in Table 10.8. Using Eq. (10.93), the results are converted to the stress intensity factors defined in Cho et al. (1992). When $\theta^{(1)} = 0$ and 90°, material 1 is orthotropic and W_2 vanishes. The results plotted in Cho et al. (1992) are digitized and listed for comparison. Good agreement is observed. When $\theta^{(1)} = 30$ and 60°, material 1 is anisotropic with reference to the crack faces ($W_2 \neq 0$). The stress intensity factors in Cho et al. (1992) are not shown.

◼

◼ **Example 10.4   A V-notched Bimaterial Body**

A rectangular bimaterial body with a V-notch is shown in Figure 10.11a with its dimensions. The geometry is symmetric about the material interface. A uniform tension p is applied. The elastic properties of the two isotropic materials are E_1, E_2 and $v_1 = v_2 = 0.3$. Plane stress conditions exist. The scaling centre is selected at the tip of the V-notch. The boundary is discretized with ten 10th order elements.

The V-notch becomes a crack when the opening angle decreases to 0. This interfacial crack problem has been studied in Yuuki and Cho (1989) for various values of E_1/E_2 and crack depth a/b. The definition of the generalized stress intensity factors for this case is the same as the definition in Eq. (10.18). When $E_1/E_2 = 10$ and $a/b = 0.4$, the matrix of the order of singularity matrix along the material interface is obtained as

$$[\tilde{S}^{(s)}] = \begin{bmatrix} +0.500000 & -0.093774 \\ +0.093774 & +0.500000 \end{bmatrix}$$

It is in the form of Eq. (10.89). The oscillatory index is equal to $\epsilon = -0.093774$. The same value is obtained from the analytical solution given in Eq. (10.32). The characteristic length is chosen as $L = 2a$ for consistency with Yuuki and Cho (1989). The reference result in Yuuki and Cho (1989) is $\sqrt{K_I^2 + K_{II}^2}/(p\sqrt{\pi a}) = 2.146$, $K_{II}/K_I = -0.237$. The present result of the stress intensity factors is $K_I/(p\sqrt{\pi a}) = 2.089$, $K_{II}/(p\sqrt{\pi a}) = -0.495$ which leads to $\sqrt{K_I^2 + K_{II}^2}/(p\sqrt{\pi a}) = 2.145$, $K_{II}/K_I = -0.237$. The two sets of results are in excellent agreement.

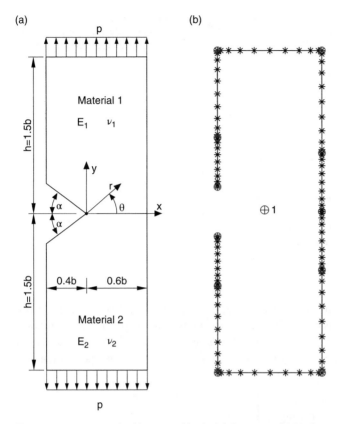

(a)

p

Material 1

E_1 ν_1

h=1.5b

y

r

α

θ

x

α

h=1.5b

0.4b | 0.6b

Material 2

E_2 ν_2

p

(b)

⊕ 1

Figure 10.11 A V-notched bimaterial body. (a) Geometry. (b) Mesh.

The V-notch is investigated with varying opening angle in the range of $0 \le \alpha \le 50°$. There are two singular stress modes, i.e. the size of $[\tilde{S}^{(s)}]$ is 2×2. The two generalized stress intensity factors normalized with $p\sqrt{\pi a}$ are plotted in Figure 10.12 as functions of α. Note that their values vary smoothly with the angle α. The entries of the matrix of orders of singularity $[\tilde{S}^{(s)}]$ are plotted in Figure 10.13a. The variation is also smooth. As shown in Eq. (10.85), the generalized stress intensity factors and the matrix of orders of singularity $[\tilde{S}^{(s)}]$ fully describe the singular stress field in the front of the notch tip ($\theta = 0$).

The two eigenvalues of $[\tilde{S}^{(s)}]$ (Eq. (10.97)) are plotted in Figure 10.13b. As the angle α increases from 0, the real part $\mathrm{Re}(\omega)$ and the positive imaginary part $\mathrm{Im}(\omega)$ of the pair of complex conjugate eigenvalues decrease. A critical angle where the pair of complex conjugate eigenvalues become two repeated real eigenvalues exists around $\alpha \approx 37.6945015°$. When the angle α increases further, the two real eigenvalues separate. In a small region close to the critical angle, the eigenvalues change rapidly.

At the critical angle, it is known that a power-logarithmic singularity $r^{-\omega} \ln r$ exists (Dempsey and Sinclair, 1979; Pageau et al., 1996; Song, 2005, 2006). It has been reported in the literature (Munz and Yang, 1993; Banks-Sills and Sherer, 2002; Hwu and Kuo, 2007) that the generalized stress intensity factors change abruptly close to this transition

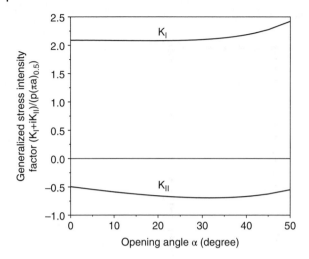

Figure 10.12 A V-notched bimaterial body of varying opening angle α: Generalized stress intensity factors.

case if the power-logarithmic singularity is not considered. It is demonstrated in the following that this anomaly is caused by separating the two nearly parallel stress modes.

The case of $\alpha = 37.6945015°$ is examined. The generalized stress intensity factors of the present definition are $K_I = 2.157p\sqrt{\pi a}$ and $K_{II} = -0.68p\sqrt{\pi a}$. The matrix of orders of singularity is equal to

$$[\tilde{S}^{(s)}] = \begin{bmatrix} +0.486970 & -0.217483 \\ +0.095435 & +0.198835 \end{bmatrix}$$

As shown in Eqs. (10.97), (10.98) and (10.99), the generalized stress intensity factors of the present definition can be converted to those of a standard definition by the eigenvalue method. The eigenvalues and eigenvectors of $[\tilde{S}^{(s)}]$, normalized as in Eq. (10.97), are equal to

$$\omega_1 = 0.342905; \quad \omega_2 = 0.342901; \quad [\Phi] = \begin{bmatrix} 1 & 1.509573 \\ 1/1.509615 & 1 \end{bmatrix}$$

The two eigenvalues are very close to each other, i.e. this case is adjacent to a power-logarithmic singularity. Two eigenvectors are nearly parallel resulting in very large numbers in the inverse of $[\Phi]$. Using Eq. (10.99), the generalized stress intensity factors for the two decoupled modes are equal to

$$K_1 = 114399.5p\sqrt{\pi a}; \quad K_2 = -75781.3p\sqrt{\pi a}$$

which are 4 orders of magnitude larger than the present results of generalized stress intensity factors. Using Eq. (10.98) (where the characteristic length has been chosen as $L = 2a$), the stresses are obtained as

$$\sigma_y^{(s)}(r,0)/p = 57199.8(r/L)^{-0.342905} - 57198.7(r/L)^{-0.342901}$$

$$\tau_{xy}^{(s)}(r,0)/p = 37890.3(r/L)^{-0.342905} - 37890.6(r/L)^{-0.342901}$$

Figure 10.13 An V-notched bimaterial body of varying opening angle α. (a) Matrix of orders of singularity. (b) Eigenvalues of matrix of orders of singularity.

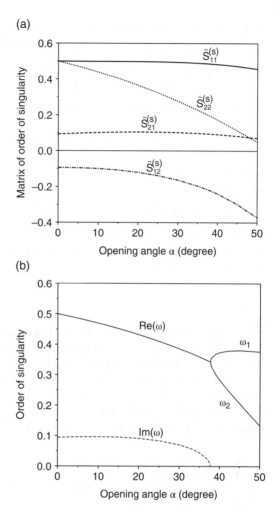

The contribution of the two modes to $\sigma_y^{(s)}(r,0)/P$ are examined. At $r/L = 10^{-4}$, which is a typical value used in determining stress intensity factors by stress extrapolation, $\sigma_y^{(s)}(r,0)/p = 1345905.3 - 1345829.9 = 75.5$ is obtained. At $r/L = 10^{-10}$, which is much smaller than a dimension of practical interest, $\sigma_y^{(s)}(r,0)/p = 153618850.2 - 153601748.1 = 17102.1$. It is obvious that neither K_1 or K_2 is suitable to represent the severity of the stress singularity. In addition, the numerical evaluation of the difference between two very large but nearly equal numbers will lead to significant loss of precision. On the other hand, the evaluation of the matrix power function in Eq. (10.85) by the 'scaling and squaring' algorithm is numerically stable. It yields $\sigma_y^{(s)}(r,0)/p = 75.07$ at $r/L = 10^{-4}$ and $\sigma_y^{(s)}(r,0)/p = 17075.6$ at $r/L = 10^{-10}$. This case demonstrates that the two singular terms should be coupled in a definition of the generalized stress intensity factors suitable for numerical analysis of singular stress field.

The anomaly caused by decoupling the stress modes in the eigenvalue method is not limited to a small range of opening angle close to the critical angle. For example, at angle $\alpha = 45°$, the generalized stress intensity factors are $K_I = 2.276P\sqrt{\pi a}$ and $K_{II} = -0.624P\sqrt{\pi a}$. The matrix of orders of singularity are

$$[\tilde{S}^{(s)}] = \begin{bmatrix} +0.473071 & -0.296053 \\ +0.083250 & +0.114594 \end{bmatrix}$$

The eigenvalues and eigenvectors are

$$\omega_1 = 0.380320; \quad \omega_2 = 0.207346; \quad [\Phi] = \begin{bmatrix} 1 & 1.114129 \\ 0.313293 & 1 \end{bmatrix}$$

Using Eq. (10.99), the generalized stress intensity factors for the two decoupled modes are equal to

$$K_1 = 4.564p\sqrt{\pi a}; \quad K_2 = -2.054p\sqrt{\pi a}$$

which are about twice as large as K_I and K_{II}. As shown in Eq. (10.99), K_1 and K_2 are equal to K_I and K_{II} only when the eigenvector matrix $[\Phi]$ is an identity matrix, i.e. $[\tilde{S}^{(s)}]$ (Eq. (10.97)) is diagonal.

◼

◼ Example 10.5   A Crack Terminating at a Material Interface

A crack terminating at a material interface is shown in Figure 10.14a with its dimension. The inclination angle of the crack is α. The crack length is $a = b/\cos \alpha$. The characteristic length is chosen as $L = 2a$. The elastic properties of the two orthotropic materials are $E_{11} = 200\,\text{MPa}$, $E_{22} = 200\,\text{MPa}$, $v_{12} = 0.4$ and $G_{12} = 29.41\text{MPa}$ for material 1 and $E_{11} = 10\,\text{MPa}$, $E_{22} = 100\,\text{MPa}$, $v_{12} = 0.02$ and $G_{12} = 28.07\text{MPa}$ for material 2. A uniform tension p is applied. Plane stress conditions are assumed.

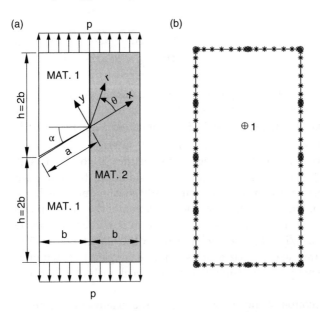

Figure 10.14 A crack terminating at a material interface. (a) Geometry. (b) Mesh.

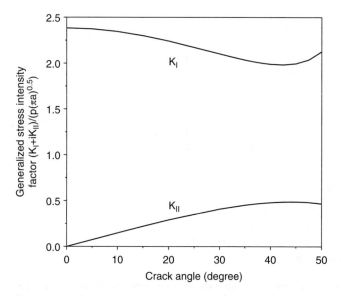

Figure 10.15 A crack terminating at a material interface for varying crack angle α: Generalized stress intensity factors.

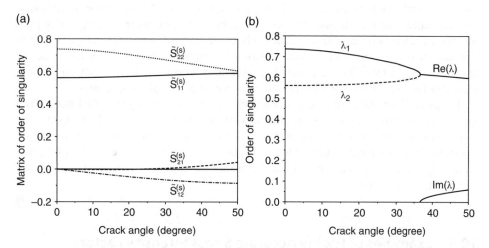

Figure 10.16 A crack terminating at a material interface for varying crack angle α. (a) Matrix of orders of singularity. (b) Eigenvalues of matrix of orders of singularity.

The scaling centre is located at the crack tip. The boundary is discretized with 12 nine-node elements (Figure 10.14b). When the crack angle changes, the scaling centre moves on the material interface accordingly and the boundary mesh is not modified.

The generalized stress intensity factors are evaluated for varying crack angle α and plotted in Figs 10.15. The matrix of orders of singularity and its eigenvalues are shown in Figures 10.16a and 10.16b, respectively. When the crack is normal to the material interface ($\alpha = 0$), the two singular stress modes are independent as $\tilde{S}_{12}^{(s)} = \tilde{S}_{21}^{(s)} = 0$

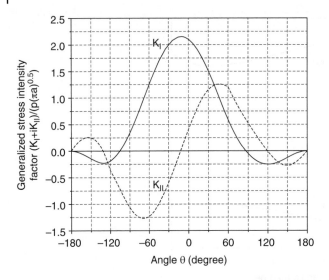

Figure 10.17 Angular variation of generalized stress intensity factors at crack inclination angle $\alpha = 30°$

(Figure 10.16a) and the two eigenvalues are real (Figure 10.16b). The generalized stress intensity factors are $K_I = 2.383p\sqrt{\pi a}$ and $K_{II} = 0$. As the angle α increases, the magnitudes of the off-diagonal entries of the matrix of orders of singularity $[\tilde{S}^{(s)}]$ increase, and the two singular stress modes become coupled. As shown in Figure 10.16b, the pair of real eigenvalues become a pair of complex conjugate eigenvalues at the angle $\alpha \approx 36.65513°$, where a power-logarithmic singularity exists. When the angle α increases further, the two eigenvalues become a pair of complex conjugates.

The generalized stress intensity factors can be obtained at any specified angle from Eq. (10.84). Figure 10.17 shows the angular variation of the generalized stress intensity factors when the inclination angle of the crack is $\alpha = 30°$. The angle θ is measured starting from the crack front as illustrated in Figure 10.14. The generalized stress intensity factors are all normalized with $p\sqrt{\pi a}$. Note that the maximum value of K_I does not occur at the crack front ($\theta = 0$).

10.6 Examples of Highly Accurate Stress Intensity Factors and *T*-stress

The stress intensity factors of many common specimens are available in the handbook of Tada et al. (1985) and other sources. The *T*-stresses are also included in Fett (2008). These results are typically evaluated at sets of discrete values of crack length and provided as tables, graphs or formulas obtained by curve-fitting. They are routinely employed in engineering design and in verifications of numerical techniques proposed to compute the stress intensity factors and the *T*-stress.

The scaled boundary finite element method is adopted to produce solutions of the stress intensity factors and the *T*-stress for several standard fracture specimens (Chowdhury et al., 2015). It has been demonstrated that highly accurate results can be obtained efficiently. This method is also very simple to use since only a boundary mesh of the S-elements is required and the crack faces are not discretized. As will be shown in Section 10.6.1, various crack lengths of a specimen can be considered using a single mesh.

After the stress intensity factors and the T-stresses of a specimen are determined at discrete values of crack length. Padé approximants of the stress intensity factors and the T-stresses are obtained by curve-fitting the data. A Padé series is a rational function defined as a ratio of two polynomials. Using the normalized crack length $\alpha = a/W$ as the independent variable, the Padé series is expressed as

$$\frac{\sum_{i=0}^{k} p_i \alpha^i}{1 + \sum_{i=1}^{l} q_i \alpha^i} = \frac{p_0 + p_1\alpha + p_2\alpha^2 + \ldots + p_k\alpha^k}{1 + q_1\alpha + q_2\alpha^2 + \ldots + q_l\alpha^l}, \tag{10.105}$$

where p_i and q_i are the coefficients of the two polynomials to be determined by curve-fitting. k and l are the degrees of polynomials in the numerator and denominator, respectively. Generally speaking, Padé approximants require fewer coefficients than polynomials do to achieve the same accuracy (Press et al., 1992). The curve fitting tool (`cftool`) in MATLAB is used in the following examples. The degrees of the two polynomials k and l are selected interactively to achieve the desired accuracy with the least number of coefficients p_i and q_i.

The analysis process is illustrated in Section 10.6.1 using a single edge-cracked specimen. For the other specimens, only Padé approximants, together with the stress intensity factors and the T-stresses at a few crack lengths, are provided. Results of additional specimens are reported in Chowdhury et al. (2015).

The stress intensity factors and the T-stress are presented as dimensionless values. In the analysis, plane stress conditions are assumed. The Poisson's ratio is selected as $\nu = 0.2$.

10.6.1   A Single Edge-cracked Rectangular Body Under Tension

A rectangular body of height $6W$ and width W is shown in Figure 10.18a. It is subject to a tension σ_0. A horizontal crack of length a is presented at the left edge. As depicted in Figure 10.18b, the rectangle is divided into three S-elements to improve the visibility of the boundary from the scaling centres. The line elements of the S-elements are illustrated by their two end nodes. The crack is located in S-element 1 with the scaling centre chosen at the crack tip (Figure 10.18c). Varying crack length is modelled by simply moving the scaling centre. The line elements are chosen in such a way that the same mesh can be used for the selected range of crack length ($0.1 \leq a/W \leq 0.9$).

For each specified crack length, the analysis is performed with increasing order of elements starting from $p = 5$. The convergence of the results is evaluated. The analysis terminates after the first six significant digits of the stress intensity factors and the T-stresses have converged. The results of the normalized stress intensity factor $\bar{K}_I = K_I/(\sigma_0\sqrt{\pi a})$ and normalized T-stress $\bar{T} = T/\sigma_0$ are listed in Tables 10.9 and 10.10, respectively, for a few selected crack lengths. All the digits of the last number in each row have converged.

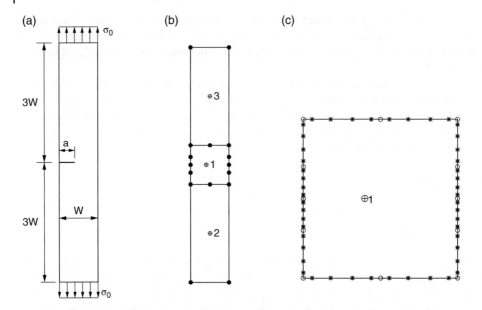

Figure 10.18 A single edge-cracked rectangular body under tension. (a) Geometry. (b) Mesh consisting of 3 S-elements with scaling centres and the S-element numbers. Only the end nodes of line elements are shown. (c) Mesh of 5th order line elements of S-element 1.

Table 10.9 A single edge-cracked rectangular body under tension: Convergence of stress intensity factor $\bar{K}_I = K_I/(\sigma_0\sqrt{\pi a})$.

$\dfrac{a}{W}$	Element order (p)			
	5	7	9	11
0.1	1.18891	1.18917	1.18918	
0.2	1.36723	1.36732	1.36733	
0.3	1.65974	1.65986	1.65987	
0.4	2.11127	2.11141		
0.5	2.82445	2.82457	2.82458	
0.6	4.03302	4.03311		
0.7	6.35483	6.35487	6.35488	
0.8	11.9544	11.9554	11.9553	
0.9	34.6013	34.6300	34.6335	34.6326

The curve-fitting of the stress intensity factors and T-stress is performed on $\bar{K}_I\sqrt{1-\alpha} = (K_I/(\sigma_0\sqrt{\pi a}))\sqrt{1-\alpha}$ and on $\bar{T}(1-\alpha^2) = T/\sigma_0(1-\alpha^2)$, respectively. The Padé approximant of the stress intensity factor K_I is obtained as

$$\frac{K_I}{\sigma_0\sqrt{\pi a}} = \frac{1.121871 + 2.201880\alpha + 1.017533\alpha^2}{\sqrt{1-\alpha}(1 + 2.474585\alpha - 5.202173\alpha^2 + 3.409654\alpha^3 - 2.243554\alpha^4 + 0.561625\alpha^5)}$$

$$(10.106)$$

Table 10.10 A single edge-cracked rectangular body under tension: Convergence of T-stress $\bar{T} = T/\sigma_0$.

$\dfrac{a}{W}$	Element order			
	5	7	9	11
0.1	−0.54908	−0.55019		
0.2	−0.58909	−0.58901	−0.58900	
0.3	−0.61061	−0.61034	−0.61033	
0.4	−0.57877	−0.57826	−0.57825	
0.5	−0.42246	−0.42169	−0.42168	
0.6	0.03713	0.03812	0.03814	
0.7	1.36022	1.36134	1.36139	
0.8	6.00669	6.00734	6.00738	6.00739
0.9	35.6718	35.7413	35.7433	

and the Padé approximant of the T-stress is written as

$$\frac{T}{\sigma_0} = \frac{-0.526471 - 1.195582\alpha + 1.647374\alpha^2 + 8.796785\alpha^3 - 13.695006\alpha^4 + 6.960189\alpha^5}{(1-\alpha)^2(1 + 4.311435\alpha - 3.484751\alpha^2 - 0.782068\alpha^3 + 3.145563\alpha^4)}$$

$$(10.107)$$

The Padé approximants are accurate up to at least the first four significant digits or the first five digits after the decimal points.

The following MATLAB script calculates the normalized stress intensity factor and T-stress using the Padé approximants. The two Padé approximants can be obtained by scanning the QR-codes in Figure 10.19.

```
%% Single edge cracked body under tension: Normalized KI
p = [1.121871, 2.201880, 1.017533];
q = [1,        2.474585, -5.202173, 3.409654 -2.243554, 0.561625];
SET_Kbar = @(a) polyval(p(end:-1:1),a)./ ...
     polyval(q(end:-1:1),a)./sqrt(1-a);
fprintf('%12.5f\n',SET_Kbar([0.1:0.1:0.9]))

%% Single edge cracked body under tension: Normalized T-stress
p = [-0.526471 -1.195582  1.647374  8.796785 -13.695006 6.960189];
q = [1,        4.311435 -3.484751 -0.782068   3.145563];
SET_Tbar = @(a) polyval(p(end:-1:1),a)./ ...
                polyval(q(end:-1:1),a)./(1-a).^2;
fprintf('%12.5f\n',SET_Tbar([0.1:0.1:0.9]))
```

10.6.2 A Single Edge-cracked Rectangular Body Under Bending

The single edge-cracked rectangular body shown at the left side of Figure 10.20 is subject to a bending moment $1/6W^2\sigma_0$. The dimensions are indicated in Figure 10.20. A convergence study on the normalized stress intensity factor K_I and the T-stress is

(a)   (b)

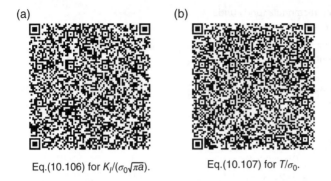

Eq.(10.106) for $K_I/(\sigma_0\sqrt{\pi a})$.   Eq.(10.107) for T/σ_0.

Figure 10.19 A single edge-cracked rectangular body under tension: QR-codes of MATLAB functions of Padé approximants of the normalized stress intensity factor and T-stress.

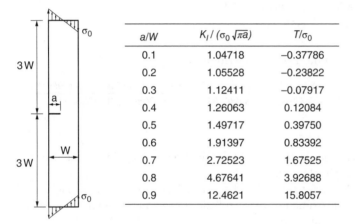

a/W	$K_I/(\sigma_0\sqrt{\pi a})$	T/σ_0
0.1	1.04718	−0.37786
0.2	1.05528	−0.23822
0.3	1.12411	−0.07917
0.4	1.26063	0.12084
0.5	1.49717	0.39750
0.6	1.91397	0.83392
0.7	2.72523	1.67525
0.8	4.67641	3.92688
0.9	12.4621	15.8057

Figure 10.20 A single edge-cracked rectangular body under bending: Geometry (left), and the normalized stress intensity factor $K_I/(\sigma_0\sqrt{\pi a})$ and T-stress T/σ_0 (right).

performed. The results at some selected crack lengths are shown in the table at the right side of Figure 10.20. The last digit has converged.

The Padé approximants of $\bar{K}_I(1-\alpha)^{1.5} = (K_I/(\sigma_0\sqrt{\pi a}))(1-\alpha)^{1.5}$ and $\bar{T}(1-\alpha^2) = T/\sigma_0(1-\alpha^2)$ are determined from the discrete data. The normalized stress intensity factor and the T-stress are expressed as

$$\frac{K_I}{\sigma_0\sqrt{\pi a}} = \frac{1.121632 + 0.474629\alpha - 1.850055\alpha^2 + 1.642133\alpha^3 - 0.466097\alpha^4}{(1-\alpha)^{1.5}\left(1 + 3.145932\alpha - 2.694734\alpha^2 + 1.015946\alpha^3\right)}$$

(10.108)

$$\frac{T}{\sigma_0} = \frac{-0.526032 + 1.005355\alpha + 3.315115\alpha^2 - 6.291778\alpha^3 + 3.272981\alpha^4}{(1-\alpha)^2\left(1 + 3.335731\alpha - 3.802655\alpha^2 + 6.458966\alpha^3 - 6.715033\alpha^4 + 4.619367\alpha^5\right)}$$

(10.109)

The MATLAB scripts of Eqs. (10.108) and (10.109) can be obtained by scanning the QR-codes in Figure 10.21.

(a)

(b)

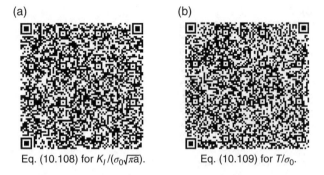

Eq. (10.108) for $K_I/(\sigma_0\sqrt{\pi a})$.   Eq. (10.109) for T/σ_0.

Figure 10.21 A single edge-cracked rectangular body under bending: QR-codes of MATLAB functions of Padé approximants of stress intensity factor and T-stress.

10.6.3   A Centre-cracked Rectangular Body Under Tension

For the centre-cracked rectangular body under tension shown in Figure 10.22, the normalized stress intensity factors and the T-stresses at a few crack lengths are listed in the table at the right side Figure 10.22. All the values have converged to the last digit.

The Padé approximants of $\bar{K}_I\sqrt{1-\alpha} = (K_I/(\sigma_0\sqrt{\pi a}))\sqrt{1-\alpha}$ and $\bar{T}(1-\alpha^2) = T/\sigma_0(1-\alpha^2)$ are obtained by curve-fitting the discrete data. The stress intensity factor and the T-stress are expressed as

$$\frac{K_I}{\sigma_0\sqrt{\pi a}} = \frac{1.000004 - 0.939533\alpha - 0.065819\alpha^2 + 0.250877\alpha^3 - 0.041437\alpha^4}{\sqrt{1-\alpha}(1 - 0.439439\alpha - 0.756218\alpha^2 + 0.442880\alpha^3)}$$

$$(10.110)$$

$$\frac{T}{\sigma_0} = \frac{-1.000011 + 1.855183\alpha - 0.082441\alpha^2 - 2.173131\alpha^3 + 1.922868\alpha^4 - 0.522434\alpha^5}{(1-\alpha)^2(1 + 0.145292\alpha - 1.328959\alpha^2 + 0.716805\alpha^3)}$$

$$(10.111)$$

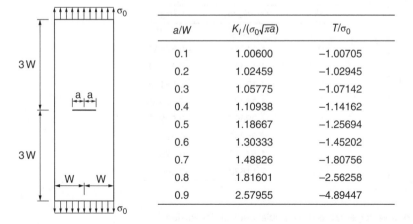

a/W	$K_I/(\sigma_0\sqrt{\pi a})$	T/σ_0
0.1	1.00600	−1.00705
0.2	1.02459	−1.02945
0.3	1.05775	−1.07142
0.4	1.10938	−1.14162
0.5	1.18667	−1.25694
0.6	1.30333	−1.45202
0.7	1.48826	−1.80756
0.8	1.81601	−2.56258
0.9	2.57955	−4.89447

Figure 10.22 Centre-cracked rectangular body under tension.

(a) (b)

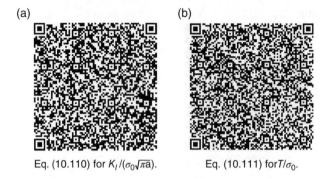

Eq. (10.110) for $K_I/(\sigma_0\sqrt{\pi a})$. Eq. (10.111) for T/σ_0.

Figure 10.23 A centre-cracked rectangular body under tension: QR-codes of MATLAB functions of Padé approximants of stress intensity factor and T-stress.

The MATLAB scripts of these two equations can be obtained by scanning the QR-codes in Figure 10.23.

10.6.4 A Double Edge-cracked Rectangular Body Under Tension

For the double edge-cracked rectangular body under tension in Figure 10.24, the normalized stress intensity factors and the T-stresses at some crack lengths (a/W) are given in the table at the right of Figure 10.24.

After determining the Padé approximants of $\bar{K}_I\sqrt{1-\alpha} = (K_I/(\sigma_0\sqrt{\pi a}))\sqrt{1-\alpha}$ and $\bar{T}(1-\alpha^2) = T/\sigma_0(1-\alpha^2)$, the stress intensity factors and the T−stress are expressed as

$$\frac{K_I}{\sigma_0\sqrt{\pi a}} = \frac{1.121652 + 2.262991\alpha + 2.008346\alpha^2 - 0.921035\alpha^3 + 0.389353\alpha^4 - 0.187798\alpha^5}{\sqrt{1-\alpha}(1 + 2.521306\alpha + 3.819594\alpha^2)}$$

(10.112)

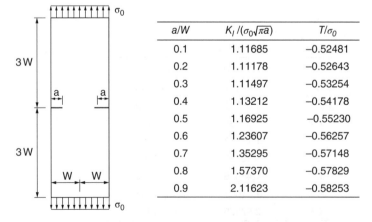

a/W	$K_I/(\sigma_0\sqrt{\pi a})$	T/σ_0
0.1	1.11685	−0.52481
0.2	1.11178	−0.52643
0.3	1.11497	−0.53254
0.4	1.13212	−0.54178
0.5	1.16925	−0.55230
0.6	1.23607	−0.56257
0.7	1.35295	−0.57148
0.8	1.57370	−0.57829
0.9	2.11623	−0.58253

Figure 10.24 A double edge-cracked rectangular body under tension.

(a)

(b)

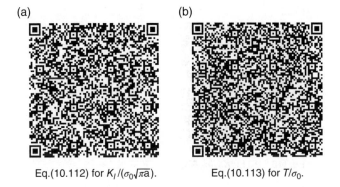

Eq.(10.112) for $K_I/(\sigma_0\sqrt{\pi a})$.　　　　Eq.(10.113) for T/σ_0.

Figure 10.25 A double edge-cracked rectangular body under tension: QR-codes of MATLAB functions of Padé approximants of stress intensity factor and T−stress.

$$\frac{T}{\sigma_0} = \frac{-0.526079 - 0.233183\alpha - 0.638753\alpha^2 + 2.341397\alpha^3 + 0.798007\alpha^4 - 1.741453\alpha^5}{(1-\alpha)^2(1 + 2.451578\alpha + 5.66626\alpha^2 - 0.2263\alpha^3 + 1.76326\alpha^4)}$$

(10.113)

and the MATLAB scripts can be obtained by scanning the QR-codes in Figure 10.25.

10.6.5 A Single Edge-cracked Rectangular Body Under End Shearing

A rectangular body of $2W \times W$ is shown in Figure 10.26. The left edge has a crack of length a. The bottom edge is fixed. A uniform shear load with intensity τ_0 is applied to its top edge. A convergence study of the normalized stress intensity factors K_I, K_{II} and the T-stress are performed. The results that have converged in the first six significant figures are obtained. Those at a few selected crack lengths are listed in the table at the right side of Figure 10.26.

The Padé approximants of $\bar{K}_{I,II}(1-\alpha)^{1.5} = (K_{I,II}/(\sigma_0\sqrt{\pi a}))(1-\alpha)^{1.5}$ and $\bar{T}(1-\alpha^2) = T/\sigma_0(1-\alpha^2)$ are obtained from curve-fitting the discrete data. The normalized stress

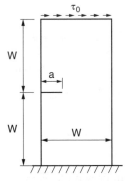

a/W	$K_I/(\tau_0\sqrt{\pi a})$	$K_{II}/(\tau_0\sqrt{\pi a})$	$T/\tau 0$
0.1	6.28271	0.37722	−2.26742
0.2	6.33210	0.69243	−1.42944
0.3	6.74555	0.95298	−0.47289
0.4	7.56453	1.17133	0.73026
0.5	8.98330	1.36269	2.39120
0.6	11.4837	1.54608	5.00765
0.7	16.3513	1.75285	10.0523
0.8	28.0585	2.05837	23.5605

Figure 10.26 A single edge-cracked rectangular body under end shearing.

(a)

(b)

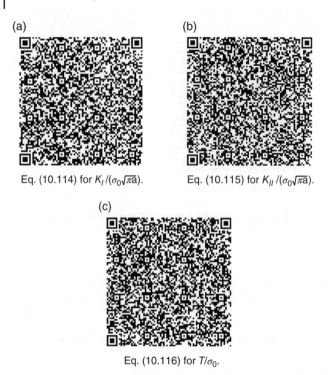

Eq. (10.114) for $K_I/(\sigma_0\sqrt{\pi a})$.

Eq. (10.115) for $K_{II}/(\sigma_0\sqrt{\pi a})$.

(c)

Eq. (10.116) for T/σ_0.

Figure 10.27 A single edge-cracked rectangular body under end shearing: QR-codes of MATLAB functions of Padé approximants of stress intensity factors and T-stress.

intensity factors K_I, K_{II} and T-stress are expressed as

$$\frac{K_I}{\sigma_0\sqrt{\pi a}} = \frac{6.728076 + 9.375094\alpha - 0.940390\alpha^2 - 5.453186\alpha^3 + 8.455276\alpha^4 - 3.240346\alpha^5}{(1-\alpha)^{1.5}(1 + 4.109803\alpha + 1.540861\alpha^2)}$$

(10.114)

$$\frac{K_{II}}{\sigma_0\sqrt{\pi a}} = \frac{0.000104 + 4.084730\alpha - 14.027485\alpha^2 + 21.086754\alpha^3 - 14.626113\alpha^4 + 3.488663\alpha^5}{(1-\alpha)^{1.5}(1 - 1.180975\alpha + 1.266745\alpha^2 - 0.678643\alpha^3 + 1.903867\alpha^4)}$$

(10.115)

$$\frac{T}{\sigma_0} = \frac{-3.155235 + 2.204099\alpha + 21.238208\alpha^2 + 11.186071\alpha^3 - 48.425158\alpha^4 + 26.545974\alpha^5}{(1-\alpha)^2(1 + 4.537868\alpha + 2.274589\alpha^3 + 2.188079\alpha^4)}$$

(10.116)

The MATLAB scripts can be obtained by scanning the QR-codes in Figure 10.27.

10.7 Modelling of Crack Propagation

The key idea of modelling crack propagation using the scaled boundary finite element method is illustrated in Figure 10.28. Because of possible complex crack paths, it becomes impossible or impractical to use a few larger S-elements, as in previous

Figure 10.28 Illustration of modelling of crack propagation by S-elements.

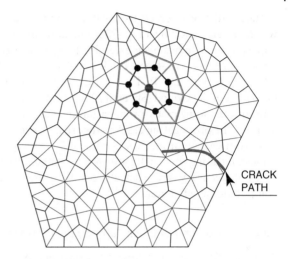

CRACK PATH

sections, to model the complete crack propagation process. A problem domain is discretized into an S-element mesh. The S-element mesh is generated from a given triangular mesh (see Section 4.3.3) without considering the presence of any cracks. The (incremental) crack path is shown by the bold solid line in Figure 10.28. This approach benefits from the following two features of the scaled boundary finite element method:

1) The arbitrary polygonal shape of S-elements greatly simplifies the modelling of crack paths that are not known *a priori*. A crack path non-conformable to the mesh is illustrated in Figure 10.28. The sizes of the S-elements with respect to the desired resolution of the crack path are chosen to be small enough so that the crack path is simple (say, a curve with a small chord height- to length-ratio) within individual S-elements. The original S-elements on the crack path are split into two pieces. Each piece is modelled as an S-element. The remeshing is handled by adding nodes to the intersections between the crack path and the edges of the original S-elements and modifying the element connectivity. Alternatively, the two S-elements resulting from splitting an S-element can be assembled to form the split S-element. The displacement discontinuity is embedded in the split S-element with the displacements of the additional nodes on the crack faces being treated as auxiliary variables. This process is similar to that in the extended finite element method (Moës et al., 1999), but does not need special numerical integrations.
2) The singular stress field around the crack tip is represented analytically using a crack S-element (Figure 10.28). Accurate solutions are obtained efficiently. The stress intensity factors are directly extracted from the semi-analytical stress solution or, if the asymptotic solution is available, from the displacement solution. The requirements on the crack S-element are:
   a) the size of the crack S-element be small enough to provide the specified resolution of the crack paths, and
   b) there are enough line elements along the circumferential direction around the crack tip so that accurate results of the stress intensity factors can be obtained. Note that this requirement is not directly related to the size of the crack S-element.

The simulation of crack propagation using the scaled boundary finite element method was first performed by Yang (2006) in static linear fracture mechanics. Since then, the application range has been extended to nonlinear cohesive and dynamic fracture problems in concrete (Yang and Deeks, 2007a,b; Ooi and Yang, 2009, 2010, 2011; Ooi et al., 2012a; Zhu et al., 2013). The techniques for tracing the crack propagation paths have been further developed (Shi et al., 2013; Ooi et al., 2013; Dai et al., 2015; Ooi et al., 2015). The simulation of crack propagation in linear fracture mechanics is presented in this section. The propagating crack paths are modelled using polygon meshes that are generated from triangular meshes (Section 4.3.3) and quadtree meshes (Chapter 9). Minimal remeshing along the crack paths is performed for primarily two reasons. One is robustness and accuracy of the simulation. The other is to allow element size gradations for computational efficiency.

10.7.1 Modelling of Crack Paths by Polygon Meshes

A methodology for modelling crack propagation using polygon meshes was developed by Ooi et al. (2012b). Polygon meshes that satisfy the scaling requirement can be generated from background triangular meshes using the algorithm and MATLAB code in Section 4.3.3. Local remeshing based on the background triangular mesh is proposed (Shi et al., 2013; Ooi et al., 2013) to improve the robustness and the quality of the mesh.

The procedure in Shi et al. (2013) and Ooi et al. (2013) is illustrated in Figure 10.29 for one crack increment. The background triangular mesh and the polygon S-element mesh are shown by thin dash lines and medium solid lines, respectively.

1) Figure 10.29a illustrates the mesh at the current step of crack propagation. The crack is shown as bold lines, and the crack tip is at Point O. The background triangular mesh conforms to the crack path. The part of crack shown by the bold dashed line is represented by scaling with respect to the crack tip serving as the scaling centre.

2) The next crack increment is shown in Figure 10.29b by the bold solid line OO'. The remeshing procedure starts from the background triangular mesh. The triangles cut by the crack increment are identified, as illustrated by the medium dash-dot lines. A group of triangles are selected by including another layer of triangles connected to the cut triangles. The boundary of this group is illustrated by the thick dash-dot lines in Figure 10.29b. The additional layer of triangles ensures that the crack increment is separated from the boundary.

3) A patch is constructed in Figure 10.29c. Its boundary is defined by the boundary of the selected group of triangles and the crack path inside it.

4) A new triangular mesh is generated on the patch, replacing the original triangular mesh (Figure 10.29d). The mesh is conformable to the crack path, and the crack tip is at a node.

5) A mesh of polygon S-elements is generated from the new triangular mesh (Section 4.3.3) as shown in Figure 10.29e. The crack tip becomes the scaling centre of the crack S-element. The other part of the crack path coincides with the edges of S-elements.

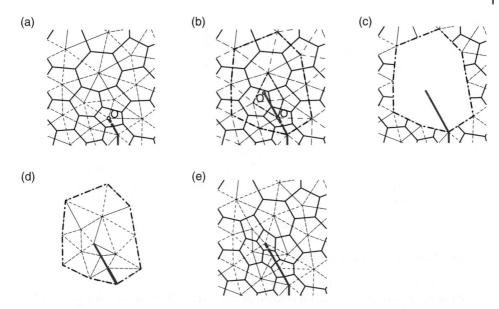

Figure 10.29 Model crack propagation by local remeshing of a polygon S-element mesh (medium solid line) generated from a triangular mesh (thin dashed line). (a) Mesh around crack tip (Point O) at current step. (b) Identification of a group of triangles surrounding the crack increment for remeshing. (c) Construction of a patch incorporating the crack increment. (d) Generate a triangular mesh on the patch conforming to the new crack path. (e) The polygon mesh of S-elements of the patch is generated from the triangular mesh incorporating the crack increment.

The crack in Figure 10.29e is modelled in the same way as that in Figure 10.29a, except that the crack has propagated by one increment. The above process is repeated to consider further increments of crack propagation.

10.7.2 Modelling of Crack Paths by Quadtree Meshes

The quadtree mesh in Chapter 9 is employed in Ooi et al. (2015, 2016, 2017) for modelling crack propagation. The remeshing process to model the crack path is depicted in Figure 10.30. The current mesh around the crack tip and the next crack increment are shown in Figure 10.30a. The size of the S-element containing the crack tip is controlled by the radius of a circle Ω_{cir}. A balanced quadtree mesh is created around the new crack tip (Figure 10.30b). The S-element containing the crack tip is obtained by merging the quadtree cells that overlap Ω_{cir}. The mesh seeds on Ω_{cir} control the size and numbers of the quadtree cells around the crack tip in a way that there are sufficient nodes on the boundary of this S-element. Each of the cells cut by the crack path is then split into two leading to the mesh in Figure 10.30c.

In this approach based on quadtree mesh, the bulk of the S-elements are square and can be obtained by scaling master S-elements. During the simulations, only the stiffness matrices from the arbitrary sided polygons need to be computed. This contributes to improve computational efficiency.

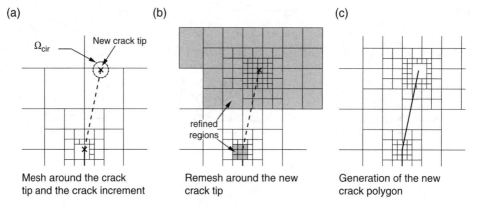

(a)   (b)   (c)

Mesh around the crack tip and the crack increment   Remesh around the new crack tip   Generation of the new crack polygon

Figure 10.30 Modelling crack propagation by local remeshing of a quadtree mesh.

10.7.3 Examples of Crack Propagation Modelling

10.7.3.1 Fatigue Crack Propagation Using Polygon Mesh

The fatigue fracture of the Arcan specimen (Gaylon et al., 2009) shown in Figure 10.31a is considered. The initial crack length a_0 at the notch tip is 2.997 mm and the thickness of the specimen is 1.6 mm. The specimen is made of aluminum (AA2024-T351) with Young's modulus $E = 73.1$ GPa and Poisson's ratio $v = 0.33$. The fatigue life is predicted by Paris's law

$$\frac{da}{dN} = C\Delta K^m$$

where da/dN (mm per cycle) is the crack growth rate due to fatigue. ΔK (MPa$\sqrt{m}$) is the range of the stress intensity factor during the fatigue cycle. The Paris constant $C = 5.3 \times 10^{-11}$ and the Paris exponent $m = 2.9$ apply. Plane stress conditions are assumed.

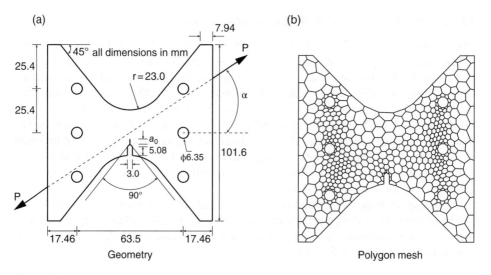

(a)   (b)

Geometry   Polygon mesh

Figure 10.31 Arcan specimen.

An external cyclic loading P with a constant amplitude ratio of $R = 0.1$ is applied at a loading angle α measured from the horizontal axis. Two loading angles $\alpha = 30°$ and $\alpha = 60°$ are considered. The magnitude of the external load is first determined by a static analysis so that the initial equivalent stress intensity factor of the crack is $K_{eq}^{(ini)} = \sqrt{K_I^2 + K_{II}^2} = 7.59 \text{MPa} \sqrt{\text{m}}$ consistent with the experiments performed by Gaylon et al. (2009).

The specimen is discretized with polygon S-elements. Figure 10.31b shows the mesh. The initial mesh has 483 polygons and 1043 nodes.

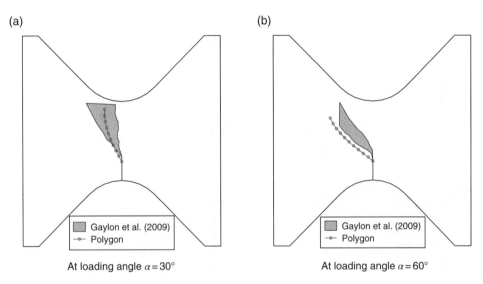

Figure 10.32 Predicted crack paths for Arcan specimen.

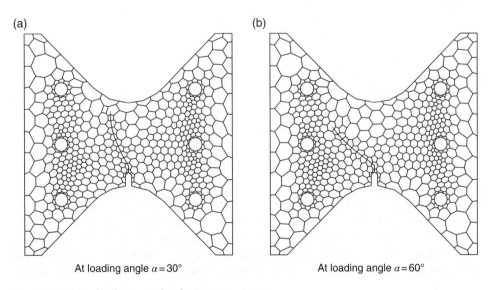

At loading angle $\alpha = 30°$ At loading angle $\alpha = 60°$

Figure 10.33 Final polygon meshes for Arcan specimen.

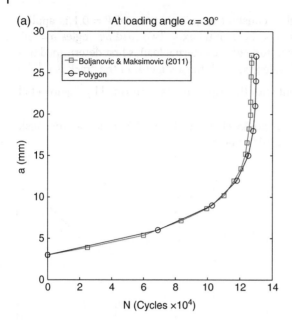

(a)

Figure 10.34 Predicted fatigue life for Arcan specimen.

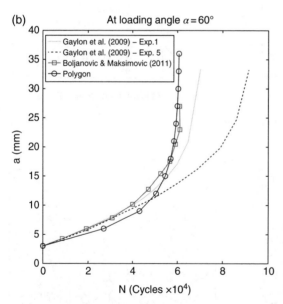

(b)

The fatigue fracture analysis is carried out using a crack incremental length of $\Delta a = 3$mm. Figures 10.32a and b compare the predicted crack paths with the experimental measurements of Gaylon et al. (2009). At the loading angle $\alpha = 30°$, the predicted crack path compares very well with the experimental result. For the loading case with $\alpha = 60°$, barring the initially vertical crack path observed in the experiments but not captured by the present method, the predicted crack path is almost parallel to the experimental measurements. Figure 10.33 shows the final polygon meshes for the load cases $\alpha = 30°$ and $\alpha = 60°$.

Figures 10.34a and b compare the predicted fatigue life of the Arcan specimen with the finite element simulations of Boljanovic and Maksimovic (2011) for $\alpha = 30°$ and $\alpha = 60°$, respectively. Good agreement is observed. The fatigue life for the load case $\alpha = 30°$ is estimated to be 129, 859 cycles, whereas it is 126, 608 cycles in Boljanovic and Maksimovic (2011). For the load case $\alpha = 60°$, the estimated fatigue life is 60, 621 cycles whereas it is 60, 855 cycles in Boljanovic and Maksimovic (2011).

10.7.3.2 Crack Propagation in a Beam with Three Holes

A cracked beam with three holes under three-point bending condition is shown in Figure 10.35. This problem has been studied in, e.g. Bittencourt et al. (1996); Phongthanapanich and Dechaumphai (2004); Azocar et al. (2010). The vertical load is $P = 4.45\text{N}$. The dimensions are given in Figure 10.35. The beam is made of polymethyl-crylate (PMMA). The material properties are: Young's modulus $E = 29 \times 10^6\text{kPa}$ and Poisson's ratio $\nu = 0.3$. Two cases of crack location and depth as indicated in the table in Figure 10.35 are considered. Plane stress conditions are assumed.

The quadtree meshes as shown in Figure 10.36. The mesh of Case I (Figure 10.36) has 438 S-elements with 320 square ones and 78 polygonal ones. The mesh of Case II

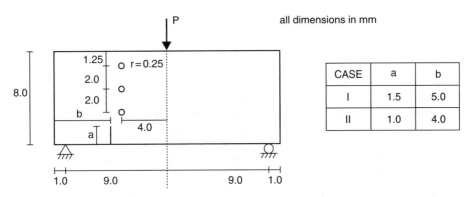

CASE	a	b
I	1.5	5.0
II	1.0	4.0

Figure 10.35 Cracked beam with three holes. The two cases of crack location and depth are indicated in the table.

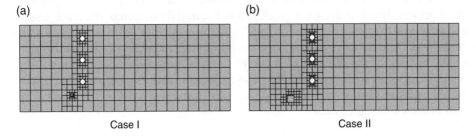

Figure 10.36 Initial meshes of the cracked beam with three holes.

(a) (b)

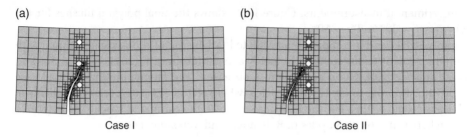

Case I   Case II

Figure 10.37 Final crack paths for cracked beam with three holes.

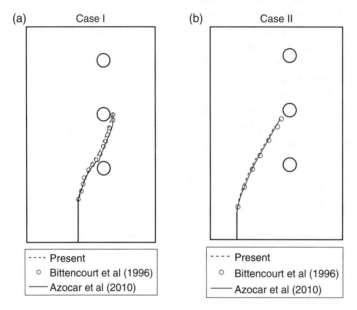

Figure 10.38 Comparison of predicted crack paths for the cracked beam with three holes with published results in the literature.

(Figure 10.36b) has 476 S-elements with 354 square ones and 122 polygonal ones. A crack incremental length of $\Delta a = 0.5$mm is adopted in both cases.

The final meshes for Case I and Case II are shown in Figure 10.37a and Figure 10.37b, respectively. It is observed the bulk of the S-elements are still square cells. polygon S-elements mostly occur along the crack paths. The initial position of the crack affects the trajectory of the crack propagation. In Case I, the crack initially curved towards the bottom hole. The crack then deviates from its initial path and deflects towards the middle hole. In Case II, the crack steadily curves towards the middle hole. The simulated crack trajectories in both cases showed good agreement with the experimental results reported by Bittencourt et al. (1996) and the finite element simulation of Azocar et al. (2010) as shown in Figure 10.38.

Appendix A

Governing Equations of Linear Elasticity

The governing equations of linear elasticity used in this book are summarized. A unified expression in two and three dimensions is adopted by using matrix notations and different operators. A thorough treatment of linear elasticity can be found in many textbooks ranging from those on fundamentals of mechanics, for example, Hibbeler (2017), to those on the finite element method, for example, Bathe (1996); Cook et al. (2002); Reddy (2005).

A.1 Three-dimensional Problems

The governing equations in three dimensions (3D) are formulated in the Cartesian coordinates x, y, z. The displacement field $\{u\} = \{u(x, y, z)\}$ of a solid body occupying domain V is expressed as

$$\{u\} = [u_x,\ u_y,\ u_z]^T \tag{A.1}$$

A.1.1 Strain

The deformation of the solid body is described by strains derived from the displacement field. In linear elastic theory, the strains are assumed to be small. In the engineering definition of strains, the normal strains describing the changes in the length along the coordinate axes are expressed as

$$\varepsilon_x = \frac{\partial u_x}{\partial x}; \qquad \varepsilon_y = \frac{\partial u_y}{\partial y}; \qquad \varepsilon_z = \frac{\partial u_z}{\partial z}; \tag{A.2}$$

The shear strains describing the changes in the angle between the coordinate axes are expressed as

$$\gamma_{yz} = \frac{\partial u_y}{\partial z} + \frac{\partial u_z}{\partial y}; \qquad \gamma_{zx} = \frac{\partial u_z}{\partial x} + \frac{\partial u_x}{\partial z}; \qquad \gamma_{xy} = \frac{\partial u_x}{\partial y} + \frac{\partial u_y}{\partial x} \tag{A.3}$$

In the finite element method, the strains are arranged in a strain vector $\{\varepsilon\} = \{\varepsilon(x, y, z)\}$ as

$$\{\varepsilon\} = [\varepsilon_x,\ \varepsilon_y,\ \varepsilon_z,\ \gamma_{yz},\ \gamma_{zx},\ \gamma_{xy}]^T \tag{A.4}$$

The Scaled Boundary Finite Element Method: Introduction to Theory and Implementation,
First Edition. Chongmin Song.
© 2018 John Wiley & Sons Ltd. Published 2018 by John Wiley & Sons Ltd.
Companion website: www.wiley.com/go/author/scaledboundaryfiniteelementmethod

The strain-displacement relationships given in Eq. (A.2) to (A.4) are expressed concisely as

$$\{\varepsilon\} = [L]\{u\} \tag{A.5}$$

by introducing the matrix of differential operator $[L]$ formulated as

$$[L] = \begin{bmatrix} \dfrac{\partial}{\partial x} & 0 & 0 \\[2mm] 0 & \dfrac{\partial}{\partial y} & 0 \\[2mm] 0 & 0 & \dfrac{\partial}{\partial z} \\[2mm] 0 & \dfrac{\partial}{\partial z} & \dfrac{\partial}{\partial y} \\[2mm] \dfrac{\partial}{\partial z} & 0 & \dfrac{\partial}{\partial x} \\[2mm] \dfrac{\partial}{\partial y} & \dfrac{\partial}{\partial x} & 0 \end{bmatrix} \tag{A.6}$$

A.1.2 Stress and Equilibrium Equation

The stresses describe the internal forces per unit area at a point in the solid body. The general stress state has three normal stresses σ_x, σ_y, σ_z and three independent shear stresses τ_{yz}, τ_{zx}, τ_{xy}. The stress field is expressed in vector form $\{\sigma\} = \{\sigma(x, y, z)\}$ as

$$\{\sigma\} = [\sigma_x, \sigma_y, \sigma_z, \tau_{yz}, \tau_{zx}, \tau_{xy}]^T \tag{A.7}$$

The equation of motion is formulated by considering an infinitesimal volume of the solid body as

$$\frac{\partial \sigma_x}{\partial x} + \frac{\partial \tau_{xy}}{\partial y} + \frac{\partial \tau_{zx}}{\partial z} + p_x = \rho \ddot{u}_x \tag{A.8a}$$

$$\frac{\partial \tau_{xy}}{\partial x} + \frac{\partial \sigma_y}{\partial y} + \frac{\partial \tau_{yz}}{\partial z} + p_y = \rho \ddot{u}_y \tag{A.8b}$$

$$\frac{\partial \tau_{zx}}{\partial x} + \frac{\partial \tau_{yz}}{\partial y} + \frac{\partial \sigma_z}{\partial z} + p_z = \rho \ddot{u}_z \tag{A.8c}$$

where p_x, p_y and p_z are the body forces. ρ is the mass density. The double dots ($^{..}$) indicate the accelerations, i.e. the second time derivatives of displacements. Equation (A.8) is expressed in matrix notation using the differential operator $[L]$ in Eq. (A.6) as

$$[L]^T \{\sigma\} + \{p\} = \rho \{\ddot{u}\} \tag{A.9}$$

with the body force vector $\{p\} = [p_x, p_y, p_z]^T$.

In statics and with vanishing body forces, the equilibrium equations are obtained by simplifying Eq. (A.9)

$$[L]^T \{\sigma\} = 0 \tag{A.10}$$

A.1.3   Stress-strain Relationship and Material Elasticity Matrix

The stress-strain relation in linear elastic conditions is expressed in matrix format as

$$\{\sigma\} = [D]\{\varepsilon\} + \{\sigma_0\} \tag{A.11}$$

where $[D]$ is the elasticity matrix of the material. $\{\sigma_0\}$ denotes the initial stresses. It can be formulated to include initial strains $\{\varepsilon_0\}$

$$\{\sigma_0\} = -[D]\{\varepsilon_0\} \tag{A.12}$$

For example, the initial strains $\{\varepsilon_0\}$ produced by temperature change T are expressed as

$$\{\varepsilon_0\} = T\{\beta\} \tag{A.13}$$

For an isotropic linear-elastic material with a coefficient of thermal expansion α, the proportionality $\{\beta\}$ is written as

$$\{\beta\} = [\,\alpha \quad \alpha \quad \alpha \quad 0 \quad 0 \quad 0\,]^T \tag{A.14}$$

The elasticity matrix $[D]$ of an isotropic material is considered. Hooke's law is applied. Under the normal stress σ_x, the strain ε_x along the direction of the stress, and the strains ε_y and ε_z along the directions perpendicular to the direction of the stress are expressed as

$$\varepsilon_x = \frac{\sigma_x}{E}; \qquad \varepsilon_y = -\frac{v\sigma_x}{E}; \qquad \varepsilon_z = -\frac{v\sigma_x}{E}; \tag{A.15a}$$

where E is the Young's modulus and v the Poisson's ratio. Similarly, the strains under the normal stresses σ_y and σ_z are expressed as

$$\varepsilon_y = \frac{\sigma_y}{E}; \qquad \varepsilon_x = -\frac{v\sigma_y}{E}; \qquad \varepsilon_z = -\frac{v\sigma_y}{E} \tag{A.15b}$$

$$\varepsilon_z = \frac{\sigma_z}{E}; \qquad \varepsilon_x = -\frac{v\sigma_z}{E}; \qquad \varepsilon_y = -\frac{v\sigma_z}{E} \tag{A.15c}$$

The relationships between shear stresses and shear strains are decoupled

$$\gamma_{yz} = \frac{\tau_{yz}}{G}; \qquad \gamma_{zx} = \frac{\tau_{zx}}{G}; \qquad \gamma_{xy} = \frac{\tau_{xy}}{G} \tag{A.16}$$

where G is the shear modulus. It is related to the Young's modulus E and Poisson's ratio v

$$G = \frac{E}{2(1+v)} \tag{A.17}$$

Applying the superposition property of linear elasticity, the stress-strain relationship under a general stress state is expressed in compliance form as

$$
\begin{Bmatrix} \varepsilon_x \\ \varepsilon_y \\ \varepsilon_z \\ \gamma_{yz} \\ \gamma_{zx} \\ \gamma_{xy} \end{Bmatrix}
=
\begin{bmatrix}
1/E & -v/E & -v/E & 0 & 0 & 0 \\
 & 1/E & -v/E & 0 & 0 & 0 \\
 & & 1/E & 0 & 0 & 0 \\
 & & & 1/G & 0 & 0 \\
 & & & & 1/G & 0 \\
 & \text{Symmetric} & & & & 1/G
\end{bmatrix}
\begin{Bmatrix} \sigma_x \\ \sigma_y \\ \sigma_z \\ \tau_{yz} \\ \tau_{zx} \\ \tau_{xy} \end{Bmatrix}
\tag{A.18}
$$

Inversely, the relationship is written in stiffness form as

$$
\begin{Bmatrix} \sigma_x \\ \sigma_y \\ \sigma_z \\ \tau_{yz} \\ \tau_{zx} \\ \tau_{xy} \end{Bmatrix} = \frac{E}{(1+v)(1-2v)}
\begin{bmatrix}
1-v & v & v & 0 & 0 & 0 \\
 & 1-v & v & 0 & 0 & 0 \\
 & & 1-v & 0 & 0 & 0 \\
 & & & \dfrac{1-2v}{2} & 0 & 0 \\
 & & & & \dfrac{1-2v}{2} & 0 \\
 & \text{Symmetric} & & & & \dfrac{1-2v}{2}
\end{bmatrix}
\begin{Bmatrix} \varepsilon_x \\ \varepsilon_y \\ \varepsilon_z \\ \gamma_{yz} \\ \gamma_{zx} \\ \gamma_{xy} \end{Bmatrix}
$$

$$(A.19)$$

Using the stress vector in Eq. (A.7) and strain vector in Eq. (A.4), Eq. (A.19) is rewritten as

$$\{\sigma\} = [D]\{\varepsilon\} \tag{A.20}$$

with the elasticity matrix

$$
[D] = \frac{E}{(1+v)(1-2v)}
\begin{bmatrix}
1-v & v & v & 0 & 0 & 0 \\
 & 1-v & v & 0 & 0 & 0 \\
 & & 1-v & 0 & 0 & 0 \\
 & & & \dfrac{1-2v}{2} & 0 & 0 \\
 & & & & \dfrac{1-2v}{2} & 0 \\
 & & \text{Symmetric} & & & \dfrac{1-2v}{2}
\end{bmatrix}
\tag{A.21}
$$

Note that the elasticity matrix is symmetric.

When the material is orthotropic, the stress-strain relationship is written in compliance form as

$$
\begin{Bmatrix} \varepsilon_x \\ \varepsilon_y \\ \varepsilon_z \\ \gamma_{yz} \\ \gamma_{zx} \\ \gamma_{xy} \end{Bmatrix} =
\underbrace{
\begin{bmatrix}
\dfrac{1}{E_x} & -\dfrac{v_{yx}}{E_y} & -\dfrac{v_{zx}}{E_z} & 0 & 0 & 0 \\
-\dfrac{v_{xy}}{E_x} & \dfrac{1}{E_y} & -\dfrac{v_{zy}}{E_z} & 0 & 0 & 0 \\
-\dfrac{v_{xz}}{E_x} & -\dfrac{v_{yz}}{E_y} & \dfrac{1}{E_z} & 0 & 0 & 0 \\
0 & 0 & 0 & \dfrac{1}{G_{yz}} & 0 & 0 \\
0 & 0 & 0 & 0 & \dfrac{1}{G_{zx}} & 0 \\
0 & 0 & 0 & 0 & 0 & \dfrac{1}{G_{xy}}
\end{bmatrix}
}_{[D]^{-1}}
\begin{Bmatrix} \sigma_x \\ \sigma_y \\ \sigma_z \\ \tau_{yz} \\ \tau_{zx} \\ \tau_{xy} \end{Bmatrix}
\tag{A.22}
$$

where the matrix is symmetric, i.e.

$$\frac{v_{yx}}{E_y} = \frac{v_{xy}}{E_x}; \qquad \frac{v_{zx}}{E_z} = \frac{v_{xz}}{E_x}; \qquad \frac{v_{zy}}{E_z} = \frac{v_{yz}}{E_y} \tag{A.23}$$

E_i ($i = x, y, z$) is the Young's modulus along axis i. G_{ij} ($i, j = x, y, z$) is the shear modulus in direction j on the plane whose normal state is in direction i. v_{ij} is the Poisson's ratio that corresponds to a contraction in direction j when an extension is applied in direction i. In total, 9 material constants are required.

Inverting Eq. (A.22), the stress-strain relationship is expressed in stiffness form (Eq. (A.20)). The elasticity matrix follows from inverting $[D]^{-1}$ and is expressed as

$$[D] = \begin{bmatrix} d_{11} & d_{12} & d_{13} & 0 & 0 & 0 \\ & d_{22} & d_{23} & 0 & 0 & 0 \\ & & d_{33} & 0 & 0 & 0 \\ & & & G_{yz} & 0 & 0 \\ & & & & G_{zx} & 0 \\ & \text{Symmetric} & & & & G_{xy} \end{bmatrix} \tag{A.24}$$

A.1.4 Boundary Conditions

On the surface S of the solid body, either displacements or surface tractions are prescribed as boundary conditions. At a point, the corresponding components of displacement and surface traction are considered as a pair, i.e. (u_x, t_x), (u_y, t_y) or (u_z, t_z). At any point of a well-posed problem, one and only one out of each pair must be prescribed as a boundary condition. A mixed type of boundary condition providing a relationship of the two variables in a pair is also permissible.

The surface tractions $\{t\} = [t_x, t_y, t_z]^T$ at a point on the boundary are related to the stresses. Denoting the outward unit normal vector as $\{n\} = \{n_x, n_y, n_z\}^T$, the equilibrium of an infinitesimal volume on the boundary leads to

$$t_x = n_x \sigma_x + n_y \tau_{xy} + n_z \tau_{zx} \tag{A.25a}$$
$$t_y = n_x \tau_{xy} + n_y \sigma_y + n_z \tau_{yz} \tag{A.25b}$$
$$t_z = n_x \tau_{zx} + n_y \tau_{yz} + n_z \sigma_z \tag{A.25c}$$

Using the stress vector in Eq. (A.7), it is expressed as

$$\{t\} = [n]\{\sigma\} \tag{A.26}$$

with the matrix $[n]$ formed by the outward unit normal vector

$$[n] = \begin{bmatrix} n_x & 0 & 0 & 0 & n_z & n_y \\ 0 & n_y & 0 & n_z & 0 & n_x \\ 0 & 0 & n_z & n_y & n_x & 0 \end{bmatrix} \tag{A.27}$$

It can be identified that the matrix $[n]$ has the same structure as the transpose of the differential operator $[L]$ in Eq. (A.6), where the partial differential operators $\partial/\partial x$, $\partial/\partial y$ and $\partial/\partial z$ are replaced by the vector components n_x, n_y and n_z, respectively.

A.2 Two-dimensional Problems

Under certain conditions as discussed below, some 3D problems are simplified as 2D problems in plane stress or plane strain condition. For 2D elasticity, the governing equations are formulated in the Cartesian coordinates x, y. As becomes apparent later, it is sufficient to consider the in-plane quantities only. The in-plane displacements $\{u\} = \{u(x, y)\}$ are expressed as

$$\{u\} = [u_x, u_y]^T \tag{A.28}$$

Only the in-plane strains in Eq. (A.4) are considered. The strain vector $\{\varepsilon\} = \{\varepsilon(x, y)\}$ is expressed as

$$\{\varepsilon\} = [\varepsilon_x, \varepsilon_y, \gamma_{xy}]^T \tag{A.29}$$

In matrix notation, the strain-displacement relationship is expressed in Eq. (A.5) (repeated here for easy reference)

$$\{\varepsilon\} = [L]\{u\} \tag{A.30}$$

with the differential operator $[L]$ in Eq. (A.6) simplified as

$$[L] = \begin{bmatrix} \dfrac{\partial}{\partial x} & 0 \\[2mm] 0 & \dfrac{\partial}{\partial y} \\[2mm] \dfrac{\partial}{\partial y} & \dfrac{\partial}{\partial x} \end{bmatrix} \tag{A.31}$$

The in-plane stresses are extracted from Eq. (A.7) and repressed in vector notation

$$\{\sigma\} = [\sigma_x, \sigma_y, \tau_{xy}]^T \tag{A.32}$$

The equation of motion in three dimensions (Eq. (A.8)) is simplified as

$$\frac{\partial \sigma_x}{\partial x} + \frac{\partial \tau_{xy}}{\partial y} + p_x = \rho \ddot{u}_x \tag{A.33a}$$

$$\frac{\partial \tau_{xy}}{\partial x} + \frac{\partial \sigma_y}{\partial y} + p_y = \rho \ddot{u}_y \tag{A.33b}$$

It is again expressed as (Eq. (A.9))

$$[L]^T \{\sigma\} + \{p\} = \rho\{\ddot{u}\} \tag{A.34}$$

using the differential operator in Eq. (A.31).

The surface tractions $\{t\} = [t_x, t_y]^T$ at a point on the boundary with the outward unit normal vector $\{n_x, n_y\}^T$ are equal to

$$t_x = n_x \sigma_x + n_y \tau_{xy} \tag{A.35a}$$

$$t_y = n_x \tau_{xy} + n_y \sigma_y \tag{A.35b}$$

and expressed in matrix form as

$$\{t\} = [n]\{\sigma\} \tag{A.36}$$

with the matrix $[n]$ formed by the normal vector

$$[n] = \begin{bmatrix} n_x & 0 & n_y \\ 0 & n_y & n_x \end{bmatrix} \tag{A.37}$$

The matrix $[n]$ has the same structure as the transpose of the differential operator $[L]$ in Eq. (A.31).

A.2.1 Elasticity Matrix in Plane Stress

When a thin flat plate is subject to in-plane (xy plane) forces only, the stresses on the z−plane vanish, i.e.

$$\sigma_z = \tau_{yz} = \tau_{zx} = 0 \tag{A.38}$$

The 2D stress vector $\{\sigma\} = \{\sigma(x, y)\}$ are constant over the thickness of the plate (independent of coordinate z).

Considering Eq. (A.38), the stress-strain relationship of an isotropic material is expressed in compliance form as

$$\begin{Bmatrix} \varepsilon_x \\ \varepsilon_y \\ \gamma_{xy} \end{Bmatrix} = \begin{bmatrix} 1/E & -v/E & 0 \\ -v/E & 1/E & 0 \\ 0 & 0 & 1/G \end{bmatrix} \begin{Bmatrix} \sigma_x \\ \sigma_y \\ \tau_{xy} \end{Bmatrix} \tag{A.39}$$

with the other strain components equal to

$$\varepsilon_z = -\frac{v}{E}(\sigma_x + \sigma_y); \qquad \gamma_{yz} = \gamma_{zx} = 0 \tag{A.40}$$

Although the normal strain ε_z perpendicular to the plane is non-zero, the in-plane quantities are independent of it. The solution of ε_z follows from the solution of σ_x and σ_y.

Inverting Eq. (A.39), the stress-strain relationship is written as

$$\begin{Bmatrix} \sigma_x \\ \sigma_y \\ \tau_{xy} \end{Bmatrix} = \frac{E}{1 - v^2} \begin{bmatrix} 1 & v & 0 \\ v & 1 & 0 \\ 0 & 0 & \frac{1-v}{2} \end{bmatrix} \begin{Bmatrix} \varepsilon_x \\ \varepsilon_y \\ \gamma_{xy} \end{Bmatrix} \tag{A.41}$$

It is expressed as Eq. (A.20) with the stress vector in Eq. (A.32), the strain vector in Eq. (A.29) and the elasticity matrix

$$[D] = \frac{E}{1 - v^2} \begin{bmatrix} 1 & v & 0 \\ v & 1 & 0 \\ 0 & 0 & \frac{1-v}{2} \end{bmatrix} \tag{A.42}$$

The elasticity matrix in plane stress is related to the elasticity matrix in plane strain as described below.

For an orthotropic material, the stress-strain relationship in compliance form is expressed in the plane stress condition as

$$
\begin{Bmatrix} \varepsilon_x \\ \varepsilon_y \\ \gamma_{xy} \end{Bmatrix} = \underbrace{\begin{bmatrix} \dfrac{1}{E_x} & -\dfrac{\nu_{yx}}{E_y} & 0 \\ -\dfrac{\nu_{xy}}{E_x} & \dfrac{1}{E_y} & 0 \\ 0 & 0 & \dfrac{1}{G_{xy}} \end{bmatrix}}_{[D]^{-1}} \begin{Bmatrix} \sigma_x \\ \sigma_y \\ \tau_{xy} \end{Bmatrix}
\tag{A.43}
$$

with

$$
\frac{\nu_{yx}}{E_y} = \frac{\nu_{yx}}{E_y}
\tag{A.44}
$$

The normal strain ε_z perpendicular to the plane and the other strain components are equal to

$$
\varepsilon_z = -\frac{\nu_{xz}}{E_x}\sigma_x - \frac{\nu_{yz}}{E_y}\sigma_y; \qquad \gamma_{yz} = \gamma_{zx} = 0
\tag{A.45}
$$

Inverting Eq. (A.43) results in the elasticity matrix

$$
[D] = \begin{bmatrix} \dfrac{E_x}{1 - \nu_{xy}\nu_{yx}} & \dfrac{\nu_{yx}E_x}{1 - \nu_{xy}\nu_{yx}} & 0 \\ \dfrac{\nu_{xy}E_y}{1 - \nu_{xy}\nu_{yx}} & \dfrac{E_y}{1 - \nu_{xy}\nu_{yx}} & 0 \\ 0 & 0 & G_{xy} \end{bmatrix}
\tag{A.46}
$$

Four material constants E_x, E_y, ν_{yx} and G_{xy} are needed in the elasticity matrix.

A.2.2 Elasticity Matrix in Plane Strain

An infinitely long prismatic structure is considered. The external forces are uniform along the axis and also perpendicular to the axis. Every section of the structure behaves identically. It is sufficient in the analysis to consider a section of the structure as a 2D problem. Owing to symmetry, only in-plane strains exist. Setting the coordinate z along the axis of the prismatic structure, the plane strain condition is expressed as

$$
\varepsilon_z = \gamma_{yz} = \gamma_{zx} = 0
\tag{A.47}
$$

The 3D stress-strain relationship in Eq. (A.19) is written in the plane strain condition (Eq. (A.47)) as

$$
\begin{Bmatrix} \sigma_x \\ \sigma_y \\ \tau_{xy} \end{Bmatrix} = \frac{E}{(1+\nu)(1-2\nu)} \begin{bmatrix} 1-\nu & \nu & 0 \\ \nu & 1-\nu & 0 \\ 0 & 0 & \dfrac{1-2\nu}{2} \end{bmatrix} \begin{Bmatrix} \varepsilon_x \\ \varepsilon_y \\ \gamma_{xy} \end{Bmatrix}
\tag{A.48}
$$

and the normal stress

$$\sigma_z = \frac{Ev(\sigma_x + \sigma_y)}{(1+v)(1-2v)} \tag{A.49}$$

is a function of x, y and independent of z. In deriving the equation of motion in 2D (Eq. A.33), $\partial\sigma_z/\partial z = 0$ has been used.

Expressing Eq. (A.48) as Eq. (A.20), the elasticity matrix follows as

$$[D] = \frac{E}{(1+v)(1-2v)} \begin{bmatrix} 1-v & v & 0 \\ v & 1-v & 0 \\ 0 & 0 & \dfrac{1-2v}{2} \end{bmatrix} \tag{A.50}$$

Similarly, the elasticity matrix for orthotropic materials in Eq. (A.24) is reduced to

$$[D] = \begin{bmatrix} d_{11} & d_{12} & 0 \\ d_{12} & d_{22} & 0 \\ 0 & 0 & G_{xy} \end{bmatrix} \tag{A.51}$$

In this case, seven material constants (E_x, E_y, E_z, v_{yx}, v_{zx}, v_{zy} and G_{xy}) are needed in the elasticity matrix.

For isotropic materials, the elasticity matrices in plane stress and plane strain conditions are related. Replacing the Young's modulus E and Poisson's ratio v in Eq. (A.50) (the elasticity matrix in plane strain) with $v/(1+v)$ and $E(1+2v)/(1+v)^2$, respectively, leads to Eq. (A.42) (the elasticity matrix in plane stress). Inversely, the elasticity matrix in plane strain is obtained from that in plane stress by replacing E and v with $v/(1-v)$ and $E/(1-v^2)$.

Introducing Kolosov's constant

$$\kappa = \begin{cases} 3 - 4v & \text{for plane strain} \\ \dfrac{3-v}{1+v} & \text{for plane stress} \end{cases} \tag{A.52}$$

the elasticity matrices in plane stress (Eq. (A.42)) and plane strain (Eq. (A.50)) are both written as

$$[D] = \frac{G}{\kappa - 1} \begin{bmatrix} 1+\kappa & 3-\kappa & 0 \\ 3-\kappa & 1+\kappa & 0 \\ 0 & 0 & \kappa-1 \end{bmatrix} \tag{A.53}$$

A.3 Unified Expressions of Governing Equations

Using matrix notations, the governing equations of linear elasticity in two and three dimensions are written in a set of unified expressions. The strain-displacement relationship is expressed as

$$\{\varepsilon\} = [L]\{u\} \tag{A.54}$$

where the matrix of differential operator $[L]$ is defined in Eq. (A.6) for 3D problems and in Eq. (A.31) for 2D problems. The equation of motion is written, using the matrix of differential operator, as

$$[L]^T\{\sigma\} + \{b\} = \rho\{\ddot{u}\} \tag{A.55}$$

The stress-strain relationship is given by

$$\{\sigma\} = [D]\{\varepsilon\} + \{\sigma_0\} \tag{A.56}$$

with the material elasticity matrix $[D]$ chosen according to the type of problem (3D, plane stress or plane strain). The surface tractions on boundary are expressed as

$$\{t\} = [n]\{\sigma\} \tag{A.57}$$

with the matrix $[n]$ found in Eq. (A.27) for 2D problems and in Eq. (A.37) for 3D problems.

Appendix B

Matrix Power Function

Comprehensive textbooks on the theory of matrix functions, for example, Gantmacher (1977); Golub and van Loan (1996), are available. In this appendix, only the theory directly employed in this book is summarized.

B.1 Definition of Matrix Power Function

The matrix power function is closely related to the matrix exponential function defined by a Taylor series

$$e^{[A]} = [I] + \sum_{k=1} \frac{1}{k!} [A]^k \tag{B.1}$$

where $[A]$ is a square matrix, and $[I]$ is an identity matrix of the same dimension. The corresponding matrix power function is expressed as

$$x^{[A]} = e^{[A]\ln x} = [I] + \sum_{k=1} \frac{1}{k!} ([A]\ln x)^k \tag{B.2}$$

which is defined for $x > 0$.

The following property can be identified from Eq. (B.2):

1) At $x = 1$,

$$x^{[A]} = 1^{[A]} = [I] \tag{B.3}$$

where $[I]$ is a identity matrix having the same size as $[A]$.

2) When $[A] = 0$,

$$x^{[A]} = x^0 = [I] \tag{B.4}$$

3) The matrix $[A]$ and the function $x^{[A]}$ are commutative, i.e.

$$[A]x^{[A]} = x^{[A]}[A] \tag{B.5}$$

which can be shown using Eq. (B.2).

4) The matrix

$$[A] = \begin{bmatrix} 0 & a \\ 0 & 0 \end{bmatrix} \qquad (a \neq 0) \tag{B.6}$$

The Scaled Boundary Finite Element Method: Introduction to Theory and Implementation,
First Edition. Chongmin Song.
© 2018 John Wiley & Sons Ltd. Published 2018 by John Wiley & Sons Ltd.
Companion website: www.wiley.com/go/author/scaledboundaryfiniteelementmethod

has two zero eigenvalues and a nonzero off-diagonal entry a. Since

$$[A]^k = 0 \qquad \text{when } k \geq 2 \tag{B.7}$$

applies, the matrix power function $x^{[A]}$ is expressed as

$$x^{[A]} = [I] + \begin{bmatrix} 0 & a \\ 0 & 0 \end{bmatrix} \ln x = \begin{bmatrix} 1 & a \ln x \\ 0 & 1 \end{bmatrix} \tag{B.8}$$

A logarithmic function appears.

5) The matrix

$$[A] = -\epsilon \begin{bmatrix} 0 & +1 \\ -1 & 0 \end{bmatrix} \tag{B.9}$$

has two pure imaginary eigenvalues $\pm\epsilon i$. The matrix power function $x^{[A]}$ is expressed as

$$x^{[A]} = \begin{bmatrix} 1 & 0 \\ 0 & 1 \end{bmatrix} - \begin{bmatrix} 0 & 1 \\ -1 & 0 \end{bmatrix} \epsilon \ln x - \begin{bmatrix} 1 & 0 \\ 0 & 1 \end{bmatrix} \frac{(\epsilon \ln x)^2}{2!}$$

$$- \begin{bmatrix} 0 & -1 \\ 1 & 0 \end{bmatrix} \frac{(\epsilon \ln x)^3}{3!} + \begin{bmatrix} 1 & 0 \\ 0 & 1 \end{bmatrix} \frac{(\epsilon \ln x)^4}{4!} - \begin{bmatrix} 0 & +1 \\ -1 & 0 \end{bmatrix} \frac{(\epsilon \ln x)^5}{5!} + \cdots$$

$$= \begin{bmatrix} \cos(\epsilon \ln x) & -\sin(\epsilon \ln x) \\ \sin(\epsilon \ln x) & \cos(\epsilon \ln x) \end{bmatrix} \tag{B.10}$$

considering the following Taylor series

$$\cos \theta = \sum_{i=0}^{\infty} \frac{(-1)^i}{(2i)!} \theta^{2n} = 1 - \frac{\theta^2}{2!} + \frac{\theta^4}{4!} + \cdots \tag{B.11}$$

$$\sin \theta = \sum_{i=0}^{\infty} \frac{(-1)^i}{(2i+1)!} \theta^{2n+2} = x - \frac{\theta^3}{3!} + \frac{\theta^5}{5!} + \cdots \tag{B.12}$$

B.2 Application to Solution of System of Ordinary Differential Equations

One application of the matrix power function is to solve the system of linear ordinary differential equations

$$x\{y'(x)\} = [A]\{y(x)\} \tag{B.13}$$

with the initial condition

$$\{y(1)\} = \{y_1\} \tag{B.14}$$

The solution is given by

$$\{y(x)\} = x^{[A]}\{y_1\} \tag{B.15}$$

The matrix power function $x^{[A]}$ is the fundamental matrix of the system of differential equation (B.13). $x^{[A]}$ itself satisfies

$$x(x^{[A]})' = [A]x^{[A]} = x^{[A]}[A] \tag{B.16}$$

Its derivative is expressed as

$$(x^{[A]})' = [A]x^{[A]-[I]} = x^{[A]-[I]}[A] \tag{B.17}$$

The matrix power function satisfies the following properties:

1) The inverse of the power function of a matrix $[A]$ is the power function of the negative of the matrix

$$x^{[A]}x^{-[A]} = [I] \tag{B.18}$$

2) If a and b are two positive numbers, then

$$(ab)^{[A]} = a^{[A]}b^{[A]} \tag{B.19}$$

3) If two matrices $[A]$ and $[B]$ commute (i.e. $[A][B] = [B][A]$), then

$$x^{[A]}x^{[B]} = x^{[B]}x^{[A]} = x^{[A]+[B]} \tag{B.20}$$

4) If $[B]$ is a matrix similar to $[A]$, i.e. there exists an invertible matrix $[T]$ such that

$$[B] = [T]^{-1}[A][T] \tag{B.21a}$$

then

$$x^{[B]} = [T]^{-1}x^{[A]}[T] \tag{B.21b}$$

5) The integration of $x^{[A]}$, which none of the eigenvalues of $[A]$ is equal to -1, is expressed as

$$\int x^{[A]}dx = ([A] + [I])^{-1}x^{[A]+[I]} + [c] = x^{[A]+[I]}([A] + [I])^{-1} + [c] \tag{B.22}$$

which $[c]$ is a constant matrix. This can be shown by considering Eq. (B.17) formulated for the constant matrix $([A] + [I])$

$$(x^{[A]+[I]})' = ([A] + [I])x^{[A]} = x^{[A]}([A] + [I]) \tag{B.23}$$

where the matrix commutativity in Eq. (B.5) is used.

B.3 Computation of Matrix Power Function by Eigenvalue Method

When all the eigenvectors of matrix $[A]$ are independent, its eigenvalue decomposition is expressed as

$$[A] = [\phi]\text{diag}(\lambda_i)[\phi]^{-1} \tag{B.24}$$

where λ_i are the eigenvalues and $[\phi]$ consists of eigenvectors. Using Eq. (B.21), the matrix power function is expressed explicitly as

$$x^{[A]} = [\phi]\text{diag}(x^{\lambda_i})[\phi]^{-1} \tag{B.25}$$

The eigenvalue method breaks down when multiple eigenvalues with parallel eigenvectors exist. The matrix in Example 1, Section B.1

$$[A] = \begin{bmatrix} 0 & a \\ 0 & 0 \end{bmatrix} \qquad (a \neq 0)$$

is considered. It has double eigenvalues with parallel eigenvectors

$$\lambda_1 = \lambda_2 = 0; \qquad [\phi] = \begin{bmatrix} \phi_{11} & \phi_{12} \\ 0 & 0 \end{bmatrix}$$

As the matrix $[\phi]$ is singular, the solution in Eq. B.25 cannot be defined.

A robust approach to compute the matrix exponential function is the 'scaling and square' algorithm (Section 11.3, Golub and van Loan, 1996). It is based on Padé approximation of the exponential function. This algorithm is implemented in the function expm(x) of MATLAB.

Bibliography

Al-Mohy, A. H., Higham, N. J., 2009. A new scaling and squaring algorithm for the matrix exponential. *SIAM Journal on Matrix Analysis and Applications* 31, 970–989.

Anderson, E., Bai, Z., Bischof, C., et al., 1999. LAPACK Users' Guide, 3rd Edition. Society for Industrial and Applied Mathematics, Philadelphia.

Anderson, T. L., 2005. *Fracture Mechanics: Fundamentals and Applications*, 3rd Edition. Taylor & Francis.

Atluri, N. S., Zhu, T., 1998. A new Meshless Local Petrov-Galerkin (MLPG) approach in computational mechanics. *Computational Mechanics* 22, 117–127.

Attene, M., Campen, M., Kobbelt, L., Mar. 2013. Polygon mesh repairing: An application perspective. *ACM Computing Surveys* 45 (2), 15:1–15:33.

Azocar, D., Elgueta, M., Rivara, M. C., 2010. Automatic LEFM crack propagation method based on local Lepp-Delaunay mesh refinement. *Advances in Engineering Software* 41, 111–119.

Banks-Sills, L., Hershkovitz, I., Wawrzynek, P. A., Eliasi, R., Ingraffea, A. R., 2005. Methods for calculating stress intensity factors in anisotropic materials: Part I–z = 0 is a symmetric plane. *Engineering Fracture Mechanics* 72 (15), 2328–2358.

Banks-Sills, L., Sherer, A., 2002. A conservative integral for determining stress intensity factors of a bimaterial notch. *International Journal of Fracture* 115, 1–26.

Barsoum, R. S., 1974. Application of quadratic isoparametric finite elements in linear fracture mechanics. *International Journal of Fracture* 10, 603–605.

Bartels, R. H., Stewart, J. L., 1972. Solution of the matrix equation $AX + XB = C$. *Communications of the ACM* 15, 820–826.

Bassoli, E., Gatto, A., Iuliano, L., Violante, M. G., 2007. 3D printing technique applied to rapid casting. *Rapid Prototyping Journal* 13 (3), 148–155.

Bathe, J. K., 1996. *Finite Element Procedures*. Prentice Hall, New Jersey.

Bavely, C. A., Stewart, G. W., 1979. An algorithm for computing reducing subspaces by block diagonalization. *SIAM Journal on Numerical Analysis* 16, 359–367.

Bazyar, M. H., Song, C., 2008. A continued-fraction-based high-order transmitting boundary for wave propagation in unbounded domains of arbitrary geometry. *International Journal for Numerical Methods in Engineering* 74, 209–237.

The Scaled Boundary Finite Element Method: Introduction to Theory and Implementation,
First Edition. Chongmin Song.
© 2018 John Wiley & Sons Ltd. Published 2018 by John Wiley & Sons Ltd.
Companion website: www.wiley.com/go/author/scaledboundaryfiniteelementmethod

Bazyar, M. H., Song, C., 2017. Analysis of transient wave scattering and its applications to site response analysis using the scaled boundary finite-element method. *Soil Dynamics and Earthquake Engineering* 98, 191–205.

Béchet, E., Cuilliere, J. C., Trochu, F., 2002. Generation of a finite element MESH from stereolithography (STL) files. *Computer-Aided Design* 34 (1), 1–17.

Behnke, R., Mundil, M., Birk, C., Kaliske, M., 2014. A physically and geometrically non-linear scaled-boundary-based finite element formulation for fracture in elastomers. *International Journal for Numerical Methods in Engineering* 99 (13), 966–999.

Belytschko, T., Lu, Y., Gu, L., 1994. Element-free Galerkin methods. *International Journal for Numerical Methods in Engineering* 37, 229–256.

Bern, M., Eppstein, D., Gilbert, J., 1994. Provably good mesh generation. *Journal of Computer and System Sciences* 48 (3), 384–409.

Bird, G., Trevelyan, J., Augarde, C., 2010. A coupled bem/scaled boundary FEM formulation for accurate computations in linear elastic fracture mechanics. *Engineering Analysis with Boundary Elements* 34 (6), 599–610.

Birk, C., Chen, D., Song, C., Du, C., 2014. A high-order approach for modelling transient wave propagation problems using the scaled boundary finite element method. *International Journal for Numerical Methods in Engineering* 97, 937–959.

Birk, C., Song, C., 2009. A continued-fraction approach for transient diffusion in unbound medium. *Computer Methods in Applied Mechanics and Engineering* 198 (33-36), 2576–2590.

Bishop, J., 2014. A displacement-based finite element formulation for general polyhedra using harmonic shape functions. *International Journal for Numerical Methods in Engineering* 97, 1–31.
URL http://dx.doi.org/10.1002/nme.4562

Bittencourt, T. N., Wawrzynek, P. A., Ingraffea, A. R., 1996. Quasi-automatic simulation of crack propagation for 2D LEFM problems. *Engineering Fracture Mechanics* 55, 321–334.

Blacker, T. D., Meyers, R. J., 1993. Seams and wedges in plastering: A 3-D hexahedral mesh generation algorithm. *Engineering with Computers* 9 (2), 83–93.

Boljanovic, S., Maksimovic, S., 2011. Analysis of crack growth propagation under mixed-mode loading. *Engineering Fracture Mechanics* 78, 1565–1576.

Bower, A. F., 2009. *Applied Mechanics of Solids*. CRC Press.

Brebbia, C. A., Telles, J. C. F., Wrobel, L. C., 1984. *Boundary Element Techniques*. Springer-Verlag.

Canuto, C., Hussaini, M. Y., Quarteroni, A., Zeng, T. A., 1988. *Spectral Methods in Fluid Dynamics*. Springer-Verlag.

Chen, L., Dornisch, W., Klinkel, S., 2015a. Hybrid collocation-Galerkin approach for the analysis of surface represented 3D-solids employing SB-FEM. *Computer Methods in Applied Mechanics and Engineering* 295, 268–289.

Chen, X., Birk, C., Song, C., 2015b. Transient analysis of wave propagation in layered soil by using the scaled boundary finite element method. *Computers and Geotechnics* 63, 1–12.

Chidgzey, S. R., Deeks, A. J., 2005. Determination of coefficients of crack tip asymptotic fields using the scaled boundary finite element method. *Engineering Fracture Mechanics* 72, 2019–2036.

Chiong, I., Ooi, E., Song, C., Tin-Loi, F., 2014. Scaled boundary polygons with application to fracture analysis of functionally graded materials. *International Journal for Numerical Methods in Engineering* 98, 562–589.

Cho, S. B., Lee, K. R., Choy, Y. S., Yuuki, R., 1992. Determination of stress intensity factors and boundary element analysis for interface cracks in dissimilar anisotropic materials. *Engineering Fracture Mechanics* 43, 603–614.

Chowdhury, M., Song, C., Gao, W., 2014. Shape sensitivity analysis of stress intensity factors by the scaled boundary finite element method. *Engineering Fracture Mechanics* 116, 13–30.

Chowdhury, M., Song, C., Gao, W., 2015. Highly accurate solutions and Padé approximants of the stress intensity factors and T-stress for standard specimens. *Engineering Fracture Mechanics* 144, 46–67.

Collini, L., 2005. Micromechanical modeling of the elasto-plastic behavior of heterogeneous nodular cast iron. Ph.D. thesis.

Cook, R. D., Malkus, D. S., Plesha, M. E., Witt, R. J., 2002. *Concepts and Applications of Finite Element Analysis.* John Wiley & Sons.

Cotterell, B., 1966. Notes on the paths and stability of cracks. *International Journal of Fracture Mechanics* 2, 526–533.

Dai, S., Augarde, C., Du, C., Chen, D., 2015. A fully automatic polygon scaled boundary finite element method for modelling crack propagation. *Engineering Fracture Mechanics* 133, 163–178.

Deeks, A. J., 2004. Prescribed side-face displacements in the scaled boundary finite-element method. *Computers & Structures* 82, 1153–1165.

Deeks, A. J., Augarde, C. E., 2005. A meshless local Petrov-Galerkin scaled boundary method. *Computational Mechanics* 36 (3), 159–170.

Deeks, A. J., Wolf, J. P., 2002a. Semi-analytical elastostatic analysis of unbounded two-dimensional domains. *International Journal for Numerical and Analytical Methods in Geomechanics* 26, 1031–1057.

Deeks, A. J., Wolf, J. P., 2002b. Stress recovery and error estimation for the scaled boundary finite-element method. *International Journal for Numerical Methods in Engineering* 54, 557–583.

Deeks, A. J., Wolf, J. P., 2002c. An *h*-hierarchical adaptive procedure for the scaled boundary finite-element method. *International Journal for Numerical Methods in Engineering* 54, 585–605.

Deeks, A. J., Wolf, J. P., 2002d. A virtual work derivation of the scaled boundary finite-element method for elastostatics. *Computation Mechanics* 28, 489–504.

Dempsey, J. P., Sinclair, G. B., 1979. On the stress singularities in the plane elasticity of the composite wedge. *Journal of Elasticity* 9, 373–391.

Dominguez, J., 1993. *Boundary Elements in Dynamics.* Computational Mechanics Publications, Southampton.

Ebeida, M. S., Davis, R. L., Freund, R. W., 2010. A new fast hybrid adaptive grid generation technique for arbitrary two-dimensional domains. *International Journal for Numerical Methods in Engineering* 84, 305–329.

Fett, T., 2008. *Stress Intensity Factors, T-stresses, Weight Functions.* Univ.-Verlag Karlsruhe.

Frey, P. J., George, P. L., 2008. *Mesh Generation.* ISTE, London.

Gadi, K. S., Josepha, P. F., Zhang, N. S., Kaya, A. C., 2000. Thermally induced logarithmic stress singularities in a composite wedge and other anomalies. *Engineering Fracture Mechanics* 65, 645–664.

Gantmacher, F. R., 1977. *The Theory of Matrices*. Chelsea, New York.

Gaylon, S. E., Arunachalam, S. R., Greer, J., Hammond, M., Fawaz, S. A., 2009. Three-dimensional crack growth prediction. In: Bos, M. J. (Ed.), *Proceedings of the 25th Symposium of the International Committee on Aeronautical Fatigue (ICAF 2009), Bridging the Gap Between Theory and Operational Practice*. Rotterdam, the Netherlands, pp. 1035–1068.

Golub, G. H., van Loan, C. F., 1996. *Matrix Computations*, 3rd Edition. The Johns Hopkins University Press.

Goswami, S., Becker, W., 2012. Computation of 3-D stress singularities for multiple cracks and crack intersections by the scaled boundary finite element method. *International Journal of Fracture* 175 (1), 13–25.

Gravenkamp, H., Prager, J., Saputra, A., Song, C., 2012. The simulation of Lamb waves in a cracked plate using the scaled boundary finite element method. *The Journal of the Acoustical Society of America* 132 (3), 1358.

Greaves, D. M., Borthwick, A. G. L., 1999. Hierarchical tree-based finite element mesh generation. *International Journal for Numerical Methods in Engineering* 45, 447–471.

Gupta, A. K., 1978. A finite element for transition from a fine to a coarse grid. *International Journal for Numerical Methods in Engineering* 12, 35–45.

GVS, P. R., Montas, H. J., Samet, H., Shirmohammadi, A., 2001. Quadtree-based triangular mesh generation for finite element analysis of heterogeneous spatial data. In: The 2001 ASAE Annual International Meeting (Paper ID:01-3072). Sacramento, CA.

Hell, S., Becker, W., 2015. The scaled boundary finite element method for the analysis of 3D crack interaction. *Journal of Computational Science* 9, 76–81.

Henshell, R. D., Shaw, K. G., 1975. Crack tip finite elements are unnecessary. *International Journal for Numerical Methods in Engineering* 9, 495–507.

Hibbeler, R. C., 2017. *Mechanics of Materials*, 10th Edition. Pearson.

Huang, Y., Yang, Z., Ren, W., Liu, G., Zhang, C., 2015. 3D meso-scale fracture modelling and validation of concrete based on in-situ X-ray Computed Tomography images using damage plasticity model. *International Journal of Solids and Structures* 67-68, 340–352.

Hughes, T. J. R., Cottrell, J. A., Bazilevs, Y., 2005. Isogeometric analysis: CAD, finite elements, NURBS, exact geometry and mesh refinement. *Computer Methods in Applied Mechanics and Engineering* 194 (39-41), 4135–4195.

Hwu, C., Kuo, T., 2007. A unified definition for stress intensity factors of interface corners and cracks. *International Journal of Solids and Structures* 44 (18-19), 6340–6359.

Kanninen, Melvin F.and Popelar, C. H., 1985. *Advanced Fracture Mechanics*. Oxford University Press.

Keyak, J., Meagher, J., Skinner, H., Mote, C., 1990. Automated three-dimensional finite element modelling of bone: a new method. *Journal of Biomedical Engineering* 12 (5), 389–397.

Klinkel, S., Chen, L., Dornisch, W., 2015. A {NURBS} based hybrid collocation-Galerkin method for the analysis of boundary represented solids. *Computer Methods in Applied Mechanics and Engineering* 284, 689–711, isogeometric Analysis Special Issue. URL http://www.sciencedirect.com/science/article/pii/S0045782514003995

Kreyszig, E., 2011. *Advanced Engineering Mathematics*, 10th Edition. John Wiley & Sons, New York.

Labossiere, P. E. W., Dunn, M. L., 1999. Stress intensities at interface corners in anisotropic bimaterials. *Engineering Fracture Mechanics* 62, 555–575.

Laub, A. J., 1979. A Schur method for solving algebraic Riccati equations. *IEEE Transactions on Automatic Control* AC-24, 913–921.

Lengsfeld, M., Schmitt, J., Alter, P., Kaminsky, J., Leppek, R., 1998. Comparison of geometry-based and CT voxel-based finite element modelling and experimental validation. *Medical Engineering & Physics* 20 (7), 515–522.

Li, C., Man, H., Song, C., Gao, W., 2013a. Analysis of cracks and notches in piezoelectric composites using scaled boundary finite element method. *Composite Structures* 101, 191–203.

Li, C., Song, C., Man, H., Ooi, E., Gao, W., 2014. 2D dynamic analysis of cracks and interface cracks in piezoelectric composites using the SBFEM. *International Journal of Solids and Structures* 51 (11-12), 2096–2108.

Li, J., Liu, J., Lin, G., 2013b. Dynamic interaction numerical models in the time domain based on the high performance scaled boundary finite element method. *Earthquake Engineering and Engineering Vibration* 12 (4), 541–546.

Li, S. M., 2009. Diagonalization procedure for scaled boundary finite element method in modeling semi-infinite reservoir with uniform cross-section. *International Journal for Numerical Methods in Engineering* 80, 596–608.

Lian, W. D., Legrain, G., Cartraud, P., 2013. Image-based computational homogenization and localization: comparison between X-FEM/levelset and voxel-based approaches. *Computational Mechanics* 51 (3), 279–293.

Lin, G., Liu, J., Li, J., Hu, Z., 2015. A scaled boundary finite element approach for sloshing analysis of liquid storage tanks. *Engineering Analysis with Boundary Elements* 56, 70–80.

Lindemann, J., Becker, W., 2002. Free-edge stresses around holes in laminates by the boundary finite-element method. *Mechanics of Composite Materials* 38, 407–416.

Liu, J., Lin, G., Wang, F., Li, J., 2010. The scaled boundary finite element method applied to electromagnetic field problems. In: *IOP Conference Series: Materials Science and Engineering*. Vol. 10. IOP Publishing, p. 012245.

Liu, J. Y., Lin, G., 2007. Evaluation of stress intensity factors subjected to arbitrarily distributed tractions on crack surfaces. *China Ocean Engineering* 21, 293–304.

Löhner, R., Parikh, P., 1988. Generation of three-dimensional unstructured grids by the advancing-front method. *International Journal for Numerical Methods in Fluids* 8 (10), 1135–1149.

Long, X., Jiang, C., Han, X., Gao, W., Bi, R., 2014. Sensitivity analysis of the scaled boundary finite element method for elastostatics. *Computer Methods in Applied Mechanics and Engineering* 276, 212–232.

Lorensen, W. E., Cline, H. E., 1987. Marching cubes: A high resolution 3D surface construction algorithm. In: *ACM Siggraph Computer Graphics*. Vol. 21. ACM, pp. 163–169.

Lucy, L. B., 1977. A numerical approach to the testing of the fission hypothesis. *Astronomical Journal* 82, 1013–1024.

Ma, C. C., Luo, J. J., 1996. Plane solutions of interface cracks in anisotropic dissimilar media. *Journal of Engineering Mechanics, ASCE* 122, 30–38.

Man, H., Song, C., Gao, W., Tin-Loi, F., 2012. A unified 3D-based technique for plate bending analysis using scaled boundary finite element method. *International Journal for Numerical Methods in Engineering* 91 (5), 491–515.

Man, H., Song, C., Gao, W., Tin-Loi, F., 2014a. Semi-analytical analysis for piezoelectric plate using the scaled boundary finite-element method. *Computers & Structures* 137, 47–62.

Man, H., Song, C., Natarajan, S., Ooi, E. T., Birk, C., 2014b. Towards automatic stress analysis using scaled boundary finite element method with quadtree mesh of high-order elements. arXiv:1402.5186, https://arxiv.org/pdf/1402.5186v1.pdf.

Man, H., Song, C., Xiang, T., Gao, W., Tin-Loi, F., 2013. High-order plate bending analysis based on the scaled boundary finite element method. *International Journal for Numerical Methods in Engineering* 95 (4), 331–360.

Mcdill, J. M., Goldak, J. A., Oddy, A. S., Bibby, M. J., 1987. Isoparametric quadrilaterals and hexahedrons for mesh-grading algorithms. *Communications in Applied Numerical Methods* 3, 155–163.

Meagher, D., 1982. Geometric modeling using octree encoding. *Computer Graphics and Image Processing* 19 (2), 129–147.

Mittelstedt, C., Becker, W., 2006. Efficient computation of order and mode of three-dimensional stress singularities in linear elasticity by the boundary finite element method. *International Journal of Solids and Structures* 43 (10), 2868–2903.

Moës, N., Dolbow, J., Belytschko, T., 1999. A finite element method for crack growth without remeshing. *International Journal for Numerical Methods in Engineering* 46 (1), 131–150.

Moler, C. B., van Loan, C., 2003. Nineteen dubious ways to compute the exponential of a matrix, twenty-five years later. *SIAM Review* 45, 3–49.

Müller, A., Wenck, J., Goswami, S., Lindemann, J., Hohe, J., Becker, W., 2005. The boundary finite element method for predicting directions of cracks emerging from notches at bimaterial junctions. *Engineering Fracture Mechanics* 72, 373–386.

Müller, R., Rüegsegger, P., 1995. Three-dimensional finite element modelling of non-invasively assessed trabecular bone structures. *Medical Engineering & Physics* 17 (2), 126–133.

Munz, D., Yang, Y., 1993. Stresses near the edge of bonded dissimilar materials described by two stress intensity factors. *International Journal of Fracture* 60, 169–177.

Natarajan, S., Ooi, E. T., Saputra, A., Song, C., 2017. A scaled boundary finite element formulation over arbitrary faceted star convex polyhedra. *Engineering Analysis with Boundary Elements* 80, 218–229.

Natarajan, S., Wang, J., Song, C., Birk, C., 2015. Isogeometric analysis enhanced by the scaled boundary finite element method. *Computer Methods in Applied Mechanics and Engineering* 283, 733–762.

Newman, T. S., Yi, H., 2006. A survey of the marching cubes algorithm. *Computers & Graphics* 30 (5), 854–879.

Nguyen-Xuan, H., 2017. A polytree-based adaptive polygonal finite element method for topology optimization. *International Journal for Numerical Methods in Engineering* 110, 972–1000.

Ooi, E., Shi, M., Song, C., Tin-Loi, F., Yang, Z., Jul. 2013. Dynamic crack propagation simulation with scaled boundary polygon elements and automatic remeshing technique. *Engineering Fracture Mechanics* 106 (2012), 1–21.

Ooi, E., Song, C., Tin-Loi, F., Yang, Z., Oct. 2012a. Automatic modelling of cohesive crack propagation in concrete using polygon scaled boundary finite elements. *Engineering Fracture Mechanics* 93, 13–33.

Ooi, E., Yang, Z., 2010. A hybrid finite element-scaled boundary finite element method for crack propagation modelling. *Computer Methods in Applied Mechanics and Engineering* 199 (17), 1178–1192.

Ooi, E. T., Man, H., Natarajan, S., Song, C., 2015. Adaptation of quadtree meshes in the scaled boundary finite element method for crack propagation modelling. *Engineering Fracture Mechanics* 144, 101–117.

Ooi, E. T., Natarajan, S., Song, C., Ooi, E. H., 2016. Dynamic fracture simulations using the scaled boundary finite element method on hybrid polygon-quadtree meshes. *International Journal of Impact Engineering* 90, 154–164.

Ooi, E. T., Natarajan, S., Song, C., Ooi, E. H., 2017. Crack propagation modelling in concrete using the scaled boundary finite element method with hybrid polygon-quadtree meshes. *International Journal of Fracture* 203 (1-2), 135–157.

Ooi, E. T., Song, C., Tin-Loi, F., 2014. A scaled boundary polygon formulation for elasto-plastic analyses. *Computer Methods in Applied Mechanics and Engineering* 268, 905–937.

Ooi, E. T., Song, C., Tin-loi, F., Yang, Z., 2012b. Polygon scaled boundary finite elements for crack propagation modelling. *International Journal for Numerical Methods in Engineering* 91, 319–342.

Ooi, E. T., Yang, Z. J., 2009. Modelling multiple cohesive crack propagation using a finite element-scaled boundary finite element coupled method. *Engineering Analysis with Boundary Elements* 33 (7), 925–929.

Ooi, E. T., Yang, Z. J., 2011. Modelling crack propagation in reinforced concrete using a hybrid finite element–scaled boundary finite element method. *Engineering Fracture Mechanics* 78 (2), 252–273.

Owen, S., 1998. A survey of unstructured mesh generation technology. In: *7th International Meshing Roundtable*. pp. 26–28.

Pageau, S. S., Gadi, K., Biggers Jr., S. B., Joseph, P. F., 1996. Standardized complex and logarithmic eigensolutions for n-material wedges and junctions. *International Journal of Fracture Mechanics* 77, 51–76.

Parvizian, J., Düster, A., Rank, E., 2007. Finite cell method. *Computational Mechanics* 41 (1), 121–133.

Persson, P.-O., Strang, G., 2004. A simple mesh generator in MATLAB. *SIAM Review* 46.

Phongthanapanich, S., Dechaumphai, P., 2004. Adaptive Delaunay triangulation with object-oriented programming for crack propagation analysis. *Finite Elements in Analysis and Design* 40, 1753–1771.

Preparata, F. P., Shamos, M., 1985. *Computational Geometry: An Introduction*. Springer-Verlag.

Press, W. H., Teukolsky, S. A., Vetterling, W. T., Flannery, B., 1992. *Numerical Recipes in Fortran*. Cambridge University Press, Cambridge.

Reddy, J. N., 2005. *An Introduction to the Finite Element Method*, 3rd Edition. McGraw-Hill Education.

Ren, W., Yang, Z. J., Withers, P., 2013. Meso-scale fracture modelling of concrete based on X-ray computed tomography images. In: *Proceedings of the 5th Asia Pacific Congress on Computational Mechanics, Singapore*, 11-14th December 2013.

Rengier, F., Mehndiratta, A., von Tengg-Kobligk, H., et al., 2010. 3D printing based on imaging data: review of medical applications. *International Journal of Computer Assisted Radiology and Surgery* 5 (4), 335–341.

Rice, J. R., 1968a. *Fracture: An Advanced Treatise.* (ed. H. Liebowitz), Academic Press, N.Y., Ch. Mathematical Analysis in the Mechanics of Fracture, Chapter 3, pp. 191–311.

Rice, J. R., 1968b. A path independent integral and the approximate analysis of strain concentration by notches and cracks. *Journal of Applied Mechanics*, ASME 35, 379–386.

Rice, J. R., 1988. Elastic fracture mechanics concepts for interfacial cracks. *Journal of Applied Mechanics, ASME* 55, 98–103.

Saputra, A. A., Birk, C., Song, C., 2015. Computation of three-dimensional fracture parameters at interface cracks and notches by the scaled boundary finite element method. *Engineering Fracture Mechanics* 148, 213–242.

Shi, M., Zhong, H., Ooi, E. T., Zhang, C., S, C., 2013. Modelling of crack propagation of gravity dams by scaled boundary polygons and cohesive crack model. *International Journal of Fracture* 183 (1), 29–48.

Sladek, J., Sladek, V., 1997. Evaluations of the T-stress for interface cracks by the boundary element method. *Engineering Fracture Mechanics* 56, 813–825.

Sladek, J., Sladek, V., Krahulec, K., Song, C., 2016. Crack analyses in porous piezoelectric brittle materials by the SBFEM. *Engineering Fracture Mechanics* 160, 78–94.

Sladek, J., Sladek, V., Stanak, P., 2015. Scaled boundary finite element method for thermoelasticity in voided materials. *CMES: Computer Modeling in Engineering & Sciences* 106, 229–262.

Song, C., 2004a. A matrix function solution for the scaled boundary finite-element equation in statics. *Computer Methods in Applied Mechanics and Engineering* 193, 2325–2356.

Song, C., 2004b. A super-element for crack analysis in the time domain. *International Journal for Numerical Methods in Engineering* 61, 1332–1357.

Song, C., 2004c. Weighted block-orthogonal base functions for static analysis of unbounded domains. In: *Proceedings of The 6th World Congress on Computational Mechanics*, Beijing, China, 5–10 September 2004, pp. 615–620.

Song, C., 2005. Evaluation of power-logarithmic singularities, T-stresses and higher order terms of in-plane singular stress fields at cracks and multi-material corners. *Engineering Fracture Mechanics* 72, 1498–1530.

Song, C., 2006. Analysis of singular stress fields at multi-material corners under thermal loading. *International Journal for Numerical Methods in Engineering* 65 (5), 620–652.

Song, C., 2009. The scaled boundary finite element method in structural dynamics. *International Journal for Numerical Methods in Engineering* 77 (8), 1139–1171.

Song, C., Tin-Loi, F., Gao, W., Aug. 2010a. A definition and evaluation procedure of generalized stress intensity factors at cracks and multi-material wedges. *Engineering Fracture Mechanics* 77, 2316–2336.

Song, C., Tin-Loi, F., Gao, W., 2010b. Transient dynamic analysis of interface cracks in anisotropic bimaterials by the scaled boundary finite-element method. *International Journal of Solids and Structures* 47, 978–989.

Song, C., Wolf, J. P., 1995. Consistent infinitesimal finite-element–cell method: out-of-plane motion. *Journal of Engineering Mechanics, ASCE* 121, 613–619.

Song, C., Wolf, J. P., 1996. Consistent infinitesimal finite-element cell method: three-dimensional vector wave equation. *International Journal for Numerical Methods in Engineering* 39, 2189–2208.

Song, C., Wolf, J. P., 1997. The scaled boundary finite-element method – alias consistent infinitesimal finite-element cell method – for elastodynamics. *Computer Methods in Applied Mechanics and Engineering* 147, 329–355.

Song, C., Wolf, J. P., 1999. Body loads in the scaled boundary finite-element method. *Computer Methods in Applied Mechanics and Engineering* 180, 117–135.

Song, C., Wolf, J. P., 2002. Semi-analytical representation of stress singularity as occurring in cracks in anisotropic multi-materials with the scaled boundary finite-element method. *Computers & Structures* 80, 183–197.

Staten, M. L., Canann, S. A., Owen, S. J., 1999. BMSweep: Locating interior nodes during sweeping. *Engineering with Computers* 15 (3), 212–218.

Sternberg, E., Koiter, W., 1958. The wedge under a concentrated couple: a paradox in the two-dimensional theory of elasticity. *Journal of Applied Mechanics, ASME* 4, 575–581.

Stewart, I., Tall, D., 1983. *Complex Analysis*. Cambridge University Press.

Suo, Z., 1990. Singularities, interfaces and cracks in dissimilar anisotropic media. *Proceedings of the Royal Society London Series A-Mathematical, Physical & Engineering Sciences* 427, 331–358.

Tabarraei, A., Sukumar, N., 2005. Adaptive computations on conforming quadtree meshes. *Finite Elements in Analysis and Design* 41 (7-8), 686–702.

Tada, H., Paris, P., Irwin, G., 1985. *The Stress Analysis of Cracks Handbook*, 2nd Edition. Paris Productions Inc., St. Louis, MO.

Talischi, C., Paulino, G. H., Pereira, A., Menezes, I. F. M., Jan. 2012. PolyMesher: a general-purpose mesh generator for polygonal elements written in Matlab. *Structural and Multidisciplinary Optimization* 45, 309–328.

Tan, C. L., Gao, Y. L., 1990. Treatment of bimaterial interface crack problems using the boundary element method. *Engineering Fracture Mechanics* 36, 919–932.

Tautges, T. J., Blacker, T., Mitchell, S. A., 1996. The whisker weaving algorithm: A connectivity-based method for constructing all-hexahedral finite element meshes. *International Journal for Numerical Methods in Engineering* 39 (19), 3327–3349.

Tisseur, F., Meerbergen, K., 2001. The quadratic eigenvalue problem. *SIAM Review* 43, 235–286.

Tong, P., Pian, T. H. H., Lasary, S. J., 1973. A hybrid-element approach to crack problems in plane elasticity. *International Journal for Numerical Methods in Engineering* 7, 297–308.

Ulrich, D., van Rietbergen, B., Weinans, H., Rüegsegger, P., 1998. Finite element analysis of trabecular bone structure: a comparison of image-based meshing techniques. *Journal of Biomechanics* 31 (12), 1187–1192.

Vu, T. H., Deeks, A. J., 2006. Use of higher-order shape functions in the scaled boundary finite element method. *International Journal for Numerical Methods in Engineering* 65, 1714–1733.

Vu, T. H., Deeks, A. J., 2008. A p-adaptive scaled boundary finite element method based on maximization of the error decrease rate. *Computational Mechanics* 41 (3), 441–455.

Wang, C. C. L., 2006. Incremental reconstruction of sharp edges on mesh surfaces. *Computer-Aided Design* 38 (6), 689–702.

Wang, D., Hassan, O., Morgan, K., Weatherill, N., 2007. Enhanced remeshing from STL files with applications to surface grid generation. *Communications in Numerical Methods in Engineering* 23 (3), 227–239.

Wang, Y., Lin, G., Hu, Z., 2015. Novel nonreflecting boundary condition for an infinite reservoir based on the scaled boundary finite-element method. *Journal of Engineering Mechanics* 141 (5), 04014150.

Wang, Z., Teo, J. C.-M., Chui, C.-K., Ong, S. H., Yan, C. H., Wang, S., Wong, H.-K., Teoh, S.-H., 2005. Computational biomechanical modelling of the lumbar spine using marching-cubes surface smoothened finite element voxel meshing. *Computer Methods and Programs in Biomedicine* 80 (1), 25–35.

Watson, D. F., 1981. Computing the n-dimensional Delaunay tessellation with application to Voronoi polytopes. *The Computer Journal* 24 (2), 167–172.

Williams, M. L., 1957. On the stress distribution at the base of a stationary crack. *Journal of Applied Mechanics, ASME* 24, 109–114.

Wolf, J. P., Song, C., 1995. Consistent infinitesimal finite-element cell method: in-plane motion. *Computer Methods in Applied Mechanics and Engineering* 123, 355–370.

Wolf, J. P., Song, C., 1996. *Finite-Element Modelling of Unbounded Media*. John Wiley & Sons, Chichester.

Xiao, Q. Z., Karihaloo, B. L., Liu, X. Y., 2004. Direct determination of SIF and higher order terms of mixed mode cracks by a hybrid crack element. *International Journal of Fracture* 125, 207–225.

Yang, Y., Munz, D., 1995. Stress distribution in a dissimilar materials joint for complex singular eigenvalues under thermal loading. *Journal of Thermal Stresses* 18, 407–419.

Yang, Z., 2006. Fully automatic modelling of mixed-mode crack propagation using scaled boundary finite element method. *Engineering Fracture Mechanics* 73 (12), 1711–1731.

Yang, Z., Deeks, A., 2007a. Fully-automatic modelling of cohesive crack growth using a finite element-scaled boundary finite element coupled method. *Engineering Fracture Mechanics* 74 (16), 2547–2573.

Yang, Z. J., Deeks, A. J., 2007b. Modelling cohesive crack growth using a two-step finite element-scaled boundary finite element coupled method. *International Journal of Fracture* 143 (4), 333–354.

Yang, Z., Deeks, A., 2008. Calculation of transient dynamic stress intensity factors at bimaterial interface cracks using a SBFEM-based frequency-domain approach. *Science in China Series G: Physics, Mechanics and Astronomy* 51 (5), 519–531.

Yang, Z. J., Zhang, Z. H., Liu, G. H., Ooi, E. T., 2011. An *h*-hierarchical adaptive scaled boundary finite element method for elastodynamics. *Computers & Structures* 89 (13), 1417–1429.

Yerry, M., Shephard, M., 1983. A modified quadtree approach to finite element mesh generation. *IEEE Computer Graphics and Applications* 3 (1), 39–46.

Yerry, M. A., Shephard, M. S., 1984. Automatic three-dimensional mesh generation by the modified-octree technique. *International Journal for Numerical Methods in Engineering* 20 (11), 1965–1990.

Young, P., Beresford-West, T., Coward, S., Notarberardino, B., Walker, B., Abdul-Aziz, A., 2008. An efficient approach to converting three-dimensional image data into highly accurate computational models. *Philosophical Transactions of the Royal Society A: Mathematical, Physical and Engineering Sciences* 366 (1878), 3155–3173.

Yuuki, R., Cho, S. B., 1989. Efficient boundary element analysis of stress intensity factors for interface cracks in dissimilar materials. *Engineering Fracture Mechanics* 34, 179–188.

Zhong, H., Ooi, E., Song, C., Ding, T., Lin, G., Li, H., 2014. Experimental and numerical study of the dependency of interface fracture in concrete-rock specimens on mode mixity. *Engineering Fracture Mechanics* 124-125, 287–309.

Zhu, C., Lin, G., Li, J., 2013. Modelling cohesive crack growth in concrete beams using scaled boundary finite element method based on super-element remeshing technique. *Computers and Structures* 121, 76–86.

Zienkiewicz, O. C., Taylor, R. L., Zhu, J. Z., 2005. *The Finite Element Method: Its Basis and Fundamentals*, 6th Edition. Butterworth-Heinemann.

Zou, D., Chen, K., Kong, X., Liu, J., 2017. An enhanced octree polyhedral scaled boundary finite element method and its applications in structure analysis. *Engineering Analysis with Boundary Elements* 84, 87–107.

Index

The Scaled Boundary Finite Element Method: Introduction to Theory and Implementation,
First Edition. Chongmin Song.
© 2018 John Wiley & Sons Ltd. Published 2018 by John Wiley & Sons Ltd.
Companion website: www.wiley.com/go/author/scaledboundaryfiniteelementmethod